Industrial Noise Control and Acoustics

MECHANICAL ENGINEERING
A Series of Textbooks and Reference Books

Founding Editor

L. L. Faulkner

*Columbus Division, Battelle Memorial Institute
and Department of Mechanical Engineering
The Ohio State University
Columbus, Ohio*

1. *Spring Designer's Handbook*, Harold Carlson
2. *Computer-Aided Graphics and Design*, Daniel L. Ryan
3. *Lubrication Fundamentals*, J. George Wills
4. *Solar Engineering for Domestic Buildings*, William A. Himmelman
5. *Applied Engineering Mechanics: Statics and Dynamics*, G. Boothroyd and C. Poli
6. *Centrifugal Pump Clinic*, Igor J. Karassik
7. *Computer-Aided Kinetics for Machine Design*, Daniel L. Ryan
8. *Plastics Products Design Handbook, Part A: Materials and Components; Part B: Processes and Design for Processes*, edited by Edward Miller
9. *Turbomachinery: Basic Theory and Applications*, Earl Logan, Jr.
10. *Vibrations of Shells and Plates*, Werner Soedel
11. *Flat and Corrugated Diaphragm Design Handbook*, Mario Di Giovanni
12. *Practical Stress Analysis in Engineering Design*, Alexander Blake
13. *An Introduction to the Design and Behavior of Bolted Joints*, John H. Bickford
14. *Optimal Engineering Design: Principles and Applications*, James N. Siddall
15. *Spring Manufacturing Handbook*, Harold Carlson
16. *Industrial Noise Control: Fundamentals and Applications*, edited by Lewis H. Bell
17. *Gears and Their Vibration: A Basic Approach to Understanding Gear Noise*, J. Derek Smith
18. *Chains for Power Transmission and Material Handling: Design and Applications Handbook*, American Chain Association
19. *Corrosion and Corrosion Protection Handbook*, edited by Philip A. Schweitzer
20. *Gear Drive Systems: Design and Application*, Peter Lynwander
21. *Controlling In-Plant Airborne Contaminants: Systems Design and Calculations*, John D. Constance
22. *CAD/CAM Systems Planning and Implementation*, Charles S. Knox
23. *Probabilistic Engineering Design: Principles and Applications*, James N. Siddall
24. *Traction Drives: Selection and Application*, Frederick W. Heilich III and Eugene E. Shube
25. *Finite Element Methods: An Introduction*, Ronald L. Huston and Chris E. Passerello

26. *Mechanical Fastening of Plastics: An Engineering Handbook*, Brayton Lincoln, Kenneth J. Gomes, and James F. Braden
27. *Lubrication in Practice: Second Edition*, edited by W. S. Robertson
28. *Principles of Automated Drafting*, Daniel L. Ryan
29. *Practical Seal Design*, edited by Leonard J. Martini
30. *Engineering Documentation for CAD/CAM Applications*, Charles S. Knox
31. *Design Dimensioning with Computer Graphics Applications*, Jerome C. Lange
32. *Mechanism Analysis: Simplified Graphical and Analytical Techniques*, Lyndon O. Barton
33. *CAD/CAM Systems: Justification, Implementation, Productivity Measurement*, Edward J. Preston, George W. Crawford, and Mark E. Coticchia
34. *Steam Plant Calculations Manual*, V. Ganapathy
35. *Design Assurance for Engineers and Managers*, John A. Burgess
36. *Heat Transfer Fluids and Systems for Process and Energy Applications*, Jasbir Singh
37. *Potential Flows: Computer Graphic Solutions*, Robert H. Kirchhoff
38. *Computer-Aided Graphics and Design: Second Edition*, Daniel L. Ryan
39. *Electronically Controlled Proportional Valves: Selection and Application*, Michael J. Tonyan, edited by Tobi Goldoftas
40. *Pressure Gauge Handbook*, AMETEK, U.S. Gauge Division, edited by Philip W. Harland
41. *Fabric Filtration for Combustion Sources: Fundamentals and Basic Technology*, R. P. Donovan
42. *Design of Mechanical Joints*, Alexander Blake
43. *CAD/CAM Dictionary*, Edward J. Preston, George W. Crawford, and Mark E. Coticchia
44. *Machinery Adhesives for Locking, Retaining, and Sealing*, Girard S. Haviland
45. *Couplings and Joints: Design, Selection, and Application*, Jon R. Mancuso
46. *Shaft Alignment Handbook*, John Piotrowski
47. *BASIC Programs for Steam Plant Engineers: Boilers, Combustion, Fluid Flow, and Heat Transfer*, V. Ganapathy
48. *Solving Mechanical Design Problems with Computer Graphics*, Jerome C. Lange
49. *Plastics Gearing: Selection and Application*, Clifford E. Adams
50. *Clutches and Brakes: Design and Selection*, William C. Orthwein
51. *Transducers in Mechanical and Electronic Design*, Harry L. Trietley
52. *Metallurgical Applications of Shock-Wave and High-Strain-Rate Phenomena*, edited by Lawrence E. Murr, Karl P. Staudhammer, and Marc A. Meyers
53. *Magnesium Products Design*, Robert S. Busk
54. *How to Integrate CAD/CAM Systems: Management and Technology*, William D. Engelke
55. *Cam Design and Manufacture: Second Edition*; with cam design software for the IBM PC and compatibles, disk included, Preben W. Jensen
56. *Solid-State AC Motor Controls: Selection and Application*, Sylvester Campbell
57. *Fundamentals of Robotics*, David D. Ardayfio
58. *Belt Selection and Application for Engineers*, edited by Wallace D. Erickson
59. *Developing Three-Dimensional CAD Software with the IBM PC*, C. Stan Wei
60. *Organizing Data for CIM Applications*, Charles S. Knox, with contributions by Thomas C. Boos, Ross S. Culverhouse, and Paul F. Muchnicki

61. *Computer-Aided Simulation in Railway Dynamics*, by Rao V. Dukkipati and Joseph R. Amyot
62. *Fiber-Reinforced Composites: Materials, Manufacturing, and Design*, P. K. Mallick
63. *Photoelectric Sensors and Controls: Selection and Application*, Scott M. Juds
64. *Finite Element Analysis with Personal Computers*, Edward R. Champion, Jr., and J. Michael Ensminger
65. *Ultrasonics: Fundamentals, Technology, Applications: Second Edition, Revised and Expanded*, Dale Ensminger
66. *Applied Finite Element Modeling: Practical Problem Solving for Engineers*, Jeffrey M. Steele
67. *Measurement and Instrumentation in Engineering: Principles and Basic Laboratory Experiments*, Francis S. Tse and Ivan E. Morse
68. *Centrifugal Pump Clinic: Second Edition, Revised and Expanded*, Igor J. Karassik
69. *Practical Stress Analysis in Engineering Design: Second Edition, Revised and Expanded*, Alexander Blake
70. *An Introduction to the Design and Behavior of Bolted Joints: Second Edition, Revised and Expanded*, John H. Bickford
71. *High Vacuum Technology: A Practical Guide*, Marsbed H. Hablanian
72. *Pressure Sensors: Selection and Application*, Duane Tandeske
73. *Zinc Handbook: Properties, Processing, and Use in Design*, Frank Porter
74. *Thermal Fatigue of Metals*, Andrzej Weronski and Tadeusz Hejwowski
75. *Classical and Modern Mechanisms for Engineers and Inventors*, Preben W. Jensen
76. *Handbook of Electronic Package Design*, edited by Michael Pecht
77. *Shock-Wave and High-Strain-Rate Phenomena in Materials*, edited by Marc A. Meyers, Lawrence E. Murr, and Karl P. Staudhammer
78. *Industrial Refrigeration: Principles, Design and Applications*, P. C. Koelet
79. *Applied Combustion*, Eugene L. Keating
80. *Engine Oils and Automotive Lubrication*, edited by Wilfried J. Bartz
81. *Mechanism Analysis: Simplified and Graphical Techniques, Second Edition, Revised and Expanded*, Lyndon O. Barton
82. *Fundamental Fluid Mechanics for the Practicing Engineer*, James W. Murdock
83. *Fiber-Reinforced Composites: Materials, Manufacturing, and Design, Second Edition, Revised and Expanded*, P. K. Mallick
84. *Numerical Methods for Engineering Applications*, Edward R. Champion, Jr.
85. *Turbomachinery: Basic Theory and Applications, Second Edition, Revised and Expanded*, Earl Logan, Jr.
86. *Vibrations of Shells and Plates: Second Edition, Revised and Expanded*, Werner Soedel
87. *Steam Plant Calculations Manual: Second Edition, Revised and Ex panded*, V. Ganapathy
88. *Industrial Noise Control: Fundamentals and Applications, Second Edition, Revised and Expanded*, Lewis H. Bell and Douglas H. Bell
89. *Finite Elements: Their Design and Performance*, Richard H. MacNeal
90. *Mechanical Properties of Polymers and Composites: Second Edition, Revised and Expanded*, Lawrence E. Nielsen and Robert F. Landel
91. *Mechanical Wear Prediction and Prevention*, Raymond G. Bayer

92. *Mechanical Power Transmission Components,* edited by David W. South and Jon R. Mancuso
93. *Handbook of Turbomachinery,* edited by Earl Logan, Jr.
94. *Engineering Documentation Control Practices and Procedures,* Ray E. Monahan
95. *Refractory Linings Thermomechanical Design and Applications,* Charles A. Schacht
96. *Geometric Dimensioning and Tolerancing: Applications and Techniques for Use in Design, Manufacturing, and Inspection,* James D. Meadows
97. *An Introduction to the Design and Behavior of Bolted Joints: Third Edition, Revised and Expanded,* John H. Bickford
98. *Shaft Alignment Handbook: Second Edition, Revised and Expanded,* John Piotrowski
99. *Computer-Aided Design of Polymer-Matrix Composite Structures,* edited by Suong Van Hoa
100. *Friction Science and Technology,* Peter J. Blau
101. *Introduction to Plastics and Composites: Mechanical Properties and Engineering Applications,* Edward Miller
102. *Practical Fracture Mechanics in Design,* Alexander Blake
103. *Pump Characteristics and Applications,* Michael W. Volk
104. *Optical Principles and Technology for Engineers,* James E. Stewart
105. *Optimizing the Shape of Mechanical Elements and Structures,* A. A. Seireg and Jorge Rodriguez
106. *Kinematics and Dynamics of Machinery,* Vladimír Stejskal and Michael Valášek
107. *Shaft Seals for Dynamic Applications,* Les Horve
108. *Reliability-Based Mechanical Design,* edited by Thomas A. Cruse
109. *Mechanical Fastening, Joining, and Assembly,* James A. Speck
110. *Turbomachinery Fluid Dynamics and Heat Transfer,* edited by Chunill Hah
111. *High-Vacuum Technology: A Practical Guide, Second Edition, Revised and Expanded,* Marsbed H. Hablanian
112. *Geometric Dimensioning and Tolerancing: Workbook and Answerbook,* James D. Meadows
113. *Handbook of Materials Selection for Engineering Applications,* edited by G. T. Murray
114. *Handbook of Thermoplastic Piping System Design,* Thomas Sixsmith and Reinhard Hanselka
115. *Practical Guide to Finite Elements: A Solid Mechanics Approach,* Steven M. Lepi
116. *Applied Computational Fluid Dynamics,* edited by Vijay K. Garg
117. *Fluid Sealing Technology,* Heinz K. Muller and Bernard S. Nau
118. *Friction and Lubrication in Mechanical Design,* A. A. Seireg
119. *Influence Functions and Matrices,* Yuri A. Melnikov
120. *Mechanical Analysis of Electronic Packaging Systems,* Stephen A. McKeown
121. *Couplings and Joints: Design, Selection, and Application, Second Edition, Revised and Expanded,* Jon R. Mancuso
122. *Thermodynamics: Processes and Applications,* Earl Logan, Jr.
123. *Gear Noise and Vibration,* J. Derek Smith
124. *Practical Fluid Mechanics for Engineering Applications,* John J. Bloomer
125. *Handbook of Hydraulic Fluid Technology,* edited by George E. Totten
126. *Heat Exchanger Design Handbook,* T. Kuppan

127. *Designing for Product Sound Quality,* Richard H. Lyon
128. *Probability Applications in Mechanical Design,* Franklin E. Fisher and Joy R. Fisher
129. *Nickel Alloys,* edited by Ulrich Heubner
130. *Rotating Machinery Vibration: Problem Analysis and Troubleshooting,* Maurice L. Adams, Jr.
131. *Formulas for Dynamic Analysis,* Ronald L. Huston and C. Q. Liu
132. *Handbook of Machinery Dynamics,* Lynn L. Faulkner and Earl Logan, Jr.
133. *Rapid Prototyping Technology: Selection and Application,* Kenneth G. Cooper
134. *Reciprocating Machinery Dynamics: Design and Analysis,* Abdulla S. Rangwala
135. *Maintenance Excellence: Optimizing Equipment Life-Cycle Decisions,* edited by John D. Campbell and Andrew K. S. Jardine
136. *Practical Guide to Industrial Boiler Systems,* Ralph L. Vandagriff
137. *Lubrication Fundamentals: Second Edition, Revised and Expanded,* D. M. Pirro and A. A. Wessol
138. *Mechanical Life Cycle Handbook: Good Environmental Design and Manufacturing,* edited by Mahendra S. Hundal
139. *Micromachining of Engineering Materials,* edited by Joseph McGeough
140. *Control Strategies for Dynamic Systems: Design and Implementation,* John H. Lumkes, Jr.
141. *Practical Guide to Pressure Vessel Manufacturing,* Sunil Pullarcot
142. *Nondestructive Evaluation: Theory, Techniques, and Applications,* edited by Peter J. Shull
143. *Diesel Engine Engineering: Thermodynamics, Dynamics, Design, and Control,* Andrei Makartchouk
144. *Handbook of Machine Tool Analysis,* Ioan D. Marinescu, Constantin Ispas, and Dan Boboc
145. *Implementing Concurrent Engineering in Small Companies,* Susan Carlson Skalak
146. *Practical Guide to the Packaging of Electronics: Thermal and Mechanical Design and Analysis,* Ali Jamnia
147. *Bearing Design in Machinery: Engineering Tribology and Lubrication,* Avraham Harnoy
148. *Mechanical Reliability Improvement: Probability and Statistics for Experimental Testing,* R. E. Little
149. *Industrial Boilers and Heat Recovery Steam Generators: Design, Applications, and Calculations,* V. Ganapathy
150. *The CAD Guidebook: A Basic Manual for Understanding and Improving Computer-Aided Design,* Stephen J. Schoonmaker
151. *Industrial Noise Control and Acoustics,* Randall F. Barron
152. *Mechanical Properties of Engineering Materials,* Wolé Soboyejo
153. *Reliability Verification, Testing, and Analysis in Engineering Design,* Gary S. Wasserman
154. *Fundamental Mechanics of Fluids: Third Edition,* I. G. Currie

Additional Volumes in Preparation

HVAC Water Chillers and Cooling Towers: Fundamentals, Application, and Operations, Herbert W. Stanford III

Handbook of Turbomachinery: Second Edition, Revised and Expanded, Earl Logan, Jr., and Ramendra Roy

Progressing Cavity Pumps, Downhole Pumps, and Mudmotors, Lev Nelik

Gear Noise and Vibration: Second Edition, Revised and Expanded, J. Derek Smith

Intermediate Heat Transfer, Kau-Fui Vincent Wong

Mechanical Engineering Software

Spring Design with an IBM PC, Al Dietrich

Mechanical Design Failure Analysis: With Failure Analysis System Software for the IBM PC, David G. Ullman

Industrial Noise Control and Acoustics

Randall F. Barron
Louisiana Tech University
Ruston, Louisiana, U.S.A.

Marcel Dekker, Inc. New York · Basel

Library of Congress Cataloging-in-Publication Data
A catalog record for this book is available from the Library of Congress.

ISBN: 0-8247-0701-X

This book is printed on acid-free paper.

Headquarters
Marcel Dekker, Inc.
270 Madison Avenue, New York, NY 10016
tel: 212-696-9000; fax: 212-685-4540

Eastern Hemisphere Distribution
Marcel Dekker AG
Hutgasse 4, Postfach 812, CH-4001 Basel, Switzerland
tel: 41-61-260-6300; fax: 41-61-260-6333

World Wide Web
http://www.dekker.com

The publisher offers discounts on this book when ordered in bulk quantities, For more information, write to Special Sales/Professional Marketing at the headquarters address above.

Copyright © 2003 by Marcel Dekker, Inc. All Rights Reserved.

Neither this book nor any part may be reproduced or transmitted in any form or by any means, electronic or mechanical, including photocopying, microfilming, and recording, or by any information storage retrieval system, without permission in writing from the publisher.

Current printing (last digit):
10 9 8 7 6 5 4 3 2 1

PRINTED IN THE UNITED STATES OF AMERICA

Preface

Since the Walsh-Healy Act of 1969 was amended to include restrictions on the noise exposure of workers, there has been much interest and motivation in industry to reduce noise emitted by machinery. In addition to concerns about air and water pollution by contaminants, efforts have also been directed toward control of environmental noise pollution.

In response to these stimuli, faculty at many engineering schools have developed and introduced courses in noise control, usually at the senior design level. It is generally much more effective to design "quietness" into a product than to try to "fix" the noise problem in the field after the product has been put on the market. Because of this, many engineering designs in industry take into account the noise levels generated by a system.

Industrial Noise Control and Acoustics was developed as a result of my 30 years of experience teaching senior-level undergraduate mechanical engineering courses in noise control, directing graduate student research projects, teaching continuing education courses on industrial noise control to practicing engineers, and consulting on various industrial projects in noise assessment and abatement. The book reflects this background, including problems for engineering students to gain experience in applying the principles presented in the text, and examples for practicing engineers to illustrate the material. Several engineering case studies are included to illustrate practical solutions of noise problems in industry. This book is designed to integrate the theory of acoustics with the practice of noise control engineering.

I would like to express my most sincere appreciation to those students in my classes who asked questions and made suggestions that helped make the text more clear and understandable. My most heartfelt thanks are reserved for my wife, Shirley, for her support and encouragement during the months of book preparation, and especially during the years before I even considered writing this book.

Randall F. Barron

Contents

Preface   *iii*

1 Introduction   1
    1.1 Noise Control   1
    1.2 Historical Background   3
    1.3 Principles of Noise Control   7
       1.3.1 Noise Control at the Source   8
       1.3.2 Noise Control in the Transmission Path   9
       1.3.3 Noise Control at the Receiver   9
    References   10

2 Basics of Acoustics   12
    2.1 Speed of Sound   12
    2.2 Wavelength, Frequency, and Wave Number   13
    2.3 Acoustic Pressure and Particle Velocity   15
    2.4 Acoustic Intensity and Acoustic Energy Density   17
    2.5 Spherical Waves   21
    2.6 Directivity Factor and Directivity Index   24
    2.7 Levels and the Decibel   27
    2.8 Combination of Sound Sources   31

	2.9	Octave Bands	33
	2.10	Weighted Sound Levels	34
		Problems	37
		References	40

3 Acoustic Measurements 41
    3.1 Sound Level Meters 42
    3.2 Intensity Level Meters 46
    3.3 Octave Band Filters 49
    3.4 Acoustic Analyzers 50
    3.5 Dosimeter 50
    3.6 Measurement of Sound Power 51
        3.6.1 Sound Power Measurement in a Reverberant Room 52
        3.6.2 Sound Power Measurement in an Anechoic or Semi-Anechoic Room 58
        3.6.3 Sound Power Survey Measurements 62
        3.6.4 Measurement of the Directivity Factor 66
    3.7 Noise Measurement Procedures 69
        Problems 73
        References 76

4 Transmission of Sound 78
    4.1 The Wave Equation 78
    4.2 Complex Number Notation 83
    4.3 Wave Equation Solution 84
    4.4 Solution for Spherical Waves 88
    4.5 Changes in Media with Normal Incidence 91
    4.6 Changes in Media with Oblique Incidence 96
    4.7 Sound Transmission Through a Wall 101
    4.8 Transmission Loss for Walls 107
        4.8.1 Region I: Stiffness-Controlled Region 108
        4.8.2 Resonant Frequency 111
        4.8.3 Region II: Mass-Controlled Region 112
        4.8.4 Critical Frequency 113
        4.8.5 Region III: Damping-Controlled Region 113
    4.9 Approximate Method for Estimating the TL 117
    4.10 Transmission Loss for Composite Walls 120
        4.10.1 Elements in Parallel 121
        4.10.2 Composite Wall with Air Space 122
        4.10.3 Two-Layer Laminate 127
        4.10.4 Rib-Stiffened Panels 131

	4.11	Sound Transmission Class	134
	4.12	Absorption of Sound	139
	4.13	Attenuation Coefficient	143
		Problems	153
		References	160
5	**Noise Sources**		**162**
	5.1	Sound Transmission Indoors and Outdoors	162
	5.2	Fan Noise	164
	5.3	Electric Motor Noise	169
	5.4	Pump Noise	171
	5.5	Gas Compressor Noise	173
	5.6	Transformer Noise	177
	5.7	Cooling Tower Noise	178
	5.8	Noise from Gas Vents	182
	5.9	Appliance and Equipment Noise	185
	5.10	Valve Noise	186
		5.10.1 Sources of Valve Noise	186
		5.10.2 Noise Prediction for Gas Flows	188
		5.10.3 Noise Prediction for Liquid Flows	190
	5.11	Air Distribution System Noise	192
		5.11.1 Noise Attenuation in Air Distribution Systems	193
		5.11.2 Noise Generation in Air Distribution System Fittings	195
		5.11.3 Noise Generation in Grilles	198
	5.12	Traffic Noise	207
	5.13	Train Noise	211
		5.13.1 Railroad Car Noise	211
		5.13.2 Locomotive Noise	213
		5.13.3 Complete Train Noise	214
		Problems	217
		References	222
6	**Acoustic Criteria**		**225**
	6.1	The Human Ear	226
	6.2	Hearing Loss	229
	6.3	Industrial Noise Criteria	231
	6.4	Speech Interference Level	235
	6.5	Noise Criteria for Interior Spaces	238
	6.6	Community Reaction to Environmental Noise	243
	6.7	The Day-Night Level	247

	6.7.1	EPA Criteria	247
	6.7.2	Estimation of Community Reaction	250
6.8	HUD Criteria		253
6.9	Aircraft Noise Criteria		255
	6.9.1	Perceived Noise Level	256
	6.9.2	Noise Exposure Forecast	257
	Problems		262
	References		267

7 Room Acoustics — 269

7.1	Surface Absorption Coefficients		269
	7.1.1	Values for Surface Absorption Coefficients	269
	7.1.2	Noise Reduction Coefficient	270
	7.1.3	Mechanism of Acoustic Absorption	271
	7.1.4	Average Absorption Coefficient	274
7.2	Steady-State Sound Level in a Room		274
7.3	Reverberation Time		281
7.4	Effect of Energy Absorption in the Air		289
	7.4.1	Steady-State Sound Level with Absorption in the Air	289
	7.4.2	Reverberation Time with Absorption in the Air	291
7.5	Noise from an Adjacent Room		293
	7.5.1	Sound Source Covering One Wall	293
	7.5.2	Sound Transmission from an Adjacent Room	295
7.6	Acoustic Enclosures		299
	7.6.1	Small Acoustic Enclosures	300
	7.6.2	Large Acoustic Enclosures	304
	7.6.3	Design Practice for Enclosures	311
7.7	Acoustic Barriers		312
	7.7.1	Barriers Located Outdoors	313
	7.7.2	Barriers Located Indoors	317
	Problems		321
	References		328

8 Silencer Design — 330

8.1	Silencer Design Requirements		330
8.2	Lumped Parameter Analysis		332
	8.2.1	Acoustic Mass	332
	8.2.2	Acoustic Compliance	335
	8.2.3	Acoustic Resistance	338
	8.2.4	Transfer Matrix	339

Contents ix

    8.3 The Helmholtz Resonator 341
           8.3.1 Helmholtz Resonator System 341
           8.3.2 Resonance for the Helmholtz Resonator 342
           8.3.3 Acoustic Impedance for the Helmholtz Resonator 343
           8.3.4 Half-Power Bandwidth 344
           8.3.5 Sound Pressure Level Gain 348
    8.4 Side Branch Mufflers 350
           8.4.1 Transmission Loss for a Side-Branch Muffler 351
           8.4.2 Directed Design Procedure for Side-Branch Mufflers 357
           8.4.3 Closed Tube as a Side-Branch Muffler 361
           8.4.4 Open Tube (Orifice) as a Side Branch 365
    8.5 Expansion Chamber Mufflers 368
           8.5.1 Transmission Loss for an Expansion Chamber Muffler 368
           8.5.2 Design Procedure for Single-Expansion Chamber Mufflers 371
           8.5.3 Double-Chamber Mufflers 373
    8.6 Dissipative Mufflers 377
    8.7 Evaluation of the Attenuation Coefficient 381
           8.7.1 Estimation of the Attenuation Coefficient 381
           8.7.2 Effective Density 383
           8.7.3 Effective Elasticity Coefficient 384
           8.7.4 Effective Specific Flow Resistance 385
           8.7.5 Correction for Random Incidence End Effects 387
    8.8 Commercial Silencers 389
    8.9 Plenum Chambers 391
        Problems 397
        References 405

9 Vibration Isolation for Noise Control 406
    9.1 Undamped Single-Degree-of-Freedom (SDOF) System 407
    9.2 Damped Single-Degree-of-Freedom (SDOF) System 410
           9.2.1 Critically Damped System 411
           9.2.2 Over-Damped System 412
           9.2.3 Under-Damped System 412
    9.3 Damping Factors 413
    9.4 Forced Vibration 419
    9.5 Mechanical Impedance and Mobility 424
    9.6 Transmissibility 427
    9.7 Rotating Unbalance 431

	9.8	Displacement Excitation	436
	9.9	Dynamic Vibration Isolator	439
	9.10	Vibration Isolation Materials	446
		9.10.1 Cork and Felt Resilient Materials	446
		9.10.2 Rubber and Elastomer Vibration Isolators	450
		9.10.3 Metal Spring Isolators	457
	9.11	Effects of Vibration on Humans	464
		Problems	469
		References	474
10	**Case Studies in Noise Control**		**475**
	10.1	Introduction	475
	10.2	Folding Carton Packing Station Noise	476
		10.2.1 Analysis	476
		10.2.2 Control Approach Chosen	479
		10.2.3 Cost	479
		10.2.4 Pitfalls	480
	10.3	Metal Cut-Off Saw Noise	480
		10.3.1 Analysis	480
		10.3.2 Control Approach Chosen	481
		10.3.3 Cost	482
		10.3.4 Pitfalls	482
	10.4	Paper Machine Wet End	482
		10.4.1 Analysis	483
		10.4.2 Control Approach Chosen	487
		10.4.3 Cost	487
		10.4.4 Pitfalls	488
	10.5	Air Scrap Handling Duct Noise	488
		10.5.1 Analysis	488
		10.5.2 Control Approach Chosen	491
		10.5.3 Cost	492
		10.5.4 Pitfalls	492
	10.6	Air-Operated Hoist Motor	492
	10.7	Blanking Press Noise	494
		10.7.1 Analysis	495
		10.7.2 Control Approach Chosen	497
		10.7.3 Cost	497
		10.7.4 Pitfalls	497
	10.8	Noise in a Small Meeting Room	498
		10.8.1 Analysis	499
		10.8.2 Control Approach Chosen	502
		10.8.3 Cost	503

Contents

        10.8.4   Pitfalls   503
     Problems   503
     References   504

Appendix A   Preferred Prefixes in SI   **506**

Appendix B   Properties of Gases, Liquids, and Solids   **507**

Appendix C   Plate Properties of Solids   **509**

Appendix D   Surface Absorption Coefficients   **510**

Appendix E   Nomenclature   **514**

Index   *525*

Industrial Noise Control and Acoustics

1
Introduction

1.1 NOISE CONTROL

Concern about problems of noise in the workplace and in the living space has escalated since the amendment of the Walsh–Healy Act of 1969. This act created the first set of nationwide occupational noise regulations (Occupational Safety and Health Administration, 1983). There is a real danger of permanent hearing loss when a person is exposed to noise above a certain level. Most industries are strongly motivated to find an effective, economical solution to this problem.

The noise level near airports has become serious enough for some people to move out of residential areas near airports. These areas were considered pleasant living areas before the airport was constructed, but environmental noise has changed this perception. The airport noise in the areas surrounding the airport is generally not dangerous to a person's health, but the noise may be unpleasant and annoying.

In the design of many appliances, such as dishwashers, the designer must be concerned about the noise generated by the appliance in operation; otherwise, prospective customers may decide to purchase other quieter models. It is important that noise control be addressed in the design stage for many mechanical devices.

Lack of proper acoustic treatment in offices, apartments, and classrooms may interfere with the effective functioning of the people in the rooms. Even though the noise is not dangerous and not particularly annoying, if the person cannot communicate effectively, then the noise is undesirable.

Much can be done to reduce the seriousness of noise problems. It is often not as simple as turning down the volume on the teenager's stereo set, however. Effective silencers (mufflers) are available for trucks and automobiles, but there are other significant sources of noise, such as tire noise and wind noise, that are not affected by the installation of a silencer. Household appliances and other machines may be made quieter by proper treatment of vibrating surfaces, use of adequately sized piping and smoother channels for water flow, and including vibration isolation mounts. Obviously, the noise treatment must not interfere with the operation of the appliance or machine. This stipulation places limitations on the noise control procedure that can be used.

In many instances, the quieter product can function as well as the noisier product, and the cost of reducing the potential noise during the design stage may be minor. Even if the reduction of noise is somewhat expensive, it is important to reduce the level of noise to an acceptable value. There are more than 1000 local ordinances that limit the community noise from industrial installations, and there are legal liabilities associated with hearing loss of workers in industry.

The designer can no longer ignore noise when designing an industrial plant, an electrical generating system, or a commercial complex. In this book, we will consider some of the techniques that may be used by the engineer in reduction of noise from existing equipment and in design of a quieter product, in the case of new equipment.

We will begin with an introduction to the basic concepts of acoustics and acoustic measurement. It is important for the engineer to understand the nomenclature and physical principles involved in sound transmission in order to suggest a rational procedure for noise reduction.

We will examine methods for predicting the noise generated by several common engineering systems, such as fans, motors, compressors, and cooling towers. This information is required in the design stage of any noise control project. Information about the characteristics of the noise source can allow the design of equipment that is quieter in operation through adjustment of the machine speed or some other parameter.

How quiet should the machine be? This question may be answered by consideration of some of the design criteria for noise, including the OSHA, EPA, and HUD regulations, for example. We will also consider some of the

Introduction 3

criteria for noise transmitted outdoors and indoors, so that the anticipated community response to the noise may be evaluated.

A study of the noise control techniques applicable to rooms will be made. These procedures include the use of acoustic treatment of the walls of the room and the use of barriers and enclosures. It is important to determine if acoustic treatment of the walls will be effective or if the offending noise source must be enclosed to reduce the noise to an acceptable level.

The acoustic design principles for silencers or mufflers will be outlined. Specific design techniques for several muffler types will be presented.

Some noise problems are associated with excessive vibration of portions of the machine or transmission of machine vibration to the supporting structure. We will consider some of the techniques for vibration isolation to reduce noise radiated from machinery. The application of commercially available vibration isolators will be discussed.

Finally, several case studies will be presented in which the noise control principles are applied to specific pieces of equipment. The noise reduction achieved by the treatment will be presented, along with any pitfalls or caveats associated with the noise control procedure.

1.2 HISTORICAL BACKGROUND

Because of its connection with music, acoustics has been a field of interest for many centuries (Hunt, 1978). The Greek philosopher Pythagoras (who also stated the Pythagorean theorem of triangles) is credited with conducting the first studies on the physical origin of musical sounds around 550 BC (Rayleigh, 1945). He discovered that when two strings on a musical instrument are struck, the shorter one will emit a higher pitched sound than the longer one. He found that if the shorter string were half the length of the longer one, the shorter string would produce a musical note that was 1 octave higher in pitch than the note produced by the longer string: an *octave* difference in frequency (or pitch) means that the upper or higher frequency is two times that of the lower frequency. For example, the frequency of the note "middle C" is 262.6 Hz (cycles/sec), and the frequency of the "C" 1 octave higher is 523.2 Hz. Today, we may make measurements of the sound generated over standard octave bands or frequency ranges encompassing one octave. The knowledge of the frequency distribution of the noise generated by machinery is important in deciding which noise control procedure will be most effective.

The Greek philosopher Crysippus (240 BC) suggested that sound was generated by vibration of parts of the musical instrument (the strings, for example). He was aware that sound was transmitted by means of vibration of the air or other fluid, and that this motion caused the sensation of "hearing" when the waves strike a person's ear.

Credit is usually given to the Franciscan friar, Marin Mersenne (1588–1648) for the first published analysis of the vibration of strings (Mersenne, 1636). He measured the vibrational frequency of an audible tone (84 Hz) from a long string; he was also aware that the frequency ratio for two musical notes an octave apart was 2:1.

In 1638 Galileo Galilei (1939) published a discussion on the vibration of strings in which he developed quantitative relationships between the frequency of vibration of the string, the length of the string, its tension, and the density of the string. Galileo observed that when a set of pendulums of different lengths were set in motion, the oscillation produced a pattern which was pleasant to watch if the frequencies of the different pendulums were related by certain ratios, such as 2:1, 3:2, and 5:4 or octave, perfect fifth, and major third on the musical scale. On the other hand, if the frequencies were not related by simple integer ratios, the resulting pattern appeared chaotic and jumbled. He made the analogy between vibrations of strings in a musical instrument and the oscillating pendulums by observint that, if the frequencies of vibration of the strings were related by certain ratios, the sound would be pleasant or "musical." If the frequencies were not related by simple integer ratios, the resulting sound would be discordant and considered to be "noise."

In 1713 the English mathematician Brook Taylor (who also invented the Taylor series) first worked out the mathematical solution of the shape of a vibrating string. His equation could be used to derive a formula for the frequency of vibration of the string that was in perfect agreement with the experimental work of Galileo and Mersenne. The general problem of the shape of the wave in a string was fully solved using partial derivatives by the young French mathematician Joseph Louis Lagrange (1759).

There are some great blunders along the scientific route to the development of modern acoustic science. The French philosopher Gassendi (1592–1655) insisted that sound was propagated by the emission of small invisible particles from the vibrating surface. He claimed that these particles moved through the air and struck the ear to produce the sensation of sound.

Otto von Guericke (1602–1686) said that he doubted sound was transmitted by the vibratory motion of air, because sound was transmitted better when the air was still than when there was a breeze. Around the mid-1600s, he placed a bell in a vacuum jar and rang the bell. He claimed that he could hear the bell ringing inside the container when the air had been evacuated from the container. From this observation, von Guericke concluded that the air was not necessary for the transmission of sound. He did not recognize that the sound was being transmitted through the solid support structure of

Introduction

the bell. This story emphasized that we must be careful to consider *all* paths that noise may take, if we are to reduce noise effectively.

In 1660 Robert Boyle (who discovered Boyle's law for gases) repeated the experiment of von Guericke with a more efficient vacuum pump and more careful attention to the support. He observed a pronounced decrease in the intensity of the sound emitted from a ticking watch in the vacuum chamber as the air was pumped out. He correctly concluded that the air was definitely involved as a medium for sound transmission, although the air was not the only path that sound could take.

Sir Isaac Newton (1687) compared the transmission of sound and the motion of waves on the surface of water. By analogy with the vibration of a pendulum, Newton developed an expression for the speed of sound based on the assumption that the sound wave was transmitted *isothermally*, when in fact sound is transmitted *adiabatically* for small-amplitude sound waves. His incorrect expression for the speed of sound in a gas was:

$$c = (RT)^{1/2} \qquad (incorrect!) \tag{1-1}$$

R is the gas constant for the gas and T is the absolute temperature of the gas. For air (gas constant $R = 287$ J/kg-K) at 15°C (288.2K or 59°F), Newton's equation would predict the speed of sound to be 288 m/s (944 ft/sec), whereas the experimental value for the speed of sound at this temperature is 340 m/s (1116 ft/sec). Newton's expression was about 16% in error, compared with the experimental data. This was not a bad order of magnitude difference at the time; however, later more accurate measurements of the speed of sound consistently produced values larger than that predicted by Newton's relationship.

It wasn't until 1816 that the French astronomer and mathematician Pierre Simon Laplace suggested that sound was actually transmitted adiabatically because of the high frequency of the sound waves. Laplace proposed the correct expression for the speed of sound in a gas:

$$c = (\gamma RT)^{1/2} \tag{1-2}$$

where γ is the specific heat ratio for the gas. For air, $\gamma = 1.40$.

In 1877 John William Strutt Rayleigh published a two-volume work, *The Theory of Sound*, which placed the field of acoustics on a firm scientific foundation. Rayleigh also published 128 papers on acoustics between 1870 and 1919.

Between 1898 and 1900 Wallace Clement Sabine (1922) published a series of papers on reverberation of sound in rooms in which he laid the foundations of architectural acoustics. He also served as acoustic consultant for several projects, including the Boston Symphony Hall and the chamber of the House of Representatives in the Rhode Island State Capitol Building.

Sabine initially tried several optical devices, such as photographing a sensitive manometric gas flame, for measuring the sound intensity, but these measurements were not consistent. He found that the human ear, along with a suitable electrical timepiece, gave sensitive and accurate measurements of the duration of audible sound in the room.

One of the early acoustic "instruments" was a stethoscope developed by the French physician Rene Laennee. He used the stethoscope for clinical purposes in 1819. In 1827 Sir Charles Wheatstone, a British physicist who invented the famous Wheatstone bridge, developed an instrument similar to the stethoscope, which he called a "microphone." Following the invention of the triode vacuum tube in 1907 and the initial development of radio broadcasting in the 1920s, electric microphones and loudspeakers were produced. These developments were followed by the production of sensitive instruments designed to measure sound pressure levels and other acoustic quantities with a greater accuracy than could be achieved by the human ear.

Research was conducted during the 1920s on the concepts of subjective loudness and the response of the human ear to sound. Between 1930 and 1940, noise control principles began to be applied to buildings, automobiles, aircraft and ships. Also, during this time, researchers began to investigate the physical processes involved in sound absorption by porous acoustic materials.

With the advent of World War II, there was a renewed emphasis on solving problems in speech communication in noisy environments, such as in tanks and aircraft (Beranek, 1960). The concern for this problem area was so critical that the National Defense Research Committee (which later became the Office of Scientific Research and Development) established two laboratories at Harvard University. The Psycho-Acoustic Laboratory was involved in studies on sound control techniques in combat vehicles, and the Electro-Acoustic Laboratory conducted research on communication equipment for operation in a noisy environment and acoustic materials for noise control. After World War II ended, research in noise control and acoustics was continued at several other universities.

Noise problems in architecture and in industry were addressed in the post-war period. Research was directed toward solution of residential, workplace, and transportation noise problems. The amendment of the Walsh–Healy Act in 1969 gave rise to even more intense noise control activity in industry. This law required that the noise exposure of workers in the industrial environment be limited to a specific value (90 dBA for an 8-hour period). If this level of noise exposure could not be prevented, the law required that the workers be provided with and trained in the use of personal hearing protection devices.

Introduction

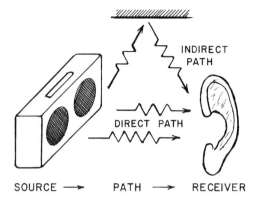

FIGURE 1-1 Three components of a general noise system: source of noise, path of the noise, and the receiver. The path may be direct from the source to the receiver, or the path may be indirect.

1.3 PRINCIPLES OF NOISE CONTROL

There are three basic elements in any noise control system, as illustrated in Fig. 1-1:

1. The source of the sound
2. The path through which the sound travels
3. The receiver of the sound (Faulkner, 1976).

In many situations, of course, there are several sources of sound, various paths for the sound, and more than one receiver, but the basic principles of noise control would be the same as for the more simple case. The objective of most noise control programs is to reduce the noise at the receiver. This may be accomplished by making modifications to the source, the path, or the receiver, or to any combination of these elements.

The source of noise or undesirable sound is a vibrating surface, such as a panel in an item of machinery, or small eddies with fluctuating velocities in a fluid stream, such as the eddies in a jet stream leaving an air vent pipe.

The path for the sound may be the air between the source and receiver, as is the case for machinery noise transmitted directly to the operator's ears. The path may also be indirect, such as sound being reflected by a wall to a person in the room. Solid surfaces, such as piping between a vibrating pump and another machine element, may also serve as the path for the noise propagation. It is important that the acoustic engineer identify all possible acoustic paths when considering a solution for a noise problem.

The receiver in the noise control system is usually the human ear, although the receiver could be sensitive equipment that would suffer impaired operation if exposed to excessively intense sound. It is important that the acoustic designer specify the "failure mode" for the receiver in any noise control project. The purpose of the noise control procedure may be to prevent hearing loss for personnel, to allow effective face-to-face communication or telephone conversation, or to reduce noise so that neighbors of the facility will not become intensely annoyed with the sound emitted by the plant. The engineering approach is often different in each of these cases.

1.3.1. Noise Control at the Source

Modifications at the source of sound are usually considered to be the best solution for a noise control problem. Components of a machine may be modified to effect a significant change in noise emission. For example, in a machine used to manufacture paper bags, by replacing the impact blade mechanism used to cut off the individual bags from the paper roll with a rolling cutter blade, a severe noise problem was alleviated.

Noise at the source may indicate other problems, such as a need for maintenance. For example, excessive noise from a roller bearing in a machine may indicate wear failure in one of the rollers in the bearing. Replacement of the defective bearing may solve the noise problem, in addition to preventing further mechanical damage to the machine.

There may be areas, such as panel coverings, that vibrate excessively on a machine. These panels are efficient sound radiators at wavelengths on the order of the dimensions of the panel. The noise generated by large vibrating panels can be reduced by applying damping material to the panel surface or by uncoupling the panel from the vibrating force, if possible. Making the panel stiffer by increasing the panel thickness or reducing the panel dimensions or using stiffening ribs may also reduce the amplitude of vibration. In most cases, reducing the amplitude of vibratory motion of elements in a machine will reduce the noise generated by the machine element.

In some cases, using two units with the same combined capacity as one larger unit may reduce the overall source noise. To determine whether this approach is feasible, the engineer would need information about the relationship of the machine capacity (power rating, flow rate capacity, etc.) and the sound power level for the generated noise from the machine. This information is presented in Chapter 5 for several noise sources.

A change in the process may also be used to reduce noise. Instead of using an air jet to remove debris from a manufactured part, rotating clean-

Introduction

ing brushes may be used. A centrifugal fan may replace a propellor-type fan to reduce the fan noise.

1.3.2 Noise Control in the Transmission Path

Modifying the path through which the noise is propagated is often used when modification of the noise source is not possible, not practical, or not economically feasible. For noise sources located outdoors, one simple approach for noise control would be to move the sound source farther away from the receiver, i.e. make the noise path longer.

For noise sources located outdoors or indoors, the transmission path may be modified by placing a wall or barrier between the source and receiver. Reduction of traffic noise from vehicles on freeways passing near residential areas and hospitals has been achieved by installation of acoustic barriers along the roadway.

The use of a barrier will not be effective in noise reduction indoors when the sound transmitted directly from the source to receiver is much less significant than the sound transmitted indirectly to the receiver through reflections on the room surfaces. For this case, the noise may be reduced by applying acoustic absorbing materials on the walls of the room or by placing additional acoustic absorbing surfaces in the room.

A very effective, although sometimes expensive, noise control procedure is to enclose the sound source in an acoustic enclosure or enclose the receiver in a personnel booth. The noise from metal cut-off saws has been reduced to acceptable levels by enclosing the saw in an acoustically treated box. Provision was made to introduce stock material to the saw through openings in the enclosure without allowing a significant amount of noise to be transmitted through the openings. If the equipment or process can be remotely operated, a personnel booth is usually an effective solution in reducing the workers' noise exposure. An air-conditioned control booth is also more comfortable for the operator of a paper machine than working in the hot, humid area surrounding the wet end of the paper machine, for example.

The exhaust noise from engines, fans, and turbines is often controlled by using mufflers or silencers in the exhaust line for the device. The muffler acts to reflect acoustic energy back to the noise source (the engine, for example) or to dissipate the acoustic energy as it is transmitted through the muffler.

1.3.3 Noise Control at the Receiver

The human ear is the usual "receiver" for noise, and there is a limited amount of modification that can be done for the person's ear. One possible approach to limit the noise exposure of a worker to industrial noise is to limit the time during which the person is exposed to high noise levels. As discussed in Chapter 6, a person can be exposed to a sound level of 95 dBA for 4 hours during each working day, and encounter a risk of "only" 10% of suffering significant permanent hearing loss, if the person remains in a much more quiet area during the remainder of the day. The 95 dBA sound level is typical of the noise from printing and cutting presses for folding cartons, for example (Salmon et al., 1975).

Hearing protectors (earplugs or acoustic muffs) can be effective in preventing noise-induced hearing loss in an industrial environment. In some cases, the use of hearing protectors may be the only practical means of limiting the workers' noise exposure, as is the case for workers who "park" airplanes at large air terminals. Because of inherent problems with hearing protectors, however, it is recommended that they should be used only as a last resort after other techniques have been reviewed. For example, the worker may not be able to hear warning horns or shouts of co-workers when wearing earplugs. One can get accustomed to wearing hearing protectors, but the earplugs are often less comfortable than wearing nothing at all. This characteristic of earplugs and people introduces some difficulty in enforcement of the use of hearing protection devices. In cases where earplugs are the only feasible solution to a noise exposure problem, an education, training, and monitoring program should be in place to encourage strongly the proper and effective use of the protective devices.

REFERENCES

Beranek, L. L. 1960. *Noise Reduction*, pp. 1–10. McGraw-Hill, New York.
Faulkner, L. L. 1976. *Handbook of Industrial Noise Control*, pp. 39–42. Industrial Press, New York.
Galilei, G. H. 1939. *Dialogues Concerning Two New Sciences* [translated from the Italian and Latin by Henry Crew and Alfonso de Salvio]. Evanston and Chicago. See also Lindsay, R. B. 1972. *Acoustics: Historical and Philosophical Development*, pp. 42–61. Dowden, Hutchison, and Ross, Stroudsburg, PA.
Hunt, F. V. 1978. *Origins of Acoustics*, p. 26. Yale University Press, New Haven, CT.
Mersenne, M. 1636. *Harmonicorum Liber*. Paris [see also Rayleigh, J. W. S. 1945. *The Theory of Sound*, 2nd ed., pp. xiii–xiv. Dover Publications, New York.]
Newton, I. 1687. *Principia*, 2nd book [see Cajori, F. 1934. *Newton's Principia: Motte's Translation Revised*. University of California Press, Berkeley, CA.]
Occupational Safety and Health Administration. 1983. Occupational noise exposure: hearing conservation amendment. *Fed. Reg.* 48(46): 9738–9785.
Rayleigh, J. W. S. 1945. *The Theory of Sound*, 2nd ed, pp. xi–xxii. Dover Publications, New York.

Sabine, W. C. 1922. *Collected Papers on Acoustics*, pp. 3–68. Peninsula Publishing, Los Altos, CA.

Salmon, V., Mills, J. S., and Peterson, A. C. 1975. *Industrial Noise Control Manual*. HEW Report (NIOSH) 75-183, p. 146. US Government Printing Office, Washington, DC.

2
Basics of Acoustics

2.1 SPEED OF SOUND

Sound is defined as a pressure disturbance that moves through a material at a speed which is dependent on the material (Beranek and Vér, 1992). Sound waves in fluids are often produced by vibrating solid surfaces in the fluid, as shown in Fig. 2-1. As the vibrating surface moves to the right, the fluid adjacent to the surface is compressed. This compression effect moves outward from the vibrating surface as a sound wave. Similarly, as the surface moves toward the left, the fluid next to the surface is rarefied. The vibratory motion of the solid surface causes pressure variations above and below the fluid bulk pressure (atmospheric pressure, in many cases) to be transmitted into the surrounding fluid.

Noise is usually defined as any perceived sound that is objectionable or damaging for a human. Noise is somewhat subjective, because one person's "music" may be another person's "noise." Some sounds that could be classified as noise, such as the warning whistle on a train, are actually beneficial by warning people of potential dangerous situations.

The speed of sound in various materials is given in Appendix B. For an ideal gas, the speed of sound is a function of the absolute temperature of the gas:

$$c = (g_e \gamma RT)^{1/2} \qquad (2\text{-}1)$$

Basics of Acoustics

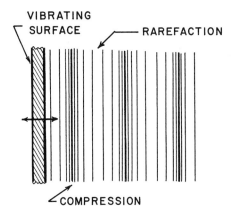

FIGURE 2-1 Sound waves in materials.

where g_c is the units conversion factor, $g_c = 1\text{ kg-m/N-s}^2 = 32.174\text{ lb}_m\text{-ft/lb}_f\text{-sec}^2$; γ is the specific heat ratio, $\gamma = c_p/c_v$; R is the specific gas constant for the gas, $R = 287\text{ J/kg-K} = 53.35\text{ ft-lb}_f/\text{lb}_m\text{-}°\text{R}$ for air; and T is the absolute temperature, K or °R.

The speed of sound (or c^2) in a fluid (liquid or gas), in general, is given by:

$$c^2 = \frac{\gamma B}{\rho} \tag{2-2}$$

where B is the isothermal bulk modulus and ρ is the fluid density. For transverse (bulk) sound waves in a solid, the speed of sound is given by (Timoshenko, 1970):

$$c^2 = \frac{(1-v)E}{(1+v)(1-2v)\rho} \tag{2-3}$$

where E is Young's modulus and v is Poisson's ratio for the material. For sound transmitted through a thin bar, the speed of sound expression reduces to:

$$c = (E/\rho)^{1/2} \tag{2-4}$$

2.2 WAVELENGTH, FREQUENCY, AND WAVE NUMBER

There is a single frequency (f) associated with a simple harmonic wave or sinusoidal wave. This frequency depends on the frequency of vibration of the source of sound and is independent of the material through which the sound is transmitted for non-dissipative sound transmission. The *period* (τ) for a wave is defined as the time elapsed during one complete cycle for the wave, or the time elapsed between the passage of the successive peaks for a simple harmonic wave, as shown in Fig. 2-2. The frequency is the reciprocal of the period, $f = 1/\tau$. The unit for the frequency is hertz (Hz), named in honor of the German physicist Heinrich Rudolph Hertz, who conducted pioneering studies in electromagnetism and in elasticity (Timoshenko, 1983). The unit hertz is the same as the unit cycle/sec.

To get a physical understanding of the magnitude of the frequency of sound waves usually considered in noise control, we may note that the range of audibility for the undamaged human ear is from about 16 Hz to about 16 kHz. Frequencies below about 16 Hz are considered *infrasound*, and frequencies above 16 kHz are *ultrasound*. The standard musical pitch (frequency) is A-440, or the note "A" above middle "C" has an assigned frequency of 440 Hz. The soprano voice usually ranges from about middle "C" ($f = 261.6$ Hz) to approximately "C" above the staff ($f = 1046.5$ Hz). Thus, the female voice has a frequency on the order of 500 Hz. The baritone voice usually ranges from about 90 Hz to 370 Hz, so the male voice has a frequency on the order of 200 Hz.

The *wavelength* (λ) of the sound wave is an important parameter in determining the behavior of sound waves. If we take a "picture" of the wave at a particular instant in time, as shown in Fig. 2-2, the wavelength is the distance between successive peaks of the wave. The wavelength and speed of sound for a simple harmonic wave are related by:

$$\lambda = c/f \tag{2-5}$$

Another parameter that is encountered in analysis of sound waves is the *wave number* (k), which is defined by:

$$k = \frac{2\pi}{\lambda} = \frac{2\pi f}{c} \tag{2-6}$$

Example 2-1. A sound wave having a frequency of 250 Hz is transmitted through air at 25°C (298.2 K or 77°F). The gas constant for air is 287 J/kg-K, and the specific heat ratio is $\gamma = 1.40$. Determine the speed of sound, wavelength, and wave number for this condition.

The speed of sound is found from Eq. (2-1):

$$c = (g_c \gamma R T)^{1/2} = [(1)(1.40)(287)(298.2)]^{1/2}$$
$$c = 346.1 \text{ m/s} = 1136 \text{ ft/sec} = 774 \text{ mph}$$

Basics of Acoustics

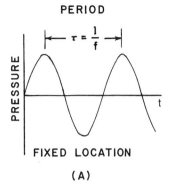

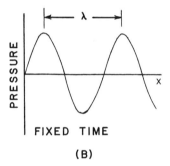

FIGURE 2-2 Wavelength and period for a simple harmonic wave: (A) pressure *vs.* time and (B) pressure *vs.* position.

The wavelength for a frequency of 250 Hz is:

$$\lambda = \frac{c}{f} = \frac{346.1}{250} = 1.385\,\text{m} = 4.543\,\text{ft}$$

The wave number is:

$$k = \frac{2\pi}{\lambda} = \frac{2\pi}{1.385} = 4.538\,\text{m}^{-1}$$

2.3 ACOUSTIC PRESSURE AND PARTICLE VELOCITY

The *acoustic pressure* (p) is defined as the instantaneous difference between the local pressure (P) and the ambient pressure (P_o) for a sound wave in the material. The acoustic pressure for a plane simple harmonic sound wave moving in the positive x-direction may be represented by the following.

$$p(x, t) = p_{\max} \sin(2\pi f t - kx) \tag{2-7}$$

The quantity $p_{\max}$ is the *amplitude* of the acoustic pressure wave.

Acoustic instruments, such as a sound level meter, generally do not measure the amplitude of the acoustic pressure wave; instead, these instruments measure the root-mean-square (rms) pressure, which is proportional to the amplitude. The relation between the pressure wave amplitude and the rms pressure is demonstrated in the following.

Suppose we define the variable $\theta = 2\pi t/\tau$, so $d\theta = 2\pi \, dt/\tau$. The rms pressure is defined as the square root of the average of the square of the instantaneous acoustic pressure over one period of vibration τ:

$$(p_{\text{rms}})^2 = \frac{1}{\tau} \int_0^\tau p^2(x, t) \, dt = \frac{(p_{\max})^2}{2\pi} \int_0^{2\pi} \sin^2(\theta - kx) \, d\theta$$

Carrying out the integration, we find:

$$(p_{\text{rms}})^2 = \frac{(p_{\max})^2}{2\pi} \left[\tfrac{1}{2}(\theta - kx) - \tfrac{1}{4}\sin(2\theta - 2kx)\right]_0^{2\pi}$$

$$(p_{\text{rms}})^2 = \tfrac{1}{2}(p_{\max})^2$$

The rms pressure is related to the pressure amplitude for a simple harmonic wave by:

$$p_{\text{rms}} = \frac{p_{\max}}{\sqrt{2}} \tag{2-8}$$

To avoid excessive numbers of subscripts, we will use the symbol p (without the subscript rms) to denote the rms acoustic pressure in the following material, except where stated otherwise.

The *instantaneous acoustic particle velocity* (u) is defined as the local motion of particles of fluid as a sound wave passes through the material. The rms acoustic particle velocity is the quantity used in engineering analysis, because it is the quantity pertinent to energy and intensity measurements.

The rms acoustic pressure and the rms acoustic particle velocity are related by the *specific acoustic impedance* (Z_s):

$$p = Z_s u \tag{2-9}$$

The specific acoustic impedance is often expressed in complex notation to display both the magnitude of the pressure–velocity ratio and the phase angle between the pressure and velocity waves. The SI units for specific acoustic impedance are Pa-s/m. This combination of units has been given

Basics of Acoustics

the special name rayl, in honor of Lord Rayleigh, who wrote the famous book on acoustics: i.e., 1 rayl ≡ 1 Pa-s/m. In conventional units, the specific acoustic impedance would be expressed in lb_f-sec/ft^3.

For plane acoustic waves, the specific acoustic impedance is a function of the fluid properties only. The specific acoustic impedance for plane waves is called the *characteristic impedance* (Z_o) and is given by:

$$Z_o = \rho c / g_c \qquad (2\text{-}10)$$

(Note that, since the quantity g_c is a units conversion factor, it is often omitted from equations, and it is assumed that consistent units will be maintained when substituting values in the equations.) Values for the characteristic impedance for several materials are given in Appendix B.

Example 2-2. A plane sound is transmitted through air ($R = 287$ J/kg-K) at 25°C (298.2K or 77°F) and 101.3 kPa (14.7 psia). The speed of sound in the air is 346.1 m/s. The sound wave has an acoustic pressure (rms) of 0.20 Pa. Determine the rms acoustic particle velocity.

The density of the air may be determined from the ideal gas equation of state:

$$\rho = \frac{P_0}{RT} = \frac{(101.3)(10^3)}{(287)(298.2)} = 1.184 \, \text{kg/m}^3$$

The characteristic impedance for the air is:

$$Z_o = \rho c / g_c = (1.184)(346.1)/(1) = 409.8 \, \text{Pa-s/m} = 409.8 \, \text{rayl} = p/u$$

The acoustic particle velocity may be evaluated:

$$u = \frac{0.20}{409.8} = 0.488 \times 10^{-3} \, \text{m/s} = 0.488 \, \text{mm/s} \, (0.0192 \, \text{in/sec})$$

We observe that the acoustic particle velocity (0.000448 m/s) is a rather small quantity and is generally much smaller than the acoustic velocity (346.1 m/s).

2.4 ACOUSTIC INTENSITY AND ACOUSTIC ENERGY DENSITY

The *acoustic intensity* (I) is defined as the average energy transmitted through a unit area per unit time, or the acoustic power (W) transmitted per unit area. The SI units for acoustic intensity are W/m^2. The conventional units ft-lb_f/sec-ft^2 are not used in acoustic work at the present time.

For plane sound waves, as shown in Fig. 2-3, the acoustic intensity is related to the acoustic power and the area (S) by:

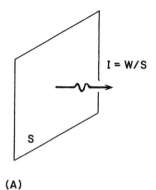

(A)

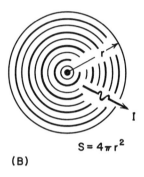

(B)

FIGURE 2-3 Intensity for (A) plane waves and (B) spherical waves.

$$I = \frac{W}{S} \qquad (2\text{-}11)$$

For a spherical sound wave (a sound wave that moves out uniformly in all directions from the source), the area through which the acoustic energy is transmitted is $4\pi r^2$, where r is the distance from the sound source, so the intensity is given by:

$$I = \frac{W}{4\pi r^2} \qquad (2\text{-}12)$$

For the general case in which the sound is not radiated uniformly from the source, but the acoustic intensity may vary with direction, the intensity is given by:

$$I = \frac{QW}{4\pi r^2} \qquad (2\text{-}13)$$

Basics of Acoustics

The quantity Q is called the *directivity factor*, which is a dimensionless quantity that generally depends on the direction and the frequency of the sound wave.

The acoustic intensity may be related to the rms acoustic pressure. The average acoustic power per unit area, averaged over one period for the acoustic wave, is given by:

$$I = \frac{1}{\tau}\int_0^\tau p(x,t)u(x,t)\,dt = \frac{1}{2\pi}\int_0^{2\pi} p(x,t)u(x,t)\,d\theta \qquad (2\text{-}14)$$

where $\theta = 2\pi f t = (2\pi/\tau)t$. Let us use the following expressions for the acoustic pressure and acoustic particle velocity for a plane wave:

$$\begin{aligned} p(x,t) &= \sqrt{2}\,p_{\text{rms}}\sin(2\pi t - kx) \\ u(x,t) &= (\sqrt{2}\,p_{\text{rms}}/\rho c)\sin(2\pi t - kx) \end{aligned} \qquad (2\text{-}15)$$

Making these substitutions into Eq. (2-14), we find:

$$I = \frac{1}{2\pi}\int_0^{2\pi}\frac{2(p_{\text{rms}})^2}{\rho c}\sin^2(\theta - kx)\,d\theta$$

$$I = \frac{2(p_{\text{rms}})^2}{2\pi\rho c}\left[\tfrac{1}{2}(\theta - kx) - \tfrac{1}{4}\sin(2\theta - 2kx)\right]_0^{2\pi}$$

The final expression for the acoustic intensity becomes:

$$I = \frac{p^2}{\rho c} \qquad (2\text{-}16)$$

where $p = p_{\text{rms}}$. We will show that this same expression also applies for a spherical sound wave and for a non-spherical sound wave.

When making sound measurements in a room or other enclosure, one parameter of interest is the *acoustic energy density* (D), which is the total acoustic energy per unit volume. The SI unit for the acoustic energy density is J/m³. The total acoustic energy is composed of two parts: the kinetic energy, associated with the motion of the vibrating fluid; and the potential energy, associated with energy stored through compression of the fluid.

The kinetic energy per unit volume, averaged over one wavelength, may be expressed in terms of the acoustic particle velocity:

$$\text{KE} = \frac{1}{\lambda}\int_0^\lambda \tfrac{1}{2}\rho u^2(x,t)\,dx = \frac{1}{2\pi}\int_0^{2\pi}\tfrac{1}{2}\rho u^2(\zeta,\theta)\,d\zeta$$

where $\zeta = kx$. If we use the acoustic particle velocity expression from Eq. (2-15) for a plane wave, we find:

$$KE = \frac{p^2}{2\rho c^2} \tag{2-17}$$

For a spherical sound wave, the acoustic pressure and acoustic particle velocity are not in-phase. We will show that the kinetic energy per unit volume for a spherical wave is dependent on the frequency (or the wave number, k) for the sound wave, and the distance from the sound source, r, as follows.

$$KE = \frac{p^2}{2\rho c^2}\left(1 + \frac{1}{k^2 r^2}\right) \tag{2-18}$$

The potential energy may also be related to the acoustic pressure. For a plane sound wave, the potential energy per unit volume, averaged over one wavelength, is given by:

$$PE = \frac{1}{\lambda}\int_0^\lambda \frac{p^2(x,t)}{2\rho c^2}\,dx = \frac{1}{2\pi}\int_0^{2\pi} \frac{p^2(\theta,\zeta)}{2\rho c^2}\,d\zeta$$

Using the expression for the acoustic pressure from Eq. (2-15), we obtain the following equation for the potential energy per unit volume:

$$PE = \frac{p^2}{2\rho c^2} \tag{2-19}$$

By comparison of Eqs (2-17) and (2-19), we see that, for a plane sound wave, the kinetic and potential contributions to the total energy are equal. The total acoustic energy is half kinetic and half potential, for a plane sound wave: this is not the case for a spherical wave.

For a plane sound wave, the acoustic energy density is found by adding the kinetic energy, Eq. (2-17), and the potential energy, Eq. (2-19):

$$D = \frac{p^2}{\rho c^2} \tag{2-20}$$

If we compare Eq. (2-20) with Eq. (2-16), we see that (for a plane sound wave) the acoustic intensity and acoustic energy density are related:

$$D = \frac{I}{c} \tag{2-21}$$

For a spherical sound wave, the acoustic energy density is given by:

$$D = \frac{p^2}{\rho c^2}\left(1 + \frac{1}{2k^2 r^2}\right) \tag{2-22}$$

Basics of Acoustics

Example 2-3. A plane sound wave is transmitted through air (speed of sound, 346.1 m/s; characteristic impedance, 409.8 rayl) at 25°C (298.2K or 77°F) and 101.3 kPa (14.7 psia). The sound wave has an acoustic pressure (rms) of 0.20 Pa. Determine the acoustic intensity and acoustic energy density for the sound wave.

The acoustic intensity is given by Eq. (2-16):

$$I = \frac{p^2}{\rho c} = \frac{(0.20)^2}{(409.8)} = 97.6 \times 10^{-6}\,\text{W/m}^2 = 97.6\,\mu\text{W/m}^2$$

The SI prefixes are listed in Appendix A.

The acoustic energy density is given by Eq. (2-20):

$$D = \frac{p^2}{\rho c^2} = \frac{p^2}{Z_o c} = \frac{(0.20)^2}{(409.8)(346.1)} = 0.282 \times 10^{-6}\,\text{J/m}^3 = 0.282\,\mu\text{J/m}^3$$

This is actually an extremely small quantity of energy. The specific heat of air at 25°C is $c_p = 1005.7\,\text{J/kg-°C}$. The thermal capacity per unit volume is:

$$\rho c_p = (1.184)(1005.7) = 1190.7\,\text{J/m}^3\text{-°C}$$

If all of the acoustic energy in this problem were dissipated into the air, the temperature of the air would rise by:

$$\Delta T = \frac{D}{\rho c_p} = \frac{(0.282)(10^{-6})}{(1190.7)} = 0.24 \times 10^{-9}\,°\text{C}\ (0.43 \times 10^{-9}\,°\text{F})$$

2.5 SPHERICAL WAVES

In many situations, the size of the source of sound is relatively small, and the sound is radiated from the source uniformly in all directions. In this case, the sound waves would not be planar; instead, the sound waves are called *spherical* waves. By combining Eq. (2-12) and Eq. (2-16) for spherical waves, we see that the acoustic pressure varies inversely with the distance from the sound sosurce, r, because the acoustic power W is constant for the case of zero energy dissipation:

$$I = \frac{p^2}{\rho c} = \frac{W}{4\pi r^2} \qquad (2\text{-}23)$$

The acoustic power is spread over a larger area as the sound wave moves away from the source, so the acoustic intensity decreases inversely proportional to r^2.

From the solution of the acoustic wave equation in Chapter 4, we find that the magnitude of the specific acoustic impedance for a spherical sound wave is given by:

$$Z_s = \frac{\rho c k r}{(1 + k^2 r^2)^{1/2}} = \frac{Z_o k r}{(1 + k^2 r^2)^{1/2}} \qquad (2\text{-}24)$$

where $k = (2\pi/\lambda) = (2\pi f/c) =$ wave number. The phase angle (ϕ) between the acoustic pressure and the acoustic particle velocity is found from:

$$\tan \phi = \frac{1}{kr} \qquad (2\text{-}25)$$

We may note two limiting cases for the acoustic impedance of spherical waves. For long wavelengths or low frequencies ($kr \ll 1$), the acoustic impedance approaches $(Z_o kr) = 2\pi \rho f r$, and the phase angle approaches $\frac{1}{2}\pi$ rad $= 90°$. This regime, $kr < 0.1$ approximately, is called the *near-field* regime. The acoustic pressure and acoustic particle velocity are almost $90°$ out of phase, and the acoustic pressure produced by a spherical source is very small near the source, for a given acoustic particle velocity.

For short wavelengths (high frequencies) or for distances far from the source ($kr \gg 1$), the specific acoustic impedance approaches the characteristic impedance ($Z_s \approx Z_o$), and the phase angle is approximately zero. This region, $kr > 5$ approximately, is called the *far-field* regime. In this regime, the spherical wave appears to behave almost as a plane sound wave.

Because the acoustic pressure and acoustic particle velocity are not in-phase for a spherical wave, the potential energy and kinetic energy of the acoustic wave are not equal, as is the case for a plane wave. The acoustic energy density for a spherical wave is given by:

$$D = \frac{p^2}{\rho c^2}\left(1 + \frac{1}{2k^2 r^2}\right) \qquad (2\text{-}26)$$

The kinetic energy contribution is given by Eq. (2-18), and the potential energy contribution is given by Eq. (2-19). For the near-field regime ($1/2k^2 r^2 \gg 1$), the kinetic energy contribution predominates; whereas, in the far-field regime ($1/2k^2 r^2 \ll 1$), the kinetic energy and potential energy contributions are equal.

Example 2-4. A spherical source of sound produces an acoustic pressure of 2 Pa at a distance of 1.20 m (3.937 ft or 47.2 in) from the source in air at 25°C (77°F) and 101.3 kPa (14.7 psia). The frequency of the sound wave is 125 Hz. Determine the rms acoustic particle velocity, the acoustic energy density, and acoustic intensity for the sound wave at 1.20 m from the source and at a distance of 2.50 m (8.202 ft) from the source.

Basics of Acoustics

The characteristic acoustic impedance is $Z_o = 409.8$ rayl from Appendix B. The wave number is:

$$k = \frac{2\pi f}{c} = \frac{(2\pi)(125)}{(346.1)} = 2.269 \text{ m}^{-1}$$

The parameter $kr = (2.269)(1.20) = 2.723$. This value is neither in the near-field nor the far-field regime. The specific acoustic impedance may be evaluated from Eq. (2-24):

$$Z_s = \frac{Z_o kr}{(1+k^2 r^2)^{1/2}} = \frac{(409.8)(2.723)}{(1+2.723^2)^{1/2}} = (409.8)(0.9387) = 384.7 \text{ rayl}$$

The acoustic particle velocity at a distance of 1.20 m from the source is:

$$u = \frac{p}{Z_s} = \frac{(2.00)}{(384.7)} = 0.00520 \text{ m/s} = 5.20 \text{ mm/s} \quad (0.205 \text{ in/sec})$$

The phase angle between the acoustic pressure and acoustic particle velocity is given by Eq. (2-25):

$$\phi = \tan^{-1}(1/kr) = \tan^{-1}(1/2.723) = 20.2°$$

The acoustic intensity is found from Eq. (2-23):

$$I = \frac{p^2}{Z_o} = \frac{(2.00)^2}{(409.8)} = 0.00976 \text{ W/m}^2 = 9.76 \text{ mW/m}^2$$

The acoustic power radiated from the source is:

$$W = 4\pi r^2 I = (4\pi)(1.20)^2 (9.76)(10^{-3}) = 0.1766 \text{ W}$$

For a distance of 1.20 m from the source, the acoustic energy density is given by Eq. (2.-26):

$$D = \frac{(2.00)^2}{(1.184)(346.1)^2} \left[1 + \frac{1}{(2)(2.723)^2}\right] = (28.2)(10^{-6})(1.0674)$$
$$D = 30.1 \times 10^{-6} \text{ J/m}^3 = 30.1 \text{ }\mu\text{J/m}^3$$

The acoustic energy dissipation is negligible for sound transmitted through a few meters in air; therefore, the acoustic power at a distance of 2.50 m from the source is also 0.1766 W. The acoustic intensity at a distance of 2.50 m from the source is:

$$I = \frac{W}{4\pi r^2} = \frac{(0.1766)}{(4\pi)(2.50)^2} = 0.00225 \text{ W/m}^2 = 2.25 \text{ mW/m}^2$$

The acoustic pressure at a distance of 2.50 m from the source is:

$$p = (Z_o I)^{1/2} = [(409.8)(0.00225)]^{1/2} = 0.960 \, \text{Pa}$$

We note that both the intensity and acoustic pressure decrease for a spherical wave as we move away from the source of sound, because the area through which the energy is distributed is increased.

The wave number is not affected by the position, so:

$$kr = (2.269)(2.50) = 5.673$$

The specific acoustic impedance becomes:

$$Z_s = \frac{(409.8)(5.673)}{(1 + 5.673^2)^{1/2}} = (409.8)(0.9848) = 403.6 \, \text{rayl}$$

The acoustic particle velocity at 2.50 m from the source is:

$$u = \frac{p}{Z_s} = \frac{(0.960)}{(403.6)} = 0.00238 \, \text{m/s} = 2.38 \, \text{mm/s} \quad (0.937 \, \text{in/sec})$$

The acoustic energy density is:

$$D = \frac{(0.960)^2}{(1.184)(346.1)^2} \left[1 + \frac{1}{(2)(5.673)^2} \right] = (6.50)(10^{-6})(1.0155)$$

$$D = 6.60 \times 10^{-6} \, \text{J/m}^3 = 6.60 \, \mu\text{J/m}^3$$

2.6 DIRECTIVITY FACTOR AND DIRECTIVITY INDEX

The acoustic energy is radiated uniformly in all directions for a spherical wave; however, other sources of sound may be highly directional. These directional sources radiate sound with different intensities in different directions. The intensity of noise radiated from a vent pipe along the axis of the vent pipe, for example, is different from the intensity perpendicular to the vent pipe axis. In fact, if a spherical source is placed near the floor or a wall, some sound will be reflected from the surface and will not be radiated in all directions.

The *directivity factor* (Q) is defined as the ratio of the intensity on a designated axis of a sound radiator at a specific distance from the source to the intensity that would be produced at the same location by a spherical source radiating the same total acoustic energy:

$$Q = \frac{4\pi r^2 I}{W} \tag{2-27}$$

Basics of Acoustics

The *directivity index* (DI) is related to the directivity factor by:

$$\text{DI} = 10\log_{10}(Q) \qquad (2\text{-}28)$$

For a spherical source, the directivity factor $Q = 1$ and the directivity index $\text{DI} = 0$.

The directivity factor may be determined from analytical or experimental values of the acoustic pressure. The *directional pressure distribution function* $H(\theta, \varphi)$ is defined by:

$$H(\theta, \varphi) = \frac{p(\theta, \varphi)}{p(0)} \qquad (2\text{-}29)$$

The quantity θ is the azimuth angle, and φ is the polar angle, as shown in Fig. 2-4: $p(0)$ is the acoustic pressure on the axis, $\theta = 0$. The directivity factor may be evaluated from the directional pressure distribution function:

$$Q = \frac{4\pi}{\displaystyle\int_0^\pi \int_0^{2\pi} H^2(\theta, \varphi) \sin\theta \, d\varphi \, d\theta} \qquad (2\text{-}30)$$

If the pressure distribution is symmetrical, or $H(\theta, \varphi) = H(\theta)$, the integration with respect to φ may be carried out directly. The directivity factor for a symmetrical source of sound is given by:

$$Q = \frac{2}{\displaystyle\int_0^\pi H^2(\theta) \sin\theta \, d\theta} \qquad (2\text{-}31)$$

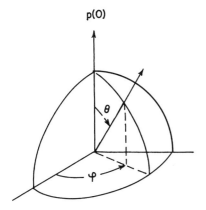

Figure 2-4 Spherical coordinates for directivity factor.

The directivity factor for locations off the axis (Q_θ) may be expressed, as follows:

$$Q_\theta(\theta, \varphi) = QH^2(\theta, \varphi) \qquad (2\text{-}32)$$

If a spherical source of sound is placed near the floor or a wall, as shown in Fig. 2-5, sound is radiated through a hemispherical area, $S = 2\pi r^2$. In this case, the intensity is:

$$I = \frac{W}{2\pi r^2} = \frac{2W}{4\pi r^2} = \frac{QW}{4\pi r^2}$$

For this case, we see that the directivity factor is $Q = 2$, and the directivity index is:

$$DI = 10\log_{10}(2) = 3.0$$

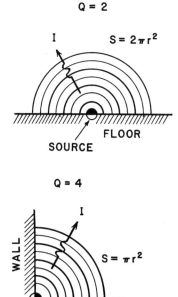

FIGURE 2-5 Sound sources near a surface for (A) directivity factor $Q = 2$ and directivity index $DI = 3$ and (B) for $Q = 4$ and $DI = 6$.

Basics of Acoustics

Similarly, if the spherical source were placed on the floor near a wall, the energy is radiated through an area $S = \pi r^2$. For this case,

$$I = \frac{W}{\pi r^2} = \frac{4W}{4\pi r^2} = \frac{QW}{4\pi r^2}$$

The directivity factor, in this case, is $Q = 4$ and the directivity index is 6.0. By going through the same reasoning, we may show that if the spherical source were placed in a corner near the floor and two walls, $Q = 8$ and $DI = 9.0$.

From a practical standpoint, these results show the importance of location of a noisy piece of machinery. If the machine is located on the floor, it will produce an intensity that is about twice that produced by the same machine away from the floor. The intensity for the machine located on the floor near a wall will be about four times that measured with the machine away from reflective surfaces.

Example 2-5. A source of sound radiates symmetrically with the following directional pressure distribution function:

$$H(\theta) = \cos \theta$$

Determine the directivity factor and directivity index in the direction $\theta = 0$.

The integral in the denominator of Eq. (2-31) may be evaluated first:

$$\int_0^\pi H^2(\theta) \sin\theta \, d\theta = \int_0^\pi \cos^2\theta \sin\theta \, d\theta = \left[-\tfrac{1}{3}\cos^3\theta\right]_0^\pi = 2/3$$

The directivity factor is evaluated from Eq. (2-31).

$$Q = \frac{2}{2/3} = 3$$

The directivity index is found from Eq. (2-28)

$$DI = 10 \log_{10}(3) = 4.8$$

The directivity factor at an angle of $\theta = 45°$ from the axis is:

$$Q_\theta = QH^2(\theta) = (3.00)\cos^2(45°) = 1.500$$

2.7 LEVELS AND THE DECIBEL

The range of the quantities used in acoustics, such as acoustic pressure, intensity, power, and energy density, is quite large. For example, the undamaged human ear can detect sounds having an acoustic pressure as small as 20 µPa, and the ear can withstand sounds for a few minutes having a sound pressure as large as 20 Pa. As a consequence of this wide range of magnitudes, there was an interest in developing a scale that could represent these quantities in a more convenient manner. In addition, it was found that the response of the human ear to sound was more dependent on the ratio of intensity of two different sounds, instead of the difference in intensity. For these reasons, a logarithmic scale called the *level* scale was defined.

The level of any quantity is defined as the logarithm to the base 10 of the ratio of an *energy-like* quantity to a standard reference value of the quantity. The common logarithms (base 10) are used, instead of the natural or napierian logarithms (base e), because the scale was developed years prior to the advent of hand-held calculators. Common logarithm tables were much more convenient to use for widely different quantities than natural logarithm tables. An energy-like quantity (for example, p^2) is used, because energy is a scalar quantity and an additive quantity. This means that all levels may be combined in the same manner, if an energy-like quantity is used.

Although the level is actually a dimensionless quantity, it is given the unit of *bel*, in honor of Alexander Graham Bell. It is general practice to use the *decibel* (dB), where 1 decibel is equal to 0.1 bel. The history of the development of the bel unit is described by Huntley (1970). The level is usually designated by the symbol L, with a subscript to denote the quantity described by the level. For example, the acoustic power level is designated by L_W. The acoustic power level is defined by:

$$L_W = 10 \log_{10}(W/W_{\text{ref}}) \tag{2-33}$$

The factor 10 converts from bels to decibels. The reference acoustic power (W_{ref}) is 10^{-12} watts or 1 pW.

The sound intensity level and sound energy density level are defined in a similar manner, since both of these quantities (I and D) are proportional to energy:

$$L_I = 10 \log_{10}(I/I_{\text{ref}}) \tag{2-34}$$

$$L_D = 10 \log_{10}(D/D_{\text{ref}}) \tag{2-35}$$

where the reference quantities are:

$$I_{\text{ref}} = 10^{-12} \text{ W/m}^2 = 1 \text{ pW/m}^2$$

$$D_{\text{ref}} = 10^{-12} \text{ J/m}^3 = 1 \text{ pJ/m}^3$$

The reference quantities were not completely arbitarily selected. At a frequency of 1000 Hz, a person with normal hearing can barely hear a sound having an acoustic pressure of 20 µPa. For this reason, the reference acoustic pressure was selected as:

Basics of Acoustics

$$p_{ref} = 20\,\mu Pa = 20 \times 10^{-6}\,Pa$$

The characteristic impedance for air at ambient temperature and pressure is approximately $Z_o \approx 400$ rayl. The acoustic intensity corresponding to a sound pressure of $20\,\mu Pa$ moving through ambient air is approximately $I_{ref} = (20 \times 10^{-6})^2/(400) = 10^{-12}\,W/m^2$ or $1\,pW$. The acoustic power corresponding to the reference intensity and a "unit" area of $1\,m^2$ is $W_{ref} = (10^{-12})(1) = 10^{-12}\,W$ or $1\,pW$. The reference acoustic energy ($D_{ref} = 1\,pJ/m^3$) was somewhat arbitarily selected, because the acoustic energy density for a plane sound wave in ambient air with the reference sound pressure level is approximately $0.003\,pJ/m^3$.

We note that the acoustic pressure is not proportional to the energy, but instead, p^2 is proportional to the energy (intensity or energy density). For this reason, the sound pressure level is defined by:

$$L_p = 10\log_{10}(p^2/p_{ref}^2) = 20\log_{10}(p/p_{ref}) \qquad (2\text{-}36)$$

The expressions for the various "levels" and the reference quantities, according to ISO and ANSI, are given in Table 2-1.

One feature of the use of the decibel notation is that many expressions involve addition or subtraction, instead of multiplication or division. This feature was advantageous before the advent of hand-held digital calculators and digital computers. If we combine Eqs (2-13) and (2-16) for the acoustic intensity, we obtain:

$$I = \frac{QW}{4\pi r^2} = \frac{p^2}{\rho c} \qquad (2\text{-}37)$$

TABLE 2-1 Reference Quantities for Acoustic Levels

Quantity	Definition, dB	Reference
Sound pressure level	$L_p = 20\log_{10}(p/p_{ref})$	$p_{ref} = 20\,\mu Pa$
Intensity level	$L_I = 10\log_{10}(I/I_{ref})$	$I_{ref} = 1\,pW/m^2$
Power level	$L_W = 10\log_{10}(W/W_{ref})$	$W_{ref} = 1\,pW$
Energy level	$L_E = 10\log_{10}(E/E_{ref})$	$E_{ref} = 1\,pJ$
Energy density level	$L_D = 10\log_{10}(D/D_{ref})$	$D_{ref} = 1\,pJ/m^3$
Vibratory acceleration level	$L_a = 20\log_{10}(a/a_{ref})$	$a_{ref} = 10\,\mu m/s^2$
Vibratory velocity level	$L_v = 20\log_{10}(v/v_{ref})$	$v_{ref} = 10\,nm/s$
Vibratory displacement level	$L_d = 20\log_{10}(d/d_{ref})$	$d_{ref} = 10\,pm$
Vibratory force level	$L_F = 20\log_{10}(F/F_{ref})$	$F_{ref} = 1\,\mu N$
Frequency level	$L_{fr} = 10\log_{10}(f/f_{ref})$	$f_{ref} = 1\,Hz$

Note: The SI prefixes are listed in Appendix A.
Source: From ISO Recommendation No. 1683 and American National Standard ANSI S1.8 (1989).

The acoustic pressure is related to the acoustic power:

$$p^2 = \frac{WQ\rho c}{4\pi r^2} \tag{2-38}$$

We may introduce the reference pressure and reference power:

$$\frac{p^2}{p_{ref}^2} = \frac{WQ\rho c W_{ref}}{4\pi W_{ref} p_{ref}^2 r^2} \tag{2-39}$$

If we take the common logarithm of both sides of Eq. (2-39) and multiply both sides by 10, we obtain:

$$L_p = L_W + DI - 20\log_{10}(r) + 10\log_{10}\left[\frac{\rho c W_{ref}}{4\pi p_{ref}^2}\right] \tag{2-40}$$

The quantity DI is the *directivity index*, defined by:

$$DI = 10\log_{10}(Q) \tag{2-41}$$

where Q is the directivity factor. If we express the radial distance r in *meters* and take the properties of air at 25°C (77°F), we may determine the numerical value for the last term in Eq. (2-40):

$$10\log_{10}\left[\frac{(4\pi)(20\times 10^{-6})^2}{(409.8)(10^{-12})}\right] = 10\log_{10}(12.266) = 10.9\,dB$$

For sound transmitted from a directional source (or spherical source, with DI = 0 or Q = 1) outdoors in air at 25°C, the sound pressure level and the sound power level are related by:

$$L_p = L_W + DI - 20\log_{10}(r) - 10.9\,dB \tag{2-42}$$

Example 2-6. The quantities in Examples 2-2 and 2-3 are as follows:

acoustic pressure, $p = 0.20\,Pa$
acoustic particle velocity, $u = 0.488\,mm/s$
acoustic intensity, $I = 97.6\,\mu W/m^2$
acoustic energy density, $D = 0.282\,\mu J/m^3$

Determine the corresponding levels for these quantities.
The sound pressure level is:

$$L_p = 20\log_{10}(0.20/20\times 10^{-6}) = 80.0\,dB$$

The velocity level is:

$$L_v = 20\log_{10}(0.488\times 10^{-3}/10\times 10^{-9}) = 93.8\,dB$$

Basics of Acoustics

The intensity level is:

$$L_I = 10\log_{10}(97.6 \times 10^{-6}/10^{-12}) = 79.9\,\text{dB}$$

We note that, for sound transmitted in air at ambient conditions, the intensity level and the sound pressure level have approximately equal values. This is a consequence of the definition of the reference quantities for acoustic pressure and intensity. For sound transmitted in water, on the other hand, the sound pressure level and intensity level have quite different values.

The energy density level is:

$$L_D = 10\log_{10}(0.282 \times 10^{-6}/10^{-12}) = 54.5\,\text{dB}$$

2.8 COMBINATION OF SOUND SOURCES

There are many situations in which we need to determine the sound level produced by several sources of sound acting at the same time. For example, we may need to determine the sound level produced by two machines in a room, but we may have only information about the sound level produced by each machine separately.

Because all levels are defined in terms of energy-like quantities, all "levels" (sound pressure level, intensity level, power level, etc.) will combine in the same manner. The total intensity, for example, is the sum of the intensities for the individual sources, if the sources produce sound waves that are not exactly in-phase or out-of-phase. It is quite likely that the noise generated by machinery is not correlated, because the noise involves a wide range of frequencies and not a single frequency only.

Suppose we have two sources of sound that produce the following intensity levels when operating alone:

$$L_{I1} = 80\,\text{dB} = 10\log_{10}(I_1/I_{\text{ref}})$$

$$L_{I2} = 85\,\text{dB} = 10\log_{10}(I_2/I_{\text{ref}})$$

We may solve for the individual intensities to obtain these values:

$$I_1 = (10^{-12})10^{(80/10)} = 10^{-4}\,\text{W/m}^2 = 0.100\,\text{mW/m}^2$$

$$I_2 = (10^{-12})10^{(85/10)} = 3.16 \times 10^{-4}\,\text{W/m}^2 = 0.316\,\text{mW/m}^2$$

The total intensity when the two sources are operating at the same time is the sum of the intensities:

$$I = I_1 + I_2 = 0.100 + 0.316 = 0.416\,\text{mW/m}^2$$

The combined intensity level is:

$$L_1 = 10\log_{10}(0.416 \times 10^{-3}/10^{-12}) = 86.2\,\text{dB}$$

The general expression for determining the combination of any set of "level" quantities is:

$$L = 10\log_{10}\left(\sum_i 10^{L_i/10}\right) \tag{2-43}$$

This expression is valid for all types of levels—including sound pressure levels—because the total pressure is not the sum of the individual pressures if the waves are not correlated. The square of the pressure is proportional to energy (the intensity, for example), so the individual sound pressures must be combined in an "energy-like" manner.

$$(p_{\text{total}})^2 = p_1^2 + p_2^2 + p_3^2 + \cdots$$

The reference quantity is the same for each intensity level, so the previous calculation could be carried out using the intensity ratio. For the values used in the previous example, we have:

$$I_1/I_{\text{ref}} = 10^{(80/10)} = 10^8$$

The total intensity ratio is the sum of the individual ratios:

$$I/I_{\text{ref}} = (I_1/I_{\text{ref}}) + (I_2/I_{\text{ref}}) = 10^8 + 3.16 \times 10^8 = 4.16 \times 10^8$$

The total intensity level is:

$$L_I = 10\log_{10}(4.16 \times 10^8) = 86.2\,\text{dB}$$

The previous calculation provides the basis for a simple routine for combining levels on the hand-held calculator. The routine is outlined in Table 2-2. Different calculators have slightly different designations on the keys; however, the routine may be adapted to any calculator. The initial level value divided by 10 is first entered. (One may easily divide a number by 10 without the use of a calculator.) The 10^x key (or the corresponding key to raise 10 to the power of the entry) is pressed. The result is then stored in memory. Each of the next levels divded by 10 is entered, the 10^x key is pressed, and the results are added to memory. To finally determine the total level, the quantity is recalled from memory, and the *log* key (or *log-base-10* key) is pressed. The display may be multiplied by 10 to obtain the total level.

Basics of Acoustics 33

TABLE 2-2 Handheld Calculator Routine for Combining Levels[a]

Keystroke entry	Key pressed	Display results
Initial level value divided by 10	[10^x]	I_1/I_{ref}, *for example*
(none)	[STO]	(value stored in memory)
Next level value divided by 10	[10^x]	I_2/I_{ref}, *for example*
(none)	[SUM]	(value added to memory)
(Repeat for the remaining levels		
(none)	[RCL]	I_{total}/I_{ref}, *for example*
(none)	[log]	Level/10
(none)	[×]	
10	[=]	Total level
Example		
8.0	[10^x]; [STO]	100000000
8.5	[10^x]; [SUM]	316227766
	[RCL]	416227766
	[log]; [×]	8.619331048
10	[=]	86.19331048

[a] The notation is that used for most "scientific" handheld calculators; however, there may be some variation for different calculator models.

2.9 OCTAVE BANDS

The human ear is sensitive to sounds having frequencies in the range from about 16 Hz to 16 kHz. Because it is not practical to measure the sound level at each of the 15,984 frequencies in this range, acoustic measuring instruments generally measure the acoustic energy included in a range of frequencies. The human ear also responds more to frequency ratios than to frequency differences, so the frequency ranges generally have terminal frequencies (upper and lower frequencies of the range) that are related by the same ratio.

The frequency interval over which measurements are made is called the *bandwidth*. The bandwidth may be described by the lower frequency of the interval (f_1) and the upper frequency of the interval (f_2). In acoustics, the bandwidths are often specified in terms of *octaves*, where an octave is a frequency interval such that the upper frequency is twice the lower frequency (Table 2-3). For an octave,

$$f_2 = 2f_1 \quad \text{or} \quad f_2/f_1 = 2$$

In some cases, a more refined division of the frequency range is used in measurement, such as 1/3-octave bands, in which $(f_2/f_1) = 2^{1/3} = 1.260$.

TABLE 2-3 Standard Octave Bands

Band No.	Lower, f_1	Center, f_0	Upper, f_2
	Frequency, Hz		
12	11	16	22
15	22	31.5	44
18	44	63	88
21	88	125	177
24	177	250	355
27	355	500	710
30	710	1,000	1,420
33	1,420	2,000	2,840
36	2,840	4,000	5,680
39	5,680	8,000	11,360
42	11,360	16,000	22,720

Source: ANSI S1.6 (1967).

The center frequency of the band (f_0) is defined as the geometric mean of the upper and lower frequencies for the interval:

$$f_0 = (f_1 f_2)^{1/2} \tag{2-44}$$

For an octave band, the upper and lower frequencies are related to the center frequency by:

$$f_1 = f_0/2^{1/2} \quad \text{and} \quad f_2 = 2^{1/2} f_0,$$

For 1/3-octave bands,

$$f_1 = f_0/2^{1/6} \quad \text{and} \quad f_2 = 2^{1/6} f_0$$

2.10 WEIGHTED SOUND LEVELS

Most sound level meters have three "weighting" networks, called the A-, B-, and C-scales (ANSI S1.4, 1971). Originally, the A-scale was designed to correspond to the response of the human ear for a sound pressure level of 40 dB at all frequencies. The B-scale was designed to correspond to the response of the human ear for a sound pressure level of 70 dB at all frequencies. The C-scale was approximately flat (constant) for frequencies between 63 Hz and 4000 Hz.

The B-scale is rarely used at present. The A-scale is widely used as a single measure of possible hearing damage, annoyance caused by noise, and

Basics of Acoustics 35

compliance with various noise regulations. The sound levels indicated by the A-scale network are denoted by L_A, and the units are designated dBA.

The weighting for the A- and C-scale is shown in Table 2-4. These values are also plotted in Fig. 2-6. The large negative weighting factor for low-frequency sounds corresponds to the fact that the human ear is not as sensitive to low-frequency sound as it is for sound at frequencies in the 1 kHz to 4 kHz range. For example, a sound having a sound pressure level of 40 dB at 63 Hz would be perceived by the human ear as having a sound pressure level of approximately $(40 - 26) = 14$ dB. Alternatively, a sound that was perceived to have a sound pressure level of 40 dB for a frequency of 63 Hz would actually have a sound pressure level of $(40 + 26) = 66$ dB. Because the human ear does not respond as significantly to low-frequency sounds, noise at low frequencies (63 Hz, for example) is generally not as damaging or annoying as sound at high frequencies (2 kHz, for example).

If the sound pressure level spectrum is measured or calculated for each octave band, the A-weighted sound level may be calculated, using the A-weighting factors (CFA) from Table 2-4:

$$L_A = 10 \log_{10}[\Sigma\, 10^{(L_p + CFA)/10}] \qquad (2\text{-}45)$$

where the summation is carried out for all octave bands. The A-scale conversion process is illustrated in the following example.

Example 2-7. The measured octave band sound pressure levels around a punch press are given in Table 2-5. Determine the A-weighted sound level and the overall sound pressure level.

TABLE 2-4 Weighting Factors for the A- and C-Scales

Octave band center frequency, Hz	A-scale CFA	C-scale CFC
31.5	−39.4	−3.0
63	−26.2	−0.8
125	−16.1	−0.2
250	−8.9	0.0
500	−3.2	0.0
1,000	0.0	0.0
2,000	+1.2	−0.2
4,000	+1.0	−0.8
8,000	−1.1	−3.0
16,000	−6.6	−8.5

Source: ANSI S1.4 (1971).

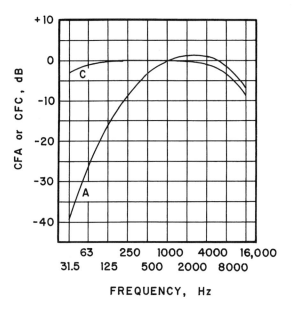

FIGURE 2-6 Weighting factors for the A- and C-scales. CFA = conversion factor to A-scale; CFC = conversion factor to C-scale.

The sound levels with the weighting factor applied are given in Table 2-5. The A-weighted sound level is calculated from Eq. (2-45).

$$L_A = 10 \log_{10}[10^{30.6/10} + 10^{54.8/10} + 10^{72.9/10} + \cdots]$$

$$L_A = 10 \log_{10}(1.3532 \times 10^{10}) = 101.3 \, \text{dBA}$$

The overall sound pressure level is obtained by adding the individual unweighted octave band sound pressure levels given in Table 2-5, using "decibel addition", Eq. (2-43):

TABLE 2-5 Data for Example 2-7

	Octave band center frequency, Hz								
	31.5	63	125	250	500	1,000	2,000	4,000	8,000
L_p(OB), dB	70	81	89	101	103	93	83	77	74
CFA, dB	−39.4	−26.2	−16.1	−8.9	−3.2	0.0	+1.2	+1.0	−1.1
L_p+ CFA, dB	30.6	54.8	72.9	9.1	99.8	93.0	84.2	78.0	72.9

Basics of Acoustics

$$L_p = 10\log_{10}[\Sigma 10^{L_p/10}]$$

$$L_p = 10\log_{10}[10^{70/10} + 10^{81/10} + 10^{89/10} + \cdots] = 10\log_{10}(3.5742 \times 10^{10})$$

$$L_p = 105.5\,\text{dB}$$

We note that the A-weighted sound level is lower than the overall sound pressure level in this problem. The reason for this difference is that the sound energy is more predominant in the lower octave bands, such as the 250 Hz band. The readings are diminished in the lower frequencies (compare L_p and $L_p + \text{CFA}$) when the A-weighting is applied. The A-weighted sound level can be larger than the overall sound pressure level if the sound energy is more concentrated in the octave bands between 1 and 4 kHz.

PROBLEMS

2-1. In the 200-meter track event, the starter is located a distance of 150 m (492.1 ft) from the timers. If the air temperature is 22°C (295.2K or 71.6°F), how long does it take the sound of the starter's gun to reach the timers? The gas constant for air is 287 J/kg-K and the specific heat ratio $\gamma = 1.400$.

2-2. One proposed type of thermometer is one that measures the speed of sound in an ideal gas. If this thermometer indicates a speed of sound in methane gas ($R = 518.4$ J/kg-K; $\gamma = 1.299$) of 425 m/s (1394 ft/sec), determine the temperature of the methane gas.

2-3. Sonic liquid level gauges are used to measure the liquid level of liquids in closed containers. Such a gauge is installed in a water container, and the transducer is placed at the bottom of the container. The sound pulse emitted from the transducer moves through the liquid, is reflected at the liquid surface, and travels back to the transducer. If the water temperature is 20°C (68°F) and the total transit time of the sound pulse is 10 ms, determine the level of liquid water in the container.

2-4. For gold, Young's modulus is 75 GPa (10.88×10^6 psi), Poisson's ratio is 0.42, and the density is 19,320 kg/m³ (1206 lb$_m$/ft³). Determine the bulk speed of sound in gold.

2-5. The wavelength of a sound wave is 305 mm (12.0 in). Determine the frequency and wave number for a plane sound wave propagated in (a) air at 20°C (293.2K or 68°F), $R = 287$ J/kg-K, $\gamma = 1.400$ and (b) helium at 20°C, $R = 2078$ J/kg-K, $\gamma = 1.667$.

2-6. A famous basketball player has a height of 2.110 m (6 ft 11 in). At what frequency would a sound wave in air at 25°C (298.2K or 77°F) have a wavelength equal to the player's height?

2-7. A trombone produces a plane sound wave having a frequency of 170 Hz and a rms acoustic pressure of 13.1 Pa in air at 25°C (298.2K or 77°F) and 101.3 kPa (14.7 psia). Determine the rms acoustic particle velocity for the sound wave.

2-8. A plane sound wave is propagated in air at 15°C (288.2K or 59°F) and 101.3 kPa (14.7 psia). The intensity of the wave is 10 mW/m². Determine the rms acoustic pressure, the rms acoustic particle velocity, and the acoustic energy density for the sound wave.

2-9. A spherical source of sound radiates uniformly into a large volume of air at 22°C (295.2K or 72°F) and 101.3 kPa (14.7 psia). The frequency of the sound wave is 274 Hz, and the acoustic power radiated from the source is 30 mW. At a radial distance of 500 mm (41.7 in) from the source, determine the intensity, the rms acoustic pressure, the rms acoustic particle velocity, and the acoustic energy density.

2-10. When Bix Beiderbecke, the talented jazz performer of the 1920s, played a high G on his cornet (concert F pitch), he produced a note having a frequency of 690.5 Hz, and the resulting sound waves could be treated as spherical waves (for this problem). At a distance of 60.5 mm (2.38 in) from his cornet, the rms acoustic particle velocity was 2.60 mm/s (0.102 in/sec). The warm air in his club was at 32°C (305.2K or 90°F). Determine the rms acoustic pressure and acoustic intensity at a distance of 60.5 mm from Bix's cornet when he played the high G.

2-11. An acoustic quadrupole source radiates sound with an acoustic pressure distribution function given by:

$$H(\theta) = \tfrac{1}{2}(3\cos^2\theta - 1)$$

Determine the expression for the directivity factor and directivity index for this symmetrical directional source along the axis, $\theta = 0$.

2-12. An acoustic triplet source radiates sound with an acoustic pressure distribution function given by:

$$H(\theta) = \tfrac{1}{2}(1 + \cos\theta)$$

Determine the expression for the directivity factor and directivity index for this symmetrical directional source along the axis $\theta = 0$, and for $\theta = 60°$.

2-13. A boiler feedwater pump radiates sound as a spherical source. The acoustic power level for the pump is 103 dB, and the frequency of the

Basics of Acoustics

sound wave is 63 Hz. The sound travels through air at 36.8°C (310K or 98.2°F) and 101.3 kPa (14.7 psia). At a distance of 1.50 m (4.92 ft) from the pump, determine (a) the intensity and intensity level and (b) the energy density and energy density level for the sound.

2-14. A plane sound wave is propagated in the water in a pipe having an internal diameter of 100 mm (3.937 in) and a length of 30 m (9.144 ft). The properties of the water are density 1000 kg/m^3 and sonic velocity 1400 m/s. The intensity level for the sound wave is 121 dB. Determine (a) the sound pressure level, (b) the acoustic particle velocity, and (c) the total acoustic energy (mJ) contained in the water in the pipe.

2-15. A chain saw produces a spherical sound wave having a frequency of 214 Hz in air at 35°C (308.2K or 95°F). At a distance of 600 mm (23.62 in), the sound pressure level is 100 dB. Determine (a) the acoustic power level and (b) the rms acoustic particle velocity and velocity level at a distance of 600 mm from the saw.

2-16. When two pumps are both operating, the sound pressure level at a distance of 10 m from the pumps is 101.8 dB. When pump A is turned off (only pump B operating), the measured sound pressure level at 10 m from the pumps is 97.0 dB. Determine the sound pressure level when pump B is turned off and only pump A is operating. The sound waves are radiated through air at 22°C (205.2K or 72°F) and 101.3 kPA (14.7 psia).

2-17. The sound pressure level spectrum around a wood chipper unit is given in Table 2-6. Determine (a) the overall sound pressure level and (b) the A-weighted sound level for the chipper noise.

2-18. A machine produces the sound pressure level spectrum in octave bands at a distance of 3 m (9.84 ft), as given in Table 2-7. Determine the overall sound pressure level and the A-weighted sound level at 3 m from the machine.

TABLE 2-6 Data for Problem 2-17

	Octave band center frequency, Hz							
	63	125	250	500	1,000	2,000	4,000	8,000
L_p(OB), dB	91	88	99	102	99	98	98	88

TABLE 2-7 Data for Problem 2-18

	Octave band center frequency, Hz							
	63	125	250	500	1,000	2,000	4,000	8,000
L_p(OB), dB	102	96	89	83	80	79	79	77

REFERENCES

ANSI S1.6. 1967. Preferred frequencies and band numbers for acoustical measurements. American National Standards Institute, Inc., New York.

ANSI S1.4. 1971. Specifications for sound level meters. American National Standards Institute, Inc., New York. Revised in 1976.

ANSI S1.8. 1989. Reference quantities for acoustical levels. American National Standards Institute, Inc., New York.

Beranek, L. L. and Vér, I. L. 1992. *Noise and Vibration Control Engineering*, p. 1. John Wiley and Sons, New York.

Huntley, R. 1970. A bel is ten decibels. *Sound and Vibration* 4(1): 22.

Timoshenko, S. P. 1970. *Theory of Elasticity*, 3rd edn, p. 488. McGraw-Hill, New York.

Timoshenko, S. P. 1983. *History of Strength of Materials*, p. 347. Dover Publications, New York.

3
Acoustic Measurements

Noise control programs often require measurements of various acoustic quantities to determine the effectiveness of the noise control procedure. There may be other reasons for the need of experimental data. Noise measurements must be made to determine compliance with noise regulations. Noise measurements may be needed for diagnostic purposes or to locate the source (or sources) of noise in a piece of machinery. The transmission path (or paths) for noise in a system may be identified through acoustic measurements.

It is important that the measuring equipment be properly selected to monitor and measure sound properties. When we have a basic situation in which we need to assess the severity of environmental noise, we may need to measure only the overall sound pressure level or the A-weighted level, using a simple sound level meter. For example, if we wish to determine if the sound level in a room exceeds 90 dBA, then a portable or hand-held sound level meter is the appropriate instrument to use.

There are cases where we require a more detailed analysis of the noise. In these cases, octave band or 1/3 octave band sound level measurements may be made. A sound level meter with octave band or 1/3 octave band filters is required. An acoustic spectrum analyzer, in which microprocessors are used to manipulate the input data, may be the instrument of choice.

For verification of compliance with noise exposure regulations, dosimeters may be used to measure and record cumulative noise exposure.

Data on the sound power generated by machines and equipment may be needed in the development of quieter mechanical systems, in making acoustic comparisons of several different machines, and in the generation of data on production machines and equipment. There is no "acoustic wattmeter" available for direct measurement of acoustic power; however, the techniques for sound power determination from sound pressure measurements will be discussed in this chapter.

3.1 SOUND LEVEL METERS

The basic parts of most sound level meters include a microphone, amplifiers, weighting networks, and a display indicating decibels. Typical sound level meters are shown in Figs 3-1 and 3-2. The microphone acts to convert the input acoustic signal (acoustic pressure) into an electrical signal (usually voltage). This signal is magnified as it passes through the electronic pream-

FIGURE 3-1 Sound level meter. This sound level meter provides manual operation and storage of results at the end of each run. The output is displayed on a screen on the meter, or the data can be downloaded to a PC. (By permission of Casella CEL Instruments Ltd.)

Acoustic Measurements

FIGURE 3-2 Sound level meter. This meter is a handheld modular precision sound analyzer system. (By permission of Brüel and Kjaer.)

plifier. The amplified signal may then be modified by the weighting network to obtain the A-, B-, or C-weighted signal. This signal is digitized to drive the display meter, where the output is indicated in decibels. The display setting may be "fast" response, "slow" response, "impact" response, or "peak" response. Unless one is interested in measuring rapid noise fluctuations, the "slow" response setting is usually used. An output jack may be provided to record or analyze the signal in an external instrument system.

Sound level meters are rated in the following categories, based on the accuracy of the meter: (a) type 1, precision; (b) type 2, general-purpose; (c) type 3, survey; and (d) special-purpose sound level meters. The type 1 or type 2 sound level meter is required for OSHA noise surveys, and is specified in most community noise ordinances.

There are several items of auxiliary equipment that are used with sound level meters, including a calibrator and a windscreen. Many sound level meters have output ports for connection to a PC for post-processing of data.

The calibrator, shown in Fig. 3-3, is a portable, battery-operated instrument that is used to calibrate the sound level meter. The microphone on the sound level meter is inserted into one end of the calibrator, and the calibrator generates a pure tone at a frequency of 1 kHz and a known level (such as 114 dB). The reading of the sound level meter is compared with the known output of the calibrator, and the sound level meter is adjusted to match the calibrator.

There are two general types of calibrators for sound level meters: the loudspeaker type and the pistonphone type. The loudspeaker type contains a small loudspeaker that produces known sound pressure levels at several frequencies, such as 125 Hz, 250 Hz, 500 Hz, 1000 Hz, and 2000 Hz. The pistonphone calibrator consists of an air cavity, in which the microphone is placed at one end and cam-driven pistons are located at the other end. The oscillation of the pistons changes the volume of the cavity and produces a known variation of the instantaneous acoustic pressure in the air cavity at the microphone diaphragm. The pistonphone usually provides calibration at one frequency, such as 250 Hz, for example.

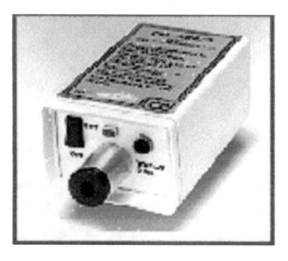

FIGURE 3-3   Acoustic calibrator. (By permission of Casella CEL Instruments Ltd.)

Acoustic Measurements

A windscreen should always be used when making sound level measurements outdoors. The windscreen consists of a spherical piece of open-cell foam material that can be fitted over the microphone of the sound level meter, as shown in Fig. 3-4. The windscreen minimizes the effect of wind turbulence over the microphone. Generally, sound level measurements are not effective when the wind speed exceeds approximately 12 mph or 19 m/s (Sköde, 1966). For acoustic measurements in ventilation ducts where the direction of the flow of air is constant, nose cones are usually attached to the microphone to alleviate the wind noise effect.

A precision sound level meter may include an impulse network or software package to measure impulsive sounds, or sounds in which the pressure level rises rapidly for short periods of time. The impulse feature has an output of maximum rms sound level or the maximum peak level (or both) for the impulsive sound. The readings may be denoted by dBA(I) or dBC(I), depending upon the weighting network used with the impulse feature.

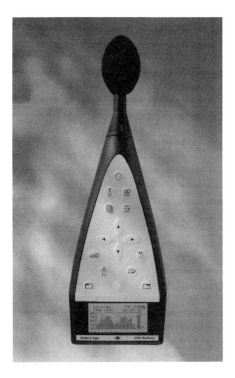

FIGURE 3-4 Sound level meter with windscreen in place on the microphone. (By permission of Brüel and Kjaer.)

46 Chapter 3

3.2 INTENSITY LEVEL METERS

There are some situations where we would like to measure the intensity directly, instead of measuring the sound pressure level and attempting to calculate the intensity from this measurement. In addition, the location of a specific noise source may be determined from the directivity pattern associated with intensity level measurements. The intensity level meter is a helpful tool for location of sources of noise problems in a machine, such as noise produced by bearing failure, internal impact problems, etc. A typical sound intensity meter is shown in Fig. 3-5.

The acoustic intensity is the time-averaged value of the product of the acoustic pressure and acoustic particle velocity.

$$I = \bar{p}\bar{u}$$

In the intensity probe, two microphones are placed face-to-face at a known spacing Δx, as shown in Fig. 3-6. Typically, the spacing is 6 mm for frequencies between 250 Hz and 12 kHz, 12 mm for the 125 Hz to 5 kHz range, and 59 mm for frequencies between 31.5 Hz and 1.25 kHz (Beranek and Vér,

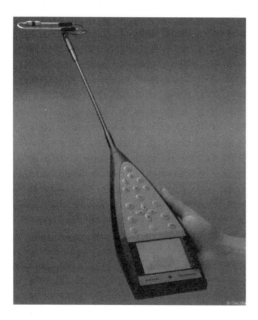

FIGURE 3-5 Sound intensity meter. The two opposed microphones may be observed in the upper part of the figure. (By permission of Brüel and Kjaer.)

Acoustic Measurements

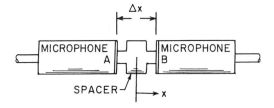

FIGURE 3-6 Sound intensity probe schematic.

1992). As we will show in Chapter 4, the particle velocity and the acoustic pressure are related by:

$$\bar{u} = -\frac{1}{\rho}\int \frac{\partial p}{\partial x}\, dt \approx -\frac{1}{\rho}\int \frac{(p_B - p_A)}{\Delta x}\, dt \tag{3-1}$$

where p_A and p_B are the sound pressures indicated by microphone A and B, respectively. The integration of the acoustic pressure is carried out in an analyzer in the intensity measuring instrument. The average acoustic pressure is:

$$\bar{p} = \tfrac{1}{2}(p_A + p_B)$$

The time-averaging of the acoustic particle velocity and average acoustic pressure may be done directly by using integrators and filters.

The accuracy of the intensity level meter in direct measurement of the intensity is a function of the wavelength of the sound and the spacing of the two microphones. Suppose the instantaneous acoustic pressure at a specific time is given by:

$$p(x) = p_m \sin(kx) \tag{3-2}$$

where p_m is the peak amplitude, k is the wave number, and the coordinate x is measured from the center of the spacer. The exact expression for the derivative in Eq. (3-1) is:

$$\frac{\partial p}{\partial x} = p_m k \cos(kx) = p_m k \quad (\text{at } x = 0) \tag{3-3}$$

The approximation in Eq. (3-1) is:

$$\frac{p_B - p_A}{\Delta x} = \frac{p_m[\sin(\tfrac{1}{2}k\Delta x) - \sin(-\tfrac{1}{2}k\Delta x)]}{\Delta x} = \frac{2p_m \sin(\tfrac{1}{2}k\Delta x)}{\Delta x} \tag{3-4}$$

The pressure p_B is measured at the position $x = +\tfrac{1}{2}\Delta x$, and the pressure p_A is measured at $x = -\tfrac{1}{2}\Delta x$.

The error in the intensity measurement is proportional to the error in the derivative approximation:

$$\text{error} = \frac{p_m k - [2 p_m \sin(\tfrac{1}{2} k \Delta x)/\Delta x]}{p_m k} = 1 - \frac{2 \sin(\tfrac{1}{2} k \Delta x)}{k \Delta x} \qquad (3\text{-}5)$$

As the dimensionless parameter $k \Delta x$ approaches zero, the second term in Eq. (3-5) approaches 1, and the error approaches zero.

The intensity probe is highly directional, so care must be taken to orient the probe such that the sound waves are incident along the probe axis. If the incident sound wave makes an angle θ with the probe axis, the indicated intensity I will deviate from the intensity along the axis I_o according to the following expression (Beranek and Vér, 1992):

$$I = I_o \cos \theta$$

One of the advantages of the use of the intensity meter is that steady background noise does not affect the meter indication. The integration procedure eliminates steady background noise components, so accuracy on the order of 1 dB may be achieved when the background noise level is as much as 10 dB higher than the noise to be measured. Another advantage of the intensity meter is that its directional characteristics may be used to identify the direction of significant noise sources relative to the instrument. For example, the intensity level meter is well suited for acoustic troubleshooting tasks to locate the specific problem causing excessive noise in a machine. A disadvantage of the intensity meter is that it is usually more expensive than a basic sound level meter.

Example 3-1. Determine the error in the intensity meter reading if the microphone spacing is 6 mm (0.236 in). The frequency of the sound wave is 12 kHz, and the speed of sound in the air around the microphone is 346 m/s (1135 fps).

The wave number is:

$$k = \frac{2\pi f}{c} = \frac{(2\pi)(12,000)}{(346)} = 217.9 \text{ m}^{-1}$$

Then,

$$\tfrac{1}{2} k \Delta x = (\tfrac{1}{2})(217.9)(0.006) = 0.6537$$

The percentage error is found from Eq. (3-5).

$$\text{error} = 1 - \frac{\sin(0.6537)}{(0.6537)} = 1 - 0.9303 = 0.070 \text{ or } 7\% \text{ error}$$

Acoustic Measurements

If we wish to limit the error to 5% or less for the conditions in this example, the microphone spacing would need to be reduced to about 5.1 mm (0.20 in).

3.3 OCTAVE BAND FILTERS

An octave band is defined as a frequency range in which the ratio of the upper and lower frequency limits for the range is equal to 2. Octave band filter sets are often included as a feature of sound level meters. If the filters were "perfect," all energy outside the octave band frequency range would be attenuated or canceled out. For a "real" filter, the attenuation is not perfect outside the octave band range, as shown in Fig. 3-7. Typically, the attenuation of the octave band filter is on the order of $-25\,\mathrm{dB}$ at a frequency one octave below or one octave above the center frequency of the octave band.

For a more detailed analysis of the frequency distribution of the acoustic energy, 1/3 octave band filters may be used. The center frequencies of the standad 1/3 octave bands are shown in Table 3-1. The ratio of the upper and lower frequency for a 1/3 octave band is $2^{1/3} = 1.260$. The lower frequency f_1 in the 1/3 octave band is related to the center frequency f_o by $f_1 = 2^{-1/6} f_o = 0.8909 f_o$. the upper frequency f_2 in the 1/3 octave band is related to the center frequency by $f_2 = 2^{1/6} f_o = 1.1225 f_o$.

Octave band (1/1) or 1/3 octave band filters are often used in basic acoustic engineering design and analysis work. By observing the frequency band in which the maximum sound pressure level occurs, the system characteristics that relate to the noise generation may be identified.

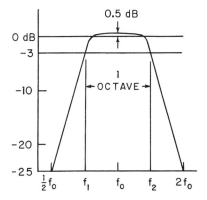

FIGURE 3-7 Typical frequency characteristics of an acoustic octave band filter.

TABLE 3-1 Standard 1/3 Octave Band Center Frequencies

Band No.	Center frequency, Hz	Band No.	Center frequency, Hz	Band No.	Center frequency, Hz
12	16	23	200	34	2,500
13	20	24	250	35	3,150
14	25	25	315	36	4,000
15	31.5	26	400	37	5,000
16	40	27	500	38	6,300
17	50	28	630	39	8,000
18	63	29	800	40	10,000
19	80	30	1,000	41	12,500
20	100	31	1,250	42	16,000
21	125	32	1,600	43	20,000
22	160	33	2,000		

3.4 ACOUSTIC ANALYZERS

For general acoustic measurements, the measurement of the sound levels on the A- and C-scales, along with octave band or 1/3 octave band data, is usually sufficient. For detailed analysis and diagnosis, however, more information may be required. In addition, other acoustic parameters or data averaging may be needed. In these cases, acoustic analyzers are often used.

Most of the commercial sound level meters are digital, microprocessor-controlled units. Various levels of acoustic analysis may be provided as features of the sound level meter. In addition, many sound level meters have the capability of transferring the data to a personal computer (PC) for post-processing using special-purpose software.

3.5 DOSIMETER

The *dosimeter* or noise-exposure meter is an instrument developed for measurement of the accumulated noise exposure of workers in an industrial environment. The dosimeter, shown in Fig. 3-8, is an integrating sound level meter system that can be carried by workers during their normal course of work. Some units have separate storage registers to allow the unit to be used to monitor several workers before downloading the data.

The output of a typical dosimeter includes the total noise exposure, in units of dBA, the percent of allowable noise dosage, and the peak (maximum) noise exposure level. In addition, such information as date, time, and duration of the data acquisition period may be included in the output. At

Acoustic Measurements

FIGURE 3-8 Personal noise dosimeter. Data can be stored in several locations to allow the monitoring of multiple inputs. (By permission of Casella CEL Instruments Ltd.)

the end of the day, the data may be downloaded for permanent storage on a disk and for export to spreadsheet software for graphing and reporting purposes.

3.6 MEASUREMENT OF SOUND POWER

The sound pressure in the vicinity of a noise source is generally dependent on the surroundings. The sound pressure level will be different for the same noise source, for example, if the source is located indoors or if it is located outdoors. The sound pressure will be different if the source is placed in a room with acoustically reflective surfaces or if the room surfaces are highly absorbent for sound. In contrast, the sound power is generally independent of the surroundings. For this reason, information about the sound power spectrum for a noise source is important to the designer interested in noise control.

There is no "acoustic wattmeter" available for direct measurement of sound power, however. The sound power must be inferred (or calculated) from measurements of sound pressure or sound intensity and other appropriate quantities, such as surface area, reverberation time, etc.

There are three broad classes of environment used in connection with sound power determination: (a) reverberant field, (b) direct or anechoic field, or (c) the actual environment to which the noise source is exposed (in-situ survey). The national and international standards for sound power measurement in a reverberant room include ANSI S1.31, ANSI S1.32, and ANSI S1.33 (Acoustical Society of America, 1986a,b,c) and ISO 3741, ISO 3742, ISO 3743 (International Organization for Standardization, 1986a,b,c). The corresponding standards for an anechoic room include ANSI S1.35 (Acoustical Society of America, 1979a) and ISO 3745 (International Organization for Standardization, 1986d). The survey method is covered by ANSI S1.36 (Acoustical Society of America, 1976b) and ISO 3746 (International Organization for Standardization, 1986e).

A *reverberant room* is a room in which the acoustic energy from sound reflected from the room surfaces (*reverberant field*) is much larger than the energy transmitted directly from the noise source to a receiver (*direct field*), as discussed in Sec. 7.2. All surfaces in a reverberant room are highly reflective or have a very low surface absorption coefficient. A reverberant room may be used to determine sound power by either comparison with a calibrated noise source or by direct measurement of sound pressure.

An *anechoic room* is a room in which practically all of the acoustic energy striking the surfaces of the room is absorbed. Because the energy reflected from the room surfaces is negligible in an anechoic room, the energy transmitted directly from the source to the receiver is predominant. Measurements in an anechoic room may be used to determine the directional characteristics (directivity factor or directivity index) of the noise source. One modification of the anechoic room is the *semi-anechoic room*, in which the floor surface is highly reflective, but the other surfaces in the room are highly absorptive. An alternative arrangement is to make measurements outdoors on a reflective surface, such as a parking lot.

In situations where the noise source cannot be moved into a reverberant or anechoic room, the sound power may be determined from measurements taken in situ, with appropriate adjustments made for surrounding surfaces and environment background noise.

3.6.1 Sound Power Measurement in a Reverberant Room

The determination of the sound power in a reverberant room requires that the diffuse or reverberant sound field in the room is much larger than the direct sound field. This requirement results in a practically uniform value of the acoustic energy density and the acoustic pressure in the room.

Acoustic Measurements

The volume of the reverberant room should be such that the wavelength of the sound waves is much smaller than the dimensions of the room. The minimum volume for the room should meet the following condition:

$$V \geq (3\lambda)^3 = 9(c/f)^3 \qquad (3\text{-}6)$$

The quantity c is the speed of sound in the air in the room, and f is the octave band (or 1/3-octave band) center frequency for the lowest-frequency band considered in the measurements. For air at 300K ($c = 347.2\,\text{m/s} = 1139\,\text{fps}$) and a frequency of 125 Hz, the corresponding wavelength is as follows:

$$\lambda = (347.2/125) = 2.78\,\text{m} \quad (9.11\,\text{ft})$$

The minimum reverberant room volume for this condition may be found from Eq. (3-6).

$$V \geq (9)(2.78)^3 = 193\,\text{m}^3 \quad (6810\,\text{ft}^3)$$

If the room dimensions (height, width, and length) are in the commonly used ratio 1 : 1.5 : 2.5, the room dimensions must be at least 3.72 m (12.2 ft) high, 5.58 m (18.3 ft) wide, and 9.30 m (30.5 ft) long. Room dimension ratios of $1 : 2^{1/3} : 4^{1/3}$ or 1 : 1.260 : 1.587 have also been used (Broch, 1971). The volume of the equipment being tested should not exceed $0.01\,V$, where V is the volume of the room.

The room surfaces in a reverberant room should have surface absorption coefficients (Sabine absorption coefficient) that are less than about 0.06. The surface absorption coefficient is discussed in Sec. 7.1.

The sound field in the region near the walls of the room will not be quite uniform or diffuse, so it is good practice to locate the microphones such that none are nearer than the smaller value of $\frac{1}{2}\lambda$ or 1 m (39 in) from the walls. For case of a wavelength of 2.78 m given in the previous example ($\frac{1}{2}\lambda = 1.36\,\text{m}$), the microphone should be located at least 1 m from the walls. If an array of microphones is used, at least three microphones should be included in the array. If a single microphone is used, measurements should be taken at three or more positions around the noise source. The positions should be spaced at a distance that is the larger of $3(\frac{1}{2}\lambda)$ or 3 m, where λ is the wavelength of the lowest-frequency sound to be measured.

The noise source should not be placed at the center of the room, because many of the resonant modes of the room would not be excited by the noise source in this position. The noise source is usually placed near the room wall at a distance not less than the major dimension of the source.

3.6.1.1 Comparison Method

The sound power may be measured by comparison of the measured sound pressure level in a reverberant room with the sound pressure level of a reference (calibrated) sound source at the same location. A reference sound power source was originally designed by a committee of the American Society of Heating, Refrigeration, and Air Conditioning Engineers (ASHRAE) in the 1960s (Baade, 1969). Reference sound power sources are commercially available with calibration accuracies of ± 0.5 dB for frequencies between 200 and 4000 Hz and ± 1.0 dB between 100 and 160 Hz and between 5 and 10 kHz.

Several microphones arranged in an array in the room or a single movable microphone may be used to measure the sound pressure level in the reverberant field. To ensure that the microphones are in the diffuse field, the distance between the microphones and the surface of the noise source d_m should meet the following condition (Beranek and Vér, 1992, p. 92):

$$20\log_{10}(1.25 d_m) \geq L_{W,cal} - L_{p,cal} \tag{3-7}$$

The quantity $L_{W,cal}$ is the sound power level of the reference source, and $L_{p,cal}$ is the measured sound pressure level produced by the calibrated sound power source.

The experimental procedure for determining the sound power level for a noise source, using the comparison method in a reverberant room, is as follows. First, the energy-averaged sound pressure level in each frequency band L_p is measured with the test source in operation. Secondly, the test source is removed and the reference sound source is placed in the same location, and energy-averaged sound pressure level in each frequency band $L_{p,cal}$ is measured with the calibrated reference source in operation. Thirdly and finally, the sound power level of the test source L_W is calculated from the measured data:

$$L_W = L_p + (L_{W,cal} - L_{p,cal}) \tag{3-8}$$

The values of the sound power level for the calibrated reference source $L_{W,cal}$ are supplied by the manufacturer of the calibrated source.

If the reverberant field is much larger than the direct sould field, Eq. (7-17), which relates the sound power level and sound pressure level, may be written in the following form:

$$L_p = L_W + 10\log_{10}(4/R) + 0.1 \tag{3-9}$$

The quantity R is the *room constant*, defined by Eq. (7-13). The value of the room constant remains constant when the test source is replaced by the calibrated source:

Acoustic Measurements

$$L_{p,\text{cal}} = L_{W,\text{cal}} + 10\log_{10}(4/R) + 0.1 \tag{3-10}$$

If we subtract Eq. (3-10) from Eq. (3-9), we obtain the expression given by Eq. (3-8).

3.6.1.2 Direct Method

If the room constant were known, Eq. (3-9) could be used to determine the sound power level directly from sound pressure level measurements. If the acoustic energy density associated with the reverberant sound field is much larger than that associated with the direct sound field, Eq. (7-14) may be written as follows:

$$\frac{4W}{4} = \frac{p^2}{\rho_0 c} \tag{3-11}$$

If the room is highly reverberant, or if the average surface absorption coefficient $\bar{\alpha}$ is small, the room constant from Eq. (7-13) may be written as follows:

$$R = \frac{\bar{\alpha} S_o}{1 - \bar{\alpha}} \approx \bar{\alpha} S_o \tag{3-12}$$

The quantity S_o is the total surface area of the room.

The reverberation time T_r may be used to determine the average surface absorption coefficient for the room surfaces. The reverberation time, adjusted for standing wave effects, is given by Eq. (7-34):

$$T_r = \frac{55.26 V}{ca}\left(1 + \frac{\lambda}{2d}\right) \tag{3-13}$$

The quantity V is the volume of the room, c is the speed of sound in the air in the room, and a is the number of absorption units, given by Eq. (7-30). The quantity $\lambda = c/f$ is the wavelength of the sound at the band center frequency, and $d = 4V/S_o$ is the mean free path for the sound waves in the room. For small values of the surface absorption coefficient, the number of absorption units may be approximated by the following, according to Eq. (7-30),

$$a = S_o \ln\left(\frac{1}{1 - \bar{\alpha}}\right) \approx \bar{\alpha} S_o \tag{3-14}$$

By comparing Eqs (3-12) and (3-14), we observe that the room constant and the number of absorption units are approximately equal.

$$R \approx a \approx \bar{\alpha} S_o \tag{3-15}$$

If we make the substitutions from Eq. (3-15) into Eq. (3-13) for the reverberation time, the following result is obtained:

$$R = \frac{55.26V}{cT_r}\left(1 + \frac{S_o c}{8Vf}\right) \qquad (3\text{-}16)$$

If we make the substitution for the room constant from Eq. (3-16) into Eq. (3-11), the following result is obtained for the acoustic power:

$$W = \frac{55.26V}{4T_r}\left(1 + \frac{S_o c}{8Vf}\right)\frac{p^2}{\rho_o c^2} \qquad (3\text{-}17)$$

We may convert Eq. (3-17) to "level" form as follows. First, introducing the reference quantities, we have the following:

$$\frac{W}{W_{ref}} = \frac{(p/p_{ref})^2 (V/V_{ref})}{(T_r/T_{ref})}\left(1 + \frac{S_o c}{8Vf}\right)\frac{13.816(p_{ref})^2 V_{ref}}{T_{ref} W_{ref} \rho_o c^2} \qquad (3\text{-}17a)$$

where $V_{ref} = 1\,\text{m}^3$ and $T_{ref} = 1$ sec. If we take log base 10 of both sides of Eq. (3-17a) and multiply both sides by 10, we obtain the final result needed to determine the sound power level of a noise source from sound pressure level measurements in a reverberant room:

$$L_W = L_p + 10\log_{10}(V/V_{ref}) - 10\log_{10}(T_r/T_{ref}) + 10\log_{10}\left(1 + \frac{S_o c}{8Vf}\right)$$

$$+ 10\log_{10}\left[\frac{13.816(p_{ref})^2 V_{ref}}{T_{ref} W_{ref} \rho_c c^2}\right]$$

$$(3\text{-}18)$$

For an ideal gas, the last term in Eq. (3-18) may be written in the following form:

$$\rho_o c^2 = \frac{P_o}{RT}(\gamma RT) = \gamma P_o = \gamma P_{o,ref}(P_o/P_{o,ref}) \qquad (3\text{-}19)$$

The quantity P_o is atmospheric pressure, γ is the specific heat ratio ($\gamma = 1.40$ for air), and $P_{o,ref} = 101.325\,\text{kPa}$ (14.696 psia). If we substitute the numerical values for the reference quantities, we obtain:

$$\frac{13.816(p_{ref})^2 V_{ref}}{T_{ref} W_{ref} \rho_o c^2} = \frac{(13.816)(20 \times 10^{-6})^2(1)}{(1)(10^{-12})(1.40)(101.325 \times 10^3)(P_o/P_{o,ref})} = \frac{0.03896}{(P_o/P_{o,ref})}$$

$$10\log_{10}\left[\frac{13.816(p_{ref})^2 V_{ref}}{T_{ref} W_{ref} \rho_o c^2}\right] = 10\log_{10}\left[\frac{0.03896}{(P_o/P_{o,ref})}\right]$$

$$= -10\log_{10}(P_o/P_{o,ref}) - 14.1$$

Acoustic Measurements

With these substitutions, Eq. (3-18) may be written in the following form:

$$L_W = L_p + 10\log_{10}(V/V_{ref}) - 10\log_{10}(T_r/T_{ref}) + 10\log_{10}\left(1 + \frac{S_o c}{8Vf}\right)$$
$$- 10\log_{10}(P_o/P_{o,ref}) - 14.1$$

(3-20)

To ensure that the diffuse or reverberant sound field predominates at the microphone location, the distance between the microphones and the surface of the noise source d_m should meet the following condition (Beranek and Vér, 1992, p. 93):

$$d_m \geq 3\left(\frac{V}{cT_r}\right)^{1/2} \tag{3-21}$$

Example 3-2. A reverberant room has dimensions of 6 m (19.69 ft) by 10 m (32.81 ft) by 4 m (13.12 ft) high. The measured reverberation time for the room is 3.50 seconds. The air in the room is at 300K (27°C or 80°F) and 101.3 kPa (14.7 psia), at which condition the speed of sound is 347.2 m/s (1139 fps). The measured sound pressure level in the 500 Hz octave band due to the noise from pump in the room is 65 dB. Determine the sound power level for the pump in the 500 Hz octave band.

The surface area of the room is as follows:

$$S_o = (2)(10+6)(4) + (2)(10)(6) = 128 + 120 = 248 \text{ m}^2 \quad (2669 \text{ ft}^2)$$

The volume of the room is:

$$V = (10)(6)(4) = 240 \text{ m}^3 \quad (8476 \text{ ft}^3)$$

The value of the following quantity for a frequency of 500 Hz may be calculated:

$$\frac{S_o c}{8Vf} = \frac{(248)(347.2)}{(8)(240)(500)} = 0.0897$$

The sound power level may now be found from Eq. (3-20):

$$L_W = 65 + 10\log_{10}(240) - 10\log_{10}(3.50) + 10\log_{10}(1 + 0.0897)$$
$$- 0 - 14.1$$
$$L_W = 65 + 23.80 - 5.44 + 0.37 - 14.1 = 69.6 \text{ dB}$$

The minimum distance of the microphone from the surface of the pump is given by Eq. (3-21):

$$d_\mathrm{m} \geq (3)\left[\frac{(240)}{(347.2)(3.50)}\right]^{1/2} = 1.333\,\mathrm{m} \quad (4.37\,\mathrm{ft})$$

3.6.2 Sound Power Measurement in an Anechoic or Semi-Anechoic Room

The acoustic power generated by a noise source may also be measured in an environment such that the direct acoustic field is much larger than the reverberant field. This situation may be achieved in an anechoic or semi-anechoic room or outdoors away from any reflecting surfaces, such as buildings, walls, etc. The directivity characteristics of the noise source may be measured in an anechoic room, as discussed in Sec. 3.6.4.

In an anechoic chamber, the room surfaces are treated with acoustic material such that the surface absorption is practically 100%. The floor of a semi-anechoic room is highly reflective, but the walls are highly absorptive, as in the case of an anechoic chamber.

An array of microphones or a single microphone moved to various positions may be used for making the measurements. The microphone measurement positions may be located at the same distance a from the center of the noise source on a spherical or hemispherical surface. The radius a of the sphere or hemisphere should be at least twice the largest dimension of the source or four times the height of the source, whichever is larger. The microphone measurement position should be farther than a distance $d_\mathrm{m} = \frac{1}{4}\lambda$ from any room surfaces, where λ is the wavelength corresponding to the lowest octave band center frequency considered in the measurements.

If the direct acoustic field is the only contributor to the acoustic pressure, the sound power W_j radiated from the noise source through an area S_j is related to the acoustic intensity I_j and measured acoustic pressure p_j as follows:

$$W_j = I_j S_j = \frac{S_j p_j^2}{\rho_0 c} \tag{3-22}$$

The total sound power radiated from the noise source is the sum of the power radiated through the entire surface around the source:

$$W = \sum_j W_j \tag{3-23}$$

The determination of the sound power may be made somewhat more convenient by dividing the total surface area of the sphere or hemisphere into N_s equal surface areas.

Acoustic Measurements

$$S_j = 2\pi a^2 / N_s \quad \text{(for a hemisphere)} \tag{3-24}$$

$$S_j = 4\pi a^2 / N_s \quad \text{(for a sphere)} \tag{3-25}$$

The surface area of the surface bounded by angles θ_1 and θ_2, for example, is illustrated in Fig. 3-9. The area is given by:

$$S_2 = \int_{\theta_1}^{\theta_2} (2\pi a \sin \theta)(a\, d\theta) = 2\pi a^2 (\cos \theta_1 - \cos \theta_2) \tag{3-26}$$

Let us consider the case for a hemispherical area or radius a, and use $N_s = 10$ equal areas. The top "cap" will be one area (S_1), and each of the next "rings" will each be divided into three areas. The surface area of the "cap", with $\theta_o = 0°$, is as follows:

$$S_1 = 2\pi a^2 (1 - \cos \theta_1) = (1/10)(2\pi a^2) \tag{3-27}$$

If we solve for the angle θ_1, we obtain the following:

$$\cos \theta_1 = 1 - 0.100 = 0.900 \quad \text{and} \quad \theta_1 = 25.84° = 0.4510\,\text{rad}$$

Using Eq. (3-26), we may obtain the following values for the other angles:

$$\cos \theta_2 = \cos \theta_1 - 0.300 = 0.600 \quad \text{and} \quad \theta_2 = 53.13° = 0.9273\,\text{rad}$$
$$\cos \theta_3 = \cos \theta_2 - 0.300 = 0.300 \quad \text{and} \quad \theta_3 = 72.54° = 1.2661\,\text{rad}$$
$$\cos \theta_4 = \cos \theta_3 - 0.300 = 0.000 \quad \text{and} \quad \theta_4 = 90.00° = \tfrac{1}{2}\pi\,\text{rad}$$

The microphone should be placed at the geometrical centroid of the surface area segment, as shown in Fig. 3-9. The centroid for the "cap" is directly at the top of the hemisphere ($\bar{\theta}_1 = 0°$). The angle locating the centroid of the "band" areas may be found from the following expression:

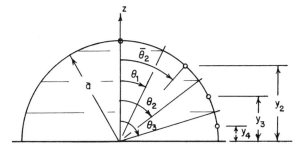

FIGURE 3-9 Coordinates for determining the microphone locations on a measuring surface of radius a. The small circles denote the points at the centroid of the area segments.

$$(\cos\theta_1 - \cos\theta_2)\bar{\theta}_2 = \int_{\theta_1}^{\theta_2} \theta \sin\theta \, d\theta \qquad (3\text{-}28)$$

$$\bar{\theta}_2 = \frac{(\sin\theta_2 - \sin\theta_1) - (\theta_2 \cos\theta_2 - \theta_1 \cos\theta_1)}{\cos\theta_1 - \cos\theta_2} \qquad (3\text{-}29)$$

If we substitute the numerical values for a 10-microphone system, the following value is obtained for the centroid of the first "band" area:

$$\bar{\theta}_2 = \frac{\sin(53.13°) - \sin(25.84°) - [(0.9273)\cos(53.13°) - (0.4510)\cos(25.84°)]}{(0.900 - 0.600)}$$

$$\bar{\theta}_2 = 0.7121 \text{ rad} = 40.80°$$

We may repeat the calculations for the other "band" areas to obtain the following result:

$$\bar{\theta}_3 = 1.1016 \text{ rad} = 63.12°$$

$$\bar{\theta}_4 = 1.4196 \text{ rad} = 81.34°$$

The vertical distance from the equator (floor for a hemisphere) y_j and the horizontal distance from the vertical axis x_j for the microphone locations may be found, as follows:

$y_2/a = \cos\bar{\theta} = 0.757$ and $x_2/a = \sin\bar{\theta}_2 = 0.653$

$y_3/a = \cos\bar{\theta}_3 = 0.452$ and $x_3/a = \sin\bar{\theta}_3 = 0.892$

$y_4/a = \cos\bar{\theta}_4 = 0.151$ and $x_4/a = \sin\bar{\theta}_4 = 0.989$

The specific locations for 10 microphones, according to ISO 3744, are shown in Fig. 3-10.

For the case in which the measuring surface subdivisions are equal in area, the total sound power may be found from Eqs (3-22) and (3-23), using the sound pressure measurements:

$$W = \Sigma I_j S_j = \frac{S_o \Sigma p_j^2}{\rho_o c N_s} \qquad (3\text{-}30)$$

The quantity S_o is the total surface area ($S_o = 2\pi a^2$ for a hemisphere; $S_o = 4\pi a^2$ for a sphere). For a spherical surface, 20 measurements could be used, for example. The 10 microphone locations below the equator would be at the same distance below the equator as those above the equator for the hemispherical surface with 10 microphone locations.

Example 3-3. The sound pressure level measurements given in Table 3-2 were obtained in a semi-anechoic room around a motor. The overall dimen-

Acoustic Measurements

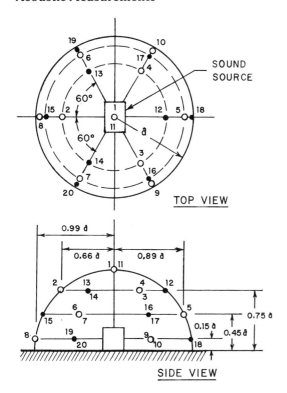

FIGURE 3-10 Microphone locations for a hemispherical measurement surface, using 10 measurement points (open circles). The closed circles denote locations for an additional 10 microphones for improved accuracy, particularly for sources having nonsymmetrical sound radiation characteristics.

TABLE 3-2 Data for Example 3-3

Point	Elevation, y, m	Angle, θ	L_p, dB
1	1.250	0.0°	86.0
2	0.9375	41.4°	81.5
3	0.9375	41.4°	82.4
4	0.9375	41.4°	81.3
5	0.5625	63.3°	70.9
6	0.5625	63.3°	72.9
7	0.5625	63.3°	68.0
8	0.1875	81.4°	79.3
9	0.1875	81.4°	78.5
10	0.1875	81.4°	80.1

sions of the motor are 500 mm (19.69 in) long by 300 mm (11.81 in) wide and 300 mm (11.81 in) high. The microphones were located at a distance of 1.250 m (49.21 in) from the center of the motor. The air in the room was at 24°C (297.2K or 75.2°F) and 101.6 kPa (14.74 psia), for which the properties are sonic velocity $c = 345.6$ m/s (1134 fps); density $\rho_o = 1.191$ kg/m^3 (0.0744 lb$_m$/ft^3), and $\rho_o c = 411.6$ rayl. Determine the overall sound power level for the motor.

The surface area for the measurement hemisphere may be calculated:

$$S_o = (2\pi)(1.250)^2 = 9.817 \text{ m}^2 \quad (105.7 \text{ ft}^2)$$

The sum of the squares of the acoustic pressure may be calculated in several different ways. Let us use the following technique:

$$\frac{\Sigma p_j^2}{(p_{\text{ref}})^2} = \Sigma 10^{L_{p_j}/10} = 1.14439 \times 10^9$$

$$\Sigma p_j^2 = (20 \times 10^{-6})^2 (1.14439 \times 10^9) = 0.45776 \text{ Pa}^2$$

The acoustic power for the motor may be calculated from Eq. (3-30):

$$W = \frac{(9.817)(0.45776)}{(411.6)(10)} = 1.092 \times 10^{-3} \text{ W} = 1.092 \text{ mW}$$

The sound power level for the motor is as follows:

$$L_W = 10 \log_{10}(1.092 \times 10^{-3}/10^{-12}) = 90.4 \text{ dB}$$

3.6.3 Sound Power Survey Measurement

There are some situations in which the noise source cannot be moved into a reverberant room or into an anechoic room. If the noise source is not located outdoors away from reflective surfaces, both the direct and diffuse or reverberant acoustic fields will influence the relationship between sound power and sound pressure. In this case, the microphone array location on a rectangular parallelepiped shown in Fig. 3-11 may be used to estimate the sound power from sound pressure measurements. The measured sound pressure levels must be "corrected" for the presence of any background noise, as discussed in Sec. 3.7.

The sound power level may be determined by the comparison method using a calibrated sound power source, as discussed in Sec. 3.6.1.1. Equation (3-8) may be used to calculate the sound power level L_W for the source from measurements of the sound pressure level L_p with the sound source in operation, the sound pressure level $L_{p,\text{cal}}$ with the calibrated sound power source in operation alone, and the sound power level $L_{W,\text{cal}}$ given by the manufacturer of the calibrated source.

Acoustic Measurements

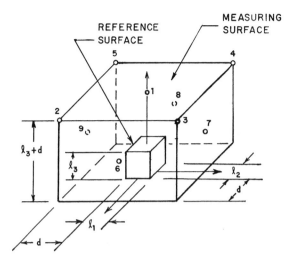

FIGURE 3-11 Microphone locations for rectangular measuring surfaces, using 9 microphone locations. Point 1 is at the center of the top surface, points 2 through 5 are at the corners of the top surface, and points 6 through 9 are at the centers of the vertical surfaces.

The sound power may also be calculated from measurements of the reverberation time for the space in which the noise source is located. If we include the acoustic energy directly transmitted through the measurement area S_m, Eq. (7-14) may be written in the following form:

$$W\left(\frac{4}{R} + \frac{1}{S_m}\right) = \frac{W}{S_m}\left(\frac{4S_m}{R} + 1\right) = \frac{p_{av}^2}{\rho_0 c} \tag{3-31}$$

The quantity p_{av} is the energy-averaged sound pressure from the measurements:

$$p_{av}^2 = \frac{\Sigma p_j^2}{N_s} \tag{3-32}$$

The room constant R may be taken from Eq. (3-16), with the second term in parenthesis neglected, since the magnitude of this term is usually smaller than the uncertainty in the sound power determination:

$$W = \frac{S_m p_{av}^2}{\rho_0 c}\left[1 + \frac{cT_r S_m}{13.816V}\right]^{-1} \tag{3-33}$$

We may convert Eq. (3-33) to "level" form by introducing the reference quantities, then taking log base 10 of both sides and multiplying through by 10:

$$L_W = L_{p,av} + 10\log_{10}(S_m/S_{ref}) - K_r - 10\log_{10}\left[\frac{\rho_0 c W_{ref}}{p_{ref}^2 S_{ref}}\right] \quad (3\text{-}34)$$

The reference area is $S_{ref} = 1\,\text{m}^2$, and the quantity K_r is defined by the following expression:

$$K_r = 10\log_{10}\left[1 + \frac{cT_r S_m}{13.816 V}\right] \quad (3\text{-}35)$$

For air at 25°C (298.2K or 77°F) and 101.325 kPa (14.696 psia), the characteristic impedance $\rho_0 c = 409.8$ rayl. Using this value, we may determine the numerical value of the last term in Eq. (3-34):

$$10\log_{10}\left[\frac{\rho_0 c W_{ref}}{p_{ref}^2 S_{ref}}\right] = 10\log_{10}\left[\frac{(409.8)(10^{-12})}{(20 \times 10^{-6})^2 (1)}\right] = 0.1\,\text{dB} \quad (3\text{-}36)$$

The sound power level expression may be written as follows, using the value of 0.1 dB for the last term in Eq. (3-34):

$$L_W = L_{p,av} + 10\log_{10}(S_m/S_{ref}) - K_r - 0.1 \quad (3\text{-}37)$$

There may be some situations in which the noise source cannot be stopped or "turned off" in order that reverberation time measurements can be made. In these cases, the sound power may be determined, with some loss in accuracy, by first estimating the room constant. Using information about the room surfaces and the techniques discussed in Chapter 7, the room constant may be estimated from Eq. (7-13):

$$R = \frac{\bar{\alpha} S_0}{1 - \bar{\alpha}} \quad (3\text{-}38)$$

The quantity $\bar{\alpha}$ is the average surface absorption coefficient for the room surfaces, and S_0 is the total surface area of the room in which the noise source is located. The factor K_r in Eq. (3-34) is given by the following expression, using the estimate for the average surface absorption coefficient:

$$K_r = 10\log_{10}\left[1 + \frac{4(1 - \bar{\alpha})S_m}{\bar{\alpha} S_0}\right] \quad (3\text{-}39)$$

The rectangular parallelepiped measuring surface and the key measuring points for the survey method of determining the sound power are illustrated in Fig. 3-11 (ISO, 1986e). The reference surface, with dimensions ℓ_1, ℓ_2, and height ℓ_3, is the smallest parallelepiped that can enclose the noise source. The measuring surface, on which the microphone measurement points are located, has dimensions of $(\ell_1 + 2d)$, $(\ell_2 + 2d)$, and height $(\ell_3 + d)$. The dimension d is somewhat flexible; however, a distance of $d = 1$ m (39.4 in) is often used for cases in which the largest dimension ℓ_{max} of the reference surface is 250 mm (9.8 in) or larger. For cases in which the largest dimension of the reference surface is less than 250 mm, the dimension d may be taken as any distance from $(4\ell_{max})$ to 1 m, but not smaller than 250 mm. For example, if the largest dimension of the reference surface is 150 mm (5.91 in), the dimension d could be chosen as any value from 0.60 m (23.6 in) to 1.00 m (39.4 in). However, if the largest dimension of the reference surface is 50 mm (1.97 in), the dimension d would be chosen as 250 mm (9.8 in) and not $(4)(50) = 200$ mm, for example.

There are nine key microphone locations, including locations at the height $h = \frac{1}{2}(\ell_3 + d)$ in the middle of the four vertical faces, at the center of the top surface, and at each of the four corners of the top surface. Eight additional microphone locations, including locations at the center of each of the four edges of the top surface, and locations at the center of each of the four vertical edges, may be used for additional accuracy.

Example 3-4. A small air compressor has envelope dimensions of 600 mm (23.6 in) wide by 800 mm (31.5 in) long by 600 mm (23.6 in) high. The compressor is located in a room having dimensions of 15 m (49.21 ft) by 18 m (59.06 ft) by 3.75 m (12.30 ft) high. The estimated average surface absorption coefficient for the room is $\bar{\alpha} = 0.15$. The measurement surface is selected with a spacing d of 1.00 m (3.28 ft or 39.4 in) from the compressor envelope surfaces, such that the dimensions of the measurement surface are 2.60 m (8.53 ft) by 2.80 m (9.19 ft) by 1.60 m (5.25ft) high. The measured sound pressure level values are given in Table 3-3 for the 500 Hz octave band. Determine the sound power level of the compressor for the 500 Hz octave band.

The energy-averaged sound pressure level may be determined from the nine data points and Eq. (3-32):

$$(p_{av}/p_{ref})^2 = (10^{8.20} + 10^{8.12} + \cdots + 10^{7.81})/(9) = 1.0972 \times 10^8$$

$$p_{av} = (20 \times 10^{-6})(1.0972 \times 10^8)^{1/2} = 0.2095 \text{ Pa}$$

$$L_{p,av} = 10\log_{10}(1.0972 \times 10^8) = 80.4 \text{ dB}$$

The measurement surface area is as follows:

$$S_m = (2)(2.60 + 2.80)(1.60) + (2.60)(2.80)$$

$$S_m = 17.28 + 7.28 = 24.56 \text{ m}^2 \quad (264.4 \text{ ft}^2)$$

TABLE 3-3 Data for Example 3-4. The measurement locations are illustrated in Fig. 3-11

Point	Location	L_p, dB
1	Top, center	82.0
2	Top corner, front	81.2
3	Top corner, front	82.6
4	Top corner, back	79.6
5	Top corner, back	80.1
6	Vertical side, center	76.7
7	Vertical side, center	79.8
8	Vertical side, center	80.6
9	Vertical side, center	78.1

The surface area of the room may be determined. We may either neglect the effect of the area covered by the compressor (less than 1% of the room surface area) or we may include the surface acoustic absorption of the compressor and exclude the floor area covered by the compressor. Let us use the first approach, since the floor area covered by the compressor is small:

$$S_o = (2)(15 + 18)(3.75) + (2)(15)(18)$$

$$S_o = 247.5 + 540.0 = 787.5 \, m^2 \quad (8477 \, ft^2)$$

The factor K_r may be calculated from Eq. (3-39) for this problem:

$$K_r = 10 \log_{10} \left[1 + \frac{(4)(1 - 0.15)(24.56)}{(0.15)(787.5)} \right] = 10 \log_{10}(1 + 0.7069)$$

$$= 2.32 \, dB$$

The sound power level for the 500 Hz octave band may be determined from Eq. (3.37):

$$L_W = 80.4 + 10 \log_{10}(24.56) - 2.32 - 0.1$$

$$L_W = 80.4 + 13.90 - 2.33 = 92.0 \, dB$$

3.6.4 Measurement of the Directivity Factor

The directivity factor Q or the directivity index DI, defined by Eqs (2-27) and (2-28), may be determined from measurements of the sound power in an anechoic or semi-anechoic room. The directivity may also be measured outdoors far away from reflecting surfaces. If the measurement is made out-

Acoustic Measurements

doors, the microphone should be located at a distance such that the sound pressure level decreases by 6 dB for each doubling of the distance from source. Generally, the number of measurement points required for effective determination of the directivity is larger than that needed for sound power determination. The measurement "mesh" should be made finer in the regions where the sound pressure varies rapidly with position.

The directivity factor Q is defined mathematically by Eq. (2-27). The directivity factor is the ratio of the sound intensity in a specified direction to the sound intensity for a spherical source having the same overall sound power:

$$Q = \frac{I}{(W/4\pi r^2)} = \frac{4\pi r^2 I}{W} \qquad (3\text{-}40)$$

The quantity r is the radial distance from the center of the source to the point at which the intensity I is determined. The directivity index is the directional characteristics expressed in "level" form, and is defined by Eq. (2-28):

$$\text{DI} = 10\log_{10}(Q) \qquad (3\text{-}41)$$

The acoustic power for the source must be measured first before the directivity factor can be determined. Using measurements of the acoustic pressure in an anechoic or semi-anechoic room on a spherical or hemispherical surface of radius a, the directivity factor may be calculated from Eq. (3-40), using Eqs (3-22) and (3-23) for the sound power:

$$Q = \frac{4\pi a^2 p_j^2}{\rho_0 c \Sigma S_j I_j} = \frac{4\pi a^2 p_j^2}{\Sigma S_j p_j^2} \qquad (3\text{-}42)$$

If the total surface area is divided into N_s equal areas, Eq. (3-42) may be simplified by making the substitutions from Eq. (3-24) or (3-25):

$$Q = \frac{2N_s p_j^2}{\Sigma p_j^2} = \frac{2p_j^2}{p_{av}^2} \quad \text{(sound source on a reflective surface)} \qquad (3\text{-}43)$$

$$Q = \frac{N_s p_j^2}{\Sigma p_j^2} = \frac{p_j^2}{p_{av}^2} \quad \text{(sound source suspended freely)} \qquad (3\text{-}44)$$

The quantity p_{av} is the energy-averaged sound pressure obtained from the sound power measurements:

$$p_{av}^2 = \frac{\Sigma p_j^2}{N_s} \qquad (3\text{-}45)$$

The directivity index may be found from Eq. (3-43) or (3-44) by introducing the reference sound pressure ($p_{\text{ref}} = 20\,\mu\text{Pa}$), taking log base 10 of both sides of the equations, and multiplying through by 10:

$$\text{DI} = L_{p_j} - L_{p,\text{av}} + 3\,\text{dB} \quad \text{(sound source on a reflective surface)}$$
(3-46)

$$\text{DI} = L_{p_j} - L_{p,\text{av}} \quad \text{(sound source suspended freely)} \quad (3\text{-}47)$$

Example 3-5. Determine the directivity factor and directivity index for the sound source given in Example 3-3 for an angle of $\theta = 0°$ with the vertical and for $\theta = 41.4°$.

The square of the energy-averaged sound pressure is:

$$p_{\text{av}}^2 = (\Sigma p_j^2)/N_s = (0.45776)/(10) = 0.045776\,\text{Pa}^2$$

The measured acoustic pressure at $\theta = 0°$ may be found from the data in Table 3-2.

$$p_1 = (20 \times 10^{-6})(10^{86/20}) = 0.3991\,\text{Pa}$$

The measurements were taken in a semi-anechoic room, so Eq. (3-43) may be used to evaluate the directivity factor:

$$Q = \frac{(2)(0.3991)^2}{(0.045776)} = 6.96 \quad \text{for } \theta = 0°$$

The directivity index is found from Eq. (3-41):

$$\text{DI} = 10\log_{10}(6.96) = 8.43\,\text{dB} \quad \text{for } \theta = 0°$$

We may use an alternative method to determine the directivity index. The average sound pressure level is as follows:

$$L_{p,\text{av}} = 10\log_{10}[(0.045776)/(20 \times 10^{-6})^2] = 80.6\,\text{dB}$$

The directivity index may be calculated from Eq. (3-46) for measurements taken in a semi-anechoic room:

$$\text{DI} = 86.0 - 80.6 + 3 = 8.4\,\text{dB}$$

From the data given in Table 3-2, we observe that the sound pressure level at 41.4° does not vary more than about 1 dB with the angle φ. Let us treat the source as an approximately symmetrical source. The average acoustic pressure level is found by averaging the data from points 2, 3, and 4:

$$L_{p2,av} = 10\log_{10}[(1/3)(10^{8.15} + 10^{8.24} + 10^{8.13})] = 81.8\,\text{dB}$$

$$p_{2,av} = (20 \times 10^{-6})(10^{81.8/20}) = 0.2461\,\text{Pa}$$

We may use Eq. (3-43) to evaluate the directivity factor for $\theta = 41.4°$:

$$Q_\theta = \frac{(2)(0.2461)^2}{(0.045776)} = 2.65 \quad \text{for } \theta = 41.4°$$

The directivity index for $\theta = 41.4°$ is as follows:

$$DI_\theta = 10\log_{10}(2.65) = 4.23\,\text{dB} \quad \text{for } \theta = 41.4°$$

The directivity index could also be determined from Eq. (3-46):

$$L_{p2,av} = 20\log_{10}(0.2461/20 \times 10^{-6}) = 81.8\,\text{dB}$$

$$DI_\theta = 81.8 - 80.6 + 3 = 4.2\,\text{dB}$$

3.7 NOISE MEASUREMENT PROCEDURES

As in any data-taking situation, the engineer should carefully define the purpose of the experimental study before taking data (Figliola and Beasley, 1991). The reason for making the measurements (hearing damage considerations, community reaction to noise, reduction of machinery noise, etc.) will determine, to a large extent, the type of instrumentation required.

It is good practice to make in initial visual and aural survey of the environment to be studied. This preliminary survey should answer such questions as (Beranek, 1971):

(a) What are the suspected or obvious sources of noise?
(b) What are the operating characteristics of the noise source?
(c) What is the physical size of the noise source?
(d) Does the source operate continuously or intermittently?
(e) What are the directional characteristics of the noise source?
(f) Are there any special environmental considerations?

The more familiar the engineer becomes with the problem prior to making measurements, the more likely he will arrive at an optimum choice of measuring instruments and obtain worthwhile data to help solve the noise problem.

After the preliminary survey has been completed, a specific strategy or plan for the experimental program should be developed. This plan should ensure that the acquired data meets the objectives for the test: a well-developed plan for the experimental program eliminates many "surprises" and errors of omission.

During the planning stage, consideration of the accuracy level required by the instrumentation and data acquisition procedure to meet the experimental objective should be made. An uncertainty analysis should be used to select an effective measurement method and instrumentation.

Portable battery-operated instruments are generally used for field measurements: Before commencing, it is important to check the batteries in all instruments. Sound level meters usually have a "battery check" function key. Spare batteries should be included with the instruments, and the batteries should be checked again when the instrumentation is set up in the field.

The sound level meter should be calibrated prior to taking measurements and the calibration checked again after taking all measurements. The "electrical noise floor" of the instrumentation should be checked to determine the lower limit on the levels of signals that can be measured accurately. This check may be accomplished by replacing the microphone with an equivalent electrical impedance or by reference to the specifications provided by the instrument manufacturer.

When the initial equipment check has been completed, several environmental factors should be considered. Barometric pressure and ambient temperature should be recorded, because some instruments are sensitive to ambient pressure and/or ambient temperature. Some calculations require knowledge of ambient air density, which can be determined from pressure and temperature readings. It is good practice to measure the ambient air humidity, because some microphones, such as condenser microphones, are sensitive to moisture in the ambient air. If the measurements are to be taken outdoors, a windscreen should be used on the microphone.

All microphones exhibit directivity effects at high frequencies, and the microphone should be positioned with these effects in mind. For example, if the response of the microphone is "flat" (very little change in gain as frequency of the noise input is changed) at $0°$ sound incidence, then the microphone should be positioned such that the microphone diaphragm faces the sound source. On the other hand, if the unit has "flat" response at $90°$ sound incidence, the microphone should be oriented with the microphone axis perpendicular to the source direction.

The sound level meter should be positioned such that the meter and the operator have a negligible effect on the sound field being measured. Obviously, the operator should not stand between the noise source and the microphone. As a rule of thumb, the meter should be placed at least 500 mm (20 in) from the operator's body or a tripod support should be used to minimize reflections from the operator.

When making measurements outdoors, the microphone should be located 1.20—1.50 m (48–60 in) above the ground and at least 3.50 m

Acoustic Measurements

(12 ft) from any reflecting surfaces, if possible. Indoor sound measurements should be taken with the microphone located 1.20–1.50 m above the floor, at least 1.00 m (40 in) from walls, and 1.50 m (60 in) from any windows. To overcome the effect of standing waves in indoor measurements, one should average at least three readings made at positions about 500 mm (20 in) apart.

If at all possible, the background noise level in octave bands should be measured with the source of noise turned off. If the background noise level is 10 dB or more below the noise level produced by the source, the error due to neglecting background noise is less than 0.5 dB. If the background noise level is 20 dB or more below the source noise level, the error is less than 0.1 dB and background noise will have negligible effect on the noise measurement of the source. If the difference between the measured and background levels is less than about 3 dB, the noise level from the source alone becomes quite difficult to measure accurately.

The noise level measurements may be "corrected" for background noise by subtracting (in decibel fashion) the background from the total noise measurement:

$$L(\text{corrected}) = 10 \log_{10}[10^{L(\text{measured})/10} - 10^{L(\text{background})/10}] \quad (3\text{-}48)$$

The readings may also be corrected using:

$$L(\text{corrected}) = L(\text{measured}) - A_\Delta \quad (3\text{-}49)$$

The factor A_Δ is a function of the difference between the measured and background noise levels. Values for this factor are given in Table 3-4.

If the background noise level is excessive, there are several approaches that can be used to obtain better measurements of the noise generated by the source. The measurements may be taken at a time when the background noise is lower (at night, for example). Moving the microphone closer to the noise source may increase the difference between the source and background noise levels, and thereby allow more accurate measurements of the source noise.

The following list includes the more important items of data that should be recorded when making most acoustic measurements. Many of the items of data are automatically recorded by some acoustic analyzers and programmable sound level meters:

(a) Date and time of measurement.
(b) Types, models, serial numbers, and other identification for all instruments and equipment used. This data allows one to replicate the data with the same instrument, if needed.

TABLE 3-4 Background noise correction factors:
$L(\text{corrected}) = L(\text{measured}) - A_\Delta$
$\Delta L = L(\text{measured}) - L(\text{background})$

ΔL, dB	A_Δ, dB	ΔL, dB	A_Δ, dB
1.0	6.9	6.5	1.1
1.5	5.3	7.0	1.0
2.0	4.3	7.5	0.9
2.5	3.6	8.0	0.7
3.0	3.0	9.0	0.6
3.5	2.6	10	0.5
4.0	2.2	12	0.3
4.5	1.9	14	0.2
5.0	1.7	16	0.1
5.5	1.4	18	0.1
6.0	1.3	20	0.0

(c) Description of the area where measurements were made. This data is important when writing the report on the experimental study.
(d) Description of the noise source, including dimensions, type of mounting, location within the room or relative to other objects, nameplate data, speed and power ratings, etc. Auxiliary information, such as surface area of the noise source, etc., may be determined from this data.
(d) Description of secondary noise sources, including location, type, dimensions, etc. This information is helpful when assessing the accuracy of the data, effect of background noise, etc.
(f) Location of observers during the time of the measurements.
(g) Orientation of the microphone axis relative to the direction of the source from the microphone.
(h) Barometric pressure, ambient temperature, wind speed and direction, and relative humidity. It is much more accurate to measure this data at the time of the experiment than to try to reconstruct the data from weather records several days after the measurements were taken.
(i) Measured frequency-band (1/1 or 1/3 octave data) levels at each microphone position.
(j) Measured frequency-band (1/1 or 1.3 octave data) levels for the background noise.

Acoustic Measurements

(k) Results of calibration tests, and the fact that the calibration tests were made prior to making noise measurements. This step is important in establishing credibility of the experimental data.

With previously mentioned data at hand, one can usually make an effective analysis of the noise problem and suggest one or more approaches that can result in a solution of the problem.

Example 3-6. The measured overall sound pressure level around a fan is 83 dB. The measured overall sound pressure level for the background (ambient) noise in the room where the fan is located is 77 dB. Determine the overall sound pressure level produced by the fan alone.

First, let us determine the fan sound pressure level from Eq. (3-48):

$$L_p(\text{corrected}) = 10 \log_{10}[10^{8.3} - 10^{7.7}]$$

$$L_p(\text{corrected}) = 10 \log_{10}(1.4941 \times 10^8) = 81.7 \, \text{dB}$$

Next, let us determine the fan sound pressure level from Eq. (3-49) and Table 3-4. The difference between the measured and background levels is:

$$\Delta L = L_p(\text{measured}) - L_p(\text{background}) = 83 - 77 = 6.0 \, \text{dB}$$

From Table 3-4, we find that: $A_\Delta = 1.3 \, \text{dB}$. The corrected fan sound pressure level is:

$$L_p(\text{corrected}) = L_p(\text{measured}) - A_\Delta = 83 - 1.3 = 81.7 \, \text{dB}$$

We do obtain the same answer by using either method.

PROBLEMS

3-1. Determine the intensity meter error when measuring the intensity of a sound wave having a frequency of 2 kHz in air with a sonic velocity of 344 m/s (1129 fps). The spacer thickness is 12 mm (0.472 in).

3-2. The measured overall sound pressure level around an electric motor (including background noise) is 73.6 dB. The measured background sound pressure level is 71.1 dB. Determine the sound pressure level for the electric motor alone (background noise removed).

3-3. The experimental data shown in Table 3-5 were measured around a steam valve. The readings are the octave band sound pressure levels with the steam flow stopped (background noise) and with the steam flowing (data). Determine the overall sound pressure level associated with the steam valve noise alone (with the background noise removed).

TABLE 3-5 Data for Problem 3-3

	Octave band center frequency, Hz							
	63	125	250	500	1,000	2,000	4,000	8,000
Background L_p, dB	74	71	67	63	60	57	54	54
Data L_p, dB	75	77	78	81	85	90	95	92

3-4. The experimental data shown in Table 3-6 were measured around an air vent. The readings are the octave band sound pressure levels with the air flow stopped (background noise) and with the air flowing (data). Determine the octave band sound pressure levels and overall sound pressure level for the vent noise alone (with the background noise removed).

3-5. A reverberant room having dimensions of 8.00 m (26.25 ft) by 9.00 m (29.53 ft) by 4.50 m (14.76 ft) high has a reverberation time of 3.80 seconds in the 1 kHz octave band. A food blender produces an octave band sound pressure level of 85 dB in the 1 kHz octave band during a test in the reverberant room. The air in the room is at 24°C (297 K or 75°F) and 104 kPa (15.08 psia), at which condition the speed of sound is 345.6 m/s (1134 fps). Determine the octave band sound power level for the 1 kHz octave band for the food blender.

3-6. A commercial leaf blower is suspended in an anechoic chamber to determine the sound power generated by the piece of equipment. Sound pressure measurements for the 500 Hz octave band are taken at 10 locations above the equator of a sphere having a radius of 800 mm (31.5 in) and 10 measurements are taken at corresponding points below the equator. The data points are given in Table 3-7. The location of the measurement points is the same as that for the points designated in Fig. 3-10. The point 11 corresponds to the point 1

TABLE 3-6 Data for Problem 3-4

	Octave band center frequency, Hz							
	63	125	250	500	1,000	2,000	4,000	8,000
Background L_p, dB	79.0	76.1	73.6	70.7	68.1	66.0	64.3	63.0
Data L_p, dB	79.1	76.6	76.3	78.7	81.2	83.1	81.2	78.1

Acoustic Measurements

TABLE 3-7 Data for Problem 3-6

Point	Elevation, y, m	Angle, θ	L_p, dB	Point	Elevation, y, m	Angle, θ	L_p, dB
1	0.800	0.0°	81.6	11	−0.800	180°	78.3
2	0.600	41.4°	80.0	12	−0.600	138.6°	77.1
3	0.600	41.4°	79.7	13	−0.600	138.6°	77.6
4	0.600	41.4°	80.9	14	−0.600	138.6°	76.5
5	0.360	63.3°	77.3	15	−0.360	116.7°	76.0
6	0.360	63.3°	76.2	16	−0.360	116.7°	75.3
7	0.360	63.3°	76.8	17	−0.360	116.7°	75.1
8	0.120	81.4°	74.6	18	−0.120	98.6°	73.2
9	0.120	81.4°	75.0	19	−0.120	98.6°	74.1
10	0.120	81.4°	75.5	20	−0.120	98.6°	74.7

location, but below the equator; point 12 corresponds to the point 2 location, etc. The air in the anechoic chamber is at 300K (80°F) and 101.3 kPa (14.7 psia), for which the density $\rho_o = 1.177\,\text{kg/m}^3$ (0.0735 lb$_m$/ft^3), sonic velocity $c = 347.2\,\text{m/s}$ (1139 fps), and the characteristic impedance $Z_o = 408.6$ rayl. Determine the sound power level for the 500 Hz octave band for the leaf blower.

3-7. Determine the directivity factor and directivity index for the leaf blower in Problem 3-6 for an angle of 0° and for an angle of 81.4°.

REFERENCES

Acoustical Society of America. 1979a. Precision methods for the determination of sound power levels of noise sources in anechoic and hemi-anechoic rooms, ANSI S1.35-1979. Acoustical Society of America, Woodbury, NY.

Acoustical Society of America. 1979b. Survey methods for the determination of sound power levels of noise sources, ANSI S1.36-1979. Acoustical Society of America, Woodbury, NY.

Acoustical Society of America. 1986a. Precision methods for the determination of sound power levels of broad-band noise sources in reverberation rooms, ANSI S1.31-1980 (R1986). Acoustical Society of America, Woodbury, NY.

Acoustical Society of America. 1986b. Precision methods for the determination of sound power levels of discrete frequency and narrow-band noise sources in reverberation rooms, ANSI S1.32-1980 (R1986). Acoustical Socieity of America, Woodbury, NY.

Acoustical Society of America. 1986c. Engineering methods for the determination of sound power levels of noise sources in a special reverberation test room, ANSI S1.33-1982 (R1986). Acoustical Society of America, Woodbury, NY.

Baade, P. K. 1969. Standardization of machinery sound measurements. ASME Paper 69-WA/FE-30. American Society of Mechanical Engineers, New York.

Beranek, L. L. 1971. *Noise and Vibration Control*, pp. 80–83. McGraw-Hill, New York.

Beranek, L. L. and Vér, I. L. 1992. *Noise and Vibration Control Engineering*, p. 102. John Wiley, New York.

Broch, J. T. 1971. *Acoustic Noise Measurements*, 2nd edn, p. 89. Brüel and Kjaer, Naerum, Denmark.

Figliola, R. S. and Beasley, D. E. *Theory and Design for Mechanical Measurements*, pp. 25–26. John Wiley, New York.

ISO. 1986a. Acoustics—determination of sound power levels of noise sources—broad-band sources in reverberation rooms, ISO 3741. International Organization for Standardization, Geneva, Switzerland.

ISO. 1986b. Acoustics—determination of sound power levels of noise sources—discrete-frequency and narrow-band sources in reverberation rooms, ISO 3742. International Organization for Standardization, Geneva, Switzerland.

ISO. 1986c. Acoustics—determination of sound power levels of noise sources—special reverberation test rooms, ISO 3743. International Organization for Standardization, Geneva, Switzerland.

ISO. 1986d. Acoustics—determination of sound power levels of noise sources—anechoic and semi-anechoic rooms, ISO 3745. International Organization for Standardization, Geneva, Switzerland.

ISO. 1986e. Acoustics—determination of sound power levels of noise sources—survey method, ISO 3746. International Organization for Standardization, Geneva, Switzerland.

Sköde, F. 1966. Windscreening of outdoor microphones. Brüel and Kjaer Technical Review, No. 1. Brüel & Kjaer, Denmark.

4
Transmission of Sound

Noise may be controlled by modification of the source of sound, the transmission path of the sound, or the receiver of the sound. In this chapter, the transmission of sound is considered, along with some techniques for controlling the level of sound transmitted. Sound transmission without attenuation or dissipation will be presented first. Then, the effect of attenuation of the acoustic energy during transmission will be examined.

4.1 THE WAVE EQUATION

To understand fully the principles governing various noise control procedures, one should be familiar with the governing equation for acoustic wave transmission, or the *wave equation*, which may be written in cartesian coordinates as follows:

$$\frac{\partial^2 p}{\partial x^2} + \frac{\partial^2 p}{\partial y^2} + \frac{\partial^2 p}{\partial z^2} = \frac{1}{c^2}\frac{\partial^2 p}{\partial t^2} \tag{4-1}$$

where p is the instantaneous acoustic pressure, c is the speed of sound, and t is the time coordinate (Norton, 1989).

The wave equation has two important restrictions: (a) energy dissipation effects are neglected, and (b) the pressure wave amplitude must be relatively small in comparison with atmospheric pressure. One may employ elaborate solution techniques to the wave equation, but if these two conditions are not met in the physical situation analyzed, the results of the analysis will be worthless.

As we will show in this chapter, energy dissipation effects are most pronounced for high-frequency sound and for sound transmission through large distances. For example, if the frequency of the sound being transmitted through atmospheric air is 500 Hz or less, the attenuation of acoustic energy is less than 0.1 dB for transmission distances smaller than about 25 m (82 ft). We see that there are many practical situations in which the effect of attenuation is negligible. On the other hand, if the frequency of the sound being transmitted through atmospheric air is 8 kHz and the distance transmitted is 125 m, the attenuation of acoustic energy can be as large as 40 dB.

For most noise control work, the amplitude of the sound wave is small. For example, for a sound pressure level of 150 dB, the rms acoustic pressure is 632 Pa (0.092 psi) or the peak pressure amplitude is 894 Pa (0.13 psi). These values are less than 1% of atmospheric pressure (101.3 kPa or 14.7 psia). Unless one is dealing with low-level sonic booms or sound from nearby blasts, for example, the sound pressure levels encountered in industrial or environmental conditions are usually less than 150 dB, and the restriction of "small" pressure amplitude is met.

Let us develop the one-dimensional wave equation for plane sound waves transmitted in the x-direction. An elemental layer of fluid initially having thickness dx and surface area S is shown in Fig. 4-1. After a small increment of time dt, one face moves from position x to a new position $(x + \xi)$, where ξ is the *instantaneous particle displacement*. The other face moves from $(x + dx)$ to a new position:

$$\left(x + dx + \xi + \frac{\partial \xi}{\partial x} dx \right)$$

The fluid moves from one position to another as a result of forces applied to the element. Now, we introduce the first restriction:

Restriction 1: frictional forces are negligible, so that the only forces acting on the element of fluid are the pressure forces. The net force acting on the element becomes:

$$F_{net} = pS - \left(p + \frac{\partial p}{\partial x} dx \right) S = -\frac{\partial p}{\partial x} dx\, S \qquad (4\text{-}2)$$

The particle velocity for the layer of material is the change in displacement per unit time:

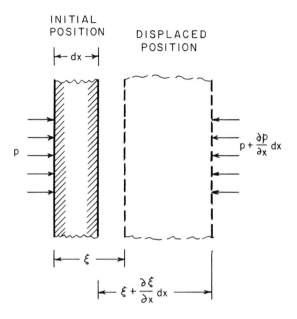

FIGURE 4-1 Initial and displaced positions of a fluid element as a sound wave passes through the element.

$$u = \frac{\partial \xi}{\partial t}$$

The acceleration of the layer of material is the change in velocity per unit time:

$$a = \frac{\partial u}{\partial t} = \frac{\partial^2 \xi}{\partial t^2}$$

The mass of the small layer is its mass per unit volume (or density, ρ) times the element volume ($S\,dx$), or $dm = \rho S\,dx$.

If we make these substitutions into Newton's second law of motion, $F_{\text{net}} = ma$, we obtain the following:

$$\frac{\partial p}{\partial x} = -\rho \frac{\partial^2 \xi}{\partial t^2} \tag{4-3}$$

The final form of the wave equation may be written in terms of several different variables. In the following, let us develop the wave equation in terms of the acoustic pressure.

Transmission of Sound

The speed of sound in any fluid may be determined from the following derivative, taken at constant entropy (van Wylen et al., 1994):

$$c^2 = \frac{\partial p}{\partial \rho} = \frac{\partial p/\partial t}{\partial \rho/\partial t} \quad \text{or} \quad \frac{\partial p}{\partial t} = c^2 \frac{\partial \rho}{\partial t} \tag{4-4}$$

Let us define the property *condensation* $\mathscr{C}$ as the fractional change in the fluid density:

$$\mathscr{C} = \frac{\rho - \rho_o}{\rho_o} = \frac{\rho}{\rho_o} - 1 \tag{4-5}$$

The quantity ρ_o is the density of the undisturbed fluid (usually at atmospheric pressure). For a fixed mass of the fluid element, we may write the condensation in the following form, in terms of the particle displacement:

$$\mathscr{C} = \frac{V_o - V}{V} = \frac{S\,dx - S\left(dx + \frac{\partial \xi}{\partial x}\,dx\right)}{\left(dx + \frac{\partial \xi}{\partial x}\,dx\right)S} = -\frac{\frac{\partial \xi}{\partial x}}{1 + \frac{\partial \xi}{\partial x}} \tag{4-6}$$

The quantity V_o is the initial volume of the element.

Now, we will introduce the second important restriction:

Restriction 2: the displacement of the fluid particles is very small, such that:

$$\frac{\partial \xi}{\partial x} \ll 1$$

With this condition, the condensation may be written in the following form:

$$\mathscr{C} = -\frac{\partial \xi}{\partial x} \tag{4-7}$$

The fluid density may be written in terms of the condensation, using Eq. (4-5):

$$\rho = \rho_o(1 + \mathscr{C})$$

And,

$$\frac{\partial \rho}{\partial t} = \rho_o \frac{\partial \mathscr{C}}{\partial t} \tag{4-8}$$

If we use this result in Eq. (4-4), we obtain the following:

$$\frac{\partial p}{\partial t} = c^2 \frac{\partial \rho}{\partial t} = \rho_o c^2 \frac{\partial \mathscr{C}}{\partial t} \tag{4-9}$$

Taking the second partial derivative both sides of Eq. (4-9),

$$\frac{\partial^2 p}{\partial t^2} = \rho_0 c^2 \frac{\partial^2 \mathscr{C}}{\partial t^2} \tag{4-10}$$

Similarly, taking the second partial derivative of Eq. (4-3), the force balance equation, assuming that the displacement is very small, such that the density is practically constant,

$$\rho = \rho_0(1 + \mathscr{C}) \approx \rho_0 = \text{constant}$$

we obtain:

$$\frac{\partial^2 p}{\partial x^2} = -\rho_0 \frac{\partial^3 \xi}{\partial t^2 \partial x} = -\rho_0 \frac{\partial^2}{\partial t^2}\left(\frac{\partial \xi}{\partial x}\right) \tag{4-11}$$

Using Eq. (4-7),

$$\frac{\partial^2 p}{\partial x^2} = +\rho_0 \frac{\partial^2 \mathscr{C}}{\partial t^2} \tag{4-12}$$

By comparing Eqs (4-10) and (4-12), we obtain the wave equation in terms of the instantaneous acoustic pressure p:

$$\frac{\partial^2 p}{\partial x^2} = \frac{1}{c^2} \frac{\partial^2 p}{\partial t^2} \tag{4-13}$$

This equation, when solved subject to the pertinent initial and boundary conditions, yields expressions for all of the acoustic quantities in a particular situation. The wave equation may also be written in terms of the instantaneous particle displacement (Randall, 1951):

$$\frac{\partial^2 \xi}{\partial x^2} = \frac{1}{c^2} \frac{\partial^2 \xi}{\partial t^2} \tag{4-14}$$

The displacement formulation may be more convenient to use in the solution of problems in which displacements are known at the boundaries.

After the wave equation has been solved for either the acoustic pressure or the particle displacement, the other quantities may be found by operating mathematically on the solution of the wave equation. For example, suppose we have solved the wave equation for the instantaneous acoustic pressure, $p(x, t)$. By integrating both sides of Eq. (4-9), we obtain the condensation expression:

$$p = \rho_0 c^2 \mathscr{C} \tag{4-15}$$

By integrating Eq. (4-7) and using Eq. (4-15), we obtain the expression for the instantaneous particle displacement:

$$\xi = -\int \mathscr{C}\, dx = -\frac{1}{\rho_0 c^2} \int p(x, t)\, dx \tag{4-16}$$

Transmission of Sound 83

Finally, if we integrate the force balance, Eq. (4-3), with respect to time and use the velocity-displacement relation, we may evaluate the instantaneous particle velocity:

$$u(x, t) = -\frac{1}{\rho_o} \int \frac{\partial p(x, t)}{\partial x} dx \qquad (4\text{-}17)$$

On the other hand, if we have obtained the solution in terms of the instantaneous particle displacement, $\xi(x, t)$, we may evaluate the other acoustic quantities from the following expressions:

$$u(x, t) = \frac{\partial \xi}{\partial t} \qquad (4\text{-}18)$$

$$\mathscr{C}(x, t) = -\frac{\partial \xi}{\partial x} \qquad (4\text{-}19)$$

$$p(x, t) = \rho_o c^2 \mathscr{C} = -\rho_o c^2 \frac{\partial \xi}{\partial x} \qquad (4\text{-}20)$$

4.2 COMPLEX NUMBER NOTATION

In working with the wave equation and its solution, it is convenient to use the complex number notation, because we are often interested in simple harmonic waves (sinusoidal waves). Even if the wave is not simple harmonic, the waveform may be expanded in a Fourier series, which involves a series of sinusoidal terms. In addition, the complex notation provides information about both the magnitude of a quantity and its phase angle in compact form.

A complex number may be written in cartesian form:

$$z = x + jy = \text{Re} + j\text{Im} \qquad (4\text{-}21)$$

where $x = \text{Re} =$ "real" and $y = \text{Im} =$ "imaginary" part of the complex quantity. We have used the symbol, $j = \sqrt{-1}$, because the symbol i is often used to represent electric current. The complex quantity may also be written in polar form:

$$z = |z| e^{j\phi} \qquad (4\text{-}22)$$

where $|z|$ is the magnitude of the quantity and ϕ is the phase angle. We may convert from one form to the other by using the following relations:

$$|z| = (\text{Re}^2 + \text{Im}^2)^{1/2} \qquad (4\text{-}23)$$

$$\tan \phi = \text{Im}/\text{Re} \qquad (4\text{-}24)$$

$$\text{Re} = |z|\cos\phi \tag{4-25}$$

$$\text{Im} = |z|\sin\phi \tag{4-26}$$

Mathematical manipulations (derivatives, integrations, etc.) are somewhat simpler to handle in terms of exponentials, because we do not need to utilize the trigonometric identities. If we have a sinusoidal expression,

$$y(t) = Y\cos\omega t = Y\cos(2\pi f t) \tag{4-27}$$

then, we may also write:

$$y(t) = Y e^{j\omega t} = Y(\cos\omega t + j\sin\omega t) = \text{Re}(y) + j\,\text{Im}(y) \tag{4-28}$$

Equations (4-27) and (4-28) are identical if we adopt the convention that only the "real" part of the complex expression is representative of the "real" physical quantity. This procedure is illustrated by an example in the following section.

4.3 WAVE EQUATION SOLUTION

Let us consider the solution of the wave equation for the case of sound waves being generated by motion of a plane wall, as shown in Fig. 4-2. The motion of the wall (at $x = 0$) may be represented by:

$$X(t) = X_m \cos\omega t \tag{4-29}$$

where X_m is the peak amplitude of motion, and $\omega = 2\pi f$ is the circular frequency for the motion. Using the complex notation, remembering that

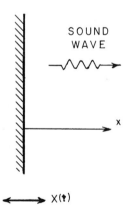

FIGURE 4-2 Sound wave generated by a vibrating wall. $X(t)$ is the instantaneous velocity of the wall.

Transmission of Sound

we are actually using only the *real* part, Eq. (4-29) may be written in an alternative form:

$$X(t) = X_m e^{j\omega t} \tag{4-30}$$

The resulting sound wave generated by the simple harmonic motion of the plane wall should also have a simple harmonic form. Let us consider the following solution form for the displacement:

$$\xi(x, t) = \psi(x) e^{j\omega t} \tag{4-31}$$

The quantity $\psi(x)$ is the *amplitude function*, which is dependent on the x-coordinate only, for one-dimensional waves.

The spatial and time derivatives from Eq. (4-31) are found, as follows:

$$\frac{\partial^2 \xi}{\partial t^2} = j^2 \omega^2 \psi(x) e^{j\omega t} = -\omega^2 \psi(x) e^{j\omega t} \tag{4-32}$$

$$\frac{\partial^2 \xi}{\partial x^2} = \frac{d^2 \psi}{dx^2} e^{j\omega t} \tag{4-33}$$

If we make these substitutions into the wave equation, Eq. (4-14), we obtain the following ordinary differential equation:

$$\frac{d^2 \psi}{dx^2} = -\frac{\omega^2}{c^2} \psi(x) \tag{4-34}$$

If we introduce the wave number, $k = \omega/c = 2\pi f/c = 2\pi/\lambda$, we obtain the final form of the equation to be solved:

$$\frac{d^2 \psi}{dx^2} + k^2 \psi = 0 \tag{4-35}$$

The general solution of Eq. (4-35) is:

$$\psi(x) = A e^{-j\omega t} + B e^{j\omega t} \tag{4-36}$$

where A and B are constants of integration to be determined from the boundary conditions. The general solution for plane waves is found by making the substitution from Eq. (4-36) into Eq. (4-31):

$$\xi(x, t) = A e^{j(\omega t - kx)} + B e^{j(\omega t + kx)} \tag{4-37}$$

The general solution involves two terms:

(a) One term containing $(\omega t - kx)$, which represents a sound wave moving in the +x-direction (*positive* x-direction), and

(b) Another term containing $(\omega t + kx)$, which represents a sound wave moving in the $-x$-direction (*negative x-direction*).

This behavior may be easily recognized if we consider the motion of the peaks in the sound waves. Consider the term containing $(\omega t - kx)$, for which the real part is $\cos(\omega t - kx)$. One peak of the wave at an arbitrary time t_o occurs at position x_o, where:

$$\omega t_o - k x_o = 0$$

A short time later (at $t = t_o + \Delta t$), the position x_1 of the peak is described by:

$$\omega(t_o + \Delta t) - k x_1 = 0$$

If we subtract these two relations, we find the location to which the peak of the sound wave has moved during the small time Δt:

$$x_1 = x_o + (\omega/k)\,\Delta t = x_o + c\,\Delta t$$

We note that $x_1 > x_o$, so the peak of the sound wave has moved in the $+x$-direction. By going through a similar procedure, we may show that the other part of the particle displacement expression, involving $(\omega t + kx)$, corresponds to a sound wave moving in the $-x$-direction.

For this example, suppose that the sound wave is transmitted into a very large (infinite) space so that there is no sound reflected back to the vibrating plate. In this case, we will have no sound wave moving in the $-x$-direction back toward the plate. To achieve this condition, we must have $B = 0$.

At the surface of the vibrating plate, the fluid must follow the motion of the plate:

$$\xi(0, t) = X(t) = X_m\, e^{j\omega t} = A\, e^{j(\omega t - kx)}|_{x=0} = A\, e^{j\omega t} \qquad (4\text{-}38)$$

Thus, the other constant of integration is $A = X_m$. The final expression for the instantaneous particle displacement caused by the vibrating plate is:

$$\xi(x, t) = X_m\, e^{j(\omega t - kx)} \qquad (4\text{-}39)$$

According to the complex notation convention, only the real part actually represents the particle displacement:

$$\xi(x, t) = X_m \cos(\omega t - kx) = X_m \cos[2\pi f t - 2\pi(x/\lambda)] \qquad (4\text{-}40)$$

We note that the peak amplitude of the particle displacement is $|\xi| = X_m$, and the rms particle displacement is $\tilde{\xi} = X_m/\sqrt{2}$.

Transmission of Sound

Let us evaluate the other quantities for a plane sound wave. The instantaneous particle velocity may be found from the velocity–displacement relation, Eq. (4-18):

$$u(x, t) = \frac{\partial \xi}{\partial t} = j\omega X_m e^{j(\omega t - kx)} = \omega X_m e^{j(\omega t - kx + \pi/2)} \quad (4\text{-}41)$$

Note that we have written $j = e^{j\pi/2}$. If we take the real part of Eq. (4-41), we obtain the actual variation of the instantaneous particle velocity:

$$u(x, t) = \omega X_m \cos(\omega t - kx + \pi/2) \quad (4\text{-}42)$$

The magnitude of the particle velocity is related to the magnitude of the particle displacement:

$$|u| = \omega X_m = \omega |\xi| = 2\pi f |\xi| \quad (4\text{-}43)$$

By comparing Eq. (4-42) and Eq. (4-40), we observe that the particle velocity leads the particle displacement by $\phi = \tfrac{1}{2}\pi$ radians $= 90°$.

Next, let us find the expression for the instantaneous acoustic pressure by applying Eq. (4-20):

$$p(x, t) = -\rho_o c^2 \frac{\partial \xi}{\partial x} = +jk\rho_o c^2 X_m e^{j(\omega t - kx)} \quad (4\text{-}44)$$

We note that $\omega = kc$ and $j = e^{j\pi/2}$:

$$p(x, t) = \rho_o c \omega X_m e^{j(\omega t - kx + \pi/2)} \quad (4\text{-}45)$$

Taking the real part of Eq. (4-45), we obtain:

$$p(x, t) = \rho_o c \omega X_m \cos(\omega t - kx + \pi/2) \quad (4\text{-}46)$$

If we compare Eq. (4-46) and Eq. (4-42), we observe that the acoustic pressure and particle velocity are related:

$$p(x, t) = \rho_o c u(x, t) \quad (4\text{-}47)$$

The characteristic acoustic impedance is defined by:

$$Z_o = \rho_o c \quad (4\text{-}48)$$

The magnitudes of the acoustic pressure and particle velocity for a plane wave are related by the following:

$$|p| = Z_o |u| \quad (4\text{-}49)$$

We also note that the phase angle between the acoustic pressure and particle velocity for a plane wave is $\phi = 0°$, or these quantities are *in-phase*.

4.4 SOLUTION FOR SPHERICAL WAVES

If we develop the wave equation for spherical coordinates, in which the quanitities are functions of the radial coordinate r and time t only, the following expression is obtained:

$$\frac{1}{r^2}\frac{\partial}{\partial r}\left(r^2\frac{\partial p}{\partial r}\right) = \frac{1}{c^2}\frac{\partial^2 p}{\partial t^2} \qquad (4\text{-}50)$$

The wave equation for one-dimensional spherical waves may be written in the following alternative form:

$$\frac{\partial^2(rp)}{\partial r^2} = \frac{1}{c^2}\frac{\partial^2(rp)}{\partial t^2} \qquad (4\text{-}51)$$

Let us try a solution for Eq. (4-51) in the form:

$$p(r, t) = \psi(r)\,e^{j\omega t} \qquad (4\text{-}52)$$

where $\psi(r)$ is the amplitude function. If we substitute the expression from Eq. (4-52) into the wave equation, Eq. (4-51), we obtain the following ordinary differential equation:

$$\frac{d^2(r\psi)}{dr^2} = \frac{j^2\omega^2(r\psi)}{c^2} = -k^2(r\psi) \qquad (4\text{-}53)$$

The general solution for the amplitude function is:

$$r\psi(r) = A\,e^{-jkr} + B\,e^{jkr} \qquad (4\text{-}54)$$

The general solution for the instantaneous acoustic pressure may be found by substituting the amplitude function expression from Eq. (4-54) into Eq. (4-52).

$$p(r, t) = \frac{1}{r}(A\,e^{-jkr} + B\,e^{jkr})\,e^{j\omega t} \qquad (4\text{-}55)$$

We observe that the amplitude of the acoustic pressure is not constant for a spherical wave; instead, the amplitude varies inversely with distance from the source, r.

If we consider only waves moving radially outward from the source or the case for no waves reflected back toward the origin, we must have $B = 0$.

$$p(r, t) = (A/r)\,e^{j(\omega t - kr)} \qquad (4\text{-}56)$$

Transmission of Sound

The real component from Eq. (4-56) is:

$$p(r, t) = (A/r)\cos(\omega t - kr) \quad (4\text{-}57)$$

The rms acoustic pressure for a spherical wave is given by:

$$p_{rms} = A/\sqrt{2}\, r \quad \text{or} \quad A = \sqrt{2}\, p_{rms}\, r \quad (4\text{-}58)$$

Next, let us determine the expression for the instantaneous particle velocity for a spherical wave. Integrating the Newton's law expression, as given by Eq. (4-17), we may evaluate the particle velocity:

$$u(r, t) = -\frac{1}{\rho_o}\int \frac{\partial p(r, t)}{\partial r}\, dt = \frac{1}{j\rho_o\omega}\frac{A}{r}\left(\frac{1}{r} + kj\right) e^{j(\omega t - kr)} \quad (4\text{-}59)$$

If we introduce $\omega = kc$, and the acoustic pressure expression, Eq. (4-56), we may write Eq. (4-59) as follows:

$$u(r, t) = -\frac{j p(r, t)}{\rho_o c k r}(1 + jkr) \quad (4\text{-}60)$$

The specific acoustic impedance, which is a complex quantity for a spherical wave, may be found from Eq. (4-60):

$$Z_s = \frac{p}{u} = \frac{j\rho_o c k r}{1 + jkr} = \frac{\rho_o c k r(kr + j)}{1 + (kr)^2} = |Z_s|\, e^{j\phi} \quad (4\text{-}61)$$

According to Eq. (4-23), the magnitude of the specific acoustic impedance is the square root of the sum of the squares of the real and imaginary parts of the complex expression in Eq. (4-61):

$$|Z_s| = \frac{|p|}{|u|} = \frac{\rho_o c k r (k^2 r^2 + 1)^{1/2}}{1 + k^2 r^2} = \frac{Z_o k r}{(1 + k^2 r^2)^{1/2}} \quad (4\text{-}62)$$

The tangent of the phase angle between the acoustic pressure and particle velocity is the ratio of the imaginary to the real parts of the complex quantity, as given by Eq. (4-24):

$$\tan \phi = 1/kr \quad (4\text{-}63)$$

We note that the magnitude of the specific acoustic impedance approaches ($Z_o kr$) and the phase angle approaches 90° for kr very small (less than about 0.15). This condition occurs for positions very near the spherical source or for very low frequencies. On the other hand, we note that the specific acoustic impedance approaches the characteristic impedance and the phase angle approaches 0° for kr large (greater than about 7). This condition occurs for positions far away from the spherical source or for high frequencies.

The expressions for the acoustic intensity and energy density may be found for a spherical wave. Using the complex notation, the intensity may be evaluated from the following:

$$I = \text{Re}(p_{rms} u^*_{rms}) \tag{4-64}$$

The quantity u^*_{rms} is the complex conjugate of the rms particle velocity. The complex conjugate of a complex quantity $z = x + jy = r e^{j\theta}$ is:

$$z^* = x - jy = r e^{-j\theta} \tag{4-65}$$

Let us write the rms quantities (with time integrated out) as follows:

$$p_{rms} = |p| e^{-jkr} \tag{4-66}$$

$$u_{rms} = \frac{|p|}{|Z_s|} e^{-j(kr+\phi)} \tag{4-67}$$

$$u^*_{rms} = \frac{|p|}{|Z_s|} e^{+j(kr+\phi)} \tag{4-68}$$

Making these substitutions into Eq. (4-64), we obtain the expressions for the intensity of a spherical wave:

$$I = \frac{|p|^2}{|Z_s|} \text{Re}(e^{j\phi}) \tag{4-69}$$

$$I = \frac{|p|^2 \cos\phi}{|Z_s|} \tag{4-70}$$

The tangent of the phase angle ϕ is given by Eq. (4-63). The cosine of the phase angle is given by the following:

$$\cos\phi = \frac{kr}{(1 + k^2 r^2)^{1/2}} \tag{4-71}$$

Using the expression for the magnitude of the specific acoustic impedance from Eq. (4-62), we obtain the final expression for the intensity of a spherical wave:

$$I = \frac{|p|^2}{Z_o} = \frac{p^2}{\rho_o c} \tag{4-72}$$

The kinetic energy per unit volume for a sound wave is given by:

$$KE = \text{Re}(\tfrac{1}{2} \rho_o u_{rms} u^*_{rms}) \tag{4-73}$$

Transmission of Sound 91

We may evaluate this expression, using Eqs (4-67) and (4-68):

$$\text{KE} = \text{Re}[\tfrac{1}{2}\rho_o(|p|^2/|Z_s|^2)\,e^{-j(kr+\phi)}\,e^{+j(kr+\phi)}] \qquad (4\text{-}74)$$

$$\text{KE} = \frac{|p|^2(1+k^2r^2)}{2\rho_o c^2 k^2 r^2} = \frac{|p|^2}{2\rho_o c^2}\left(1 + \frac{1}{k^2 r^2}\right) \qquad (4\text{-}75)$$

The potential energy per unit volume for a sound wave is given by:

$$\text{PE} = \text{Re}(\tfrac{1}{2}\rho_o p_{\text{rms}} p^*_{\text{rms}}/K_s) \qquad (4\text{-}76)$$

The quantity K_s is the adiabatic compressibility given by the following expression for an ideal gas:

$$K_s = \gamma P_o = \gamma \rho_o RT = \rho_o c^2 \qquad (4\text{-}77)$$

The quantity $\gamma = c_p/c_v$ is the specific heat ratio. The potential energy may be evaluated as follows:

$$\text{PE} = \text{Re}[(|p|^2/2\rho_o c^2)\,e^{-jkr}\,e^{+jkr}] = \frac{|p|^2}{2\rho_o c^2} \qquad (4\text{-}78)$$

The acoustic energy density or the total energy per unit volume is the sum of the kinetic and potential energies:

$$D = \text{KE} + \text{PE} = \frac{p^2}{\rho_o c^2}\left(1 + \frac{1}{2k^2 r^2}\right) \qquad (4\text{-}79)$$

4.5 CHANGES IN MEDIA WITH NORMAL INCIDENCE

We will analyze the transmission of sound from one material to another material in this section. As shown in Fig. 4-3, when a sound wave moving in one fluid strikes the surface or interface of a different material, a portion of the acoustic energy is reflected, and a portion is transmitted into the second medium. Let us consider the case of transmission from one material into another one in which the sound wave strikes the interface at *normal incidence* or with zero angle between the direction of the sound wave and the normal drawn to the interface.

From the previous discussion of plane sound waves, we know that the acoustic pressure may be written in the following form for a wave moving in material 1:

$$p_1(x,t) = A_1\,e^{j(\omega t - k_1 x)} + B_1\,e^{j(\omega t + k_1 x)} \qquad (4\text{-}80)$$

(Incident wave) + (Reflected wave)

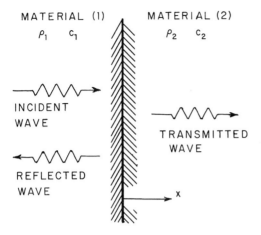

FIGURE 4-3 Transmission of sound from one material into another for normal incidence of the sound wave.

Similarly, assuming no reflections in material 2 or that material 2 is very large in extent, the instantaneous acoustic pressure for the transmitted wave in material 2 may be written as follows:

$$p_2(x, t) = A_2 \, e^{j(\omega t - k_2 x)} \qquad (4\text{-}81)$$

The instantaneous particle velocities in the two materials may be written from Eq. (4-47), noting that the reflected wave is traveling in the $-x$-direction:

$$u_1(x, t) = (1/Z_1)[A_1 \, e^{j(\omega t - k_1 x)} - B_1 \, e^{j(\omega t + k_1 x)}] \qquad (4\text{-}82)$$

$$u_2(x, t) = (1/Z_2) A_2 \, e^{j(\omega t - k_2 x)} \qquad (4\text{-}83)$$

The quantities $Z_1 = \rho_1 c_1$ and $Z_2 = \rho_2 c_2$ are the characteristic impedances for materials 1 and 2, respectively.

At the interface between the two materials, the instantaneous acoustic pressure in material 1 must be equal to the instantaneous acoustic pressure in material 2. Using this fact, at $x = 0$, we find the following relation between the coefficients:

$$A_1 + B_1 = A_2 \qquad (4\text{-}84)$$

Similarly, the instantaneous particle velocities must be the same in each media at the interface ($x = 0$):

Transmission of Sound

$$\frac{(A_1 - B_1)}{Z_1} = \frac{A_2}{Z_2} \tag{4-85}$$

We may use Eqs (4-84) and (4-85) to solve for the ratio of the two constants:

$$\frac{A_2}{A_1} = \frac{2Z_2}{Z_1 + Z_2} \tag{4-86}$$

We note that the magnitudes of the rms acoustic pressure for the transmitted wave and for the incident wave are given by:

$$p_{tr} = A_2/\sqrt{2} \quad \text{and} \quad p_{in} = A_1/\sqrt{2} \tag{4-87}$$

Therefore, $A_2/A_1 = p_{tr}/p_{in}$.

The *sound power transmission coefficient* a_t is defined as the ratio of the transmitted acoustic power to the incident acoustic power. This is a significant parameter in selecting the materials for controlling sound transmission:

$$a_t \equiv \frac{W_{tr}}{W_{in}} = \frac{SI_{tr}}{SI_{in}} = \frac{(p_{tr})^2/Z_2}{(p_{in})^2/Z_1} = \frac{A_2^2 Z_1}{A_1^2 Z_2} \tag{4-88}$$

Making the substitution for the coefficient ratio from Eq. (4-86), we obtain the following expression for the sound power transmission coefficient for a sound wave in material 1 striking material 2:

$$a_t = \frac{4Z_1 Z_2}{(Z_1 + Z_2)^2} \tag{4-89}$$

An alternative way of expressing the transmission of acoustic energy from one material to another is in terms of the *transmission loss* TL. The transmission loss expresses the sound power transmission coefficient in decibel units:

$$\text{TL} \equiv 10 \log_{10}(W_{in}/W_{tr}) = 10 \log_{10}(1/a_t) \tag{4-90}$$

We note from Eq. (4-89) that the sound power transmission coefficient is unity if the characteristic impedances of the two materials are the same, i.e., the impedances are *matched*. This result means that all of the acoustic energy is transmitted through the interface and none is reflected. On the other hand, if the acoustic impedances are quite different from each other, then the sound power transmission coefficient will be small. This result means that little acoustic energy is transmitted through the interface and most of the energy is reflected.

Another term that is not as widely used in noise control work as is the sound power transmission coefficient is the *sound power reflection coefficient* a_r:

$$a_r \equiv W_r/W_{in} \tag{4-91}$$

For this case, the acoustic energy is either reflected or transmitted, so $W_{in} = W_{tr} + W_r$, and the sound power reflection coefficient is related to the sound power transmission coefficient by the following relation:

$$a_r = 1 - a_t \qquad (4\text{-}92)$$

Making the substitution from Eq. (4-89), we obtain the following:

$$a_r = \left(\frac{Z_2 - Z_1}{Z_1 + Z_2}\right)^2 \qquad (4\text{-}93)$$

Example 4-1. A sound wave in air at 25°C (77°F) strikes a concrete wall at normal incidence, as shown in Fig. 4-4. The intensity level of the incident sound wave is 90 dB. Determine the transmission loss, and the sound pressure level for the transmitted wave.

We find the following values for the characteristic impedance for air and concrete in Appendix B:

air $\quad Z_1 = 409.8 \,\text{rayl}$

concrete $\quad Z_2 = 7.44 \times 10^6 \,\text{rayl}$

The sound power transmission coefficient is found from Eq. (4-89):

$$a_t = \frac{(4)(409.8)(7.44)(10^6)}{(409.8 + 7.44 \times 10^6)^2} = 2.203 \times 10^{-4}$$

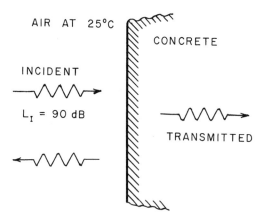

Figure 4-4 Physical system for Example 4-1.

Transmission of Sound

The transmission loss is found from Eq. (4-90):

$$\text{TL} = 10\log_{10}(1/2.203 \times 10^{-4}) = 36.6\,\text{dB}$$

The intensity of the incident wave is found from the following:

$$L_{\text{I,in}} = 10\log_{10}(I_{\text{in}}/I_{\text{ref}})$$

$$I_{\text{in}} = (10^{-12})10^{90/10} = 0.00100\,\text{W/m}^2 = 1.00\,\text{mW/m}^2$$

The intensity of the transmitted wave is given by Eq. (4-88):

$$I_{\text{tr}} = a_t I_{\text{in}} = (2.203)(10^{-4})(0.00100) = 0.2203 \times 10^{-6}\,\text{W/m}^2$$
$$= 0.2203\,\mu\text{W/m}^2$$

The intensity level of the transmitted wave is given by:

$$L_{\text{I,tr}} = 10\log_{10}(0.2203 \times 10^{-6}/10^{-12}) = 53.4\,\text{dB}$$

We note that, in this case, we could also have calculated the intensity level for the transmitted wave from:

$$L_{\text{I,tr}} = L_{\text{I,in}} - \text{TL} = 90 - 36.6 = 53.4\,\text{dB}$$

The intensity and rms acoustic pressure magnitude are related by the following:

$$I = \frac{p^2}{\rho c} = \frac{p^2}{Z_o}$$

The acoustic pressure for the incident and transmitted waves is found, as follows:

$$p_{\text{in}} = (Z_1 I_{\text{in}})^{1/2} = [(409.8)(0.00100)]^{1/2} = 0.640\,\text{Pa}$$

$$p_{\text{tr}} = (Z_2 I_{\text{tr}})^{1/2} = [(7.44)(10^6)(0.2203)(10^{-6})]^{1/2} = 1.280\,\text{Pa}$$

The sound pressure levels are found from the definition of sound pressure level:

$$L_p = 20\log_{10}(p/p_{\text{ref}})$$

For the incident sound wave,

$$L_{p,\text{in}} = 20\log_{10}(0.640/20 \times 10^{-6}) = 90.1\,\text{dB}$$

For the transmitted wave,

$$L_{p,\text{tr}} = 20\log_{10}(1.280/20 \times 10^{-6}) = 96.1\,\text{dB}$$

Although the acoustic pressure of the transmitted wave is greater than the acoustic pressure of the incident wave, in this example, there is no violation of any physical principle. The acoustic energy is conserved (the conservation of energy principle is valid), because we find that $I_{in} = I_{tr} + I_r$. The acoustic pressures are different because the media in which the two waves (incident and transmitted waves) are transmitted are different. The intensity of the reflected wave is:

$$I_r = 1.00 \times 10^{-3} - 0.2203 \times 10^{-6} = 0.999780 \times 10^{-3} \text{ W/m}^2$$

In this example, most of the energy is reflected and only about 0.02% is transmitted, because of the large difference in the characteristic impedances of the two materials.

4.6 CHANGES IN MEDIA WITH OBLIQUE INCIDENCE

In the previous section, we examined the case in which the sound wave strikes the interface at normal incidence. In most cases in noise control, we find that sound waves may strike a surface at various angles of incidence. Let us now consider the case of a plane sound wave which strikes the interface between two materials at an angle of incidence θ_i with the normal to the interface, as shown in Fig. 4-5.

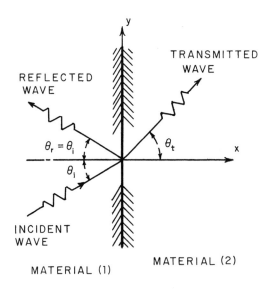

FIGURE 4-5 Transmission of sound from one material into another for oblique incidence of the sound wave.

Transmission of Sound

The expression for a wave moving at an angle θ with the normal to the interface may be written in the following form:

$$p(x, y, t) = A\, e^{j(\omega t - kL)} \tag{4-94}$$

The quantity L is related to the coordinates by the following expression, as illustrated in Fig. 4-6:

$$L = x\cos\theta + y\sin\theta \tag{4-95}$$

The corresponding expression for the acoustic wave in medium 1 may be written as follows:

$$p_1(x, y, t) = A_1\, e^{j[\omega t - k_1(x\cos\theta_i + y\sin\theta_i)]} + B_1\, e^{j[\omega t + k_1(x\cos\theta_i - y\sin\theta_i)]} \tag{4-96}$$

We have used the fact that the angle of reflection is equal to the angle of incidence, or $\theta_r = \theta_i$. The expression for the wave moving in medium 2 is:

$$p_2(x, y, t) = A_2\, e^{j[\omega t - k_2(x\cos\theta_t + y\sin\theta_t)]} \tag{4-97}$$

The angle of transmission θ_t is related to the angle of incidence θ_i through *Snell's law*. For the sound wave to remain a plane sound wave, the wave must travel the distance L_1 in the same time as it travels the distance L_2, as illustrated in Fig. 4-7:

$$d\sin\theta_i = L_1 = c_1\,\Delta t$$

$$d\sin\theta_t = L_2 = c_2\,\Delta t$$

By dividing the first expression by the second, we obtain *Snell's law*:

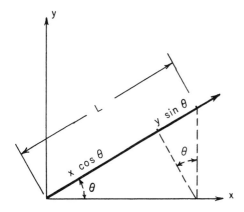

FIGURE 4-6 Relationship between the x- and y-coordinates and the coordinate L in the direction of propagation of an oblique sound wave.

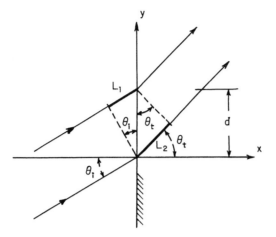

FIGURE 4-7 Wave front striking an interface at oblique incidence.

$$\frac{\sin \theta_i}{\sin \theta_t} = \frac{c_1}{c_2} = \frac{k_2}{k_1} \tag{4-98}$$

The expressions for the particle velocity may be written as follows:

$$u_1(x, y, t) = (A_1/Z_1)\,e^{j[\omega t - k_1(x\cos\theta_i + y\sin\theta_i)]} - (B_1/Z_1)\,e^{j[\omega t + k_1(x\cos\theta_i - y\sin\theta_i)]} \tag{4-99}$$

$$u_2(x, y, t) = (A_2/Z_2)\,e^{j[\omega t - k_2(x\cos\theta_t + y\sin\theta_t)]} \tag{4-100}$$

The acoustic pressure at the interface ($x = 0$) is the same in each medium:

$$p_1(0, y, t) = p_2(0, y, t)$$

The normal component (x-component) of the particle velocities is also the same in each medium at the interface:

$$u_1(0, y, t)\cos\theta_i = u_2(0, y, t)\cos\theta_t$$

If we use the previous expressions for the sound pressure and particle velocity in these two conditions, we obtain the following expression for the pressure magnitude ratio:

$$\frac{A_2}{A_1} = \frac{p_{tr}}{p_{in}} = \frac{2Z_2 \cos\theta_i}{Z_1 \cos\theta_t + Z_2 \cos\theta_i} \tag{4-101}$$

The sound power transmission coefficient for oblique incidence may be found from its definition and Eq. (4-101):

Transmission of Sound

$$a_t = \frac{W_{tr}}{W_{in}} = \frac{I_{tr} S_{tr}}{I_{in} S_{in}} \tag{4-102}$$

As shown in Fig. 4-7, the area through which the incidence and transmitted waves travel is related to the surface area of the interface S by the expressions:

$$S_{in} = S \cos \theta_i$$
$$S_{tr} = S \cos \theta_t$$

Using the expression for the intensity of the plane wave, $I = p^2/Z_o$, Eq. (4-102) may be written in the form:

$$a_t = \frac{p_{tr}^2 Z_1 \cos \theta_t}{p_{in}^2 Z_2 \cos \theta_i} \tag{4-103}$$

If we make the substitution for the pressure ratio from Eq. (4-101) into Eq. (4-103), we obtain the final expression for the sound power transmission coefficient for oblique incidence:

$$a_t = \frac{4 Z_1 Z_2 \cos \theta_i \cos \theta_t}{(Z_1 \cos \theta_t + Z_2 \cos \theta_i)^2} \tag{4-104}$$

From the Snell law expression, Eq. (4-98), we note that there can be a *critical angle of incidence* θ_{cr} for which the transmitted wave will make an angle of 90° with the normal to the interface. For this condition, $\sin \theta_t = \sin 90° = 1$. The expression for the critical angle of incidence may be found by making this substitution into the Snell law expression:

$$\sin \theta_{cr} = c_1/c_2 \tag{4-105}$$

A critical angle of incidence exists only if ($c_1 < c_2$). If the actual angle of incidence is equal to or greater than the critical angle of incidence, then no acoustic energy will be transmitted into the second material.

$$a_t = 0 \quad \text{for } \theta_i \geq \theta_{cr} \tag{4-106}$$

Example 4-2. A sound wave having a sound pressure level of 70 dB is incident at an angle of 45° with the normal to the interface between oil and water. The properties of the oil and water are as follows:

oil (material 1) $\rho_1 = 850 \text{ kg/m}^3$; $c_1 = 1350 \text{ m/s}$;
$Z_1 = 1.148 \times 10^6$ rayl

water (material 2) $\rho_2 = 998 \text{ kg/m}^3$; $c_2 = 1481 \text{ m/s}$;
$Z_2 = 1.478 \times 10^6$ rayl

Determine the sound pressure level for the transmitted wave in the water.
The angle of transmission may be found from Snells' law, Eq. (4-98):

$$\sin\theta_t = (c_2/c_1)\sin\theta_i = (1.481/1.350)\sin(45°) = 0.7757$$

$$\theta_t = 50.87°$$

We note that there is a critical angle of incidence. In this case:

$$\sin\theta_{cr} = c_1/c_2 = (1.350)/(1.481) = 0.9115$$

$$\theta_{cr} = 65.72°$$

If the angle of incidence were greater than 65.72°, there would be total reflection and no acoustic energy transmission.

The sound power transmission coefficient is found from Eq. (4-104):

$$a_t = \frac{(4)(1.478)(10^6)(1.148)(10^6)\cos(45°)\cos(50.87°)}{[(1.478)(10^6)\cos(45°) + (1.148)(10^6)\cos(50.87°)]^2}$$

$$a_t = 0.9649$$

The transmission loss is fairly small in this example:

$$TL = 10\log_{10}(1/0.9649) = 0.2\,dB$$

The magnitude of the incident pressure wave is found from the sound pressure level:

$$p_{in} = (20)(10^{-6})10^{70/20} = 0.0632\,Pa = 63.2\,mPa$$

The intensity of the incident wave is:

$$I_{in} = \frac{p_{in}^2}{Z_1} = \frac{(0.0632)^2}{(1.148)(10^6)} = 3.484 \times 10^{-9}\,W/m^2 = 3.484\,nW/m^2$$

The intensity of the transmitted wave is found from Eq. (4-102):

$$I_{tr} = \frac{\cos\theta_i}{\cos\theta_t} a_t I_{in} = \frac{\cos(45°)}{\cos(50.87°)}(0.9649)(3.484) = 3.767\,nW/m^2$$

Although the intensity of the transmitted wave is somewhat greater than the intensity of the incident wave, there is no violation of any physical principle. The energy of the transmitted wave is "squeezed" into a smaller area than that of the incident wave, so the intensity of the transmitted wave is increased. This phenomenon is analogous to the fact that flowing water will experience an increase in velocity (volume flow per unit area) when the flow enters a smaller size pipe.

The sound pressure for the transmitted wave may be found from the intensity of the wave:

Transmission of Sound

$$p_{tr} = (Z_2 I_{tr})^{1/2} = [(1.478)(10^6)(3.767)(10^{-9})]^{1/2}$$

$$p_{tr} = 0.0746\,\text{Pa} = 74.6\,\text{mPa}$$

The sound pressure level for the transmitted wave is:

$$L_{p,tr} = 20\log_{10}(0.0746/20 \times 10^{-6}) = 71.4\,\text{dB}$$

4.7 SOUND TRANSMISSION THROUGH A WALL

One of the more important problems in noise control is the determination of the energy transmitted through a wall, as shown in Fig. 4-8. The following analysis is valid if the wall is not too thin, in which case, vibrations of the wall as a whole can occur. Also, the analysis is valid if the frequency is not high enough that energy dissipation can occur. These effects will be examined later in this chapter. The sound wave is considered to strike the wall at normal incidence.

The expressions for the acoustic pressure in each of the three media may be written as follows:

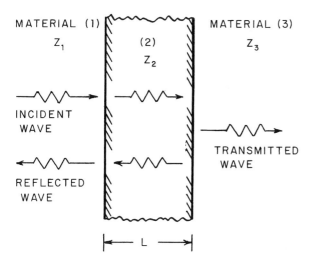

FIGURE 4-8 Sound transmission from one material through a second material into a third material for normal incidence.

$$p_1(x,t) = A_1 e^{j(\omega t - k_1 x)} + B_1 e^{j(\omega t + k_1 x)} \quad (4\text{-}107)$$
(Incident wave) + (Reflected wave)

$$p_2(x,t) = A_2 e^{j(\omega t - k_2 x)} + B_2 e^{j(\omega t + k_2 x)} \quad (4\text{-}108)$$

$$p_3(x,t) = A_3 e^{j[\omega t - k_3(x-L)]} \quad (4\text{-}109)$$
(Transmitted wave)

The constants A_1, B_1, etc., are complex quantities in this case.

For a plane wave at normal incidence, the instantaneous particle velocity in each material may be written as follows:

$$u_1(x,t) = (1/Z_1)[A_1 e^{j(\omega t - k_1 x)} - B_1 e^{j(\omega t + k_1 x)}] \quad (4\text{-}110)$$

$$u_2(x,t) = (1/Z_2)[A_2 e^{j(\omega t - k_2 x)} - B_2 e^{j(\omega t + k_2 x)}] \quad (4\text{-}111)$$

$$u_3(x,t) = (1/Z_3) A_3 e^{j[\omega t - k_3(x-L)]} \quad (4\text{-}112)$$

At the first interface ($x = 0$), the pressure in medium 1 and the pressure in medium 2 are equal, and the particle velocities in mediums 1 and 2 are also the same at the interface. Using these conditions in Eqs (4-107), (4-108), (4-110), and (4-111), we find the following relations:

$$A_1 + B_1 = A_2 + B_2 \quad (4\text{-}113)$$

$$\frac{A_1 - B_1}{Z_1} = \frac{A_2 - B_2}{Z_2} \quad (4\text{-}114)$$

At the second interface ($x = L$), the pressures and particle velocities are also equal. Using this condition in Eqs (4-108), (4-109), (4-111), and (4-112), we obtain a second set of relationships between the coefficients:

$$A_2 e^{-jk_2 L} + B_2 e^{jk_2 L} = A_3 \quad (4\text{-}115)$$

$$\frac{A_2 e^{-jk_2 L} - B_2 e^{jk_2 L}}{Z_2} = \frac{A_3}{Z_3} \quad (4\text{-}116)$$

We may combine Eqs (4-113) through (4-116) to obtain the expression for the following complex number ratio:

$$\frac{A_1}{A_3} = \frac{1}{4}\left(1 + \frac{Z_1}{Z_2}\right)\left(1 + \frac{Z_2}{Z_3}\right) e^{jk_2 L} + \frac{1}{4}\left(1 - \frac{Z_1}{Z_2}\right)\left(1 - \frac{Z_2}{Z_3}\right) e^{-jk_2 L}$$
$$(4\text{-}117)$$

The exponential terms may be written as follows:

$$e^{jk_2 L} = \cos(k_2 L) + j \sin(k_2 L) \quad (4\text{-}118)$$

$$e^{-jk_2 L} = \cos(k_2 L) - j \sin(k_2 L) \quad (4\text{-}119)$$

Transmission of Sound

Substituting the results from Eqs (4-118) and (4-119) into Eq. (4-117), the following expression is obtained:

$$\frac{A_1}{A_3} = \frac{1}{2}\left(1 + \frac{Z_1}{Z_3}\right)\cos(k_2 L) + j\frac{1}{2}\left(\frac{Z_1}{Z_2} + \frac{Z_2}{Z_3}\right)\sin(k_2 L) \qquad (4\text{-}120)$$

For any complex quantity, the magnitude is given by Eq. (4-23). The magnitude of the ratio A_1/A_3 may be written from Eq. (4-120):

$$\left|\frac{A_1}{A_3}\right| = \frac{1}{2}\left[\left(1 + \frac{Z_1}{Z_3}\right)^2 \cos^2(k_2 L) + \left(\frac{Z_1}{Z_2} + \frac{Z_2}{Z_3}\right)^2 \sin^2(k_2 L)\right]^{1/2} \qquad (4\text{-}121)$$

The sound power transmission coefficient for transmission of acoustic energy from medium 1 through medium 2 into medium 3 is given by:

$$a_t = \frac{I_{tr}}{I_{in}} = \frac{|p_3|^2/Z_3}{|p_{in}|^2/Z_1} = \left|\frac{A_3}{A_1}\right|^2 \frac{Z_1}{Z_3} \qquad (4\text{-}122)$$

Eliminating the ratio $|A_3/A_1|$ by using Eq. (4-121), we obtain the final expression for the sound power transmission coefficient:

$$a_t = \frac{4(Z_1/Z_3)}{\left(1 + \frac{Z_1}{Z_3}\right)^2 \cos^2(k_2 L) + \left(\frac{Z_1}{Z_2} + \frac{Z_2}{Z_3}\right)^2 \sin^2(k_2 L)} \qquad (4\text{-}123)$$

Note that when the trigonometric terms are evaluated numerically, the term $k_2 L$ must be expressed in *radians*.

The tangent of the phase angle between the transmitted wave and the incident wave is found from Eq. (4-120), using Eq. (4-24):

$$\tan\phi = \frac{[(Z_1/Z_2) + (Z_2/Z_3)]\tan(k_2 L)}{1 + (Z_1/Z_3)} \qquad (4\text{-}124)$$

There are several special cases of practical importance for Eq. (4-123). First, suppose the materials are the same on both sides of the wall, i.e., materials 1 and 3 are the same. This corresponds to the transmission of sound from air (1) through a solid wall (2) into air (3) on the other side of the wall. For this special case, $Z_1 = Z_3$, and Eq. (4-123) reduces to:

$$a_t = \frac{4}{4\cos^2(k_2 L) + [(Z_1/Z_2) + (Z_2/Z_1)]^2 \sin^2(k_2 L)} \qquad (4\text{-}125)$$

Next, we observe that the characteristic impedance of most solids is much larger than that of air. For example, concrete: $Z_2 = 7,440,000$ rayl

air (25°C): $Z_1 = 409.8$ rayl

$Z_1/Z_2 = 0.0000551$ and $Z_2/Z_1 = 18,200$

For this special case, we may neglect the term (Z_1/Z_2) in Eq. (4-125).

For the frequency range of interest in analysis of transmission of sound through walls, the term $k_2 L$ is often small. For example, for a 100 mm (3.94 in) thick wall of concrete ($c_2 = 3100$ m/s) at a frequency of 1000 Hz, we find the following numerical value:

$$k_2 L = \frac{2\pi f L}{c_2} = \frac{(2\pi)(1000)(0.100)}{(3100)} = 0.203 \text{ rad}$$

Using this value, we find:

$\sin(k_2 L) = 0.201 \approx k_2 L = 0.203$ (within 1%)

$\cos(k_2 L) = 0.980 \approx 1$ (within 2%)

Based on this observation, we see that for $(k_2 L) \leq 0.25$ rad, we may approximate:

$\sin(k_2 L) \approx k_2 L$ and $\cos(k_2 L) \approx 1$

within about 3% error. With these approximations and for $Z_1 = Z_3$, Eq. (4-125) reduces to the following:

$$a_t = \frac{1}{1 + (Z_2/2Z_1)^2 (k_2 L)^2} \tag{4-126}$$

If we make the substitution for the wave number, $k_2 = 2\pi f/c_2$, we obtain the following important relationship:

$$\frac{1}{a_t} = 1 + \left(\frac{\pi \rho_2 L f}{\rho_1 c_1}\right)^2 \tag{4-127}$$

If we introduce the quantity, $M_s = \rho_2 L$, called the *specific mass*, Eq. (4-127) may be written in a form often called the *mass law*:

$$\frac{1}{a_t} = 1 + \left(\frac{\pi M_s f}{\rho_1 c_1}\right)^2 \tag{4-128}$$

Another special case of interest is when $k_2 L = n\pi$, where $n =$ an integer (1, 2, 3, ...). For this case, $\cos^2(k_2 L) = 1$ and $\sin(k_2 L) = 0$. If we make these substitutions into Eq. (4-123), we find the following expression for the sound power transmission coefficient:

$$a_t = \frac{4 Z_1 Z_3}{(Z_1 + Z_3)^2} \tag{4-129}$$

Transmission of Sound

If we have the same material on both sides of the wall ($Z_1 = Z_3$), then the sound power transmission from Eq. (4-129) becomes unity, i.e., $a_t = 1$. The sound is transmitted through the wall with no attenuation!

The wall is also transparent to sound waves having a frequency given by the following relationship, obtained from $\sin(k_2 L) = 0$:

$$k_2 L = \frac{2\pi f L}{c_2} = n\pi \quad \text{or} \quad f = \frac{nc_2}{2L} \quad (4\text{-}130)$$

This condition from Eq. (4-130) may also be written in the following form:

$$2\pi L/\lambda = n\pi \quad \text{or} \quad L = \tfrac{1}{2}n\lambda \quad (4\text{-}131)$$

When the thickness of the wall is a half-integer multiple of the wavelength, the sound wave is transmitted directly through the wall. This principle has been used in the design of free-flooding streamlined domes for housing sonar transducers (Kinsler et al., 1982). For other applications, the condition described by Eq. (4-130) may not be practical to achieve. For example, for a 100 mm (3.94 in) thick concrete wall and with $n = 1$, the corresponding frequency is as follows:

$$f = \frac{(1)(3100)}{(2)(0.100)} = 15{,}500\,\text{Hz} = 15.5\,\text{kHz}$$

At this high frequency, dissipation effects within the material and bending wave effects tend to become significant, and Eq. (4-123) is no longer valid.

Example 4-3. A sound wave having a frequency of 250 Hz and an intensity level of 90 dB strikes a wooden (oak) door (material 2) at normal incidence, as shown in Fig. 4-9. The air in which the incident wave moves (material 1) is at 0°C (32°F), and the air on the other side of the door (material 3) is at 25°C (77°F). The thickness of the door is 40 mm (1.575 in). Determine the sound pressure level of the transmitted wave.

The properties of the materials are found in Appendix B:

air at 0°C $\quad \rho_1 = 1.292\,\text{kg/m}^3; \quad c_1 = 331.3\,\text{m/s}; \quad Z_1 = 428.1\,\text{rayl}$

oakwood $\quad \rho_2 = 770\,\text{kg/m}^3; \quad c_2 = 4300\,\text{m/s}; \quad Z_2 = 3.30 \times 10^6\,\text{rayl}$

air at 25°C $\quad \rho_3 = 1.184\,\text{kg/m}^3; \quad c_3 = 346.1\,\text{m/s}; \quad Z_3 = 409.8\,\text{rayl}$

The wave number for the wood is:

$$k_2 = \frac{2\pi f}{c_2} = \frac{(2\pi)(250)}{(4300)} = 0.3653\,\text{m}^{-1}$$

$$k_2 L = (0.3653)(0.040) = 0.01461\,\text{rad}$$

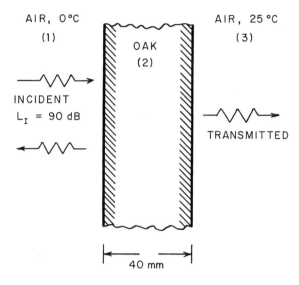

FIGURE 4-9 Physical system for Example 4-3.

Let us evaluate the sound power transmission coefficient from the general expression, Eq. (4-123):

$$a_t = \frac{(4)(428.1/409.8)}{\left(1 + \frac{428.1}{409.8}\right)^2 \cos^2(0.01461) + \left(\frac{428.1}{3.30 \times 10^6} + \frac{3.30 \times 10^6}{409.8}\right)^2 \sin^2(0.01461)}$$

$$a_t = \frac{(4)(1.0447)}{4.18 + (0.000130 + 8052.7)^2(0.01461)^2} = 3.017 \times 10^{-4}$$

The transmission loss is:

$$\text{TL} = 10\log_{10}(1/a_t) = 10\log_{10}(1/3.017 \times 10^{-4}) = 35.2\,\text{dB}$$

The intensity for the incident wave is given by:

$$I_{in} = (10^{-12})10^{90/10} = 0.0010\,\text{W/m}^2 = 1.00\,\text{mW/m}^2$$

The intensity of the transmitted wave is found from the definition of the sound power transmission coefficient:

$$I_{tr} = a_t I_{in} = (3.017)(10^{-4})(0.0010) = 0.3017 \times 10^{-6}\,\text{W/m}^2$$
$$= 0.3017\,\mu\text{W/m}^2$$

The intensity level of the transmitted wave is:

$$L_{I,tr} = 10\log_{10}(0.3017 \times 10^{-6}/10^{-12}) = 54.8\,\text{dB}$$

Transmission of Sound

We could also have calculated the transmitted wave intensity level from:

$L_{\text{I,tr}} = L_{\text{I,in}} - \text{TL} = 90 - 35.2 = 54.8\,\text{dB}$

The acoustic pressure for a plane wave may be evaluated from:

$p_{\text{tr}} = (Z_3 I_{\text{tr}})^{1/2} = [(409.8)(0.3017)(10^{-6})]^{1/2} = 0.01112\,\text{Pa} = 11.12\,\text{mPa}$

The sound pressure level for the transmitted wave is:

$L_{\text{p,tr}} = 20\log_{10}(0.01112/20 \times 10^{-6}) = 54.9\,\text{dB}$

The phase angle between the transmitted wave and the incident wave may be found from Eq. (4-124):

$$\tan\phi = \frac{[(428.1/3.30 \times 10^6) + (3.30 \times 10^6/409.8)]\tan(0.01461)}{1 + (428.1/409.8)}$$

$\tan\phi = (3938.4)(0.01461) = 57.54$

$\phi = 1.553\,\text{rad} = 89.0°$

The transmitted wave is almost 90° out of phase with the incident wave.

Let us check the accuracy of the approximate equation, Eq. (4-127), for this problem. We note that this expression is strictly valid only if $Z_1 = Z_3$; however, in this example, $Z_1/Z_3 = (428.1/409.8) = 1.045$. Using Eq. (4-127), we find:

$$\frac{1}{a_t} = 1 + \left[\frac{(\pi)(770)(0.040)(250)}{(428.1)}\right]^2 = 1 + 3192.9 = 3193.9$$

$a_t = 3.131 \times 10^{-4}$

The error in using Eq. (4-127) instead of the general expression for the sound transmission loss is approximately the same as the error in assuming the characteristic impedances are the same on both sides of the door:

$(3.131 - 3.017)/(3.017) = 0.038 = 3.8\%$

4.8 TRANSMISSION LOSS FOR WALLS

One procedure for noise control is to provide an acoustic barrier or wall to reduce the transmission of sound. For design purposes, one must be able to predict the transmission loss for the wall over a wide range of frequencies. In this section, we will examine the more general case of transmission of sound through a panel or partition.

The general variation of the transmission loss with frequency for a homogeneous wall is shown in Fig. 4-10. We note that there are three general regions of behavior for the wall or panel:

(a) Region I: stiffness-controlled region
(b) Region II: mass-controlled region
(c) Region III: wave-coincidence region (damping-controlled region)

Techniques for prediction of the transmission loss for each of these regions are given in the following material.

4.8.1 Region I: Stiffness-Controlled Region

At low frequencies, the wall or panel vibrates as a whole, and sound transmission through the panel is determined primarily by the stiffness of the panel. Let us consider a panel, as shown in Fig. 4-11, in which the medium is the same on both sides of the panel, and the panel is very thin. The expressions for the acoustic pressure and particle velocity on each side of the panel may be written as follows:

$$p_1(x, t) = A_1 e^{j(\omega t - kx)} + B_1 e^{j(\omega t + kx)} \qquad (4\text{-}132)$$

(Incident wave) + (Reflected wave)

$$p_2(x, t) = A_2 e^{j(\omega t - kx)} \qquad (4\text{-}133)$$

(Transmitted wave)

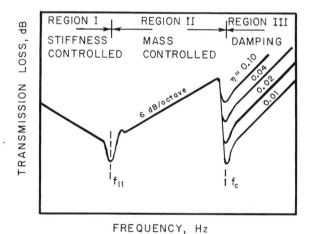

FIGURE 4-10 General variation of the transmission loss with frequency for a homogeneous wall or panel.

Transmission of Sound

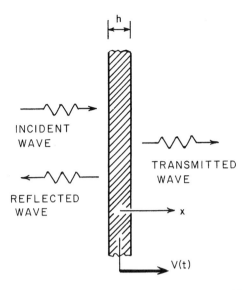

FIGURE 4-11 Vibration of a panel in the stiffness-controlled region, Region I. $V(t)$ is the vibrational velocity of the panel.

$$u_1(x, t) = (1/\rho_0 c)[A_1 e^{j(\omega t - kx)} - B_1 e^{j(\omega t + kx)}] \quad (4\text{-}134)$$

$$u_2(x, t) = (1/\rho_0 c) A_2 e^{j(\omega t - kx)} \quad (4\text{-}135)$$

At the surface of the panel (for a very thin panel), the particle velocities are both equal to the instantaneous velocity of the panel, $V(t)$. We may write the following expressions from Eqs (4-134) and (4-135) for $x = 0$:

$$A_1 - B_1 = A_2 \quad (4\text{-}136)$$

$$V(t) = \frac{A_2 e^{j\omega t}}{\rho_0 c} \quad (4\text{-}137)$$

If the panel has a finite stiffness, the net force acting on the panel is equal to the "spring-force" of the panel. The *specific mechanical compliance* or mechanical compliance per unit area will be denoted by the symbol C_S. The compliance is the reciprocal of the spring constant. If we make a force balance at the surface of the thin panel, we obtain the following expression:

$$p_1(0, t) - p_2(0, t) = -\frac{1}{C_S} \int V(t)\,dt = -\frac{A_2 e^{j\omega t}}{j\omega c \rho_0 C_S} \quad (4\text{-}138)$$

Making the substitutions from Eqs (4-132) and (4-133) for the acoustic pressure force, we obtain the following expression for the coefficients:

$$A_1 + B_1 - A_2 = +\frac{jA_2}{\omega \rho_o c C_S} \qquad (4\text{-}139)$$

Combining Eqs (4-136), (4-137), and (4-139), we obtain the following expression for the ratio of coefficients:

$$\frac{A_2}{A_1} = \frac{1 - \dfrac{j}{2\omega \rho_o c C_S}}{1 + (1/2\omega \rho_o c C_S)^2} \qquad (4\text{-}140)$$

The sound power transmission coefficient for normal incidence may be determined from Eq. (4-140):

$$a_{tn} = \frac{I_{tr}}{I_{in}} = \frac{|p_{tr}|^2}{|p_{in}|^2} = \left|\frac{A_2}{A_1}\right|^2 = \frac{1}{1 + (1/2\omega \rho_o c C_S)^2} \qquad (4\text{-}141)$$

Substituting for the frequency, $\omega = 2\pi f$, we obtain an alternative form of Eq. (4-141):

$$1/a_{tn} = 1 + (4\pi f \rho_o c C_S)^{-2} = 1 + (K_S)^{-2} \qquad (4\text{-}142)$$

where:

$$K_S = 4\pi f \rho_o c C_S \qquad (4\text{-}143)$$

If we repeat the development for the case of oblique incidence of the sound wave, we obtain the following expression for the sound power transmission for an angle of incidence θ:

$$a_t(\theta) = \frac{1}{1 + (\cos \theta / K_S)^2} \qquad (4\text{-}144)$$

In many situations in noise control work, the sound waves strike the surface at all angles of incidence (random incidence). The average sound power transmission coefficient for random incidence of the sound waves is given by:

$$a_t = 2 \int_0^{\pi/2} a_t(\theta) \cos \theta \sin \theta \, d\theta \qquad (4\text{-}145)$$

If we use the expression for $a_t(\theta)$ from Eq. (4-144) in Eq. (4-145), we obtain the following expression for the sound power transmission coefficient in the stiffness-controlled region, Region I:

$$a_t = K_S^2 \ln(1 + K_S^{-2}) = K_S^2 \ln(1/a_{tn}) \qquad (4\text{-}146)$$

Transmission of Sound

The transmission loss for the stiffness-controlled region is given by the following:

$$\text{TL} = 10\log_{10}(1/a_t) = 10\log_{10}(1/K_S^2) - 10\log_{10}[\ln(1 + K_S^{-2})] \quad (4\text{-}147)$$

The transmission loss for normal incidence may be written as follows:

$$\text{TL}_n = 10\log_{10}(1/a_{tn}) = 10\log_{10}(1 + K_S^{-2}) \quad (4\text{-}148)$$

$$\text{TL}_n = (10)(\log_{10} e)\ln(1 + K_S^{-2}) = 4.3429\ln(1 + K_S^{-2}) \quad (4\text{-}149)$$

or

$$\ln(1 + K_S^{-2}) = 0.23026\,\text{TL}_n \quad (4\text{-}150)$$

If we substitute the expression from Eq. (4-150) into Eq. (4-147), we obtain the final expression for the transmission loss for Region I, the stiffness-controlled region:

$$\text{TL} = 20\log_{10}(1/K_S) - 10\log_{10}(0.23026\,\text{TL}_n) \quad (4\text{-}151)$$

For a rectangular panel, the expression for the specific mechanical compliance is given by the following:

$$C_S = \frac{768(1 - \sigma^2)}{\pi^8 E h^3 (1/a^2 + 1/b^2)^2} \quad (4\text{-}152)$$

The quantities a and b are the width and height of the panel; h is the thickness of the panel; and E and σ are the Young's modulus and Poisson's ratio for the panel material, respectively. For a circular panel with a diameter D and thickness h, the specific mechanical compliance is given by:

$$C_S = \frac{3D^4(1 - \sigma^2)}{256 E h^3} \quad (4\text{-}153)$$

Some properties of various panel materials are given in Appendix C.

4.8.2 Resonant Frequency

As the frequency of the incident wave is increased, the plate will resonate at a series of frequencies, called the *resonant frequencies*. The lowest resonant frequency marks the transition between Region I and Region II behavior. The resonant frequencies are a function of the plate dimensions. For a rectangular plate having dimensions $a \times b \times h$ thick, the resonant frequencies are given by the following expression (Roark and Young, 1975):

$$f_{mn} = (\pi/4\sqrt{3})c_L h[(m/a)^2 + (n/b)^2] \quad (4\text{-}155)$$

The factors m and n are integers, 1, 2, 3, ... The quantity c_L is the speed of longitudinal sound waves in the solid panel material:

$$c_L = \left[\frac{E}{\rho_w(1-\sigma^2)}\right]^{1/2} \qquad (4\text{-}156)$$

The quantity ρ_w is the density of the panel material. Usually, the lowest resonant frequency (the *fundamental frequency*) is the most predominant frequency. This frequency corresponds to $m = n = 1$ in Eq. (4-155):

$$f_{11} = (\pi/4\sqrt{3})c_L h[(1/a)^2 + (1/b)^2] \qquad (4\text{-}157)$$

The magnitude of the transmission loss at the first few resonant frequencies is strongly dependent on the damping at the edges of the panel.

The fundamental resonant frequency for a circular plate is given by the following expressions. For a circular plate of diameter D and thickness h clamped at the edge (Roark and Young, 1975):

$$f_{11} = \frac{10.2 c_L h}{\pi\sqrt{3}\, D^2} \qquad (4\text{-}158)$$

For a circular plate with a simple supported edge, the fundamental resonant frequency is given by a similar equation:

$$f_{11} = \frac{5.25 c_L h}{\pi\sqrt{3}\, D^2} \qquad (4\text{-}159)$$

4.8.3 Region II: Mass-Controlled Region

For frequencies higher than the first resonant frequency, the transmission loss of the panel is controlled by the mass of the panel and is independent of the stiffness of the panel. In this region, some acoustic energy is transmitted through the panel and the remainder is reflected at the panel surfaces. This is the physical situation analysed in Sec. 4.7.

The sound power transmission coefficient for normal incidence is given by Eq. (4-128):

$$\frac{1}{a_{tn}} = 1 + \left(\frac{\pi f \rho_w h}{\rho_1 c_1}\right)^2 = 1 + \left(\frac{\pi f M_S}{\rho_1 c_1}\right)^2 \qquad (4\text{-}160)$$

The quantity M_S is called the *surface mass*, or the panel mass per unit surface area:

$$M_S = \rho_w h \qquad (4\text{-}161)$$

The quantity ρ_w is the density of the wall or panel, and ρ_1 and c_1 are the density and speed of sound in the air around the panel, respectively.

Transmission of Sound

The transmission loss for normal incidence is related to the sound power transmission coefficient for normal incidence:

$$TL_n = 10 \log_{10}(1/a_{tn}) \qquad (4\text{-}162)$$

For random incidence (field incidence), it has been found experimentally that the transmission loss for the mass-controlled region is related to TL_n by the following expression (Beranek, 1971):

$$TL = TL_n - 5 \qquad (4\text{-}163)$$

In many cases, the second term is Eq. (4-160) is much larger than 1. In these cases, the reciprocal of the sound power transmission coefficient for normal incident is proportional to f^2. The transmission loss is proportional to $20 \log_{10}(f)$, so that if the frequency is doubled the transmission loss will be increased by $20 \log_{10}(2)$ or 6 dB/octave for the mass-controlled region.

4.8.4 Critical Frequency

As the frequency of the impinging sound wave increases in the mass-controlled region, the wavelength of bending waves in the material, which are frequency-dependent, approaches the wavelength of the sound waves in the air. *Coincidence* (equality of the wavelengths) first occurs at grazing incidence, or for an angle of incidence of 90°. When this condition happens, the incident sound waves and the bending waves in the panel reinforce each other. The resulting panel vibration causes a sharp decrease in the panel transmission loss. This point corresponds to the transition from Region II behavior to Region III behavior.

The *critical frequency* (or *wave coincidence* frequency) is given by the following expression (Reynolds, 1981):

$$f_c = \frac{\sqrt{3}\, c^2}{\pi c_L h} \qquad (4\text{-}164)$$

If we combine Eqs (4-161) and (4-164), we find that the product ($M_S f_c$) is a function of the physical properties of the panel and the sonic velocity (c) in the air around the panel:

$$M_S f_c = \frac{\sqrt{3}\, c^2 \rho_w}{\pi c_L} \qquad (4\text{-}165)$$

4.8.5 Region III: Damping-Controlled Region

For frequencies above the critical frequency, the transmission loss is strongly dependent on the frequency of the incident sound waves and the internal damping of the panel material.

For sound waves striking the panel at all angles (random incidence) at frequencies greater than the critical frequency, the following empirical field-incidence expression applies for the transmission loss in the damping-controlled region (Beranek, 1971):

$$\text{TL} = \text{TL}_n(f_c) + 10\log_{10}(\eta) + 33.22\log_{10}(f/f_c) - 5.7 \qquad (4\text{-}166)$$

The quantity $\text{TL}_n(f_c)$ is the transmission loss for normal incidence at the critical frequency:

$$\text{TL}_n(f_c) = 10\log_{10}\left[1 + \left(\frac{\pi M_S f_c}{\rho_1 c_1}\right)^2\right] \qquad (4\text{-}167)$$

The quantity η is the damping coefficient for the panel material. Some numerical values for the damping coefficient for various materials are given in Appendix C.

For the damping-controlled region, the transmission loss is proportional to $33.22\log_{10}(f)$. If the frequency is doubled, the transmission loss is increased by $33.22\log_{10}(2) = 10\,\text{dB/octave}$.

Example 4-4. An oak door has dimensions of 0.900 m (35.4 in) wide by 1.800 m (70.9 in) high by 35 mm (1.38 in) thick. The air on both sides of the door has a temperature of 20°C (68°F), for which $c = 343.2$ m/s (1126 ft/sec), $\rho = 1.204$ kg/m³ (0.0752 lb$_m$/ft³), and $z_o = 413.3$ rayl. Determine the transmission loss for the following frequencies: (a) 63 Hz, (b) 250 Hz, and (c) 2000 Hz.

We find the following properties for the oak door from Appendix C:

Longitudinal sound wave wave speed $\qquad c_L = 3860$ m/s (12,700 ft/sec)
Density $\qquad \rho_w = 770$ kg/m³ (48.1 lb$_m$/ft³)
Critical frequency product
$\qquad M_S f_c = (11{,}900\,\text{Hz-kg/m}^2)(343.2/346.1)^2$
$\qquad\qquad = 11{,}700\,\text{Hz-kg/m}^2$ (2397 Hz-lb$_m$/ft²)
Damping factor $\qquad \eta = 0.008$
Young's modulus $\qquad E = 11.2$ GPa (1.62×10^6 psi)
Poisson's ratio $\qquad \sigma = 0.15$

The first resonant frequency is found from Eq. (4-157):

$$f_{11} = 0.4534 c_L h(1/a^2 + 1/b^2)$$

$$f_{11} = (0.4534)(3860)(0.035)[(1/0.90^2) + (1/1.80^2)] = 94.5\,\text{Hz}$$

The specific mass is:

$$M_S = \rho_w h = (770)(0.035) = 26.95\,\text{kg/m}^2\ (5.52\,\text{lb}_m/\text{ft}^2)$$

Transmission of Sound

The critical or wave coincidence frequency is found from the ratio $M_S f_c$:

$$f_c = \frac{M_S f_c}{M_S} = \frac{(11{,}700)}{(26.95)} = 434.1\,\text{Hz}$$

(a) For $f = 63$ Hz.

The frequency, $f = 63\,\text{Hz} < 94.5\,\text{Hz} = f_{11}$; therefore, this case lies in Region I, the stiffness-controlled region. The specific mechanical compliance may be evaluated from Eq. (4-152):

$$C_S = \frac{(768)(1 - 0.15^2)}{(\pi^8)(11.2)(10^9)(0.035)^3[(1/0.90)^2 + (1/1.80)^2]^2}$$
$$= 70.81 \times 10^{-9}\,\text{m}^3/\text{N}$$

$$C_S = 70.81\,\text{nm/Pa}$$

The value of the parameter defined by Eq. (4-143) is as follows:

$$K_S = 4\pi f Z_1 C_S = (4\pi)(63)(413.3)(70.81)(10^{-9}) = 0.02317$$

The sound power transmission coefficient may be calculated from Eq. (4-146):

$$a_t = K_S^2 \ln(1 + K_S^{-2}) = (0.02317)^2 \ln[1 + (0.02317)^{-2}] = 0.004042$$

The transmission loss for a frequency of 64 Hz is as follows:

$$\text{TL} = 10\log_{10}(1/0.004042) = 23.9\,\text{dB}$$

(b) For $f = 250$ Hz.

For this case, $f_{11} = 94.5\,\text{Hz} < 250\,\text{Hz} < 434.1\,\text{Hz} = f_c$; therefore, the operating region is Region II, the mass-controlled region. The sound power transmission coefficient for normal incidence is found from Eq. (4-160):

$$\frac{1}{a_{tn}} = 1 + \left(\frac{\pi f M_S}{Z_1}\right)^2 = 1 + \left[\frac{(\pi)(250)(26.95)}{(413.3)}\right]^2 = 1 + (51.21)^2$$

$$1/a_{tn} = 2623.8$$

The transmission loss for normal incidence is found from Eq. (4-162):

$$\text{TL}_n = 10\log_{10}(1/a_{tn}) = 10\log_{10}(2623.8) = 34.2\,\text{dB}$$

The transmission loss with random incidence for a frequency of 250 Hz is found from Eq. (4-163):

$$\text{TL} = 34.2 - 5 = 29.2\,\text{dB}$$

(c) For $f = 2000$ Hz.

The frequency, $f = 2000\,\text{Hz} > 434.1\,\text{Hz} = f_c$; therefore, this case lies in Region III, the damping-controlled region. The transmission loss for normal incidence at the critical frequency is found from Eq. (4-167):

$$\text{TL}_n(f_c) = 10\log_{10}\left\{1 + \left[\frac{(\pi)(11{,}700)}{(413.3)}\right]^2\right\} = 10\log_{10}(1 + 7909)$$

$$= 39.0\,\text{dB}$$

The transmission loss for a frequency of 2000 Hz is found from Eq. (4-166):

$$\text{TL} = 39.0 + 10\log_{10}(0.008) + 33.22\log_{10}(2000/434.1) - 5.7$$

$$\text{TL} = 39.0 + (-21.0) + 22.0 - 5.7 = 34.3\,\text{dB}$$

Example 4-5. A steel plate (density 7700 kg/m^3) has dimensions of 0.900 m (35.4 in) by 1.800 m (70.9 in). The air on both sides of the plate has a characteristic impedance of 413.3 rayl (at 20°C) and sonic velocity of 343.3 m/s. At a frequency of 500 Hz, it is desired to have a transmission loss of 30 dB. Determine the required thickness of the plate.

This problem involves iteration, because we do not know the region for the transmission loss. Let us begin by trying Region II, the mass-controlled region. The required transmission loss for normal incidence is given by Eq. (4-163):

$$\text{TL}_n = \text{TL} + 5 = 30 + 5 = 35\,\text{dB}$$

We may use Eqs (4-160) and (4-162) to determine the surface mass:

$$\text{TL}_n = 10\log_{10}[1 + (\pi M_S f/Z_1)^2] = 35\,\text{dB}$$

$$(\pi M_S f/Z_1)^2 = 10^{35/10} - 1 = 3161.3$$

The surface mass is:

$$M_S = \frac{(3161.3)^{1/2}(413.3)}{(\pi)(500)} = 14.79\,\text{kg/m}^2 = \rho_w h$$

The required thickness (if the TL region is Region II) is as follows:

$$h = \frac{14.79}{7700} = 0.00192\,\text{m} = 1.92\,\text{mm}\,(0.076\,\text{in})$$

Now, let us check the assumption of Region II behavior. The critical frequency is found from the $M_S f_c$ product, obtained from Appendix C for steel:

$$M_S f_c = (99{,}700)(343.2/346.1)^2 = 98{,}040\,\text{Hz-kg/m}^2$$

Transmission of Sound

The critical or wave-coincidence frequency is:
$$f_c = \frac{M_S f_c}{M_S} = \frac{(98{,}040)}{(14.79)} = 6630\,\text{Hz} > 500\,\text{Hz} = f$$

The first resonant frequency for the panel is found from Eq. (4-157):
$$f_{11} = (0.4534)(5100)(0.00192)[(1/0.900)^2 + (1/1.800)^2]$$
$$f_{11} = 6.85\,\text{Hz} < 500\,\text{Hz} = f$$

The frequency $f = 500\,\text{Hz}$ lies in Region II, because $f_{11} < f < f_c$, and the required panel thickness is:
$$h = 1.92\,\text{mm}\ (0.076\,\text{in})$$

4.9 APPROXIMATE METHOD FOR ESTIMATING THE TL

In preliminary design, it is often required to estimate the transmission loss spectrum for a panel. This section presents an outline of an approximate method for calculating the transmission loss curve for Regions II and III (Watters, 1959). If the panel dimensions a and b are at least 20 times the panel thickness h, the first resonant frequency for the panel is usually less than 125 Hz, so the major portion of the transmission loss curve will involve Regions II and III. In addition, the application of the Region II equations for Region I results in a conservative estimate for the transmission loss. When using the approximate method, one should check the importance of the Region I behavior, however.

In Region II, the mass-controlled region, the random-incidence transmission loss is given by:

$$\text{TL} = \text{TL}_n - 5 = 10\log_{10}\left[1 + \left(\frac{\pi M_S f}{\rho_1 c_1}\right)^2\right] - 5 \qquad (4\text{-}168)$$

For frequencies above about 60 Hz, the term $(\pi M_S f/\rho_1 c_1)$ is usually much larger than 1; therefore, Eq. (4-168) may be approximated by the following expression:

$$\text{TL} = 10\log_{10}\left[\frac{\pi M_S f}{\rho_1 c_1}\right]^2 - 5 \qquad (4\text{-}169)$$

Equation (4-169) may be written in the following alternative form:

$$\text{TL} = 20\log_{10}(M_S) + 20\log_{10}(f) - 20\log_{10}(\rho_1 c_1/\pi) - 5 \qquad (4\text{-}170)$$

For the case of air at 101.3 kPa (14.7 psia) and 22°C (72°F), the density and sonic velocity are:

$$\rho_1 = 1.196 \text{ kg/m}^3 \quad \text{and} \quad c_1 = 344 \text{ m/s}$$

Using these values, we find the following value for the third term in Eq. (4-170):

$$20 \log_{10}[(1.196)(344)/\pi] = 42.3 \text{ dB}$$

For frequencies below the plateau (Region II), the transmission loss may be approximated by the following:

$$\text{TL} = 20 \log_{10}(M_S) + 20 \log_{10}(f) - 47.3 \quad (4\text{-}171)$$

The specific mass M_S is in kg/m^2 and the frequency f is in Hz. We note from Eq. (4-171) that a doubling of the frequency (a frequency change of one octave) results in a change of the transmission loss of:

$$\Delta \text{TL} = 20 \log_{10}(2) = 6.02 \text{ dB/octave} \approx 6 \text{ dB/octave}$$

The approximate method replaces the transition "peaks-and-valleys" between Region II and Region III by a horizontal line or *plateau*, as shown in Fig. 4-12. The height of the plateau (TL$_P$) and the width of the plateau (Δf_P) depend on the material. Some typical values of these quantities are given in Table 4-1.

For the damping-controlled region, the only term in Eq. (4-166) that contains the frequency is the following:

$$33.22 \log_{10}(f/f_c)$$

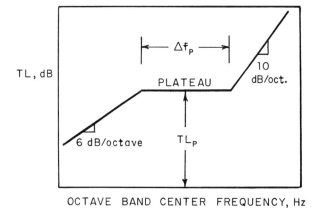

Figure 4-12 Schematic of the approximate curve for the transmission loss of a panel.

Transmission of Sound

TABLE 4-1 Values of the Plateau Height (TL$_P$) and Plateau Width (Δf_P) for the Approximate Method of Calculation of the Transmission Loss for Panels.

Material	TL$_P$, dB	$\Delta f_P = f_2 - f_1$, octaves	f_2/f_1
Aluminum	29	3.5	11
Brick	37	2.2	4.5
Concrete	38	2.2	4.5
Glass	27	3.3	10
Lead	56	2.0	4
Masonry block			
Cinder	30	2.7	6.5
Dense	32	3.0	8
Plywood	19	2.7	6.5
Sand plaster	30	3.0	8
Steel	40	3.5	11

Source: Watters (1959).

For a frequency ratio of 2 (1 octave), we find the following value for this term:

$$33.22 \log_{10}(2) = 10.0 \, \text{dB/octave}$$

In Region III, the slope of the transmission loss curve is 10 dB/octave. To be on the "safe" side or for a conservative estimate of the transmission loss, it is recommended that the TL curve in Region III be drawn with a slope of the 10 dB/octave for the first 2 octaves above the plateau. The remainder of the curve should be drawn with a slope of 6 dB/octave (Beranek, 1960).

The application of the approximate method for estimating the transmission loss curve is illustrated in the following example.

Example 4-6. Estimate the transmission loss curve for a steel plate having a thickness of 3 mm (0.118 in), using the approximate method.

The specific mass for the plate is found as follows:

$$M_S = \rho_w h = (7700)(0.003) = 23.1 \, \text{kg/m}^2$$

Let us start by calculating the transmission loss at 125 Hz, using Eq. (4-171):

$$\text{TL} = 20 \log_{10}(23.1) + 20 \log_{10}(125) - 47.3$$

$$\text{TL} = 27.27 + 41.94 - 47.3 = 21.9 \, \text{dB (at 125 Hz)}$$

The plateau transmission loss for steel is TL$_P$ = 40 dB, from Table 4-1. Using Eq. (4-171), we find the frequency at which the plateau begins:

$$20 \log_{10}(f_1) = 40 - 27.27 + 47.3 = 60.03$$

$$f_1 = 10^{60.03/20} = 1003 \text{Hz (beginning of plateau)}$$

Using the frequency ratio from Table 4-1, we may find the frequency at the end of the plateau:

$$f_2 = (f_2/f_1)f_1 = (11)(1003) = 11,040 \text{ Hz}$$

The region for the transmission loss from 63 Hz to 1003 Hz is Region II, the mass-controlled region. In this range, the approximate transmission loss values may be found by adding (or subtracting) 6 dB for each octave above (or below) 125 Hz. For frequencies above 11.04 kHz, the region is Region III, the damping-controlled region. The transmission loss at 16 kHz, may be found from the following:

$$\text{TL} = \text{TL}_\text{P} + 33.22 \log_{10}(f/f_2) = 40 + 33.22 \log_{10}(16,000/11,040)$$
$$= 45.4 \text{ dB}$$

The values for the complete TL curve are shown in Table 4-2, and a plot of the TL curve is given in Fig. 4-13.

4.10 TRANSMISSION LOSS FOR COMPOSITE WALLS

The material presented in the previous sections applies for transmission of sound through homogeneous, single-component panels, such as a plate of

TABLE 4-2 Transmission Loss Values for Example 4-6.

f, Hz	TL, dB	Explanation
63	15.9	Down by 6 dB from 125 Hz value
125	21.9	Value calculated
250	27.9	Up by 6 dB from 125 Hz value
500	33.9	Up by 6 dB
1,000	39.9	Up by 6 dB
1,003	40.0	Plateau begins
2,000	40.0	Plateau
4,000	40.0	Plateau
8,000	40.0	Plateau
11,040	40.0	Plateau ends
16,000	45.4	Value calculated

Transmission of Sound

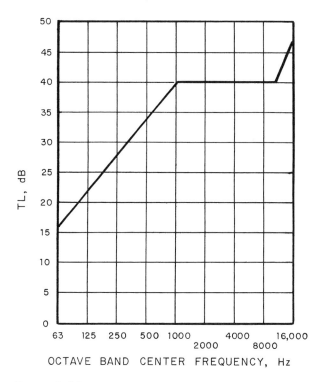

FIGURE 4-13 Solution for Example 4-6.

glass. In this section, we will consider some more complex constructions that can be analyzed analytically.

4.10.1 Elements in Parallel

One common form of construction consists of elements in parallel in a composite wall, such as a window or door in the wall. The total power transmitted through the wall is the sum of the power transmitted through each element, because the incident acoustic intensity is the same for all elements:

$$W_{tr} = \Sigma W_{tr,j} = a_t W_{in} = a_t S I_{in} = I_{in} \Sigma a_{t,j} S_j \tag{4-172}$$

The quantity $S = \Sigma S_j$ is the total surface area, and $a_{t,j}$ is the sound power transmission coefficient for each individual element. The overall sound power transmission coefficient for elements in parallel in a composite wall is given by the following:

$$a_t = \frac{\Sigma a_{t,j} S_j}{S} = \frac{a_{t,1} S_1 + a_{t,2} S_2 + \cdots}{S_1 + S_2 + \cdots} \qquad (4\text{-}173)$$

The effect of openings in a panel is generally significant, because the sound power transmission coefficient for an opening is unity (all energy is transmitted through the opening). This effect is illustrated in the following example.

Example 4-7. A wall has a transmission loss of 20 dB with no opening in the wall. If an opening having an area equal to 10% of the total wall area is added in the wall, determine the overall transmission loss for the wall with the opening included.

The sound power transmission coefficient for the wall material is:

$1/a_{t,1} = 10^{20/10} = 100$

$a_{t,1} = 0.010$

The sound power transmission coefficient for the opening is $a_{t,2} = 1$. Using Eq. (4-173), we may evaluate the overall sound power transmission coefficient:

$$a_t = \frac{(0.900 S)(0.010) + (0.10 S)(1.00)}{S} = 0.1090$$

The transmission loss for the wall with the opening included is:

$\text{TL} = 10 \log_{10}(1/0.1090) = 9.6 \,\text{dB}$

We observe that an opening of only 10% of the total wall area reduces the transmission loss from 20 dB to a value slightly less than 10 dB. If the noise reduction for a wall is to be effective, any openings must be as small as possible or completely eliminated, if practical.

4.10.2 Composite Wall with Air Space

The double-wall construction, consisting of two panels separated by an air space, is often used as a barrier to reduce noise transmission. For this construction, shown in Fig. 4-14, the overall transmission loss is influenced by the air mass in the space, in addition to the effect of the transmission loss for each separate panel. The behavior of the TL curve for the composite wall may be divided into three regimes (Beranek, 1971).

Regime A, the low-frequency regime, occurs for closely spaced panels. When the two panels are placed very close together, the panels act as one unit, as far as the sound transmission is concerned. The air space between

Transmission of Sound

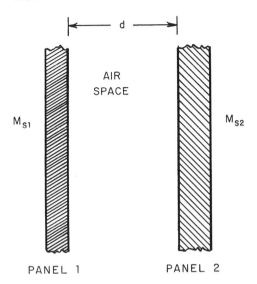

FIGURE 4-14 Composite wall with an air space between the two panels.

the panels has a negligible effect. This behavior occurs for the frequency range, as follows:

$$\frac{\rho c}{\pi(M_{S1} + M_{S2})} < f < f_o \qquad (4\text{-}174)$$

The density and speed of sound for Eq. (4-174) are the values for the air around the panel. The frequency f_o is the resonant frequency of the two panels coupled by the air space. This frequency is given by the following:

$$f_o = \frac{c}{2\pi}\left[\frac{\rho}{d}\left(\frac{1}{M_{S1}} + \frac{1}{M_{S2}}\right)\right]^{1/2} \qquad (4\text{-}175)$$

The quantities M_{S1} and M_{S2} are the specific mass for panels 1 and 2, respectively. The quantity d is the spacing between the panels.

The transmission loss for Regime A is given by the following:

$$\text{TL} = 20\log_{10}(M_{S1} + M_{S2}) + 20\log_{10}(f) - 47.3 \qquad (4\text{-}176)$$

As the panels are moved farther apart, standing waves are set up in the air space between the panels, and Regime B behavior is observed. This regime occurs for the frequency range, as follows:

$$f_o < f < (c/2\pi d) \qquad (4\text{-}177)$$

The transmission loss in Regime B is given by the following:

$$\text{TL} = \text{TL}_1 + \text{TL}_2 + 20\log_{10}(4\pi f\, d/c) \tag{4-178}$$

The quantities TL_1 and TL_2 are the transmission loss values for each of the panels acting alone.

When the panels are moved sufficiently far apart, the two panels act independently, and Regime C behavior is observed. The air space between the panels acts as a small "room." This behavior occurs for the frequency range, $f > (c/2\pi d)$. The transmission loss in Regime C is given by:

$$\text{TL} = \text{TL}_1 + \text{TL}_2 + 10\log_{10}\left[\frac{4}{1+(2/\alpha)}\right] \tag{4-179}$$

The quantity α is the surface absorption coefficient for the panels.

The transmission loss expressions given in this section apply for the sound transmitted through the airspace only. There is a second path that the sound may take, called the *structureborne flanking path*, which involves sound transmission through mechanical links between the panels. Prediction methods for this contribution to the transmission loss are given by Sharp (1973).

Example 4-8. Two panels of glass, each having a thickness of 6 mm (0.24 in), are to be used to reduce the sound transmission through an opening 1.00 m (39.4 in) high and 2.00 m (78.7 in) wide. The panels are spaced 75 mm (2.95 in) apart, and the air around the panels is at 24°C (75°F), for which $\rho = 1.188\,\text{kg/m}^3$ (0.0742 lb$_\text{m}$/ft^3) and $c = 345.6$ m/s (1134 ft/sec). The surface absorption coefficient for the glass is $\alpha = 0.03$. Determine the transmission loss at the following frequencies: (a) 250 Hz, (b) 1000 Hz, and (c) 4 kHz.

The properties of glass are found in Appendix C:

Longitudinal sound wave speed $\quad c_\text{L} = 5450$ m/s (17,880 ft/sec)
Density $\quad \rho_\text{w} = 2500$ kg/m^3 (156 lb$_\text{m}$/ft^3)
Critical frequency product
$\quad M_\text{S} f_\text{c} = (30{,}300\,\text{Hz-kg/m}^2)(345.6/346.1)^2$
$\quad\quad\quad = 30{,}210$ Hz-kg/m^2 (6190 Hz-lb$_\text{m}$/ft^2)
Damping factor $\quad \eta = 0.002$
Young's modulus $\quad E = 71.0$ GPa (10.3×10^6 psi)
Poisson's ratio $\quad \sigma = 0.21$

First, let us determine the transmission loss for a single glass panel. The first resonant frequency is found from Eq. (4-157):

Transmission of Sound

$$f_{11} = 0.4534 c_L h[(1/a)^2 + (1/b)^2]$$

$$f_{11} = (0.4534)(5450)(0.006)[(1/1.00)^2 + (1/2.00)^2] = 18.5 \text{ Hz}$$

The surface mass for one panel is:

$$M_S = \rho_w h = (2500)(0.006) = 15.0 \text{ kg/m}^2 \ (3.07 \text{ lb}_m/\text{ft}^2)$$

The critical frequency is:

$$f_c = (M_S f_c)/M_S = (30{,}210)/(15.00) = 2014 \text{ Hz}$$

For the frequencies of 250 Hz and 1000 Hz, the single panel operates in Region II, the mass-controlled region. The sound power transmission coefficient for normal incidence is found from Eq. (4-160) for a frequency of 250 Hz:

$$\frac{1}{a_{tn}} = 1 + \left[\frac{(\pi)(250)(15.0)}{(1.188)(345.6)}\right]^2 = 1 + 823.3 = 824.3$$

The transmission loss for normal incidence is:

$$TL_n = 10 \log_{10}(824.3) = 29.2 \text{ dB}$$

The transmission loss for a single panel of glass at 250 Hz is found from Eq. (4-163):

$$TL = 29.2 - 5 = 24.2 \text{ dB}$$

Repeating the calculations for a frequency of 1000 Hz, we find the following values:

$$1/a_{tn} = 1 + 13{,}174 = 13{,}175$$

$$TL_n = 41.2 \text{ dB}$$

$$TL = 41.2 - 5 = 36.2 \text{ dB}$$

The frequency of 4 kHz lies in Region III, the damping-controlled region. The transmission loss for normal incidence at the critical frequency is found from Eq. (4-167):

$$TL_n(f_c) = 10 \log_{10}\left\{1 + \left[\frac{(\pi)(30{,}210)}{(410.6)}\right]^2\right\} = 47.3 \text{ dB}$$

The transmission loss for a single panel at 4000 Hz is found from Eq. (4-166):

$$TL = 47.3 + 10 \log_{10}(0.002) + 33.22 \log_{10}(4000/2014) - 5.7$$

$$TL = 47.3 - 27.0 + 9.9 - 5.7 = 24.5 \text{ dB}$$

The complete transmission loss curve for a single panel of glass is given in Table 4-3.

Next, let us examine the case for two glass panels, each having a surface mass of $M_{S1} = M_{S2} = 15.0\,\text{kg/m}^2$. The various frequencies which divide the different regimes of behavior for the double panel may be evaluated:

$$\frac{\rho c}{\pi(M_{S1} + M_{S2})} = \frac{(410.6)}{(\pi)(15.0 + 15.0)} = 4.4\,\text{Hz}$$

Using Eq. (4-175), we find the resonant frequency for the panel:

$$f_o = \frac{(345.6)}{(2\pi)}\left[\frac{(1.188)}{(0.075)}\left(\frac{1}{15.0} + \frac{1}{15.0}\right)\right]^{1/2} = 79.9\,\text{Hz}$$

$$\frac{c}{2\pi d} = \frac{(345.6)}{(2\pi)(0.075)} = 733\,\text{Hz}$$

We note that the frequency $f = 63\,\text{Hz}$ lies in Regime A.

(a) For $f = 250\,\text{Hz}$, we find that $79.9\,\text{Hz} < f = 250\,\text{Hz} < 733\,\text{Hz}$. This case lies in Regime B. The transmission loss may be calculated from Eq. (4-178):

$$\text{TL} = 24.2 + 24.2 + 20\log_{10}\left[\frac{(4\pi)(250)(0.075)}{(345.6)}\right] = 48.4 + (-3.3)\,\text{dB}$$

$$\text{TL} = 45.1\,\text{dB}$$

(b) For $f = 1000\,\text{Hz}$, we find that $733\,\text{Hz} < f = 1000\,\text{Hz}$. This case lies in Regime C. The transmission loss may be calculated from Eq. (4-179):

$$\text{TL} = 36.2 + 36.2 + 10\log_{10}\left[\frac{4}{1 + (2/0.03)}\right] = 72.4 - 12.3 = 60.1\,\text{dB}$$

TABLE 4-3 Tabular Results for Example 4-8.

	Frequency, Hz							
	63	125	250	500	1,000	2,000	4,000	8,000
Single panel:								
Region	II	II	II	II	II	II	III	III
TL, dB	12.3	18.2	24.2	30.2	36.2	42.2	24.5	34.5
Double panel:								
Regime	A	B	B	B	C	C	C	C
TL, dB	18.2	27.1	45.1	63.1	60.1	72.1	36.7	56.7

Transmission of Sound

(c) For $f = 4000\,\text{Hz}$. As far as the behavior of the double panel is concerned, the regime is Regime C; however, the individual panels are operating in Region III, the damping-controlled region. The transmission loss may be determined from Eq. (4-179), using the single panel transmission loss values calculated for Region III at 4000 Hz:

$$\text{TL} = 24.5 + 24.5 + (-12.3) = 36.7\,\text{dB}$$

The complete transmission loss curve is tabulated in Table 4-3 and plotted in Fig. 4-15.

4.10.3 Two-Layer Laminate

Panels composed of two or more solid layers are often used as partitions for enclosures and other acoustic structures. If the layers are bonded at the interface with no air space, as shown in Fig. 4-16, then the composite panel bends about an overall neutral axis. If we let χ be the distance from the interface to the overall neutral axis, positive toward material 1 side, we may find the quantity in terms of the properties of the individual layers:

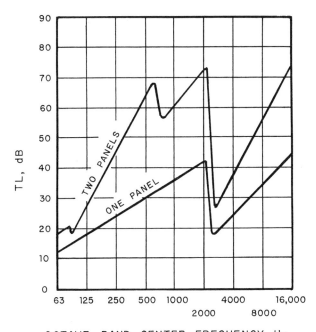

FIGURE 4-15 Solution for Example 4-8.

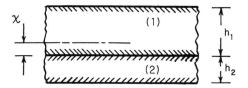

FIGURE 4-16 Two-ply laminated panel: χ is the distance from the interface of the two materials to the overall neutral axis of the composite panel in bending, with positive values measured toward material 1.

$$\chi = \frac{E_1 h_1^2 - E_2 h_2^2}{2(E_1 h_1 + E_2 h_2)} \qquad (4\text{-}180)$$

The transmission loss for Region II, the mass-controlled region, may be determined from Eq. (4-163):

$$\text{TL} = 10 \log_{10}\left[1 + \left(\frac{\pi f M_S}{\rho_0 c}\right)^2\right] - 5 \qquad (4\text{-}181)$$

The specific mass for the layered panel is given by the following:

$$M_S = \rho_1 h_1 + \rho_2 h_2 \qquad (4\text{-}182)$$

The critical or wave coincidence frequency for the layered panel may be found from the following expression:

$$f_c = \frac{c^2}{2\pi}\left(\frac{M_S}{B}\right)^{1/2} \qquad (4\text{-}183)$$

The quantity c is the speed of sound in the air around the panel, and B is the *flexural rigidity* of the panel, given by the following expression:

$$B = \frac{E_1 h_1^3}{12(1 - \sigma_1^2)}[1 + 3(1 - 2\chi/h_1)^2] + \frac{E_2 h_2^3}{12(1 - \sigma_2^2)}[1 + 3(1 + 2\chi/h_2)^2]$$

$$(4\text{-}184)$$

Note that the algebraic sign for χ must be maintained in Eq. (4-184). The quantity χ is positive when the overall neutral axis is on the material 1 side of the interface.

The transmission loss for a layered panel may be determined from Eq. (4-166) with the overall damping coefficient η calculated from the following:

$$\eta = \frac{(\eta_1 E_1 h_1 + \eta_2 E_2 h_2)(h_1 + h_2)^2}{E_1 h_1^3[1 + 3(1 - 2\chi/h_1)^2] + E_2 h_2^3[1 + 3(1 + 2\chi/h_2)^2]} \qquad (4\text{-}185)$$

Transmission of Sound

Example 4-9. An aluminum plate (material 1) having a thickness of 1.6 mm (0.063 in) is bonded to a rubber sheet (material 2) having a thickness of 4.8 mm (0.189 in). The panel dimensions are 400 mm (15.75 in) by 750 mm (29.53 in). The air around the panel is at 21°C (70°F), for which the density and speed of sound are $\rho_o = 1.200$ kg/m^3 (0.0749 lb$_m$/ft^3) and $c = 343.8$ m/s (1128 ft/sec), respectively. Determine the transmission loss for the panel at (a) 500 Hz and (b) 8 kHz.

The properties of the aluminum (subscript 1) and rubber (subscript 2) are found from Appendix C:

density $\rho_1 = 2800$ kg/m^3 (174.5 lb$_m$/ft^3);
$\rho_2 = 950$ kg/m^3 (59.3 lb$_m$/ft^3)
Young's modulus $E_1 = 73.1$ GPa (10.6×10^6 psi);
$E_2 = 2.30$ GPa (0.334×10^6 psi)
Poisson's ratio $\sigma_1 = 0.33$; $\sigma_2 = 0.400$
damping factor $\eta_1 = 0.001$; $\eta_2 = 0.080$

The specific mass for the composite panel is found from Eq. (4-182):
$M_S = (2800)(0.0016) + (950)(0.0048)$
$M_S = 4.48 + 4.56 = 9.04$ kg/m^2 (1.852 lb$_m$/ft^2)

The location of the neutral axis for the composite panel is found from Eq. (4-180):

$$\chi = \frac{(73.1)(0.0016)^2 - (2.30)(0.0048)^2}{(2)[(73.1)(0.0016) + (2.30)(0.0048)]} = 0.000524 \text{ m} = 0.524 \text{ mm}$$

Let us calculate the following parameters for Eq. (4-184):
$1 + 3(1 - 2\chi/h_1)^2 = 1 + (3)[1 - (2)(0.524)/(1.60)]^2 = 1.357$
$1 + 3(1 + 2\chi/h_2)^2 = 1 + (3)[1 + (2)(0.524)/(4.80)]^2 = 5.453$

The flexural rigidity for the composite panel is found from Eq. (4-184):

$$B = \frac{(73.1)(10^9)(0.0016)^3(1.357)}{(12)(1 - 0.33^2)} + \frac{(2.30)(10^9)(0.0048)^3(5.453)}{(12)(1 - 0.40^2)}$$

$B = 38.0 + 137.6 = 175.6$ Pa-m^3 = 175.6 N-m (129.5 lb$_f$-ft)

The critical or wave-coincidence frequency for the composite panel is found from Eq. (4.183):

$$f_c = \frac{(343.8)^2}{2\pi} \left(\frac{9.04}{175.6} \right)^2 = 4268 \text{ Hz}$$

If the panel were constructed of aluminum only, the critical frequency would be found from Eq. (4-164):

$$f_c(1) = \frac{\sqrt{3}(343.8)^2}{(\pi)(5420)(0.0016)} = 7515\,\text{Hz}$$

(a) For a frequency of 500 Hz. This frequency is less than the critical frequency, so the panel behavior falls in Region II, the mass-controlled region. The transmission loss may be found from Eq. (4-181) for the composite panel:

$$\frac{1}{a_{tn}} = 1 + \left[\frac{(\pi)(500)(9.04)}{(1.20)(343.8)}\right]^2 = 1186$$

$$\text{TL} = 10\log_{10}(1186) - 5 = 30.7 - 5 = 25.7\,\text{dB}$$

For a single aluminum sheet, the transmission loss is as follows:

$$1/a_{tn} = 1 + 291 = 292$$

$$\text{TL} = 24.7 - 5 = 19.7\,\text{dB}$$

The addition of the mass of the rubber sheet increases the transmission loss by 6.0 dB.

(b) For a frequency of 8 kHz. This frequency is greater than the critical frequency, so the panel behavior falls in Region III, the damping-controlled region. The composite panel damping factor is found from Eq. (4-185):

$$\eta = \frac{[(0.001)(73.1)(0.0016) + (0.008)(2.30)(0.0048)](0.0016 + 0.0048)^2}{(73.1)(0.0016)^3(1.357) + (2.30)(0.0048)^3(5.453)}$$

$$\eta = 0.00469$$

Using Eq. (4-167), we find the transmission loss at the critical frequency for normal incidence:

$$\text{TL}_n(f_c) = 10\log_{10}\left\{1 + \left[\frac{(\pi)(9.04)(4268)}{(1.20)(343.8)}\right]^2\right\} = 10\log_{10}(86{,}320)$$

$$= 49.4\,\text{dB}$$

The transmission loss for the composite panel is found from Eq. (4-166):

$$\text{TL} = 49.4 + 10\log_{10}(0.00469) + 33.22\log_{10}(8000/4268) - 5.7$$

$$\text{TL} = 49.4 + (-23.3) + 9.1 - 5.7 = 29.5\,\text{dB}$$

For a single layer of aluminum, we find the following value, using Eq. (4-167):

$$\text{TL}_n(f_c) = 10\log_{10}\left\{1 + \left[\frac{(\pi)(4.48)(7515)}{(1.20)(343.8)}\right]^2\right\} = 48.2\,\text{dB}$$

Transmission of Sound

The transmission loss for the aluminum alone at 8000 Hz is as follows:

$$TL(1) = 48.2 + 10 \log_{10}(0.001) + 33.22 \log_{10}(8000/7515) - 5.7$$
$$TL(1) = 48.2 + (-30.0) + 0.9 - 5.7 = 13.4 \, dB$$

The addition of the rubber layer increases the transmission loss at 8 kHz by about 16 dB.

4.10.4 Rib-Stiffened Panels

Panels may have ribs attached to increase the stiffness of the panel and to reduce stress levels for a given applied load. The rib-stiffened panel shown in Fig. 4-17 has a stiffness that is different in the direction parallel to the ribs (the more stiff direction) than in the direction perpendicular to the ribs. This difference in stiffness has an influence on the transmission loss for the panel (Maidanik, 1962).

In the mass-controlled region (for $f < f_{c1}$), the transmission loss may be calculated from Eq. (4-163), using the following expression for the surface mass or mass per unit surface area:

$$M_S = \rho_w h [1 + (h_r/h)(t/d)] \quad (4\text{-}186)$$

There are two different wave coincidence or critical frequencies for an orthotropic plate, such as a rib-stiffened panel, corresponding to the different stiffness of the panel. The two critical frequencies are given by expressions similar to Eq. (4-183):

$$f_{c1} = \frac{c^2}{2\pi} \left(\frac{M_S}{B_1} \right)^{1/2} \quad (4\text{-}187)$$

$$f_{c2} = \frac{c^2}{2\pi} \left(\frac{M_S}{B_2} \right)^{1/2} \quad (4\text{-}188)$$

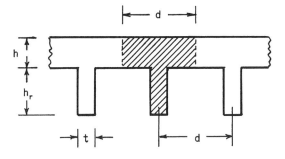

FIGURE 4-17 Dimensions for a rib-stiffened panel.

The flexural rigidity in the two perpendicular directions is given by the following expressions (Ugural, 1999):

$$B_1 = EI/d \qquad (4\text{-}189)$$

where I is the moment of inertia about the neutral axis of the T-section shown shaded in Fig. 4-17 and d is the center-to-center spacing of the ribs.

$$B_2 = \frac{Eh^3}{12\{1 - (t/d) + (t/d)/[1 + (h_r/h)]^3\}} \qquad (4\text{-}190)$$

For the intermediate frequency range, $f_{c1} < f < f_{c2}$, the transmission loss may be calculated from the following expression (Beranek and Vér, 1992):

$$\text{TL} = \text{TL}_n(f_{c1}) + 10\log_{10}(\eta) + 30\log_{10}(f/f_{c1}) - 40\log_{10}[\ln(4f/f_{c1})] \\ + 10\log_{10}[2\pi^3(f_{c2}/f_{c1})^{1/2}] \qquad (4\text{-}191)$$

The quantity $\text{TL}_n(f_{c1})$ is the transmission loss from Eq. (4-167) evaluated at the first critical frequency f_{c1}.

For the high-frequency range, $f > f_{c2}$, the transmission loss may be found from the following expression:

$$\text{TL} = \text{TL}_n(f_{c2}) + 10\log_{10}(\eta) + 30\log_{10}(f/f_{c2}) - 2 \qquad (4\text{-}192)$$

Example 4-10. A pine wood sheet, 1.22 m (48 in) by 2.44 m (96 in) with a thickness of 12.7 mm (0.500 in), has pine wood ribs attached. The dimensions of the ribs are 25.4 mm (1.000 in) high and 19.1 mm (0.750 in) thick. The ribs are spaced 101.6 mm (4.000 in) apart on centers and are oriented parallel to the long dimension of the sheet. Air at 25°C (77°F) and 101.3 kPa (14.7 psia) is on both sides of the panel. Determine the transmission loss for the panel for a frequency of 500 Hz.

The location of the centroid axis for the T-section shown in Fig. 4-17 may be found, as follows. The cross-sectional areas of the sheet portion (1) and the rib (2) are first determined, along with the distances from the interface to the individual centers of the areas, with the positive direction toward the sheet:

$$A_1 = (101.6)(12.7) = 1290 \text{ mm}^2 \quad \text{and} \quad y_1 = (\tfrac{1}{2})(12.7) = 6.35 \text{ mm}$$

$$A_2 = (19.1)(25.4) = 485 \text{ mm}^2 \quad \text{and} \quad y_2 = -(\tfrac{1}{2})(25.4) = -12.7 \text{ mm}$$

The distance from the interface between the sheet and the rib to the overall centroid axis is as follows:

Transmission of Sound

$$\chi = \frac{(1290)(6.35) + (485)(-12.7)}{(1290 + 485)} = 1.145\,\text{mm}\ (0.045\,\text{in})$$

The area moments of inertia of the individual areas are as follows:

$$I_1 = (101.6)(12.7)^3/(12) = 17{,}343\,\text{mm}^4\ (0.0417\,\text{in}^4)$$

$$I_2 = (19.1)(25.4)^3/(12) = 26{,}083\,\text{mm}^4\ (0.0627\,\text{in}^4)$$

The distances from the individual centroids to the overall centroid axis are as follows:

$$r_1 = 6.35 - 1.145 = 5.205\,\text{mm} \quad \text{and}$$
$$r_2 = -12.7 - 1.145 = -13.845\,\text{mm}$$

The area moment of inertia of the T-section about the overall centroid axis may be calculated from the following expression:

$$I = \Sigma(I_j + A_j r_j^2)$$

$$I = 17{,}343 + (1290)(5.205)^2 + 26{,}083 + (485)(-13.845)^2$$

$$I = 171{,}380\,\text{mm}^4 = 17.138\,\text{cm}^4\ (0.4117\,\text{in}^4)$$

The flexural rigidity of the stiffened panel in the direction parallel to the ribs is found from Eq. (4-189). The properties of pine wood are found in Appendix C.

$$B_1 = (13.7)(10^9)(17.138)(10^{-8})/(0.1016) = 23{,}109\,\text{Pa-m}^3$$

$$B_1 = 23{,}109\,\text{N-m} = 23.109\,\text{kN-m}\ (17{,}040\,\text{lb}_\text{f}\text{-ft})$$

The flexural rigidity in the direction perpendicular to the ribs is found from Eq. (4-190):

$$B_2 = \frac{(13.7)(10^9)(0.0127)^3}{(12)\{1 - (19.1/101.6) + (19.1/101.6)/[1 + (25.4/12.7)]^3\}}$$

$$B_2 = 2856\,\text{Pa-m}^3 = 2856\,\text{N-m} = 2.856\,\text{kN-m}\ (2110\,\text{lb}_\text{f}\text{-ft})$$

The surface mass or mass per unit surface area for the rib-stiffened panel may be found using Eq. (4-186):

$$M_S = (640)(0.0127)[1 + (25.4/12.7)(19.1/101.6)] = 11.184\,\text{kg/m}^2$$

The two critical frequencies may be determined from Eqs (4-187) and (4-188):

$$f_{c1} = \frac{(346.1)^2}{(2\pi)} \left[\frac{11.184}{23,109}\right]^{1/2} = 419.4 \, \text{Hz}$$

$$f_{c2} = \frac{(346.1)^2}{(2\pi)} \left[\frac{11.184}{2856}\right]^{1/2} = 1193.1 \, \text{Hz}$$

For the panel without the ribs, the surface mass is as follows:

$$M_S^o = \rho_w h = (640)(0.0127) = 8.128 \, \text{kg/m}^2$$

The critical frequency for the panel without stiffening ribs is determined as follows:

$$f_c^o = (M_S f_c)/M_S^o = (8160)/(8.128) = 1004 \, \text{Hz}$$

For the rib-stiffened panel, the frequency of 500 Hz falls in the intermediate region, so the transmission loss may be found from Eq. (4-191). The transmission loss for normal incidence at the lower critical frequency is calculated from Eq. (4-167):

$$\text{TL}_n(f_{c1}) = 10 \log_{10}\left\{1 + \left[\frac{(\pi)(11.184)(419.4)}{(409.8)}\right]^2\right\}$$

$$\text{TL}_n(f_{c1}) = 10 \log_{10}(1294) = 31.1 \, \text{dB}$$

The transmission loss for the rib-stiffened panel at 500 Hz is as follows:

$$\text{TL} = 31.1 + 10 \log_{10}(0.020) + 30 \log_{10}(500/419.4)$$
$$- 40 \log_{10}\{\ln[(4)(500)/419.4]\} + 10 \log_{10}[(2\pi^3)(1193.1/419.4)^{1/2}]$$

$$\text{TL} = 31.1 + (-17.0) + 2.3 - 7.7 + 20.2 = 28.9 \, \text{dB}$$

For the pine wood panel without stiffening ribs, the frequency of 500 Hz falls in the transmission loss Region II, the mass-controlled region. The sound power coefficient for normal incidence may be found from Eq. (4-160):

$$1/a_{tn} = 1 + [(\pi)(500)(8.128)/(409.8)]^2 = 971.7$$

The transmission loss for the panel without stiffening ribs is calculated from Eq. (4-163):

$$\text{TL} = 10 \log_{10}(971.7) - 5 = 29.9 - 5 = 24.9 \, \text{dB}$$

4.11 SOUND TRANSMISSION CLASS

In the previous sections, we considered some relatively simple panel constructions and presented techniques for estimation of the transmission loss

Transmission of Sound

curve. In practice, partitions separating two spaces are often much more complicated in construction than those previously discussed. In these cases, the estimation of the transmission loss curve by analytical or numerical techniques may not be practical. For this reason, the transmission loss curve must be measured experimentally. The standard test technique is described by the American Society of Testing Materials standard (ASTM, 1983).

It is often convenient to have available a single-figure rating that can be used to compare the performance of partitions in reducing noise transmission. Prior to 1970, the arithmetic average of the transmission loss at nine test frequencies was used to rate partitions (Faulkner, 1976). This technique has the weakness inherent in all "averaging" systems, i.e., two specimens may have the same average transmission loss but quite different frequency curves. The use of average transmission loss values does not present an overall picture of the ability of the material to reduce noise transmission over the entire range of frequencies of interest to the acoustic designer.

The sound transmission class (STC) rating was developed to provide a single-number rating of partitions, and yet provide additional information about the frequency spectrum of the transmission loss (ASTM, 1984). The STC rating generally correlates the impressions of the sound insulation characteristics of walls for transmission of such sounds as speech, radio, television, and other broadband noise sources in buildings. The STC rating is defined as the value of the transmission loss at 500 Hz which approximates a standard TL curve, measured in sixteen 1/3 octave band intervals from 125 Hz through 4000 Hz. The standard curve has three portions: (a) from 125 Hz to 400 Hz, in which the curve increases 3 dB for each 1/3 octave increase; (b) from 400 Hz to 1250 Hz, in which the curve increases 1 dB for each 1/3 octave increase; and (c) from 1250 Hz to 4000 Hz, in which the curve is constant at a value 4 dB higher than the value at 500 Hz.

The STC rating is determined by comparison of the experimental measurements of the transmission loss with the standard TL curve, subject to two conditions:

1. No single value of the experimental TL may be more than 8 dB below (less than) the standard curve.
2. The sum of the deviations below the standard curve cannot exceed 32 dB. The STC rating is generally specified to 1 dB significant figures.

A graphical technique may be used to determine the STC rating; however, the following procedure is much more adaptable to computer utilization.

Step 1. Measure the TL value for the sixteen 1/3 octave bands having center frequencies from 125 Hz through 4000 Hz.

Step 2. Calculate the corresponding difference Δ between the STC-50 standard curve S_{50} and the TL at each 1/3 octave band center frequency, where:

$$\Delta = S_{50} - TL \tag{4-193}$$

The standard STC-50 curve is given in Table 4-4.

Step 3. Calculate the *first* estimate for the STC rating (STC_1) from the following expression:

$$STC_1 = 50 - (\Delta)_{max} + 8 = 58 - (\Delta)_{max} \tag{4-194}$$

The quantity $(\Delta)_{max}$ is the largest (algebraically) value of the differences Δ. The calculation in Step 3 meets the first requirement of the STC rating; i.e., none of the experimental points will lie more than 8 dB below the standard curve.

Step 4. Determine the deficiencies at each experimental point, where the deficiency Def is defined by the following:

$$Def = S_{50} - (50 - STC_1) - TL = \Delta - (50 - STC_1) \tag{4-195}$$

The *deficiency* is the difference between the standard STC curve for STC_1 and the experimental data points for the transmission loss. A *positive* value of the deficiency corresponds to a point *below* the standard curve.

Step 5. Add all of the *positive* values of the deficiencies, which are the values that lie below the first estimate (STC_1) curve.

Step 6. (A) if the sum of the positive values of the deficiencies is 32 dB or less, the STC is equal to the first estimate value, STC_1:

If $\quad \Sigma(+Def) \leq 32\,dB,\quad$ then $\quad STC = STC_1$

TABLE 4-4 Values for the Standard STC-50 Curve.

1/3 Octave band center frequency, Hz	S_{50}, dB	1/3 Octave band center frequency, Hz	S_{50}, dB
125	34	800	52
160	37	1,000	53
200	40	1,250	54
250	43	1,600	54
315	46	2,000	54
400	49	2,500	54
500	50	3,150	54
630	51	4,000	54

Transmission of Sound

This calculation meets the second criterion for the STC rating; i.e., the sum of the deficiencies (deviations from the STC curve) must not be greater than 32 dB.

(B) On the other hand, if the sum of the positive values of the deficiencies is greater than 32 dB, the first estimate for the STC must be adjusted to meet the STC criterion.

If $\quad \Sigma(+\text{Def}) > 32\,\text{dB}$

then calculate the adjustment:

$$\text{Adj} = \frac{\Sigma(+\text{Def}) - 32}{N_{PD}}$$

The quantity N_{PD} is the number of positive values of the deficiencies. The adjustment Adj is rounded *up* to the next whole integer. The sound transmission class is found by applying the adjustment to the initial estimate for the STC:

$$\text{STC} = \text{STC}_1 - \text{Adj}$$

The application of this technique is illustrated in the following example.

Example 4-11. The measured values of the transmission loss for a partition are given in Table 4-5. Determine the STC rating for the partition.

The calculations are summarized in Table 4-5, where the second column contains the experimental values for the TL, the third column contains the standard curve for STC-50, and the fourth column contains the values for the differences Δ, calculated from Eq. (4-193). For example, for the 125 Hz 1/3 octave band,

$$\Delta = S_{50} - \text{TL} = 34 - 22 = 12\,\text{dB}$$

The largest value of the differences is 31 dB, which occurs for the 500 Hz 1/3 octave band.

The first estimate for the STC rating is found from Eq. (4-194):

$$\text{STC}_1 = 58 - (\Delta)_{\text{max}} = 58 - 31 = 27$$

The values in the fifth column (Def) are calculated from Eq. (4-195). For example, for the 125 Hz 1/3 octave band,

$$\text{Def} = \Delta - (50 - \text{STC}_1) = \Delta - (50 - 27) = \Delta - 23$$

$$\text{Def} = 12 - 23 = -11$$

In this example, there are $N_{PD} = 10$ positive values of the deficiencies. The sum of these 10 positive values of Def is:

TABLE 4-5 Solution for Example 4-11.

1/3 Octave band center frequency, Hz	TL (expt.), dB	S_{50}, dB	Δ, dB	Def, dB	S_{25}, dB
125	22	34	12	−11	9
160	21	37	16	−7	12
200	21	40	19	−4	15
250	22	43	21	−2	18
315	18	46	28	+5	21
400	21	49	28	+5	24
500	19	50	31 = Δ_{max}	+8	25
630	22	51	29	+6	26
800	24	52	28	+5	27
1,000	23	53	30	+7	28
1,250	29	54	25	+2	29
1,600	32	54	22	−1	30
2,000	31	54	23	0	30
2,500	28	54	26	+3	30
3,150	28	54	26	+3	30
4,000	30	54	24	+1	30

$$\Sigma(+\text{Def}) = 5 + 5 + 8 + 6 + 5 + 7 + 2 + 3 + 3 + 1 = 45 > 32$$

An adjustment is required:

$$\text{Adj} = \frac{45 - 32}{10} = 1.3$$

The actual value of the adjustment is found by rounding up to the nearest integer, or

$$\text{Adj} = 2\,\text{dB}$$

The STC rating for the partition is, as follows:

$$\text{STC} = \text{STC}_1 - \text{Adj} = 27 - 2 = 25\,\text{dB}$$

The sound transmission class rating may be written as STC-25.

If we want to generate the STC-25 standard curve, we may displace the STC-50 standard curve such that the value at the 500 Hz point is equal to 25 dB. In general,

$$S_{\text{STC}} = S_{50} - \text{STC} \tag{4-196}$$

For the STC-25 standard curve, STC = S_{50} − 25. These values are shown in the last column in Table 4-5.

Transmission of Sound

The sound transmission class rating may be used as a design criterion for partitions within or between dwellings, between areas in an office building, or within schools, theaters, etc. Detailed criteria are given by the U.S. Dept. of Housing and Urban Development (HUD) (Berendt et al., 1967). The HUD recommendations are classified according to the environment in which the dwelling is located:

- Grade I: surburban and outer urban residential areas. These are considered "quiet" areas, as far as background noise is concerned. The A-weighted sound levels during nighttime would be in the range of 35 dBA to 40 dBA or lower.
- Grade II: residential urban and surburban areas. These are considered "average" areas, as far as the background noise level is concerned. The nighttime levels are generally around 40 dBA to 45 dBA. This is probably the most commonly used category for design.
- Grade III: urban areas. These are considered "noisy" areas, and this category is considered as the minimum recommended category. The sound levels during nighttime are generally 55 dBA or higher.

Selected values of the STC criteria are given in Table 4-6.

4.12 ABSORPTION OF SOUND

In the previous discussion of sound transmission, we had assumed that the energy dissipation within the medium was negligible. This assumption is generally quite valid, except for high-frequency sound waves and for sound transmitted over large distances. In this section, we will examine the effect of energy attenuation as the sound wave moves through a medium.

In the development of the wave equation in Sec. 4.1, we had considered only pressure forces acting on a fluid element. The attenuation or dissipation effects may be represented by a dissipation force, defined by:

$$F_d = C_D S \frac{\partial \mathscr{C}}{\partial t} = -C_D S \frac{\partial^2 \xi}{\partial x \partial t} \qquad (4\text{-}197)$$

where C_D is the *dissipation coefficient* with units, Pa-s, and the quantity $\mathscr{C}$ is the condensation, as defined by Eq. (4-5). With the dissipation effect included, Eq. (4-2) for the net force becomes, as follows:

$$F_{net} = -\frac{\partial p}{\partial x} S\,dx + C_D \frac{\partial^3 \xi}{\partial x^2 \partial t} S\,dx \qquad (4\text{-}198)$$

TABLE 4-6 Selected Design Values of the STC for Partitions According to HUD Criteria.

	Sound transmission class, STC, dB		
Partitions between dwellings	Grade I	Grade II	Grade III
Bedroom to bedroom	55	52	48
Corridor to bedroom	55	52	48
Kitchen to bedroom	58	55	52
Partitions within dwellings:			
Bedroom to bedroom	48	44	40
Kitchen to bedroom	52	48	45
Office areas:			
Normal office to adjacent office	37		
Normal office to building exterior	37		
Conference room to office	42		
Schools, etc.:			
Classroom to classroom	37		
Classroom to corridor	37		
Theater to similar area	52		

Source: Berendt et al. (1967).

The force–balance equation, Eq. (4-3), has the following form if dissipative forces are included:

$$\frac{\partial p}{\partial x} - C_D \frac{\partial^3 \xi}{\partial x^2 \, \partial t} = -\rho \frac{\partial^2 \xi}{\partial t^2} \qquad (4\text{-}199)$$

The development presented in Sec. 4.1 may be carried out, using Eq. (4-199), to obtain the one-dimensional wave equation with dissipative effects included:

$$\frac{\partial^2 p}{\partial x^2} + \frac{C_D}{\rho c^2} \frac{\partial^3 p}{\partial x^2 \, \partial t} = \frac{1}{c^2} \frac{\partial^2 p}{\partial t^2} \qquad (4\text{-}200)$$

The coefficient on the second term in Eq. (4-200), corresponding to dissipation effects, has time units, so we may define the *relaxation time* τ as follows:

$$\tau = \frac{C_D}{\rho c^2} \qquad (4\text{-}201)$$

Transmission of Sound

If we substitute the expression from Eq. (4-201) for the relaxation time into Eq. (4-200), we find an alternative form for the one-dimensional wave equation including dissipation effects:

$$\frac{\partial^2}{\partial x^2}\left(p + \tau \frac{\partial p}{\partial t}\right) = \frac{1}{c^2}\frac{\partial^2 p}{\partial t^2} \qquad (4\text{-}202)$$

Let us consider a simple harmonic sound wave, given by:

$$p(x, t) = \psi(x)\,e^{j\omega t} \qquad (4\text{-}203)$$

where $\psi(x)$ is the amplitude function, which is a complex quantity, in this case. Making the substitution from Eq. (4-203) into the wave equation, Eq. (4-202), we obtain the following differential equation:

$$(1 + j\omega\tau)\frac{d^2\psi}{dx^2} + \frac{\omega^2}{c^2}\psi = 0 \qquad (4\text{-}204)$$

Let us define the complex quantity κ (*complex wave number*) as follows:

$$\kappa^2 = \frac{\omega^2}{c^2(1 + j\omega\tau)} \qquad (4\text{-}205)$$

The solution of Eq. (4-204) for waves traveling in the $+x$-direction is as follows:

$$\psi(x) = A\,e^{-j\kappa x} \qquad (4\text{-}206)$$

Let us write the complex wave number in terms of its real and imaginary parts:

$$\kappa = k - j\alpha \qquad (4\text{-}207)$$

The term α is called the *attenuation coefficient*. The amplitude function may be written in terms of the attenuation coefficient, by combining Eqs (4-206) and (4-207):

$$\psi(x) = A\,e^{-j(k-j\alpha)x} = A\,e^{-\alpha x}\,e^{-jkx} \qquad (4\text{-}208)$$

If we combine Eqs (4-205) and (4-207) and solve for the real and imaginary parts, we obtain the following expressions:

$$\alpha = \frac{\omega}{c}\left[\frac{(1+\omega^2\tau^2)^{1/2} - 1}{2(1+\omega^2\tau^2)}\right]^{1/2} \qquad (4\text{-}209)$$

$$k = \frac{\omega}{c}\left[\frac{(1+\omega^2\tau^2)^{1/2} + 1}{2(1+\omega^2\tau^2)}\right]^{1/2} \qquad (4\text{-}210)$$

In many fluids, the relaxation time is quite small. For example, for monatomic gases, τ is approximately 0.2 ns. In the acoustic range of frequencies (20 Hz to 20 kHz), the term ($\omega\tau$) is much less than unity. In this case, Eqs (4-209) and (4-210) reduce to the following relationships:

$$\alpha = \frac{\omega^2 \tau}{2c} = \frac{2\pi^2 f^2 \tau}{c} \qquad (4\text{-}211)$$

$$k = \frac{\omega}{c} = \frac{2\pi f}{c} \qquad (4\text{-}212)$$

According to Eq. (4-211), if the relaxation time is independent of frequency, then the attenuation coefficient is directly proportional to the frequency squared. This behavior is observed for many gases over a wide range of frequencies.

If we denote the amplitude of the acoustic pressure at $x = 0$ by p_o, then the magnitude of the acoustic pressure at any location is given by Eq. (4-208):

$$p(x) = p_o e^{-\alpha x} \qquad (4\text{-}213)$$

For a plane acoustic wave with attenuation, the amplitude of the pressure wave decays exponentially with distance from the source of the wave. The intensity of the plane wave at any location may be written as follows:

$$I(x) = \frac{p^2}{\rho c} = \frac{p_o^2}{\rho c} e^{-2\alpha x} = I_o e^{-2\alpha x} = I_o e^{-mx} \qquad (4\text{-}214)$$

The quantity I_o is the intensity at $x = 0$ and $m = 2\alpha$. The quantity m is called the *energy attenuation coefficient*. We note from Eq. (4-214) that, for a plane wave, the intensity is not constant with position, because energy is being dissipated in this case. The change in the intensity level with distance due to energy attenuation is given by the following expression for a plane sound wave:

$$\Delta L_I = L_I(x = 0) - L_I(x) = -10 \log_{10}(e^{-2\alpha x}) \qquad (4\text{-}215)$$

$$\Delta L_I = (10)(2\alpha x)/\ln(10) = 8.6859 \alpha x \qquad (4\text{-}216)$$

For a spherical sound wave, the acoustic energy varies according to the following expression, if dissipation effects are considered:

$$W(r) = W_o e^{-2\alpha r} \qquad (4\text{-}217)$$

The intensity for a spherical sound wave with energy attenuation may be expressed by the following:

$$I = \frac{W}{4\pi r^2} = \frac{W_o}{4\pi r^2} e^{-2\alpha r} \qquad (4\text{-}218)$$

Transmission of Sound

The change in the intensity level for a spherical sound wave with dissipation effects may be written as follows:

$$\Delta L_I = L_I(r = r_o) - L_I(r) = 20 \log_{10}(r/r_o) + 8.6859\alpha(r - r_o) \quad (4\text{-}219)$$

We observe from Eq. (4-219) that the change in intensity level for a spherical wave involves two effects: (a) the spreading of the acoustic power over a larger area and (b) the dissipation of acoustic energy in the material.

The attenuation term (αx) or (αr) has been given "units" of *neper* (named after John Napier, who developed logarithms), with the abbreviation Np (Pierce, 1981). Note that the quantity (αx) is actually dimensionless. The "unit" *radian* for angular measure is also dimensionless. The attenuation coefficient α can be written with units neper per meter, Np/m. But the intensity level change has units of decibel, dB, so the term (8.6859α) should have units of dB/m in Eq. (4-219). From this observation, we see that the conversion factor between the two dimensionless "units" is:

$$8.6859 \, \text{dB/Np} = 20 \log_{10}(e) = 20/\ln(10) \quad (4\text{-}220)$$

It is important to note that the attenuation coefficient may be reported in the literature in either Np/m units or dB/m (or dB/km) units.

4.13 ATTENUATION COEFFICIENT

The attenuation or dissipation of acoustic energy as a sound wave moves through a medium may be attributed to three basic mechanisms:

(a) *Viscous effects* (dissipation of acoustic energy due to fluid friction), which result in thermodynamically irreversible propagation of sound.
(b) *Heat conduction effects* (heat transfer between high- and low-temperature regions in the wave), which result in non-adiabatic propagation of the sound.
(c) *Internal molecular energy interchanges* (molecular energy relaxation effects), which result in a time lag between changes in translational kinetic energy and the energy associated with rotation and vibration of the molecules.

The viscous energy dissipation effects result from the relative motion between different portions of the fluid during compression and expansion that occurs when a sound wave moves through the fluid. For a newtonian fluid, the magnitude of this effect is proportional to the viscosity μ of the fluid.

As the fluid is compressed and expanded during the transmission of a sound wave, changes in temperature occur in different portions of the fluid.

There is a tendency for energy to be conducted from regions of compression, where the temperature is elevated, to regions of expansion or rarefaction, where the temperature is reduced. The heat transfer effect tends to reduce the amplitude of the pressure wave and dissipate energy as the wave moves through the medium. The magnitude of this effect is proportional to the thermal conductivity k_t of the fluid and inversely proportional to the specific heat c_p or the thermal energy storage capacity of the medium.

For all fluids except monatomic gases, there is a finite time lag for conversion of energy into rotational and vibrational energy of the molecule. During this time, the acoustic wave may move past the molecule and leave behind some of the acoustic energy.

The attenuation due to the sum of the first two mechanisms, viscous and heat conduction, is called the *classical attenuation*. The classical attenuation may be written in the following form (Kinsler et al., 1982):

$$\alpha(\text{classical}) = \frac{2\pi^2 f^2 \mu}{\rho c^3}\left(\frac{4}{3} + \frac{\gamma - 1}{\text{Pr}}\right) \quad (4\text{-}221)$$

where μ = fluid viscosity, γ = specific heat ratio (1.667 for monatomic gases and 1.400 for diatomic gases), $\text{Pr} = \mu c_p / k_t$ = Prandtl number, c_p = specific heat at constant pressure, and k_t = thermal conductivity. The classical relaxation time may be found from Eqs (4-211) and (4-221):

$$\tau(\text{classical}) = \frac{\mu}{\rho c^2}\left(\frac{4}{3} + \frac{\gamma - 1}{\text{Pr}}\right) \quad (4\text{-}222)$$

The classical attenuation coefficient is proportional to the square of the frequency, and the classical relaxation time is independent of the frequency.

The expression for the classical attenuation coefficient yields good agreement with experimental values of the attenuation coefficient for monatomic gases, such as argon and helium, as shown in Table 4-7.

The third contribution to attenuation of sound in a fluid results from the finite time requried to convert translational kinetic energy into internal energies associated with rotation and vibration of the molecules. For many molecules—including CO_2, water vapor, nitrogen, and oxygen—vibrational energy transfer is predominant in the acoustic frequency range. Theoretical models, with experimental verification, have shown that the attenuation coefficient can be written in terms of the sum of the individual contributions (Pierce, 1988):

$$\alpha = \alpha(\text{classical}) + \Sigma \alpha_{v,j} \quad (4\text{-}223)$$

where the $\alpha_{v,j}$ are the contributions of the various vibrational energy relaxation effects, which may be expressed in the form:

Transmission of Sound

TABLE 4-7 Values of the Classical Attenuation Coefficient.

Gas	Viscosity, μ, µPa-s	Prandtl No., $Pr = \mu c_p / k_t$	Relaxation time, τ, ns	α/f^2(calc.), Np/m-Hz^2	α/f^2(expt.), Np/m-Hz^2
Argon	22.73	0.668	0.314	19.20×10^{-12}	19.40×10^{-12}
Helium	19.94	0.692	0.271	5.25×10^{-12}	5.35×10^{-12}
Nitrogen	17.82	0.715	0.238	13.29×10^{-12}	16.14×10^{-12}
Oxygen	20.65	0.714	0.276	16.47×10^{-12}	19.64×10^{-12}

$$\alpha_v = \frac{\alpha_\infty (\omega \tau_v)^2}{1 + (\omega \tau_v)^2} \tag{4-224}$$

The quantity $\omega = 2\pi f$ is the circular frequency of the sound wave, τ_v is the relaxation time for vibrational energy, and α_∞ is the limiting (high-frequency) value for the vibrational attenuation contribution.

The term α_∞ is related to the specific heat ratio γ for a gas and the vibrational contribution to the specific heat, c_{vib}:

$$\alpha_\infty = \frac{(c_{\text{vib}}/R)(\gamma - 1)^2}{2c\tau_v \gamma} \tag{4-225}$$

The vibrational specific heat term may be calculated from the following expression (ter Haar, 1954):

$$c_{\text{vib}}/R = y_j (\theta_v / T)^2 \exp(-\theta_v / T) \tag{4-226}$$

The term y_j is the mole fraction of the jth component in a gas mixture ($y_j = 1$ for a single component gas), θ_v is a constant, dependent on the gas, and T is the absolute temperature of the gas. For nitrogen and oxygen, the constant θ_v is as follows (ANSI, 1978):

$$\theta_v(N_2) = 3352 K$$

$$\theta_v(O_2) = 2239 K$$

The relaxation times for atmospheric air are sensitive to the amount of water vapor present in the air. An O_2 or N_2 molecule colliding with an H_2O molecule is more likely to exhibit a change in vibrational energy than when the molecules collide with other N_2 or O_2 molecules. The following expressions may be used to estimate the relaxation times for oxygen and nitrogen in atmospheric air (Sutherland et al., 1974):

$$\frac{(p_{\text{ref}}/p)}{2\pi\tau_v(O_2)} = 24 + (4.41)(10^6)h \frac{0.05 + 100h}{0.391 + 100h} \tag{4-227}$$

$$\frac{(p_{\text{ref}}/p)}{2\pi\tau_v(N_2)} = [9 + (3.5)(10^4)h\,e^{-F}](T_{\text{ref}}/T)^{1/2} \tag{4-228}$$

$$F = 6.142[(T_{\text{ref}}/T)^{1/3} - 1] \tag{4-229}$$

The reference pressure and temperature values are $p_{\text{ref}} = 101.325\,\text{kPa}$ and $T_{\text{ref}} = 293.16\,\text{K}$, respectively. The range of validity of Eqs (4-227) and (4-228) is between 0°C and 40°C (32°F and 104°F), for an accuracy within 10%. The quantity h is the fraction of molecules in the gas that are H_2O molecules. This fraction is related to the relative humidity RH, expressed as a decimal (0.40 instead of 40%), and the saturation pressure of the water vapor at the air temperature, p_{sat}:

$$h = (\text{RH})p_{\text{sat}}/p \tag{4-230}$$

The energy attenuation coefficient for atmospheric air at various temperatures and relative humidity values is presented in Table 4-8. The energy attenuation coefficient m is related to the attenuation coefficient α by $m = 2\alpha$. The values given in Table 4-8 are values of the energy attenuation coefficient averaged over the octave band with the indicated center frequency. The octave band values are about 10% different from the values

TABLE 4-8 Energy Attenuation Coefficient m (km^{-1}) for Atmospheric Air at 101.325 kPa (14.7 psia): Note that $m = 2\alpha$, where α is the Attenuation Coefficient.

Relative humidity, %	Temperature, °C	Octave band center frequency, Hz				
		500	1,000	2,000	4,000	8,000
10	10	1.28	4.30	10.6	16.3	16.4
	15	0.98	3.41	10.9	22.3	24.0
	20	0.78	2.67	9.02	25.7	34.1
	25	0.71	2.14	7.18	24.2	36.3
	30	0.69	1.80	5.84	20.4	38.9
20	10	0.63	2.04	6.98	21.1	29.9
	15	0.56	1.61	5.50	18.7	33.2
	20	0.53	1.40	4.31	14.7	29.4
	25	0.52	1.33	3.58	11.7	24.3
	30	0.52	1.30	3.23	9.80	20.7

Transmission of Sound

TABLE 4-8 (Cont'd)

Relative humidity, %	Temperature, °C	Octave band center frequency, Hz				
		500	1,000	2,000	4,000	8,000
30	10	0.50	1.35	4.60	15.1	28.4
	15	0.48	1.23	3.59	12.0	25.1
	20	0.46	1.17	3.02	9.62	20.3
	25	0.46	1.14	2.80	7.90	16.4
	30	0.46	1.13	2.76	7.14	13.4
40	10	0.45	1.13	3.37	11.3	22.8
	15	0.44	1.07	2.80	8.91	18.7
	20	0.43	1.05	2.62	7.22	15.0
	25	0.42	1.03	2.57	6.33	12.5
	30	0.42	1.02	2.56	6.20	11.1
50	10	0.41	1.01	2.77	8.93	17.8
	15	0.40	0.99	2.50	7.16	14.5
	20	0.39	0.96	2.34	6.17	11.9
	25	0.39	0.95	2.30	5.88	10.4
	30	0.38	0.94	2.26	5.76	9.88
60	10	0.38	0.94	2.51	7.92	15.0
	15	0.38	0.92	2.31	6.12	12.2
	20	0.37	0.90	2.20	5.66	10.3
	25	0.37	0.89	2.16	5.50	9.27
	30	0.37	0.88	2.14	5.43	9.01
70	10	0.36	0.89	2.30	6.45	13.4
	15	0.36	0.86	2.16	5.58	11.0
	20	0.35	0.85	2.08	5.33	9.57
	25	0.35	0.84	2.06	5.18	8.85
	30	0.35	0.84	2.05	5.14	8.71
80	10	0.35	0.84	2.14	5.80	11.6
	15	0.34	0.82	2.02	5.32	9.86
	20	0.34	0.81	1.97	5.04	9.05
	25	0.33	0.80	1.95	4.93	8.52
	30	0.33	0.80	1.95	4.88	8.47
90	10	0.33	0.80	1.97	5.37	10.1
	15	0.33	0.79	1.92	5.09	8.93
	20	0.32	0.78	1.87	4.87	8.56
	25	0.32	0.77	1.87	4.72	8.36
	30	0.32	0.77	1.86	4.68	8.34

evaluated at the center frequency, because the attenuation coefficient is not a linear function of frequency.

Example 4-12. Determine the classical attenuation coefficient for argon gas at 273.2K (0°C or 32°F) and 101.3 kPa (14.7 psia) at a frequency of 1000 Hz. The properties of argon gas at this condition are as follows:

viscosity $\quad \mu = 21.03\ \mu\text{Pa-s}\ (0.0509\ \text{lb}_m/\text{ft-hr})$
Prandtl number $\quad \text{Pr} = 0.668$
specific heat ratio $\quad \gamma = 1.667$
sonic velocity $\quad c = 307.8\ \text{m/s}\ (1010\ \text{ft/sec})$
density $\quad \rho = 1.782\ \text{kg/m}^3\ (0.1113\ \text{lb}_m/\text{ft}^3)$

The relaxation time may be found from Eq. (4-222):

$$\tau(\text{classical}) = \frac{(21.03)(10^{-6})}{(1.782)(307.8)^2}\left[\frac{4}{3} + \frac{(1.667-1)}{(0.668)}\right]$$

$$\tau(\text{classical}) = (1.246)(10^{-10})(2.332) = 0.290 \times 10^{-9}\ \text{s} = 0.290\ \text{ns}$$

Note that $\omega\tau = (2\pi)(1000)(0.290)(10^{-9}) = 1.825 \times 10^{-6} \ll 1$.

The attenuation coefficient may be calculated from Eq. (4-211), because the quantity $\omega\tau$ is so small:

$$\alpha(\text{classical}) = \frac{(2\pi^2)(1000)^2(0.290)(10^{-9})}{(307.8)} = 1.863 \times 10^{-5}\ \text{Np/m}$$

The attenuation coefficient may be expressed in decibel units.

$$8.6859\alpha = (8.6859)(1.863)(10^{-5}) = 0.162 \times 10^{-3}\ \text{dB/m} = 0.162\ \text{dB/km}$$

Example 4-13. Calculate the energy attenuation coefficient for atmospheric air at 293.2K (20°C or 68°F) and 101.3 kPa (14.7 psia) for a frequency of 4 kHz. The relative humidity of the air is 20%. The properties of air are as follows:

viscosity $\quad \mu = 18.21\ \mu\text{Pa-s}\ (0.0441\ \text{lb}_m/\text{ft-hr})$
Prandtl number $\quad \text{Pr} = 0.717$
specific heat ratio $\quad \gamma = 1.400$
sonic velocity $\quad c = 343.2\ \text{m/s}\ (1126\ \text{ft/sec})$
density $\quad \rho = 1.204\ \text{kg/m}^3\ (0.0752\ \text{lb}_m/\text{ft}^3)$

The saturation pressure for water vapor at 20°C is $p_{\text{sat}} = 2.338\ \text{kPa}$.

First, let us calculate the contribution of viscous and thermal conduction effects (the classical attenuation coefficient). The relaxation time for viscous and thermal conduction effects is given by Eq. (4-222):

Transmission of Sound

$$\tau(\text{classical}) = \frac{(18.21)(10^{-6})}{(1.204)(343.2)^2}\left[\frac{4}{3} + \frac{(1.40-1)}{(0.717)}\right]$$

$$\tau(\text{classical}) = 0.243 \times 10^{-9}\,\text{s} = 0.243\,\text{ns}$$

The classical attenuation coefficient may be calculated from Eq. (4-211):

$$\alpha(\text{classical}) = \frac{(2\pi^2)(4000)^2(0.243)(10^{-9})}{(343.2)} = (13.97)(10^{-9})(4000)^2$$

$$\alpha(\text{classical}) = 0.223 \times 10^{-3}\,\text{Np/m}$$

Next, let us calculate the effect of molecular interactions between the O_2 and H_2O molecules. The fraction of air molecules at 20% relative humidity that are water molecules is found from Eq. (4-230):

$$h = (0.20)(2.338)/(101.3) = 0.00461$$

The relaxation time for the O_2 interactions is found from Eq. (4-227):

$$\frac{(1)}{(2\pi)\tau_v(O_2)} = 24 + (4.41)(10^6)(0.00461)\frac{0.05 + 0.461}{0.391 + 0.461} = 12{,}235\,\text{s}^{-1}$$

$$\tau_v(O_2) = 1/(2\pi)(12{,}235) = 13.01 \times 10^{-6}\,\text{s} = 13.01\,\mu\text{s}$$

In atmospheric air, the mole fraction of oxygen is about $y(O_2) = 0.21$. The vibrational specific heat terms may be determined from Eq. (4-226):

$$c_{\text{vib}}/R = (0.21)(2239/293.2)^2 \exp(-2239/293.2) = 0.005904$$

The limiting attenuation coefficient for oxygen is found from Eq. (4-225), using the sonic velocity for O_2 of 326.6 m/s:

$$\alpha_\infty = \frac{(0.005904)(1.40-1)^2}{(2)(326.6)(13.01)(10^{-6})(1.40)} = 0.07941\,\text{Np/m}$$

For a frequency of 4000 Hz, we find the following:

$$\omega\tau_v(O_2) = (2\pi)(4000)(13.01)(10^{-6}) = 0.3269$$

The vibrational contribution associated with O_2–H_2O interactions may be calculated from Eq. (4-224):

$$\alpha_v(O_2) = \frac{(0.07941)(0.3269)^2}{1 + (0.3269)^2} = 7.668 \times 10^{-3}\,\text{Np/m}$$

Let us repeat the calculations for the nitrogen and water vapor interactions. The relaxation time is found from Eq. (4-228). The factor $F = 0$ because $T = T_{\text{ref}}$ in this example:

$$\frac{(1)}{2\pi\tau_v(N_2)} = (1)[9 + (3.5)(10^4)(0.00461)] = 170.5\,\text{s}^{-1}$$

$$\tau_v(N_2) = 1/(2\pi)(170.5) = 0.933 \times 10^{-3}\,\text{s} = 0.933\,\text{ms}$$

The mole fraction of nitrogen in atmospheric air is $y(N_2) = 0.79$. The vibrational specific heat term for N_2–H_2O interactions is as follows:

$$c_{vib}/R = (0.79)(3352/293.16)^2 \exp(-3352/293.16) = 0.001118$$

The limiting attenuation coefficient for nitrogen—with a specific heat ratio $\gamma = 1.40$ and a sonic velocity of 349.0 m/s—is as follows:

$$\alpha_\infty = \frac{(0.001118)(1.40-1)^2}{(2)(349.0)(0.933)(10^{-3})(1.40)} = 0.0001961\,\text{Np/m}$$

At a frequency of 4000 Hz, we find the following for N_2:

$$\omega\tau_v(N_2) = (2\pi)(4000)(0.933)(10^{-3}) = 23.45$$

The vibrational contribution associated with N_2–H_2O interactions is as follows:

$$\alpha_v(N_2) = \frac{(0.0001961)(23.45)^2}{1 + (23.45)^2} = 0.196 \times 10^{-3}$$

The attenuation coefficient is composed of the components that we have calculated:

$$\alpha = \alpha(\text{classical}) + [\alpha_v(O_2) + \alpha_v(N_2)]$$

$$\alpha = [0.223 + (7.668 + 0.196)](10^{-3}) = 8.087 \times 10^{-3}\,\text{Np/m}$$

$$2.8\% + 94.8\% + 2.4\%$$

The most important contribution to the attenuation coefficient is the internal vibrational energy interactions for the oxygen molecules (almost 95% of the total).

The attenuation coefficient may be expressed in decibel "units":

$$8.6859\alpha = 0.0702\,\text{dB/m}$$

The energy attenuation coefficient is as follows:

$$m = 2\alpha = (2)(0.008087) = 0.01617\,\text{m}^{-1}$$

$$4.3429m = 0.0702\,\text{dB/m}$$

It is noted from Eqs (4-221) and (4-222) that the viscous and thermal conduction effects result in a classical attenuation that is proportional to the frequency squared, or the attenuation coefficient increases at a rate of 6 dB/octave:

Transmission of Sound

$$\alpha(\text{classical})/f^2 = 2\pi^2\tau(\text{classical})/c = \text{constant}$$

For the case, $(2\pi f \tau_v) > 10$, the vibration contribution to the attenuation is approximately constant with frequency (with 1%), as shown by Eq. (4-224):

$$\alpha_v \to \alpha_\infty = \text{constant} \qquad (\text{for } \omega\tau_v > 10)$$

On the other hand, for the case, $(2\pi f \tau_v) < 10$, the vibrational contribution is approximately proportional to the frequency squared.

$$\alpha_v/f^2 \to 4\pi^2 \alpha_\infty \tau_v^2 = \text{constant} \qquad (\text{for } \omega\tau_v < 10)$$

Example 4-14. A gas turbine has a sound power output spectrum as given in Table 4-9. The directivity factor may be taken as $Q = 2$ for all frequencies. The noise is transmitted through atmospheric air at 25°C (77°C or 298.2K) and 101.3 kPa (14.7 psia) with a relative humidity of 50%. A residence is located 400 m (1312 ft) from the gas turbine unit. Determine the overall sound pressure level at the residence location due to the turbine noise.

The acoustic intensity for each octave band is given by Eq. (4-218), with the directivity factor included:

$$I = \frac{QW_o e^{-2\alpha r}}{4\pi r^2} = \frac{QW_o e^{-mr}}{4\pi r^2}$$

The results of the calculations are summarized in Table 4-9. Let us present the calculations for the 2000 Hz octave band. The acoustic power at the source (gas turbine) is as follows:

$$W_o = W_{\text{ref}} 10^{L_w/10} = (10^{-12})(10^{12.4}) = 2.512 \text{ W}$$

The factor involving the energy attenuation coefficient for the 2000 Hz octave band—note that the energy attenuation coefficient is given in units of km^{-1}—is as follows:

$$\exp[-mr] = \exp[-(2.30)(0.400)] = 0.3985$$

Attenuation of sound by atmospheric air reduces the acoustic power by a factor of almost 0.40. The acoustic intensity at the receiver position (at the residence) for the 2000 Hz octave band is as follows:

$$I = \frac{(2)(2.512)(0.3985)}{(4\pi)(400)^2} = 0.996 \times 10^{-6} \text{ W/m}^2 = 0.996\,\mu\text{W/m}^2$$

The calculations may be repeated for the other octave bands. The overall intensity is the sum of the intensities in each octave band:

TABLE 4-9 Solution for Example 4-14.

	Octave band center frequency, Hz							
	63	125	250	500	1,000	2,000	4,000	8,000
Turbine L_W, dB	120	124	128	128	127	124	123	123
W_o, watts	1.000	2.512	6.310	6.310	5.012	2.512	1.995	1.995
m, km^{-1}	0.0093	0.037	0.15	0.39	0.95	2.30	5.88	10.4
e^{-mr}	0.9963	0.9853	0.9418	0.8556	0.6839	0.3985	0.0952	0.0156
I, μW/m^2	0.991	2.462	5.911	5.370	3.409	0.996	0.189	0.031

Transmission of Sound

$$I_o = \Sigma I = (0.991 + 2.462 + 5.911 + \cdots)(10^{-6})$$
$$I_o = 19.359 \times 10^{-6}\,\text{W/m}^2 = 19.359\,\mu\text{W/m}^2$$

The overall sound pressure level is calculated from the intensity as follows:
$$p = (\rho_o c I_o)^{1/2} = [(409.8)(19.359)(10^{-6})]^{1/2} = 0.0891\,\text{Pa}$$

The overall sound pressure level is found as follows:
$$L_p = 20\log_{10}(0.0891/20 \times 10^{-6}) = 73.0\,\text{dB}$$

We will show how these calculations can be carried out directly in terms of decibels in Chapter 5.

It may be noted from Table 4-9 that the effect of atmospheric attenuation is practically negligible ($e^{-mr} > 0.90$ or $1 - e^{-mr} < 0.10$) in the 63 Hz, 125 Hz, and 250 Hz octave bands. On the other hand, the attenuation is significant ($e^{-mr} < 0.10$) in the 4 kHz and 8 kHz octave bands. From this result, we may conclude that we are generally justified in neglecting atmospheric air attenuation at low frequencies (below about 500 Hz), unless the distance from the source is large. For a distance of 400 m (1312 ft or about $\frac{1}{4}$ mile), the reduction in the intensity due to atmospheric attenuation is $e^{-mr} = 0.0156 \approx 1/64$. On the other hand, for a distance of 4 m (13.1 ft), the factor $e^{-mr} = 0.959$ for the 8 kHz octave band. We can conclude that the effect of attenuation in atmospheric air is also negligible when the sound is transmitted over relatively small distances. If we set the "negligible" limit at less than 0.5 dB, then atmospheric attenuation may be neglected when the following condition is valid:

$$4.3429 mr \leq 0.5\,\text{dB}$$
$$mr \leq 0.12$$

PROBLEMS

4-1. The one-dimensional wave equation in cylindrical coordinates is as follows:

$$\frac{1}{r}\frac{\partial}{\partial r}\left(r\,\frac{\partial p}{\partial r}\right) = \frac{1}{c^2}\frac{\partial^2 p}{\partial t^2}$$

[A] Determine the differential equation for the amplitude function $\psi(r)$, where the acoustic pressure is written in the following form:

$$p(r,t) = \psi(r)\,e^{j\omega t}$$

[B] Obtain the differential equation resulting from the change of variable:

$$\psi(r) = r^{-1/2}\varphi(r)$$

Your result should be the following:

$$\frac{d^2\varphi}{dr^2} + \left(\frac{1}{4r^2} + k^2\right)\varphi = 0$$

where the wave number $k = \omega/c = 2\pi/\lambda$ and λ is the wavelength. [C] when the quantity $2rk = 4\pi r/\lambda > \pi$, or $r/\lambda > \frac{1}{4}$, in the differential equation obtained in Part B, then the first term in parentheses will be less than about 10% of the second term, and the first term $(1/4r^2)$ may be neglected. This is called the *far-field limit* for the cylindrical sound waves. Solve the differential equation for the acoustic pressure $p(r, t)$ with the term $(1/4r^2)$ omitted for the sound wave moving outward (in the $+r$-direction) from a cylindrical source, with the condition that the peak magnitude of the sound wave at a distance r_o is p_o.

[D] Using the result from Part C, determine the rms acoustic pressure and sound pressure level at a distance 800 mm (31.5 in) from a cylindrical source, if $p_o = 2$ Pa and $r_o = 400$ mm. The sound wave is propagated in air at 25°C (77°F) with a frequency of 1000 Hz.

4-2. A plane sound wave in water at 20°C (68°F) has a sound pressure level for the incident wave of 105 dB and a frequency of 1000 Hz. The wave is normally incident on a very thick concrete wall. Determine [A] the transmission loss, dB; [B] the intensity of the incident wave and the intensity of the transmitted wave; and [C] the sound pressure level of the transmitted wave.

4-3. An acoustic liquid level meter is used to determine the water level in a container filled with liquid water and steam above the liquid. A plane sound wave moving in the water (density, 960 kg/m³ = 59.9 lb$_m$/ft³; sonic velocity, 1750 m/s = 5740 ft/sec) strikes the interface between the water and steam (density, 0.600 kg/m³ = 0.0375 lb$_m$/ft³; sonic velocity, 405 m/s = 1329 ft/sec) at normal incidence. The frequency of the sound wave is 500 Hz. The sound pressure level of the transmitted wave (in steam) is 46 dB. Determine [A] the intensity level of the transmitted sound wave in the steam, [B] the intensity level of the incident sound wave in the water, and [C] the rms acoustic velocity for the incident sound wave.

4-4. A plane sound wave in air ($\rho = 1.19$ kg/m³ = 0.0743 lb$_m$/ft³; $c = 345$ m/s = 1132 ft/sec) strikes a thick gypsum board panel ($\rho = 650$ kg/m³ = 40.6 lb$_m$/ft³; $c = 6750$ m/s = 22,100 ft/sec).

Determine the critical angle of incidence (if it exists) for total reflection of the sound wave.

4-5. A plane sound wave in water ($\rho = 1000 \text{ kg/m}^3 = 62.4 \text{ lb}_m/\text{ft}^3$; $c = 1440 \text{ m/s} = 4724 \text{ ft/sec}$) is incident on a thick plate of Plexiglas ($\rho = 1200 \text{ kg/m}^3 = 74.9 \text{ lb}_m/\text{ft}^3$; $c = 1800 \text{ m/s} = 5910 \text{ ft/sec}$) at an angle of 43.85°. Determine the angle of transmission in the Plexiglas and the critical angle of incidence (if it exists).

4-6. There was some concern that over-water flights of the supersonic transport (SST) would harm marine life. A plane sound wave from the aircraft in air ($\rho = 1.18 \text{ kg/m}^3 = 0.0737 \text{ lb}_m/\text{ft}^3$; $c = 347 \text{ m/s} = 1138 \text{ ft/sec}$) has a sound pressure level of 140 dB. The sound wave strikes the surface of the sea water ($\rho = 1022 \text{ kg/m}^3 = 63.8 \text{ lb}_m/\text{ft}^3$; $c = 1500 \text{ m/s} = 4920 \text{ ft/sec}$) at an angle of incidence of 12°. Determine the intensity of the transmitted wave in sea water and the magnitude of the rms acoustic pressure of the transmitted wave.

4-7. A plane sound wave in water (density, $998 \text{ kg/m}^3 = 62.3 \text{ lb}_m/\text{ft}^3$; sonic velocity, $1481 \text{ m/s} = 4859 \text{ ft/sec}$) has an rms acoustic pressure magnitude of 100 Pa and is incident at an angle of 45° on the bottom of a lake. The density of the lake bottom material is 2000 kg/m^3 ($124.9 \text{ lb}_m/\text{ft}^3$), the sonic velocity is 1000 m/s (3280 ft/sec). Determine [A] the angle at which the wave is transmitted into the lake bottom, [B] the transmission loss, and [C] the sound pressure level (dB) for the transmitted wave.

4-8. A plane sound wave in water ($Z_o = 1.48 \times 10^6$ rayl) is normally incident on a steel plate ($Z_o = 47.0 \times 10^6$ rayl; $c = 6100 \text{ m/s}$) having a thickness of 15 mm (0.591 in). The frequency of the wave is 500 Hz. The sound is transmitted through the steel plate into air ($Z_o = 407$ rayl) on the other side of the plate. The sound pressure level of the transmitted wave in the air is 60 dB. Determine [A] the transmission loss, [B] the sound pressure level of the incident wave in the water, and [C] the phase angle between the transmitted wave and the incident wave.

4-9. A large glass window, 50 mm (1.97 in) thick, is placed in the dolphin tank at Sea World. Seawater ($Z_o = 1.533 \times 10^6$ rayl) is on one side of the window and atmospheric air ($Z_o = 415$ rayl) is on the other side. A plane sound wave in the air strikes the window at normal incidence. The frequency of the sound wave is 2 kHz. The resulting sound pressure level transmitted into the seawater is 114.3 dB. Determine [A] the acoustic intensity of the incident wave and [B] the rms acoustic pressure of the incident wave.

4-10. A Lucite plate (density, $1200 \text{ kg/m}^3 = 74.9 \text{ lb}_m/\text{ft}^3$; sonic velocity, $1800 \text{ m/s} = 5910 \text{ ft/sec}$) has seawater ($Z_o = 1.539 \times 10^6$ rayl) on both sides of the plate. A sound wave having a frequency of 5 kHz is normally incident on one side of the plate. Determine the thickness of the plate such that the sound power transmission coefficient through the plate is unity.

4-11. The hull of an oil tanker is constructed of 9% Ni steel (density, $8000 \text{ kg/m}^3 = 499 \text{ lb}_m/\text{ft}^3$; sonic velocity, $5000 \text{ m/s} = 16{,}400 \text{ ft/sec}$) having a thickness of 83.3 mm (3.28 in). Air ($\rho = 1.270 \text{ kg/m}^3 = 0.0793 \text{ lb}_m/\text{ft}^3$; $c = 334 \text{ m/s} = 1096 \text{ ft/sec}$) at 5°C (41°F) is on one side of the hull and oil ($\rho = 850 \text{ kg/m}^3 = 53.1 \text{ lb}_m/\text{ft}^3$; $c = 1353 \text{ m/s} = 4439 \text{ ft/sec}$) is on the other side. A plane sound wave having a frequency of 5 kHz is normally incident on the air-side of the hull. The sound pressure level of the transmitted wave in the oil is 81.2 dB. Determine [A] the transmission loss, [B] the sound pressure level of the incident wave in the air, and [C] the phase angle between the transmitted wave and the incident wave.

4-12. The rms acoustic pressure for a sound wave in the water (density, $1000 \text{ kg/m}^3 = 62.4 \text{ lb}_m/\text{ft}^3$; sonic velocity, $1440 \text{ m/s} = 4724 \text{ ft/sec}$) in a large swimming pool is 2.233 Pa. The frequency of the sound wave is 5 kHz. The sound is normally incident on the 155 mm (6.10 in) thick concrete bottom ($\rho = 2600 \text{ kg/m}^3 = 162.3 \text{ lb}_m/\text{ft}^3$; $c = 3100 \text{ m/s} = 10{,}170 \text{ ft/sec}$) of the pool. The sound wave is transmitted through the concrete into the soil ($\rho = 1800 \text{ kg/m}^3 = 112.4 \text{ lb}_m/\text{ft}^3$; $c = 1600 \text{ m/s} = 5250 \text{ ft/sec}$) beneath the pool. Determine [A] the rms acoustic pressure of the sound wave transmitted into the soil, [B] the transmission loss, and [C] the phase angle between the transmitted and incident wave.

4-13. After an oil tanker accident, a layer of oil (density, $865 \text{ kg/m}^3 = 54.0 \text{ lb}_m/\text{ft}^3$; sonic velocity, $1600 \text{ m/s} = 5249 \text{ ft/sec}$) 16 mm (0.630 in) thick is formed over seawater ($\rho = 1025 \text{ kg/m}^3 = 64.0 \text{ lb}_m/\text{ft}^3$; $c = 1500 \text{ m/s} = 4920 \text{ ft/sec}$). The air above the oil slick is at 25°C (77°F). A plane sound wave having a frequency of 12.5 kHz and an intensity level of 93.3 dB is generated in the seawater, and the sound wave strikes the interface at normal incidence. Determine [A] the intensity of the transmitted wave and [B] the sound pressure level of the transmitted wave in air.

4-14. Determine the transmission loss, using the "exact" method, for a steel plate, 500 mm by 750 mm (19.7 in by 29.5 in) and 1.50 mm (0.059 in) thick, for the following frequencies: [A] 125 Hz,

[B] 500 Hz, and [C] 16 kHz. Atmospheric air at 25°C (77°F) is on both sides of the plate.

4-15. A circular glass plate has a diameter of 350 mm (13.78 in) and a thickness of 10 mm (0.394 in). The edge of the plate is clamped, and air at 25°C (77°F) is on both sides of the plate. Determine the transmission loss for the plate for a frequency of [A] 250 Hz and [B] 1000 Hz.

4-16. A gypsum panel has dimensions of 1.20 m by 2.4 m (3.94 ft by 7.87 ft) and a thickness of 20 mm (0.787 in). The air around the panel is at 300K (26.8°C or 80°F). Using the "exact" method, determine the transmission loss for the panel for a frequency of [A] 250 Hz and [B] 4 kHz.

4-17. It is desired to construct a concrete barrier 2.00 m (6.56 ft) high and 4.00 m (13.12 ft) wide such that the transmission loss for random incidence is 37 dB for a frequency of 100 Hz. Air at 20°C (68°F) and 101.3 kPa (14.7 psia) is present on both sides of the wall. The physical properties of the concrete are density, 2300 kg/m^3 (143.6 lb$_m$/ft^3); sonic velocity, 3400 m/s (11,150 ft/sec); Young's modulus, 20.7 GPa (3.00 × 10^6 psi); Poisson's ratio, 0.130; and damping coefficient, 0.020. Assuming that the barrier operates in the mass-controlled regime, determine the required thickness of the barrier. With the calculated value of barrier thickness, verify that the frequency of 100 Hz falls in Region II.

4-18. A glass window has a thickness of 8.30 mm (0.327 in). Using the "approximate" method, determine the transmission loss as a function of frequency in octaves over the range from 63 Hz to 8000 Hz. Present the results in a table and as a graph of TL vs. frequency (log scale).

4-19. A wall having dimensions of 3.00 m (9.84 ft) by 12 m (39.37 ft) has a transmission loss of 25 dB. If a glass window, 1.00 m (3.28 ft) by 2.00 m (6.56 ft) with a transmission loss of 5 dB, is placed in the wall, determine the transmission loss for the wall with the window installed.

4-20. A hollow panel consists of two 6 mm (0.236 in) thick plywood sheets separated by a 30 mm (1.181 in) thick air space. The properties of the plywood are density, 600 kg/m^3 (37.5 lb$_m$/ft^3); sonic velocity, 3100 m/s (10,170 ft/sec); surface absorption coefficient, $\alpha = 0.10$; and damping coefficient, $\eta = 0.03$. The air in the space between the two panels is at 25°C (77°F). Determine the transmission loss for the composite panel for a frequency of [A] 125 Hz and [B] 1000 Hz.

4-21. The transmission loss curve from 1/3 octave band measurements for a hollow concrete block wall 200 mm (7.87 in) thick is given in Table 4-10. Determine the sound transmission class (STC) rating for the wall. Make a plot of the measured transmission loss, dB, vs. frequency (log scale), along with a plot of the STC curve.

4-22. An aluminum sheet, 400 mm (15.75 in) wide by 600 mm (23.62 in) high, with a thickness of 3.0 mm (0.118 in), has aluminum ribs attached to improve heat transfer from the surface. The ribs are 4.5 mm (0.177 in) high, 3.0 mm (0.118 in) wide, and are spaced 13.5 mm (0.531 in) on centers, as shown in Fig. 4-18. The ribs are oriented parallel to the short dimension of the plate. Air at 25°C (77°F) is on both sides of the panel. Determine the transmission loss of the rib-stiffened panel at [A] 500 Hz, [B] 2000 Hz, and [C] 8000 Hz.

4-23. The following properties are found for xenon gas at 300K (26.8°C or 80°F) and 101.3 kPa (14.7 psia):
viscosity, $\mu = 23.3\,\mu\text{Pa-s}$ ($0.0564\,\text{lb}_m/\text{ft-hr}$)
specific heat, $c_p = 158.4\,\text{J/kg-K}$ ($0.0378\,\text{Btu/lb}_m\text{-}°\text{R}$)
specific heat ratio, $c_p/c_v = 1.667$
density, $\rho = 5.334\,\text{kg/m}^3$ ($0.333\,\text{lb}_m/\text{ft}^3$)
sonic velocity, $c = 178.0\,\text{m/s}$ ($584\,\text{ft/sec}$)
Prandtl number, $\text{Pr} = 0.727$
For a sound wave having a frequency of 4 kHz in xenon gas at 300K, determine [A] the classical relaxation time and [B] the classical attenuation coefficient in units of Np/m and dB/km.

4-24. The properties of neon gas at 300K (26.8°C or 80°F) and 101.3 kPa (14.7 psia) are as follows:
viscosity, $\mu = 31.74\,\mu\text{Pa-s}$ ($0.0768\,\text{lb}_m/\text{ft-hr}$)

TABLE 4-10 Data for Problem 4-21.

1/3 Octave band center frequency, Hz	Transmission loss, TL, dB	1/3 Octave band center frequency, Hz	Transmission loss, TL, dB
125	24	800	54
160	25	1,000	56
200	28	1,250	57
250	33	1,600	58
315	39	2,000	59
400	41	2,500	58
500	47	3,150	56
630	50	4,000	58

Transmission of Sound

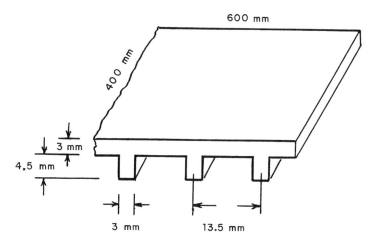

FIGURE 4-18 Physical system for Problem 4-22.

specific heat, $c_p = 1030$ J/kg-k $(0.246$ Btu/lb$_m$-°R)
specific heat ratio, $c_p/c_v = 1.667$
density, $\rho = 0.817$ kg/m^3 $(0.0510$ lb$_m$/ft$^3)$
sonic velocity, $c = 454$ m/s $(1490$ ft/sec)
Prandtl number, Pr $= 0.663$

For a sound wave having a frequency of 8 kHz in argon gas at 300K, determine [A] the classical relaxation time and [B] the classical attenuation coefficient in units of Np/m and dB/km.

4-25. A siren operates in atmospheric air (density, 1.160 kg/m^3 = 0.0724 lb$_m$/ft^3; sonic velocity, 345 m/s = 1132 ft/sec). The frequency of the sound wave emitted from the siren, which may be treated as a spherical source of sound, is 6 kHz. Determine the required acoustic power output of the siren (watts) and the acoustic power level of the siren for each of the following conditions, if the siren produces at intensity level of 60 dB at a distance of 1.00 km (3281 ft) from the siren: [A] for no attenuation ($\alpha \approx 0$) of energy in the air; [B] for only the classical contribution considered, for which $\alpha/f^2 = 20 \times 10^{-12}$ Np/m-Hz^2; and [C] for moist air at 60% relative humidity, for which the attenuation coefficient is $\alpha = 0.0056$ Np/m.

4-26. A small jet engine emits noise at a frequency of 16.34 kHz, and the engine may be treated as a spherical source of sound. A worker at a distance of 320 m (1050 ft) from the aircraft experiences a sound pressure level of 80 dB. The sound is transmitted through air at -6°C (21°F) at 101.3 kPa (14.7 psia), for which $\rho_o = 1.322$ kg/m^3 $(0.0825$ lb$_m$/ft$^3)$ and $c = 328$ m/s $(1076$ ft/sec). The attenuation coefficient for the air is given by $(\alpha/f^2) = 13.50 \times 10^{-12}$ Np/m-Hz^2. Determine [A] the sound power level at the source and [B] the sound pressure level at a distance of 450 m (1476 ft) from the engine, including the effect of attenuation in the air.

4-27. A natural gas compressor emits noise at a frequency of 8 kHz. The sound power level for the compressor is 170 dB, and the directivity factor for the compressor is $Q = 2$. The noise is transmitted through atmospheric air at 20°C (68°F) with relative humidity of 50%. Determine the sound pressure level at a distance of 535 m (1755 ft) from the compressor, including the effect of air attenuation.

4-28. A novice hunter becomes lost in a large open area, and he begins to call for help. The acoustic power level of his frantic voice is 98.6 dB at a frequency of 1000 Hz. The ambient air is at 15°C (59°F) with a relative humidity of 10%. The desperate hunter has a directivity factor of $Q = 2$. Determine the sound pressure level associated with his screams at a distance of 1.60 km (0.994 miles) from the lost hunter, including the effect of air attenuation. If the sound pressure level at the threshold of hearing is about 0 dB, what are the prospects of the hunter's shouts being heard by his companions located at a distance of 1.60 km from him?

REFERENCES

ANSI. 1978. ANSI standard method for calculating the absorption of sound in the atmosphere, ANSI S1.26/ASA23-1978. American National Standards Institute, Inc., New York.

ASTM. 1983. Laboratory measurement of airborne sound transmission loss of building partitions, ASTM E 90-83. American Society for Testing and Materials, Philadelphia.

ASTM. 1984. Measurement of airborne sound insulation in buildings, ASTM E 336-84. American Society for Testing and Materials, Philadelphia.

Beranek, L. L. 1960. *Noise Reduction*, p. 301. McGraw-Hill, New York.

Beranek, L. L. 1971. *Noise and Vibration Control*, pp. 283, 311, 317. McGraw-Hill, New York.

Beranek, L. L. and Vér, I. S. 1992. *Noise and Vibration Control Engineering*, p. 293. John Wiley and Sons, New York.

Berendt, R. D., Winzer, G. E., Burroughs, C. B. 1967. Airborne, impact, and structure borne noise control in multifamily dwellings. U.S. Dept. of Housing and Urban Development, Washington, DC.

Faulkner, L. L. 1976. *Handbook of Industrial Noise Control*, p. 159. Industrial Press, Inc., New York.

Kinsler, L. E., Frey, A. R., Coppens, A. B., and Sanders, J. V. 1982. *Fundamentals of Acoustics*, 3rd edn, p. 129. John Wiley and Sons, New York.

Maidanik, G. 1962. Response of ribbed panels to reverberant acoustic fields. *J. Acoust. Soc. Am.* 34: 640–647.

Norton, M. P. 1989. *Fundamentals of Noise and Vibration Analysis for Engineers*, pp. 111–119. Cambridge University Press, New York.

Pierce, A. D. 1981. *Acoustics: An Introduction to Its Physical Principles and Applications*, pp. 65, 558. McGraw-Hill, New York.

Randall, R. H. 1951. *An Introduction to Acoustics*. pp. 40–42. Addison-Wesley, Reading, MA.

Reynolds, D. D. 1981. *Engineering Principles of Acoustics*, p. 310. Allyn and Bacon, Boston.

Roark, R. J. and Young, W. C. 1975. *Formulas for Stress and Strain*, 5th edn, p. 579. McGraw-Hill, New York.

Sharp, B. H. 1973. A study of techniques to increase the sound insulation of building elements. Wyle Laboratories Report WR-73-5, HUD Contract No. H-1095.

Sutherland, L. C., Piercy, J. E., Bass, H. E., and Evans, L. B. 1974. A method for calculating the absorption of sound in the atmosphere. 84th Meeting of the Acoustical Society of America, St. Louis, MO.

ter Haar, D. 1954. *Elements of Statistical Mechanics*, pp. 22–25, 46–50. Rinehart, New York.

Ugural, A. C. 1999. *Stresses in Plates and Shells*, p. 257. McGraw-Hill, New York.

van Wylen, G., Sonntag, R., and Borgnakke, C. 1994. *Fundamentals of Classical Thermodynamics*, pp. 666–668. John Wiley, New York.

Watters, B. G. 1959. The transmission loss of some masonry walls. *J. Acoust. Soc. Am.* 31: 898–911.

5
Noise Sources

When one carries out an acoustic design, it is often necessary to estimate the sound pressure level that may be expected to be produced from a particular noise source. If experimental data are not available for the noise source, predictive relationships are needed. This situation arises when a new system or installation is being developed. In addition, correlations for the noise generated by a source, such as an item of machinery, provide information for design which enables the designer to reduce the noise output from the source by suitable modification of such factors as size, operational speed, etc.

In this chapter, we will examine methods that may be used to predict the noise emitted from several mechanical systems. In general, it is important to predict not only the overall noise level but also the noise spectrum or the sound pressure level in each octave band. From a knowledge of the sound pressure level spectrum (sound pressure level vs. octave band center frequency), one may predict the A-weighted sound level, which is important in determining compliance with noise regulations.

5.1 SOUND TRANSMISSION INDOORS AND OUTDOORS

The most useful correlation for noise emitted from a system is the correlation of the sound power level L_W as a function of known or measurable

Noise Sources

characteristics of the system. The corresponding sound pressure level L_p produced by the noise emission depends on distance from the source, whether the source is located indoors or outdoors, and other factors. Let us develop two general relationships that are needed for prediction of the sound pressure level when the sound power level can be determined.

For sound transmission outdoors, the acoustic intensity for a sound wave, not necessarily a spherical wave, is given by Eq. (4-128) with the directivity factor Q included:

$$I = \frac{QW_o}{4\pi r^2} e^{-2\alpha r} = \frac{p^2}{\rho_o c} \tag{5-1}$$

If we solve for the rms acoustic pressure (or p^2) and include the reference pressure and power terms, we obtain the following expression, where $m = 2\alpha$:

$$\frac{p^2}{(p_{ref})^2} = \frac{W_o Q e^{-mr} \rho_o c W_{ref}}{W_{ref} r^2 (4\pi)(p_{ref})^2} \tag{5-2}$$

Taking log base 10 of both sides of Eq. (5-2) and multiplying by 10, we obtain the following relationship in terms of levels:

$$L_p = L_W + 10\log_{10}(Q) - 10\log_{10}(r^2) + 10\log_{10}(e^{-mr})$$
$$+ 10\log_{10}\left(\frac{\rho_o c W_{ref}}{4\pi p_{ref}^2}\right) \tag{5-3}$$

If we introduce the definition of the directivity index DI from Eq. (2-41), we obtain the following expression:

$$L_p = L_W + DI - 20\log_{10}(r) + 10\log_{10}(e^{-mr}) - 10\log_{10}\left(\frac{4\pi p_{ref}^2}{\rho_o c W_{ref}}\right) \tag{5-4}$$

The characteristic impedance for atmospheric air at 300K (27°C or 80°F) and 101.325 kPa (14.696 psia) is $Z_o = \rho_o c = 408.6$ rayl. This value may be used to evaluate the last term in Eq. (5-4):

$$\frac{4\pi p_{ref}^2}{\rho_o c W_{ref}} = \frac{(4\pi)(20 \times 10^{-6})^2}{(408.6)(10^{-12})} = 12.30 \, \text{m}^{-2}$$

$$10\log_{10}(12.30) = 10.9 \, \text{dB}$$

This constant value may be used for ±0.1 dB accuracy if the air temperature is between about 293K (20°C or 68°F) and 307K (34°C or 93°F). For air temperatures outside this range or for materials other than air, the value of the constant must be calculated.

For sound transmitted outdoors in air around 300K, the following expression may be used to estimate the sound pressure level L_p for a noise source having a sound power level L_W:

$$L_p = L_W + \text{DI} - 20\log_{10}(r) - 4.343 mr - 10.9 \tag{5-5}$$

The distance from the sound source r must be expressed in meters in Eq. (5-5), and the energy attenuation coefficient must have units of m^{-1}. The term involving the energy attenuation coefficient is usually negligible for lower frequencies and smaller distances, as discussed in Sec. 4.13.

For sound transmission indoors, the sound pressure level and sound power level are related by the following expression (developed in Chapter 7):

$$L_p = L_W + 10\log_{10}\left(\frac{4}{R} + \frac{Q}{4\pi r^2}\right) + 10\log_{10}\left(\frac{\rho_o c W_{\text{ref}}}{p_{\text{ref}}^2}\right) \tag{5-6}$$

The quantity R is called the room constant and is given by:

$$R = \frac{S_o[\bar{\alpha} + (4mV/S_o)]}{1 - \bar{\alpha} - (4mV/S_o)} \tag{5-7}$$

where S_o is the total surface area of the room, m^2; $\bar{\alpha}$ is the average surface absorption coefficient; and V is the volume of the room, m^3.

For the special case of air at 101.325 kPa (14.696 psia) and 300K (27°C or 80°F), the numerical value of the last term in Eq. (5-6) may be evaluated:

$$\frac{\rho_o c W_{\text{ref}}}{p_{\text{ref}}^2} = \frac{(408.6)(10^{-12})}{(20 \times 10^{-6})^2} = 1.0215$$

$$10\log_{10}(1.0215) = 0.1 \text{ dB}$$

In the following sections, we will consider techniques for estimation of the sound power level that may be used in Eqs (5-4) and (5-6) to estimate the sound pressure level generated by various noise sources.

5.2 FAN NOISE

There are several types of fans used in industrial and residential applications. The fans may be classified according to the nature of flow through the fan and by the blade geometry (Avallone and Baumeister, 1987). Generally, the noise signature is different for each type of fan. Various fan types and typical applications are as follows:

(a) *Centrifugal fan, airfoil blades.* The airfoil blades have backward-curved chord lines, with the leading edge of the airfoil pointing forward and the trailing edge pointing backward with respect to the direction of rotation of the fan. All centrifugal fans have a volute or scroll-type housing. This fan is used in large heating, ventilating and air conditioning systems in which relatively clean air is handled.

(b) *Centrifugal fan, backward curved blades (BCB)*. The blades are flat plates with uniform thickness. The leading edge points in a direction opposite to the rotation of the fan. BCB fans are used for general ventilating and air conditioning applications. The fan efficiency is somewhat higher than that of the other types of centrifugal fans. The fan speed must be higher for a given flow rate than that of other centrifugal fans.

(c) *Centrifugal fan, radial blades*. The blades are flat plates of uniform thickness, oriented along the radial direction of the fan cage. This fan is often used in material handling systems in industrial applications in which sand, wood chips, or other small particles are present in the air.

(d) *Centrifugal fan, forward curved blades (FCB)*. The blades are shallow and curved, such that both the leading and trailing edges point in the direction of rotation. The fan efficiency is somewhat lower than that of the other centrifugal fans. As a result, this fan is used for low pressure rise, low speed applications, including domestic furnaces and packaged home air conditioning units.

(e) *Tubular centrifugal fan*. These fans use a tubular casing so that both the entering and leaving flow is in the axial direction. The blades may be either backward curved or airfoil type. The fan is often used for low-pressure return-air systems in heating and ventilating applications.

(f) *Vaneaxial fan*. These fans usually have blades of airfoil design, which allows the fan to be used in the medium to high pressure rise range at relatively high efficiency. Guide vanes are located in the annular space between the casing and the inner cylinder. Noise generation for vaneaxial fans is generally higher than that of the centrifugal fans. Typical applications for the vaneaxial fan include fume exhaust, paint spray booths, and drying ovens.

(g) *Tubeaxial fan*. This fan is similar to the vaneaxial fan, except that straightening vanes are not used. The fan is generally used in low to medium pressure rise applications.

(h) *Propeller fan*. The fan blades are usually wider than those of the vaneaxial fan. The propeller fan is mounted in a ring with no attached ductwork. The pressure rise is relatively small, but the fan can handle a large volume flow. Applications for this type of fan include roof exhaust systems and induced-draft cooling towers.

There are several paths through which noise may be radiated from a fan, including (a) sound power radiated directly from the fan outlet and/or inlet, if there is no attached ductwork to the inlet and/or outlet; (b) sound radiated through the fan housing; and (c) sound induced by vibrations transmitted from the fan through the fan supports to the adjoining structure. These paths are illustrated in Fig. 5-1. With proper vibration isolation by the support, as discussed in Chapter 9, it is possible to reduce the noise from vibrations to a negligible level, compared with the first two contributions given above.

If we denote the sound power level generated internal to the fan by L_W, then the sound power level for sound radiated out of the inlet and/or outlet openings or radiated down the attached ductwork may be estimated from the following expressions (note that if ductwork is attached to an inlet or outlet, there is no sound radiated into the surroundings from the inlet or outlet source):

$$L_W(\text{outlet}) = L_W - 3\,\text{dB} \tag{5-8}$$

$$L_W(\text{inlet}) = L_W - 3\,\text{dB} \tag{5-9}$$

The sound power level for sound transmitted through the fan housing is related to the transmission loss TL of the fan housing:

$$L_W(\text{housing}) = L_W - \text{TL} \tag{5-10}$$

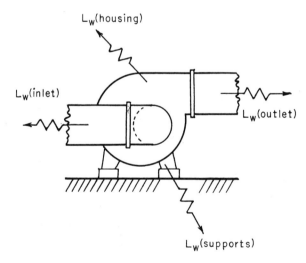

FIGURE 5-1 Fan noise paths.

Noise Sources

The noise generated internally by each of the fans mentioned previously is composed of two components: broadband noise generated by vortex shedding from the fan blades and a discrete tone (blade tone) produced as the blade passes by the inlet or outlet opening of the fan.

The sound power level of noise generated by the fan for any octave band may be estimated from the following correlation (Graham, 1972):

$$L_W = L_W(B) + 10 \log_{10}(Q/Q_o) + 20 \log_{10}(P/P_o) + B_T \qquad (5\text{-}11)$$

The term $L_W(B)$ is the basic sound power level, and is given in Table 5-1 for each of the fan types discussed previously; Q is the volumetric flow rate through the fan, and Q_o is a reference volumetric flow rate, $0.47195 \, \text{dm}^3/\text{s} = 1 \, \text{ft}^3/\text{min}$; P is the pressure rise through the fan, and P_o is a reference pressure rise, $248.8 \, \text{Pa} = 1 \, \text{in} \, \text{H}_2\text{O}$; and B_T is the blade tone component, which is zero except for the octave band in which the blade pass frequency lies. For this one octave band, the value of the blade tone component is given in Table 5-1. The blade pass frequency f_B is the number of times a blade passes one of the fan openings and is given by the following expression:

$$f_B = n_r N_b \qquad (5\text{-}12)$$

The quantity n_r is the rotational speed of the fan, rev/sec, and N_b is the number of blades on the fan.

The sound power level calculated from Eq. (5-11) is the sound power level generated internally in a particular octave band for the fan only. The

TABLE 5-1 Basic Sound Power Level Spectrum $L_W(B)$ for Fans

Fan type	Blade tone, B_T, dB	Octave band center frequency, Hz							
		63	125	250	500	1,000	2,000	4,000	8,000
Centrifugal fans:									
Airfoil blade	3	35	35	34	32	31	26	18	10
BCB	3	35	35	34	32	31	26	18	10
Radial blade	5–8	48	45	45	43	38	33	30	29
FCB	2	40	38	38	34	28	24	21	15
Tubular	4–6	46	43	43	38	37	32	28	25
Vaneaxial	6–8	42	39	41	42	40	37	35	25
Tubeaxial	6–8	44	42	46	44	42	40	37	30
Propellor	5–7	51	48	49	47	45	45	43	31

Source: Graham (1972). By permission of *Sound and Vibration*, Acoustical Publications, Inc.

additional noise due to the fan motor and drive system is not included in Eq. (5-11).

Example 5-1. A forward curved blade (FCB) centrifugal fan operates at a speed of 552 rpm against a pressure of 190 Pa (0.7626 in H$_2$O) to deliver 1.80 m^3/s (3814 cfm) of air. The fan is located outdoors (air at 300K or 80°F), and the fan has both inlet and outlet ducts, so that noise is radiated only through the housing of the fan. The transmission loss for the fan is given in Table 5-2. The directivity index for the fan may be taken as DI = 3 dB for all frequencies. Determine the overall sound pressure level produced by the fan at a distance of 3 m (9.8 ft) from the fan.

The blade pass frequency is found from Eq. (5-12):

$$f_B = (552/60)(64 \text{ blades}) = 589 \text{ Hz}$$

This frequency lies in the 500 Hz octave band (354–707 Hz), and the blade tone component for the 500 Hz octave band is found in Table 5-1 for an FCB centrifugal fan:

$B_T = 2 \text{ dB}$ in the 500 Hz octave band

$B_T = 0 \text{ dB}$ for all other octave bands

The internal noise generation sound power level can be calculated from Eq. (5-11). For the 500 Hz octave band, we obtain the following value:

$$L_W = 34 + 10 \log_{10}(1800/0.47195) + 20 \log_{10}(190/248.8) + 2$$

$$L_W = 34 + 35.8 + (-2.3) + 2 = 34 + 33.5 + 2 = 69.5 \text{ dB}$$

For the 250 Hz octave band, the sound power level is as follows:

$$L_W = 38 + 33.5 + 0 = 71.5 \text{ dB}$$

The corresponding values for the other octave bands are given in Table 5-2.

TABLE 5-2 Solution for Example 5-1

	Octave band center frequency, Hz						
	63	125	250	500	1,000	2,000	4,000	8,000
L_W, dB	73.5	71.5	71.5	69.5	61.5	57.5	54.5	48.5
TL, dB	15	21	27	33	39	40	40	40
L_W(housing), given	58.5	50.5	44.5	36.5	22.5	17.5	14.5	8.5
L_p(octave band), dB	41.1	33.1	27.1	19.1	5.1	0.1	−2.9	−8.9

Noise Sources

The sound power radiated from the fan through the housing is found for each octave band from Eq. (5-10). For the 500 Hz octave band, we find the following value:

$$L_W(\text{housing}) = L_W - \text{TL} = 69.5 - 33 = 36.5 \, \text{dB}$$

The sound pressure level in the 500 Hz octave band is found from Eq. (5-4). The atmospheric attenuation is negligible, because the distance ($r = 3$ m) is small:

$$L_p = L_W(\text{housing}) + \text{DI} - 20\log_{10}(r) - 10.9$$

$$L_p = 36.5 + 3 - 20\log_{10}(3.00) - 10.9 = 36.5 + 3 - 9.5 - 10.9$$

$$= 36.5 - 17.4$$

$$L_p = 19.1 \, \text{dB}$$

The other octave band sound pressure levels are shown in Table 5-2.

The overall sound pressure level is found by combining the octave band values of acoustic energy, as discussed in Sec. 2-8.

$$L_p(\text{overall}) = 10\log_{10}[\Sigma 10^{L(\text{OB})/10}]$$

$$L_p(\text{overall}) = 10\log_{10}(10^{4.11} + 10^{3.31} + 10^{2.71} + \cdots) = 41.9 \, \text{dB}$$

Suppose the inlet duct were removed, so that sound could be radiated out the inlet opening of the fan. For the 500 Hz octave band, the sound power radiated out the inlet is given by Eq. (5-9):

$$L_W(\text{inlet}) = 69.5 - 3 = 66.5 \, \text{dB}$$

The total power radiated from the fan in the 500 Hz octave band would have the following value:

$$L_W = 10\log_{10}(10^{3.65} + 10^{6.65}) = 66.5 \, \text{dB}$$

In this case, the effect of sound radiated through the housing is negligible.

5.3 ELECTRIC MOTOR NOISE

The noise generated by a single electric motor is usually not excessive; however, a large number of electric motors may be present in a particular location. In this case, the total noise generated by several motors may be significant.

The noise radiated from an electric motor results from several physical factors, including the following:

(a) *Windage noise generated by the motor cooling fan.* As for the case of all fans, as discussed in Sec. 5.2, the windage noise involves a pure tone component caused by the fan blades as they pass by stationary members, and broadband noise caused by turbulent eddies from the fan blades.

(b) *Rotor-slot noise generated by open slots in the motor rotor.* This noise is tonal in nature with a frequency equal to the product of the rotational speed and number of slots in the rotor. Rotor-slot noise may be made negligible by filling the slots with epoxy or other filler material.

(c) *Rotor–stator noise caused by rotor and stator slot magnetomotive force interactions.*

(d) *Noise produced by the changing magnetic flux density.* Dimensional changes produced by time-varying magnetic flux in the motor produce noise from the rotor element. The frequency of this noise component is equal to twice the power line frequency.

(e) *Dynamic unbalance noise.* This noise source indicates problems in the motor and can be corrected by dynamically balancing the motor.

(f) *Bearing noise.*

There are two primary types of electric motors, as classified by the type of motor cooling. The *drip-proof* (DRPR) *motor* cools itself by inducing a flow of air from around the motor and circulating the air over the electric conductors. The *totally enclosed fan-cooled* (TEFC) *motor* uses an internal fan to accomplish motor cooling.

Data for the A-weighted sound power level may be correlated by the following expressions. Note that when the A-weighted sound power level is used in Eq. (5-5) or Eq. (5-6), the resulting sound pressure level is the A-weighted sound level.

For drip-proof motors, the A-weighted sound power level can be correlated in terms of the rated motor horsepower (hp) and the rotational speed n_r (rpm) of the motor:

$$L_W(A) = 65 \, \text{dBA} \quad \text{(for hp} < 7 \, \text{hp)} \tag{5-13}$$

$$L_W(A) = 20 \log_{10}(\text{hp}) + 15 \log_{10}(n_r) - 3 \quad \text{(for hp} \geq 7 \, \text{hp)} \tag{5-14}$$

For TEFC motors, a similar correlation has been found:

$$L_W(A) = 78 \, \text{dBA} \quad \text{(for hp} < 5 \, \text{hp)} \tag{5-15}$$

$$L_W(A) = 20 \log_{10}(\text{hp}) + 15 \log_{10}(n_r) + 13 \quad \text{(for hp} \geq 5 \, \text{hp)} \tag{5-16}$$

The overall sound power levels may be estimated from the A-weighted values through the following conversion, which depends on the rated horsepower of the motor:

Noise Sources

TABLE 5-3 Conversion Factors CF_1 (dB) to Convert from the A-weighted Sound Power Level for an Electric Motor to the Octave Band Sound Power Levels

Motor Size	Octave band center frequency, Hz							
	63	125	250	500	1,000	2,000	4,000	8,000
1 to 250 hp	16	12	8	4	4	8	12	16
300 to 400 hp	21	15	9	3	3	8	15	22
450 hp and above	19	13	7	3	3	8	14	22

$$
\begin{aligned}
&\text{1 hp to 250 hp,} && L_W = L_W(A) + 1.1 \\
&\text{251 hp to 300 hp,} && L_W = L_W(A) + 1.2 \\
&\text{301 hp to 400 hp,} && L_W = L_W(A) + 1.3 \\
&\text{401 hp to 450 hp,} && L_W = L_W(A) + 1.5 \\
&\text{451 hp and larger,} && L_W = L_W(A) + 1.7
\end{aligned}
$$

To convert from the overall A-weighted sound power level to the octave band sound power levels, the conversion factor given in Table 5-3 may be used:

$$L_W(\text{octave band}) = L_W(A) - CF_1 \tag{5-17}$$

5.4 PUMP NOISE

Standard-line pumps are generally not severe noise sources when the pumps are operated at their rated speed and capacity. Noise from pumps involves both hydraulic and mechanical sources. Some of the sources of noise in pumps include cavitation, fluid pressure fluctuations, impact of solid surfaces, and dynamic imbalance of the rotor. The hydraulic sources are usually the more important noise generators, unless there is a mechanical problem in the pump, such as imbalance of the rotor (Heitner, 1968).

Noise may be radiated from a pump through the surrounding air or through the piping and support structure for the pump. As was the case for fans, the structureborne noise may be reduced to negligible levels by proper vibration isolation of the pump.

The sound power level for airborne noise from a pump may be estimated from the following correlation within $\pm 3\,\text{dB}$:

$$L_W = K_o + 10 \log_{10}(\text{hp}) \tag{5-18}$$

The quantity hp is the rated horsepower for the pump. For pumps having a rated speed of 1600 rpm or higher, the values of the constant K_o are as follows:

$K_o = 98\,\text{dB}$ for a centrifugal pump

$K_o = 103\,\text{dB}$ for a screw pump

$K_o = 108\,\text{dB}$ for a reciprocating pump

For pumps operating at speeds below 1600 rpm, subtract 5 dB from these values to obtain the K_o for Eq. (5-18).

To convert from the overall sound power level to the octave band sound power levels, the conversion factor given in Table 5-4 may be used:

$$L_W(\text{octave band}) = L_W - CF_2 \quad (5\text{-}19)$$

Example 5-2. An 18 kW DRPR electric motor is used to drive a 20 hp centrifugal pump at 1800 rpm, as shown in Fig. 5-2. The directivity factor for the system is $Q = 4$ for all frequencies. The system is located in a room having a room constant of 50 m² (538 ft²). Determine the overall sound pressure level at a distance of 2.00 m (6.56 ft) from the system.

The horsepower rating of the motor is found by conversion of units:

$$\text{hp} = (18\,\text{kW})(1.3410\,\text{hp/kW}) = 24.14\,\text{hp}$$

The A-weighted sound power level for the motor is found from Eq. (5-14):

$$L_W(A) = 20\log_{10}(24.14) + 15\log_{10}(1800) - 3$$

$$L_W(A) = 27.7 + 48.8 - 3 = 73.5\,\text{dBA}$$

The non-weighted sound power level for the electric motor (hp < 250 hp) is found as follows:

$$L_{W,m} = L_W(A) + 1.1 = 73.5 + 1.1 = 74.6\,\text{dB}$$

TABLE 5-4 Conversion Factors CF_2 (dB) to Convert from the Overall Sound Power Level for a Pump to the Octave Band Sound Power Levels

	Octave band center frequency, Hz							
	63	125	250	500	1,000	2,000	4,000	8,000
CF_2, dB	10	9	9	8	6	9	12	17

Noise Sources

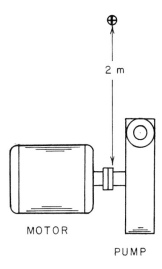

Figure 5-2 Schematic for Example 5-2.

The sound power level for the pump is found from Eq. (5-18):

$$L_{W,p} = 98 + 10\log_{10}(20) = 98 + 13.0 = 110.0\,\text{dB}$$

The total sound power level for the combined motor and pump system is found by combining the power levels:

$$L_W = 10\log_{10}[10^{7.46} + 10^{11.00}] = 10\log_{10}(1.0003 \times 10^{11}) = 110.0\,\text{dB}$$

We note that the effect of the electric motor noise is negligible in this case.

The overall sound pressure level due to the pump and motor noise may be calculated from Eq. (5-6):

$$L_p = 110.0 + 10\log_{10}\left[\frac{4}{50} + \frac{(4.0)}{(4\pi)(2.00)^2}\right] + 0.1$$

$$L_p = 110.0 + 10\log_{10}(0.0800 + 0.0796) + 0.1 = 110.0 - 8.0 + 0.1$$

$$L_p = 102.1\,\text{dB}$$

5.5 GAS COMPRESSOR NOISE

Many gas compressors are not designed with low noise emission as the primary design criterion. Such factors as high efficiency, durability and price are usually more important initially than low noise levels. Noise control procedures are often applied after the compressor has been constructed, rather than implemented during the design of the unit. There are some design factors that are usually not within the control of the designer, however. These factors include (a) compressor power input (determined by the required pressure rise and flow rate through the compressor), (b) the fluid turbulence levels, and (c) the type of gas compressed.

Some design factors that can be adjusted in the design stage for the compressor include (a) rotor–stator interaction, (b) impeller–diffuser spacing, (c) compressor rotational speed, and (d) number of compression stages (Diehl, 1972).

When a rotor blade passes a stator blade in a compressor, the gas is given an impulse, and noise is generated from this impulsive action. The peak in the noise generation curve occurs at the blade pass frequency, which is given by the following expression:

$$f_B = \frac{N_r N_s n_r}{K_f} \qquad (5\text{-}20)$$

The quantities N_r and N_s are the number of rotating and stationary blades, respectively, and n_r is the rotational speed of the compressor. The term K_f is the greatest common factor of N_r and N_s. For example, if the compressor has 6 rotating blades and 9 stationary blades, the greatest common factor between 6 and 9 is $K_f = 3$. If the compressor operates at a rotational speed of 6000 rpm, the blade pass frequency would have the following value:

$$f_B = \frac{(6)(9)(6000 \text{ rev/min}/60 \text{ s/min})}{(3)} = 1800 \text{ Hz}$$

The value of the greatest common factor ($K_f = 3$) means that 3 rotor blades line up with 3 stator blades during each revolution of the rotor. The blade passing noise is 3 times more intense than it would be if only one rotor blade was matched to one stator blade ($K_f = 1$) during each rotation.

Note that if the number of rotating blades were increased to 7 in this example, the greatest common factor would be $K_f = 1$. The blade pass frequency would be 6300 Hz. Noise at this higher frequency could be more easily controlled (by damping, for example) than noise at the lower frequency.

Increasing the radial distance between the impeller and diffuser vanes results in lower noise levels. In particular, the blade pass noise is significantly reduced by this tactic. Unfortunately, increasing the spacing also decreases the efficiency of the compressor. The designer must compromise between high noise levels and high efficiency for very close impeller–stator spacing and low noise levels and low efficiency for wide spacing.

The rotational speed of the compressor has a strong influence on the noise generated by the unit, because the sound power radiated from a com-

Noise Sources

pressor is proportional to the rotational speed raised to a power between 2 and 5, depending on the type of compressor. A design compromise must be made between a low-speed quieter unit, which generally requires a large size for a given flow rate, and a high-speed more noisy unit, which is smaller in size.

The following correlations may be used to estimate the overall sound power level for rotating compressors (Heitner 1968). For centrifugal compressors:

$$L_W = 20 \log_{10}(hp/hp_o) + 50 \log_{10}(U_t/U_o) + 81 \tag{5-21}$$

The term hp is the compressor power input, and $hp_o = 1\,hp = 745.7\,W$. The quantity U_t is the blade tip velocity, and $U_o = 800\,ft/sec = 243.8\,m/s$.

For axial compressors, the corresponding correlation is as follows:

$$L_W = 20 \log_{10}(hp/hp_o) + 76 \tag{5-22}$$

The noise spectra for centrifugal and axial compressors is broadband, with the peak (maximum) in the sound power level occurring at a frequency f_m given by the following expressions:

$$f_m = 1000(U_t/U_o) \quad \text{(centrifugal compressors)} \tag{5-23}$$

$$f_m = 2N_b n_r \quad \text{(axial compressors)} \tag{5-24}$$

The quantity N_b is the number of blades in one stage of the axial compressor, and n_r is the rotational speed of the compressor.

The sound power level in each octave band for a compressor may be calculated from the following expression:

$$L_W(\text{octave band}) = L_W - CF_3 \tag{5-25}$$

The values of the conversion factor CF_3 are given in Table 5-5.

Example 5-3. A centrifugal compressor has an overall blade tip diameter of 1.20 m (47.2 in) and operates at a rotational speed of 4800 rpm. The power input to the compressor is 500 kW. The compressor is located in a room having a room constant of 1500 m². The directivity factor for the compressor is $Q = 2.00$. Determine the sound pressure level at a distance of 15 m (49.2 ft) from the compressor and the sound level spectrum at the same point.

The compressor blade tip velocity is calculated as follows:

$$U_t = \pi n_r D = (\pi)(4800/60)(1.20) = 301.6\,m/s\ (989\,ft/sec)$$

The overall sound power level is determined from Eq. (5-21) for a centrifugal compressor:

TABLE 5-5 Conversion Factors CF_3 (dB) to Convert from the Overall Sound Power Level for a Compressor to the Octave Band Sound Power Levels

Frequency, Hz	CF_3, dB	Frequency, dB	CF_3, dB
$f_m/32$	36	f_m	4
$f_m/16$	25	$2f_m$	8
$f_m/8$[a]	18	$4f_m$	14
$f_m/4$	12	$8f_m$	21
$f_m/2$	7		

[a]The table entry $f_m/8$, for example, refers to the octave band that includes the frequency $f_m/8$, where f_m is given by Eq. (5-23) or (5-24).

$$L_W = 20 \log_{10}(500/0.7457) + 50 \log_{10}(301.6/243.8) + 81$$

$$L_W = 56.5 + 4.6 + 81 = 142.1 \, \text{dB}$$

The overall sound pressure level may be determined from Eq. (5-6) for sound propagated indoors:

$$L_p = 142.1 + 10 \log_{10}\left[\frac{4}{1500} + \frac{(2.00)}{(4\pi)(15)^2}\right] + 0.1$$

$$L_p = 142.1 + 10 \log_{10}(0.002667 + 0.0007074) + 0.1$$

$$L_p = 142.1 + (-24.7) + 0.1 = 117.5 \, \text{dB}$$

The peak in the noise level spectrum occurs at a frequency given by Eq. (5-23):

$$f_m = 1000(U_t/U_o) = (1000)(301.6/243.8) = 1237 \, \text{Hz}$$

This frequency lies in the 1000 Hz octave band (between 707 Hz and 1414 Hz); therefore, the octave band sound power level for the 1000 Hz octave band is given by Eq. (5-25):

$$L_W(\text{octave band}) = 142.1 - 4 = 138.1 \, \text{dB}$$

The octave band sound pressure level for the 1000 Hz octave band may be calculated from Eq. (5-6):

$$L_p(\text{octave band}) = 138.1 - 24.7 + 0.1 = 113.5 \, \text{dB}$$

The results for the other octave bands are given in Table 5-6.

Noise Sources

TABLE 5-6   Solution for Example 5-3

	Octave band center frequency, Hz							
	63	125	250	500	1,000	2,000	4,000	8,000
L_W, dB	142.1	142.1	142.1	142.1	142.1	142.1	142.1	142.1
CF_3, dB	25	18	12	7	4	8	14	21
L_W(octave band), dB	117.1	124.1	130.1	135.1	138.1	134.1	128.1	121.1
Equation (5-6)	−24.6	−24.6	−24.6	−24.6	−24.6	−24.6	−24.6	−24.6
L_p(octave band), dB	92.5	99.5	105.5	110.5	113.5	109.5	103.5	96.5

5.6   TRANSFORMER NOISE

Although transformer noise does not cause hearing damage, in general, the continuous hum emitted by large transformers can be quite annoying. The metal cores of transformers are usually laminated to reduce hysteresis losses. One source of transformer noise depends on the magnetic field variations in the laminations. As a result of magnetostriction effects, the laminations change in length as the magnetic field changes. The total change in length is usually on the order of micrometers, but this small dimensional change is sufficient to produce the transformer hum at frequencies of 120 Hz, 240 Hz, and 360 Hz, which are harmonics of the 60 Hz excitation current for the transformer. The harmonic frequencies fall in the 125 Hz and 250 Hz octave bands, so the hum noise tends to appear largest in these octave bands.

The level of noise generated by the transformer increases with increase of the magnetic flux density or with the kVA rating of the transformer. The flux density is determined by the design of the transformer and does not vary significantly with the load on the transformer. This means that the noise generated is practically independent of the transformer load. The hum noise is present at the no-load condition for the transformer.

One approach to reducing the noise generated by a transformer is to reduce the transmission of sound between the core, where the noise is generated, and the transformer housing, from where the sound is radiated into the surroundings. This may be accomplished by using spray cooling, evaporative cooling, or gas cooling, instead of conventional oil cooling. A resilient acoustic barrier in the oil between the core and housing could also be used to reduce the transformer hum noise.

There are two general types of transformers, depending on the method of dissipating energy from the transformer. The *self-cooled transformer* is actually cooled by natural convection. The *forced air-cooled transformer* is

TABLE 5-7 Conversion Factor CF_4 to Convert from the Overall Sound Power Level to Octave Band Sound Power Levels for Transformers

	\multicolumn{8}{c}{Octave band center frequency, Hz}
	63	125	250	500	1,000	2,000	4,000	8,000
CF_4, dB	7	3	9	13	13	19	24	30

cooled by forced convection heat transfer assisted by a fan. The fan provides an additional source of noise for this type of transformer.

Based on tests conducted by the National Electric Manufacturers Association (NEMA), the following correlations may be used to estimate the overall sound power level for transformers. For a self-cooled transformer, the sound power level is related to the kVA rating of the transformer by the following expression:

$$L_W = 45 + 12.5 \log_{10}(kVA) \tag{5-26}$$

For a forced air cooled transformer, the following expression applies:

$$L_W = 48 + 12.5 \log_{10}(kVA) \tag{5-27}$$

The octave band sound power level spectrum may be determined for either type of transformer from the following expression:

$$L_W(\text{octave band}) = L_W - CF_4 \tag{5-28}$$

Values of the conversion factor CF_4 are given in Table 5-7.

5.7 COOLING TOWER NOISE

There are several different cooling tower designs, and each has a somewhat different noise spectrum associated with it. The cooling towers may be classified as either mechanical-draft types or natural-draft types, depending on the mechanism producing motion of the air through the tower.

The mechanical-draft towers may be classified according to the type of fan used in moving the air. *Induced-draft towers* generally use a propeller fan located on the top of the tower. Air is drawn in through the intake louvers to cool the water flowing from the top of the tower over the tower packing. *Forced-draft towers* utilize a centrifugal fan located near the base of the tower. Air is exhausted from the fan into the cooling tower near the lower portion of the tower.

Noise Sources

The noise from a mechanical-draft cooling tower is produced by two primary mechanisms: (a) the fan on the tower and (b) the splashing water within the tower. The fan noise is predominant in the octave bands from 63 Hz to 1000 Hz. The splashing water contributes to noise mainly in the 2000 Hz to 8000 Hz octave bands. The fan noise is usually 15–20 dB higher than the water noise (Thumann and Miller, 1986).

The following correlations may be used to estimate the overall sound power level for mechanical-draft cooling towers. For an induced-draft tower using a propeller fan, the following expression applies:

$$L_W = 96 + 10 \log_{10}(\text{hp}/\text{hp}_o) \tag{5-29}$$

For a forced-draft tower using a centrifugal fan, the following expression may be used:

$$L_W = 87 + 10 \log_{10}(\text{hp}/\text{hp}_o) \tag{5-30}$$

The quantity hp is the power input to the tower fan, and $\text{hp}_o = 1.00 \text{ hp} = 745.7\,W$. The overall sound power level for induced-draft or forced-draft cooling towers may be converted to octave band values by using the following expression:

$$L_W(\text{octave band}) = L_W - CF_5 \tag{5-31}$$

The values for the conversion factor CF_5 are given in Table 5-8.

The only source of noise for natural-draft towers, as shown in Fig. 5-3, is the water-generated noise. If the cooling tower packing extends below the air inlet opening of the tower, noise due to water splashing over the packing material is radiated directly from the tower. In addition, the water falling from the packing material produces noise as it strikes the surface of the water in the pond at the bottom of the tower.

TABLE 5-8 Conversion Factor CF_5 to Convert from the Overall Sound Power Level to Octave Band Sound Power Levels for Induced-Draft and Forced-Draft Cooling Towers

Cooling tower type	Octave band center frequency, Hz							
	63	125	250	500	1,000	2,000	4,000	8,000
Propeller fan, induced	6	5	7	9	16	21	29	35
Centrifugal fan, forced	4	5	9	10	14	16	22	31

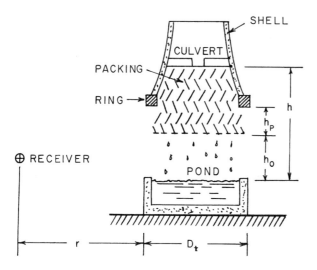

FIGURE 5-3   Natural-draft cooling tower.

The overall sound power level for noise from a natural-draft tower may be determined from the following correlation (Ellis, 1971):

$$L_W = 10\log_{10}(mgh) + 10\log_{10}[0.95(h_p/h)^2 + 1.80(h_o/h)^2] + 60.0 \tag{5-32}$$

The quantities in Eq. (5-32) and the required units are defined as follows:

m = mass flow rate of cooling water, kg/s
g = local acceleration due to gravity = 9.806 m/s^2
h = total distance that the water falls in the tower, m
h_p = depth of the packing material below the tower ring beam, m
h_o = distance between the bottom of the packing and the pond surface, m

The A-weighted sound power level may be found from the overall sound power level:

$$L_W(A) = L_W + 0.1 \, \text{dB} \tag{5-33}$$

The sound pressure level at any distance r from the edge of the pond may be determined from the following relations, depending on whether the receiver is near the tower or farther from the tower. The region *near* the tower is defined by the following relationship:

$$r < r^* = \tfrac{1}{2} D_t \{[1 + 2(h_p + h_o)/D_t]^{1/2} - 1\} \tag{5-34}$$

Noise Sources

The quantity D_t is the diameter of the tower. If the tower is rectangular with plan dimensions $a \times b$, use $D_t = (4ab/\pi)^{1/2}$. The relationship between the overall sound pressure level and the overall sound power level for the region near the tower is as follows:

$$L_p = L_W - 10\log_{10}\{\pi D_t(h_p + h_o)[1 + (2r/D_t)]\} + 10\log_{10}(\rho_o c W_{ref}/p_{ref}^2) \quad (5\text{-}35)$$

For atmospheric air around 300K (80°F), the numerical value of the last term in Eq. (5-35) is 0.1 dB.

For the region *farther* from the tower, $r \geq r^*$, the following expression may be used to determine the overall sound pressure level:

$$L_p = L_W + 10\log_{10}(Q) - 20\log_{10}(r) - 10\log_{10}(4\pi p_{ref}^2/\rho_o c W_{ref}) \quad (5\text{-}36)$$

For atmospheric air around 300K (80°F), the numerical value of the last term in Eq. (5-36) is 10.9 dB. The quantity Q is given by the following expression:

$$Q = \frac{4\tan^{-1}\{[1 + (D_t/r)]^{1/2}\}}{\pi[1 + (D_t/r)]} \quad (5\text{-}37)$$

The argument of the inverse tangent function in Eq. (5-37) must be expressed in radians when making numerical calculations.

The octave band values of the sound power level may be obtained from the overall sound power level by using the following conversion:

$$L_W(\text{octave band}) = L_W - CF_6 \quad (5\text{-}38)$$

Values of the conversion factor CF_6 are given in Table 5-9.

Example 5-4. A natural-draft cooling tower has a mass flow rate of water through the tower of 120 kg/s (196,000 lb$_m$/hr). The tower diameter is 7.50 m (24.6 ft). The packing extends 3.00 m (9.8 ft) below the tower ring, and the

TABLE 5-9 Conversion Factor CF_6 to Convert from the Overall Sound Power Level to Octave Band Sound Power Levels for Natural-Draft Cooling Towers

	Octave band center frequency, Hz						
	63	125	250	500	1,000	2,000	4,000	8,000
CF_6, dB	17.7	19.4	19.8	13.0	7.8	6.3	5.3	7.2

open height of the tower is 6.50 m (21.3 ft). The water falls a total distance of 20 m (65.6 ft) in the tower. Determine the overall sound pressure level at a distance of 25 m (82.0 ft) from the edge of the tower pond.

The overall sound power level may be determined from Eq. (5-32):

$$L_W = 10\log_{10}[(120)(9.806)(20)] + 10\log_{10}[(0.95)(3/20)^2 + (1.8)(6.5/20)^2] + 60.0$$

$$L_W = 10\log_{10}(23{,}534) + 10\log_{10}(0.0214 + 0.1901) + 60.0$$

$$L_W = 43.7 + (-6.7) + 60.0 = 97.0\,\text{dB}$$

The characteristic distance for the cooling tower may be evaluated from Eq. (5-34):

$$r^* = (\tfrac{1}{2})(7.5)\{[1 + (2)(3.0 + 6.5)/(7.5)]^{1/2} - 1\} = (\tfrac{1}{2})(7.5)(0.8797)$$

$$r^* = 3.30\,\text{m}\ (10.8\,\text{ft})$$

For this problem, the location $r = 25\,\text{m} > r^* = 3.30\,\text{m}$; therefore, the sound field corresponds to far-field conditions. We must use Eq. (5-36) to evaluate the sound pressure level.

The directivity factor may be calculated from Eq. (5-37):

$$Q = \frac{(4)\tan^{-1}\{[1 + (7.50/25)]^{1/2}\}}{(\pi)[1 + (7.50/25)]} = (0.9794)\tan^{-1}(1.1402) = 0.8333$$

The overall sound pressure level may be evaluated:

$$L_p = 97.0 + 10\log_{10}(0.8333) - 20\log_{10}(25) - 10.9$$

$$L_p = 97.0 + (-0.8) - 28.0 - 10.9 = 57.3\,\text{dB}$$

Since all factors are independent of frequency, the A-weighted sound level may be found from Eq. (5-33) in terms of the sound pressure level.

$$L_A = L_p + 0.1 = 57.3 + 0.1 = 57.4\,\text{dBA}$$

5.8 NOISE FROM GAS VENTS

One of the more serious noise problems in industrial plants is the noise produced by the discharge of air, steam, or process gas into the atmosphere. Blow-off nozzles, steam vents, and pneumatic control discharge vents are some examples of noisy venting situations. The noise generated by the jet discharged through these devices is a result of turbulent mixing in a high-shear region near the exit plane of the vent. In this region, turbulent eddies are small, and the noise radiated from the eddies is predominantly higher-frequency noise. Sound is also radiated from the fluid stream further from

Noise Sources

the jet exit plane as a result of larger turbulent eddies in this region of the jet. Lower-frequency noise is radiated from this region of the fluid.

The overall sound power level for noise radiated from a vent may be calculated from the following correlation (Burgess Industries, 1966):

$$L_W = 114.4 + 20 \log_{10}\left(\frac{P_1 T_1 M_a d}{P_o T_o M d_o}\right) \qquad (5\text{-}39)$$

The quantities in Eq. (5-39) are defined as follows:

P_1 = upstream absolute pressure of the gas
P_o = reference pressure = 101.3 kPa = 14.7 psia
T_1 = upstream absolute temperature of the gas
T_o = reference temperature = 300K = 540°R
M = gas molecular weight
M_a = molecular weight of air = 28.95 g/mol
d = inside diameter of the gas vent
d_o = reference diameter = 1.000 m = 39.37 in

The octave band sound power level spectrum may be determined for a gas vent from the following conversion.

$$L_W(\text{octave band}) = L_W - CF_7 \qquad (5\text{-}40)$$

Values of the conversion factor CF_7 are given in Table 5-10. The frequency at which the maximum sound power level occurs for the gas jet is given by the following relationship:

$$f_o = \frac{0.20c}{d} \qquad (5\text{-}41)$$

The quantity c is the sonic velocity of the flowing gas at temperature T_1.

TABLE 5-10 Conversion Factors CF_7 (dB) to Convert from the Overall Sound Power Level for a Gas Vent to the Octave Band Sound Power Levels

Frequency	CF_7	Frequency	CF_7	Frequency	CF_7
$f_o/32$	26	$f_o/2$	7	$8f_o$	17
$f_o/16$	21	f_o	5	$16f_o$	25
$f_o/8$[a]	15	$2f_o$	7	$32f_o$	31
$f_o/4$	10	$4f_o$	10	$64f_o$	37

[a] The table entry $f_o/8$, for example, refers to the octave band that includes the frequency $f_o/8$, where f_o is given by Eq. (5-41).

The noise from gas vents is highly directional, so the directivity factor is not unity in general (American Gas Association, 1969). The directivity factor Q_θ depends on the angle θ measured from the vent axis. Values of the directivity factor and the directivity index, $DI = 10 \log_{10} Q_\theta$, are given in Table 5-11.

Example 5-5. A steam vent has an inner diameter of 154 mm (6.065 in) and vents steam (molecular weight, 18.016 g/mol; sonic velocity, 500 m/s) at 615K (342°C or 647°F) and 1480 kPa (215 psia). The vent is located outdoors. Determine the overall sound pressure level and the A-weighted level at a distance of 150 m (492 ft) and at 90° from the vent axis.

The sound power level is found from Eq. (5-39):

$$L_W = 114.4 + 20 \log_{10} \left[\frac{(1480)(615)(28.95)(0.154)}{(101.3)(300)(18.016)(1.00)} \right]$$

$$L_W = 114.4 + 17.4 = 131.8 \, dB$$

The directivity index for a location 90° off the vent axis is found from Table 5-11.

$$DI = -5.5 \, dB$$

The overall sound pressure level at a distance of 150 m from the vent is found from Eq. (5-4) for negligible atmospheric air attenuation:

$$L_p = L_W + DI - 20 \log_{10}(r) - 10.9$$

$$L_p = 131.8 + (-5.5) - 20 \log_{10}(150) - 10.9$$

$$L_p = 131.8 - 5.5 - 43.5 - 10.9 = 7.19 \, dB$$

TABLE 5-11 Directivity Factor Q_θ and Directivity Index DI for a Gas Vent

θ^a	Q_θ	DI, dB	θ^a	Q_θ	DI, dB
0°	1.00	0.0	60°	0.80	−1.0
10°	1.80	2.6	75°	0.447	−3.5
15°	2.16	3.3	80°	0.381	−4.2
20°	2.52	4.0	90°	0.282	−5.5
30°	3.00	4.8	105°	0.200	−7.0
40°	2.50	4.0	120°	0.158	−8.0
45°	2.00	3.0	150°	0.118	−9.3
50°	1.53	1.8	180°	0.100	−10

[a]The angle θ is measured between the axis of the vent and the receiver.

Noise Sources

TABLE 5-12 Solution for Example 5-5[a]

	Octave band center frequency, Hz							
	63	125	250	500	1,000	2,000	4,000	8,000
L_p, dB	71.9	71.9	71.9	71.9	71.9	71.9	71.9	71.9
$-CF_7$, dB	−15	−10	−7	−5	−7	−10	−17	−25
L_p(octave band), dB	56.9	61.9	64.9	66.9	64.9	61.9	54.9	46.9
CFA, dB	−26.2	−16.1	−8.9	−3.2	0.0	+1.2	+1.0	−1.1
L_p(octave band) + CFA	30.7	45.8	56.0	63.7	64.9	63.1	55.9	45.8

[a] Because the quantities DI and r are frequency-independent, L_p(octave band) = L_p = CF_7, in this example.

The peak frequency in the vent noise spectrum is found from Eq. (5-41):

$$f_o = \frac{0.20c}{d} = \frac{(0.20)(500)}{(0.154)} = 649 \text{ Hz}$$

This frequency lies in the 500 Hz octave band (354–707 Hz). The conversion to octave band sound pressure level is found using Eq. (5-39). For example, for the 500 Hz octave band, we find the following value:

$$L_p(\text{octave band}) = 71.9 - 5 = 66.9 \text{ dB}$$

For the 250 Hz octave band, which includes the frequency $\frac{1}{2}f_o = 325$ Hz, the octave band sound pressure level is as follows:

$$L_p(\text{octave band}) = 71.9 - 7 = 64.9 \text{ dB}$$

The sound pressure levels for the other octave bands are given in Table 5-12.

The A-weighted sound pressure level may be determined from the octave band sound pressure level values using Eq. (2-46). For the 500 Hz octave band, CFA = −3.2 dBA from Table 2-4. The calculation for the A-weighted sound level is summarized in Table 5-12. The A-weighted sound level is determined, as follows:

$$L_A = 10 \log_{10}\{\Sigma 10^{[L(\text{octave band})+CFA]/10}\}$$

$$L_A = 10 \log_{10}(10^{3.07} + 10^{4.58} + 10^{5.60} + \cdots) = 69.3 \text{ dBA}$$

5.9 APPLIANCE AND EQUIPMENT NOISE

In addition to the noise sources discussed previously there are several other sources of noise that may be important in the acoustic analysis of residences,

TABLE 5-13 Median Sound Power Levels for Various
Types of Equipment and Home Appliances

Appliance	L_W, dB	Equipment	L_W, dB
Air conditioner	70	Backhoe	120
Clothes dryer	70	Concrete mixer	115
Clothes washer	70	Crane (movable)	115
Dishwasher	75	Front loader	115
Food blender	85	Jackhammer	125
Food disposal	90	Pneumatic wrench	120
Hair dryer	70	Rock drill	125
Refrigerator	50	Scraper/grader	120
Vacuum cleaner	80	Tractor	120

Source: Environmental Protection Agency (1971a).

construction sites, and offices. If one cannot obtain sound power level data from the manufacturer of the appliance or item of equipment, the median sound power level listed in Table 5-13 may be used for preliminary design (Environmental Protection Agency, 1971a). It may be noted that the sound power level from a specific item of equipment may deviate ±10 dB from the median value, so care should be exercised in using the data in Table 5-13.

5.10 VALVE NOISE

5.10.1 Sources of Valve Noise

Valves and regulators used with steam and gas lines can be a significant source of noise. There are two primary sources of noise generated by valves: (a) mechanical noise generation and (b) fluid noise generation, either hydraulic for liquids or aerodynamic for gases (Faulkner, 1976).

Mechanical vibration of the valve components results from flow-induced random pressure fluctuations in the fluid within the valve and from impingement of the fluid against flexible parts of the valve. In conventional valves, the main source of noise from mechanical vibrations arises from the sidewise motion of the valve plug within its guiding surfaces. This noise source usually produces sound at frequencies below 1500 Hz and is often classified as a metallic "rattling" sound. The noise emitted from this source is usually of less concern to the designer than the damage of the valve plug and guide surfaces resulting from the vibration. In fact, noise from valve vibration could be considered beneficial, because the noise warns of conditions in the valve (wear, excessive clearance, etc.) that could result in valve failure.

Some control valves used valve plugs that were fitted with skirts that guided the valve body ports. Flow openings were cast or machined into the skirts. Mechanical vibration of this type of valve was a serious problem, because there was a large clearance between the skirt and the body guides. The vibrations of the valve plug could be reduced by using guide posts on each end of the valve plug. One commonly used design technique for control valve noise reduction is to rigidly attach the cage member containing the flow openings to the valve body. The movable valve plug is closely guided within the inside diameter of the cage. By proper attention to the valve internal design, noise generated from mechancial vibrations can be reduced to negligible levels, compared with other sources.

Another source of mechanical vibration noise arises from valve components resonating at their natural frequencies. Resonant vibration of valve components produces a pure-tone component, usually in the frequency range between 3 kHz and 7 kHz. This vibration can cause high stresses in the component that may lead to fatigue failure. Flexible members, such as the metal seal ring of a ball valve, are subject to mechanical vibrations of this type.

The hydrodynamic flow noise from a valve handling liquids arises from several sources, including (a) turbulent velocity fluctuations in the liquid stream, (b) cavitation when bubbles of vapor collapse after being momentarily formed in the fluid within the valve, and (c) flashing (vaporization) of the liquid when the pressure within the valve falls below the vapor pressure of the liquid.

Turbulent velocity fluctuations in a liquid flow generally result in relatively low noise levels. The high turbulence levels in valves is produced by the rapid deceleration of the fluid as the velocity profile changes shape beyond the valve outlet.

Cavitation of the fluid is the major cause of hydrodynamic noise in valves. As the liquid is accelerated within the valve through valve ports, static pressure head is converted to kinetic energy, and the pressure of the liquid decreases. When the static pressure of the liquid falls below the vapor pressure of the liquid, vapor bubbles are formed within the liquid stream. As these bubbles move downstream into a region of higher pressure (greater than the vapor pressure), the bubbles collapse or implode and cavitation occurs. Noise generated by cavitation has a broad frequency range, and often has a sound similar to that produced by solid particles within the liquid stream.

Flashing of the liquid occurs when the pressure of the liquid drops below the vapor pressure of the liquid at the inlet temperature to the valve. The resulting flow from the valve is two-phase flow, a mixture of liquid and vapor. The deceleration and expansion of the two-phase flow stream produce the noise generated in a valve handling a flashing liquid.

The aerodynamic flow noise from a valve handling gases arises from turbulent fluid interactions within the flowing stream due to deceleration, expansion or impingement of the fluid (Lighthill, 1952).

5.10.2 Noise Prediction for Gas Flows

The Fisher Controls Company has developed one technique for prediction of valve noise (Stiles, 1974). The correlation was developed from A-weighted sound level measurements at a location 1.219 m (48 in) downstream of the valve outlet and 0.737 m (29 in) from the surface of the pipe connected to the valve. This location corresponds to a distance $r_o = 1.424$ m (56.1 in) from the surface of the valve. For valves located outdoors, the A-weighted sound level may be estimated from the following correlation:

$$L_A = L_A(r_o) - 20 \log_{10}(r/r_o) \tag{5-42}$$

The following correlation may be used for valves located indoors, where R is the room constant for the room in which the valve is located:

$$L_A = L_A(r_o) + 10 \log_{10}\left[\frac{4 + (R/4\pi r^2)}{4 + (R/4\pi r_o^2)}\right] \tag{5-43}$$

For valves handling a gas, the Fisher Controls equation for the A-weig'.ted noise generated by the flow through the valve may be estimated from the following:

$$L_A(r_o) = 17.4 \log_{10}(\Delta P/\Delta P_o) + 22.5 \log_{10}(C_g) \\ - 32.4 \log_{10}(\rho_p t/\rho_s t_s) + L(C) - 24.4 \tag{5-44}$$

The quantities in Eq. (5-44) are defined as follows:

ΔP = pressure drop across the valve
$\Delta P_o = 6.895$ kPa $= 1.00$ psi
C_g = valve-sizing coefficient for gas flow
ρ_p = density of the pipe material
ρ_s = density of steel = 7800 kg/m^3 = 0.282 lb$_m$/in^3
t = thickness of the pipe wall
t_s = thickness of the same nominal diameter SCH 40 pipe

The factor $L(C)$ is a function of the pressure drop across the valve ΔP and the inlet pressure to the valve P_1. This factor is given in the following expressions for some commonly used valve types:

Noise Sources

(a) *Cage-style globe valve, standard trim*

$$L(C) = \begin{cases} 0 & \text{[for } (\Delta P/P_1) \leq 0.151] \\ 17.4 \log_{10}(\Delta P/P_1) + 14.3 & \text{[for } (\Delta P/P_1) > 0.151] \end{cases} \quad (5\text{-}45)$$

(b) *Cage-style globe valve, whisper-trim*

$$L(C) = \begin{cases} -7.5 & \text{[for } (\Delta P/P_1) \leq 0.563] \\ 87.0 \log_{10}(\Delta P/P_1) + 14.2 & \text{[for } (\Delta P/P_1) > 0.563] \end{cases} \quad (5\text{-}46)$$

(c) *Guided-plug globe valve*

$$L(C) = \begin{cases} 0 & \text{[for } (\Delta P/P_1) \leq 0.216] \\ 16.1 \log_{10}(\Delta P/P_1) + 10.7 & \text{[for } (\Delta P/P_1) > 0.216] \end{cases} \quad (5\text{-}47)$$

(d) *Standard ball valve, swaged body*

$$L(C) = \begin{cases} 0 & \text{[for } (\Delta P/P_1) \leq 0.093] \\ 9.4 \log_{10}(\Delta P/P_1) + 9.7 & \text{[for } (\Delta P/P_1) > 0.093] \end{cases} \quad (5\text{-}48)$$

The valve-sizing coefficient C_g for gas or steam flow is a dimensional parameter. For gases other than steam, C_g is defined by the following dimensional relationship:

$$Q_g(\text{scfh}) = C_g P_1(\text{psia})[(M_a/M)(T_o/T_1)]^{1/2} \sin \theta \quad (5\text{-}49)$$

where:

scfh = standard cubic feet per hour, or ft^3/hr at T_o and P_o
$T_o = 519.7°\text{R} = 288.7\text{K}$ and $P_o = 14.696 \text{ psia} = 101.325 \text{ kPa}$
T_1 = absolute temperature of the gas at the inlet of the valve
P_1 = absolute pressure of the gas at the inlet of the valve
M_a = molecular weight of air = 28.95 g/mol
$\theta(\text{radians}) = (59.64/C_1)(\Delta P/P_1)^{1/2}$
$C_1 = C_g/C_V$
C_V = valve-sizing coefficient for liquids (dimensional)

Some representative values for the parameter C_1 are given in Table 5-14. The valve manufacturer should be contacted for data for a specific valve.

For mass flow of steam, the valve-sizing coefficient C_g is defined by the following relationship:

$$m(\text{lb}_m/\text{hr}) = \frac{0.0022 C_g P_1 \sin \theta}{[1 + (\rho_{\text{sat}}/\rho)]^{1/2}} \quad (5\text{-}50)$$

The quantity ρ_{sat} is the density of saturated steam at the upstream pressure P_1, and ρ is the density of the steam at the upstream pressure P_1 and temperature T_1.

TABLE 5-14 Flow Coefficients for Various Valves

Valve type	Description	Body design	$C_1 = C_g/C_V$	$C_V/[D(\text{in})]^{2\text{a}}$
Globe	Single port	A	35.0	12.90
Globe	Single port	BF	32.0	9.67
Globe	Single port	GS	35.0	14.90
Globe	Any valve plug	D	30.0	10.32
Globe	Single port	DBQ	33.0	12.90
Angle	Single port	DBAQ	34.5	10.32
Angle	Single port	461	18.0	12.90
Ball	Hi-Ball	V25	20.0	11.60
Butterfly	Swing-through vane	75° open	30.0	28.3

[a] D is the inside diameter of the pipe in inches.

5.10.3 Noise Prediction for Liquid Flows

The Fisher Controls correlation for the A-weighted sound level generated by flow of a liquid through a valve is the following expression:

$$L_A(r_o) = 10 K_L \log_{10}(\Delta P/P_o) + 20 \log_{10}(C_V) \\ - 32.4 \log_{10}(\rho_p t/\rho_s t_s) + L(\phi) + 9 \quad (5\text{-}51)$$

The quantities in Eq. (5-51) are defined as follows:

$$K_L = \begin{cases} 1 & [\text{for } 0 \le \phi \le 0.167] \\ 0.50 + 3\phi & [\text{for } \phi > 0.167] \end{cases} \quad (5\text{-}52)$$

where:

$$\phi = \frac{\Delta P}{P_1 - P_v}$$

P_1 = upstream pressure
P_v = vapor pressure of the liquid at the upstream pressure

The factor $L(\phi)$ is a function of the valve type and the pressure ratio ϕ. This factor is given by the following expressions for some commonly used valve types:

(a) *Globe valve, standard cage trim*

$$L(\phi) = \begin{cases} 0 & [\text{for } 0 \le \phi \le 0.34] \\ 48\phi(1 - 1.86\phi) & [\text{for } 0.34 < \phi \le 1.00] \end{cases} \quad (5\text{-}53)$$

(b) *Ball valve*

$$L(\phi) = \begin{cases} 0 & \text{[for } 0 \leq \phi \leq 0.50] \\ 100\phi(0.70 - \phi) & \text{[for } 0.50 < \phi \leq 1.00] \end{cases} \quad (5\text{-}54)$$

(c) *Butterfly valve*

$$L(\phi) = \begin{cases} 10 & \text{[for } 0 \leq \phi \leq 0.24] \\ 95.6\phi(1 - 1.25\phi) & \text{[for } 0.24 < \phi \leq 1.00] \end{cases} \quad (5\text{-}55)$$

The valve-sizing coefficient C_V for liquid flow with no flashing is defined by the following dimensional equation:

$$Q_L(\text{gpm}) = C_V[\Delta P(\text{psi})\rho_w/\rho_L]^{1/2} \quad (5\text{-}56)$$

where:

ρ_w = density of water at the fluid temperature
ρ_L = density of the liquid

Some representative values of the sizing coefficient are given in Table 5-14. The valve manufacturer should be contacted for data for a specific valve.

There are several other correlations that have been developed for prediction of the noise from valves. Nakano (1968) obtained the following relationship for the sound power level for a valve handling a gas flow:

$$L_W = A + B\log_{10}(mTF) \quad (5\text{-}57)$$

where:

m = mass flow rate of the gas, kg/s
T = absolute temperature of the gas at the inlet of the valve, K
$F = 1 - [1 - (\Delta P/P_1)]^{(\gamma-1)/\gamma}$
γ = specific heat ratio for the gas

The constants A and B depend on the valve type. For a globe valve, $A = 90$ and $B = 10.0$; for a gate valve, $A = 83$ and $B = 15.6$; for a ball valve, $A = 97$ and $B = 12.8$; and for an angle valve, $A = 82$ and $B = 13.1$.

The A-weighted sound level generated by valves manufactured by the Masoneilan Corporation has been correlated by an expression based on the internal conversion of fluid kinetic energy into acoustic energy (Baumann, 1970). This method has been extended to other valves also (Baumann, 1987).

Example 5-6. Natural gas (molecular weight, 20.3 g/mol; specific heat ratio, 1.252) flows through a GS globe valve with standard trim at a flow rate of 14×10^6 scfh (14 MM scfh or 110.1 std m³/s). The natural gas enters the valve at 300K (80°F) and 4.250 MPa (616.4 psia). The pressure drop across the valve is 2.40 MPa (348 psi). The gas flows through an 8-in SCH 80 steel pipe, for which the outer diameter is 8.625 in (219.1 mm) and the pipe wall thickness is 0.500 in (12.7 mm). The wall thickness of a standard 8-in SCH 40 pipe is 0.322 in (8.2 mm). Determine the A-weighted sound level at a distance of 15 m (49.2 ft) from the valve, if the valve is located outdoors.

For the GS globe valve, the parameter $C_1 = 35$, as given in Table 5-14. The factor θ is needed in Eq. (5-49) is as follows:

$$\theta = \frac{59.64}{C_1}\left(\frac{\Delta P}{P_1}\right)^{1/2} = \frac{(59.64)}{(35.0)}\left(\frac{2.400}{5.250}\right)^{1/2} = 1.2805 \text{ rad}$$

The valve-sizing coefficient for gas flow through the valve may be calculated from Eq. (5-49):

$$C_g = \frac{(14 \times 10^6)}{(616.4)\sin(1.2805)}\left[\frac{(20.3)(300)}{(28.95)(288.9)}\right]^{1/2} = 20{,}230$$

The pressure ratio term is as follows:

$$\Delta P/P_1 = (2.400)/(4.250) = 0.5647 > 0.154$$

The factor $L(C)$ is given by Eq. (5-45) for a globe valve with standard trim:

$$L(C) = 17.4\log_{10}(0.5647) + 14.3 = 10.0 \text{ dB}$$

The A-weighted sound level at the reference distance r_o is found from Eq. (5-44):

$$L_A(r_o) = 17.4\log_{10}(2400/6.895) + 22.5\log_{10}(20{,}230)$$
$$- 32.4\log_{10}(0.500/0.322) + 10.0 - 24.4$$
$$L_A(r_o) = 44.2 + 96.9 - 6.2 + 10.0 - 24.4 = 120.5 \text{ dBA}$$

The A-weighted sound level at a distance of 15 m from the valve may be found from Eq. (5-42):

$$L_A = 120.5 - 20\log_{10}(15/1.424) = 120.5 - 20.5 = 100.0 \text{ dBA}$$

5.11 AIR DISTRIBUTION SYSTEM NOISE

Noise generated in air distribution systems is one of the concerns in design of heating, ventilating, and air conditioning (HVAC) systems. Noise is transmitted from the air-handling unit (fan) into the duct system. Sound may also be generated as the air flows through elbows, fittings, and the grill or diffuser at the duct outlet into the room. The location of these components in an air distribution system is illustrated in Fig. 5-4. The sound power level of the noise introduced into the duct by the fan is given by Eq. (5-8).

Noise Sources

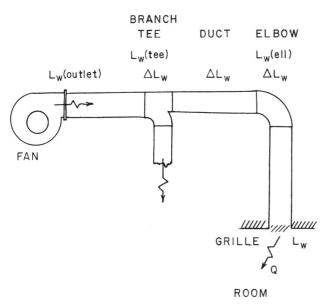

FIGURE 5-4 Air distribution system. ΔL_W is the attenuation in the element, and L_W is the noise generation within the element.

$$L_W(\text{fan outlet}) = L_W - 3\,\text{dB} \tag{5-58}$$

For industrial air distribution systems that utilize high air velocities, the noise generated by the air flowing through the distribution system is a particular concern.

Energy transmitted along the duct may be attenuated or dissipated by the interaction with the duct wall, or acoustic material may be placed inside the duct to reduce the noise transmission.

5.11.1 Noise Attenuation in Air Distribution Systems

There are three primary mechanisms responsible for noise attenuation in air distribution systems:

(a) Acoustic energy is absorbed by interaction with the duct walls.
(b) Acoustic energy is reflected at the open end of the duct.
(c) Acoustic energy is absorbed by elbows and fittings in the system.

The attenuation of sound by each of these mechanisms has been correlated in terms of the change in the sound power level produced by each element in the HVAC system (ASHRAE, 1991).

At the junction where a duct divides or has side branches or outlets, the acoustic power moving down the duct is divided at the duct branch. The acoustic power from the fan into the duct divides in proportion to the ratio of the total cross-sectional area of all branches leaving the junction to the specific branch cross-sectional area. The sound power level for the acoustic energy transmitted to a specific branch is given by the following expression:

$$L_W(i\text{th branch}) = L_W - 10\log_{10}\left(\frac{\Sigma S_j}{S_i}\right) \qquad (5\text{-}59)$$

If the duct has no external or internal lining material (a "bare" duct), the attenuation per unit length for the duct is given in Table 5-15. If the duct is externally insulated, the attenuation is approximately two times that given in Table 5-15 for the base duct. If the duct is lined internally with an absorbent material having an acoustic absorption coefficient α, the attenuation by the lining material may be estimated, as follows:

(a) For $63\,\text{Hz} \le f \le 2000\,\text{Hz}$:
$$\Delta L_W = 4.20\alpha^{1.4}(L/D_e) \qquad (5\text{-}60)$$

The quantity D_e is the equivalent or hydraulic diameter of the duct and L is the length of the duct section:

$$D_e = 4S/P_W \qquad (5\text{-}61)$$

The quantity S is the cross-sectional area of the duct and P_W is the perimeter of the duct cross section.

TABLE 5-15 Attenuation per Unit Length in Bare Ducts, $\Delta L_W/L$, dB/m

$D_e = 4S/P_w$, m[a]	Octave band center frequency, Hz				
	63	125	250	500	1,000 and greater
0.075		0.59	0.59	0.57	0.53	0.47
0.15		0.59	0.57	0.53	0.47	0.37
0.30		0.57	0.53	0.47	0.37	0.23
0.60		0.53	0.47	0.37	0.23	0.16
1.20		0.47	0.37	0.23	0.085	0.084
2.40		0.42	0.29	0.14	0.033	0.033

[a] S is the cross-sectional area of the duct and P_w is the perimeter of the duct.

Noise Sources

(b) For 2000 Hz $< f <$ 8000 Hz:

$$\Delta L_W = \tfrac{1}{2}[10 + \Delta L_W(\text{Eq. 5-60})] \quad (5\text{-}62)$$

(c) For $f \geq$ 8000 Hz:

$$\Delta L_W = 10\,\text{dB} \quad (5\text{-}63)$$

The attenuation in the 8 kHz and higher octave bands is limited to about 10 dB because of line-of-sight propagation of sound through the lined portion of the duct.

The attenuation due to reflection of the acoustic energy at the open end of the duct is presented in Table 5-16. The data apply for ducts terminating flush with the wall or ceiling.

The attenuation of elbows in the system is given in Table 5.17 (circular duct) and Table 5-18 (rectangular duct). The data presented in the tables are for unlined ducts. Using acoustic lining before or after the elbow, or both, will increase the attenuation by 7 dB to 12 dB in the 4 kHz to 8 kHz octave bands (Faulkner, 1976, p. 404).

5.11.2 Noise Generation in Air Distribution System Fittings

In addition to attenuation or dissipation of acoustic energy in fittings, flow energy may also be converted to acoustic energy in the fittings. Flow-

TABLE 5-16 Octave Band Attenuation (dB) Due to Reflection at the Open End of the Duct[a]

$f_o D$, Hz-m	Attenuation, ΔL_W, dB	$f_o D$, Hz-m	Attenuation, ΔL_W, dB
5	20.6	70	3.7
10	15.3	80	3.1
15	12.4	90	2.6
20	10.5	100	2.2
25	9.1	120	1.6
30	8.1	140	1.0
35	7.2	160	0.6
40	6.4	180	0.3
50	5.3	200 and greater	0.0
60	4.4		

[a] D is the inner diameter for a circular duct and $D = S^{1/2}$ for a rectangular duct, where S is the duct cross-sectional area; f_o is the octave band center frequency.

TABLE 5-17 Attenuation of Sound by Circular 90° Elbows[a]

$f_o D$, Hz-m	Attenuation, ΔL_W, dB	$f_o D$, Hz-m	Attenuation, ΔL_W, dB
40 or less	0.0	500	2.2
50	0.2	600	2.4
100	0.7	700	2.5
150	1.1	800	2.6
200	1.3	1,000	2.7
250	1.6	1,500	2.8
300	1.8	2,000	2.9
400	2.0	3,000 and greater	3.0

[a] The quantity D is the inner diameter of the elbow and f_o is the octave band center frequency.

induced noise generated in elbows may be estimated from the following expression (Bullock, 1970):

$$L_W = F_s + 10 \log_{10} f_o + 10 \log_{10} S + 44.4 \log_{10} u - 54 \qquad (5\text{-}64)$$

The quantity f_o is the octave band center frequency, Hz; S is the duct cross-sectional area, m^2; and u is the velocity of the air upstream of the elbow, m/s. The spectrum function F_s is given in Table 5-19. The Strouhal number N_s is defined by the following expression:

TABLE 5-18 Attenuation of Sound by Rectangular 90° Elbows[a]

$f_o S^{1/2}$, Hz-m	Attenuation, L_W, dB	$f_o S^{1/2}$, Hz-m	Attenuation, ΔL_W, dB
40 or less	0.0	300	6.8
50	0.4	350	6.5
60	1.0	400	5.9
70	1.6	450	5.4
80	2.4	500	5.0
100	3.6	600	4.5
120	4.8	700	4.1
140	5.6	800	3.8
160	6.2	1,000	3.4
200	6.8	1,200	3.2
250	7.0	1,500 or greater	3.0

[a] The quantity S is the cross-sectional area of the duct before the elbow and f_o is the octave band center frequency. The data apply to rectangular ducts with aspect ratios between 0.8 and 1.2.

Noise Sources

TABLE 5-19 Spectrum Function F_s for Noise Generation in Elbows[a]

$N_S = f_o D/u$	90° elbow, without turning vanes	90° elbow, with turning vanes
0.6	79	49
0.8	67	48
1.0	65	47
2	56	45
4	46	42
6	42	40
8	40	38
10	38	37
20	34	33
40	31	27
60	29	22
100	27	14
200	23	0

[a]The Strouhal number is $N_S = f_o D/u$, where $D = (4S/\pi)^{1/2}$; f_o is the octave band center frequency, Hz; and u is the velocity of the air before the elbow, m/s.
Note: Use logarithmic interpolation with this table;

$$\frac{F - F_1}{F_2 - F_1} = \frac{\log_{10}(N_S/N_{S1})}{\log_{10}(N_{S2}/N_{S1})}$$

$$N_S = \frac{f_o D}{u} \tag{5-65}$$

where $D = (4S/\pi)^{1/2}$.

The aerodynamic noise generation produced by 90° branch tees has been correlated in a manner similar to that used for elbows (Bullock, 1970). The noise generated and transmitted to the side branch, having a cross-sectional area S_3, m², may be estimated from the following expression:

$$L_W = F_{sb} + 10 \log_{10} f_o + 10 \log_{10} S_3 + L_{br} - 1.5 \tag{5-66}$$

For the noise generated and transmitted into the main straight branch, use the area S_2 instead of S_3 in Eq. (5-66). The spectrum function for branch tees is given in Table 5-20. The quantity L_{br} is calculated from the velocity in the main straight branch of the tee, u_2, and the velocity in the side branch of the tee, u_3, where both velocities are expressed in units of m/s:

$$L_{br} = 10 \log_{10}(10^{L_2/10} + 10^{L_3/10}) \tag{5-67}$$

$$L_2 = 46 \log_{10} u_2 - 70 \tag{5-68}$$

$$L_3 = 23 \log_{10} u_3 - 20 \tag{5-69}$$

TABLE 5-20 Spectrum Function F_{sb} for Noise Generated in Branch Duct Take-offs[a]

$N_S = f_o D/u_1$	F_{sb}, dB	$N_S = f_o D/u_1$	F_{sb}, dB
1	80	50	42
2	74	100	35
5	64	200	26
10	57	500	11
20	51	1,000	0

[a] The Strouhal number is defined by $N_S = f_o D/u_1$, where $D = (4S_1/\pi)^{1/2}$; f_o is the octave band center frequency, Hz; and u_1 is the velocity of the air before the branch, m/s:

$$\uparrow u_3$$
$$u_2 \leftarrow \perp \leftarrow u_1$$

Note: Use logarithmic interpolation with this table:

$$\frac{F - F_1}{F_2 - F_1} = \frac{\log_{10}(N_S/N_{S1})}{\log_{10}(N_{S2}/N_{S1})}$$

The attenuation values are subtracted directly from the sound power level, in decibels, because the attenuation is an exponential function. On the other hand, the acoustic energy generation in fittings must be combined by "decibel addition" with the existing sound power. The energy or power is additive, but the decibel values are not directly additive. This procedure is illustrated in the example at the end of this section.

5.11.3 Noise Generation in Grilles

Most air distribution systems are terminated by grilles or diffusers. Noise is generated by the flow of air over the grille or diffuser elements. The following correlation (Beranek and Vér, 1992) may be used to estimate the noise generated in a grille:

$$L_W(\text{grille}) = 10 + 10\log_{10} S + 30\log_{10} C_D + 60\log_{10} u \qquad (5\text{-}70)$$

where S is the cross-sectional area, m²; u is the velocity of the air before entering the grille, m/s; and C_D is the dimensionless grille pressure drop coefficient, defined by the following expression:

$$\Delta P = \tfrac{1}{2} C_D \rho u^2 \qquad (5\text{-}71)$$

Noise Sources

The quantity ΔP is the pressure drop across the grille due to the flowing air and ρ is the density of the air. Specific values for the grille pressure drop coefficient may be obtained from the grille manufacturer. Some representative values for C_D are given in Table 5-21 for various grille configurations.

The sound generated by air flow through a grille is usually broadband noise; however, some discrete frequency noise, due to vortex shedding from the solid elements of the grille, is also present. The peak in the grille noise spectrum for HVAC systems occurs at a frequency f_m given by the following dimensional relationship:

$$f_m = 150u \text{ (m/s)} \tag{5-72}$$

The octave band sound power levels may be obtained from the following expression:

$$L_W(\text{octave band}) = L_W(\text{grille}) - CF_g \tag{5-73}$$

Values for the conversion factor CF_g are given in Table 5-22.

The directivity factor is not unity for acoustic radiation from a grille, because the sound is not radiated as spherical waves. The value of the directivity factor is a function of the dimensionless ratio, $f(S)^{1/2}/c$. Values for the directivity factor Q and the directivity index, $DI = 10\log_{10} Q$, are given in Table 5-23.

Example 5-7. In the air distribution system shown in Fig. 5-5, the ducts have a square cross section and are uninsulated. There is one elbow in the

TABLE 5-21 Grille Pressure Drop Coefficient, C_D

Grille type	C_D
Rectangular grille with no dampers:	
parallel louvers	2.9
inclined louvers	2.7
Rectangular grille with dampers:	
parallel louvers, open damper	4.8
parallel louvers, partially closed damper	7.3
Circular ceiling diffuser	1.59
High side-wall diffuser (rectangular):	
zero angle of deflection of exit air	0.73
45° angle of deflection of exit air	1.93

Source: Hubert (1970) and McQuiston and Parker (1994).

TABLE 5-22 Conversion Factors CF_g to Convert from the Overall Sound Power Level for Grille Noise to the Octave Band Sound Power Levels

	Octave band center frequency, Hz							
	$f_m/16$	$f_m/8$[a]	$f_m/4$	$f_m/2$	f_m	$2f_m$	$4f_m$	$8f_m$	$16f_m$
CF_g, dB	23	17	11	6	5	7	12	18	24

[a] The table entry $f_m/8$, for example, refers to the octave band that includes the frequency $f_m/8$, where f_m is given by Eq. (5-73).

600 mm (23.6 in) branch duct and one elbow in the 900 mm (35.4 in) main duct. The grille at the duct outlet has parallel louvers and no dampers. The volumetric flow rate out of the grille (and through the 600 mm branch duct) is 1440 dm³/s (3051 cfm), and the volumetric flow rate in the 900 mm main duct from the fan is 5050 dm³/s (10,680 cfm). The internal fan sound power level spectrum is given in Table 5-24. The air in the room is at 25°C (77°F), at which condition the sonic velocity is 346.1 m/s (1136 ft/sec). Determine the steady-state sound pressure level in the room at a distance of 5 m (16.4 ft) directly in front of the grille ($\theta = 0°$).

TABLE 5-23 Directivity Factor Q and Directivity Index DI (dB) for a Duct Opening Flush with the Wall[a]

$fS^{1/2}/c$	$Q(\theta = 0°)$	DI(0°), dB	$Q(\theta = 45°)$	DI(45°), dB
0.04	2.0	3.0	2.0	3.0
0.06	2.2	3.4	2.0	3.0
0.08	2.5	4.0	2.0	3.1
0.10	2.7	4.3	2.0	3.1
0.20	3.6	5.6	2.2	3.4
0.40	4.6	6.6	2.7	4.3
0.60	5.3	7.2	3.0	4.8
0.80	5.9	7.7	3.2	5.1
1.0	6.3	8.0	3.3	5.2
2.0	7.2	8.6	3.7	5.7
4.0	7.8	8.9	3.9	5.9
∞	8.0	9.0	4.0	6.0

[a] The quantity S is the opening cross-sectional area and c is the sonic velocity in the air. The angle θ is the angle between the desired direction and the normal to the grille opening.

Noise Sources

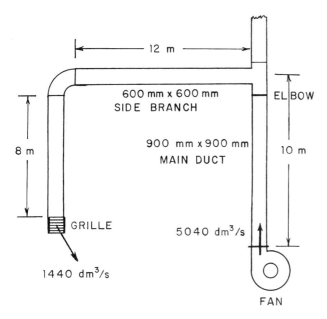

FIGURE 5-5 Diagram for Example 5-7.

A sample calculation for the 125 Hz octave band is given in the following material. The calculations for the other octave bands are summarized in Table 5-24. The calculation procedure involves beginning with the sound power level at the duct inlet and proceeding along the duct system to the outlet grille in each branch. The attenuation values are subtracted, and the noise generation values are combined by energy addition at each point where the energy is generated.

The sound power level produced internally at the fan in the 125 Hz octave band is given as 71dB. The acoustic energy transmitted through the fan outlet into the duct system is given by Eq. (5-8):

L_W(to duct) $= 71 - 3 = 68$ dB

For a square duct ($a = b$), the quantity $D_e = 4S/P_W = (4)(ab)/2(a+b) = a$, the side length of the duct cross section. For the 10 m long, 900 mm square duct, the attenuation is found from Table 5-15:

$\Delta L_W = (0.42\,\text{dB/m})(10\,\text{m}) = 4.2\,\text{dB}$

The sound power level before the elbow in the main duct is:

$L_W(a) = 68 - 4.2 = 63.8$ dB

TABLE 5-24 Solution for Example 5-7

Item	Octave band center frequency, Hz						
	63	125	250	500	1,000	2,000	4,000	8,000
L_W(fan)	73	71	71	69	61	57	54	48
L_W(to duct)	70	68	68	66	58	54	51	45
ΔL_W(10 m duct)	5.0	4.2	3.0	3.6	1.2	1.2	1.2	1.2
$L_W(a)$ before ell	65.0	63.8	65.0	62.4	56.8	52.8	49.8	43.8
N_S	10.2	20.4	40.8	81.6	163	326	652	1,304
F_s	37.8	33.9	30.9	27.8	24.2	20.2	16.2	12.2
L_W(elbow)	36.1	35.2	35.2	35.1	34.5	33.5	32.6	31.6
$L_W(b)$ in ell	65.0	63.8	65.0	62.4	56.8	52.9	49.9	44.1
$f_o D$, Hz-m	56.3	112.5	225	450	900	1,800	3,600	7,200
ΔL_W(elbow)	0.8	4.4	6.9	5.4	3.6	3.0	3.0	3.0
L_W to tee	64.2	59.4	58.1	57.0	53.2	49.9	46.9	41.1
F_{sb}	56.9	50.8	44.0	37.1	28.6	18.0	6.8	−4.2
L_W(tee)	62.8	59.7	55.9	52.0	46.5	38.9	30.7	22.7
L_W in tee	66.6	62.6	60.1	58.2	54.0	50.2	47.0	41.2
ΔL_W(branch)	5.1	5.1	5.1	5.1	5.1	5.1	5.1	5.1
L_W(branch)	61.5	57.5	55.0	53.1	48.9	45.1	41.9	36.1
ΔL_W(12 m duct)	6.4	5.6	4.4	2.8	1.9	1.9	1.9	1.9
L_W before ell	55.1	51.9	50.6	50.3	47.0	43.2	40.0	34.2
N_S	10.7	21.2	42.3	84.6	169	339	677	1,354
F_s	37.6	33.7	30.6	27.7	24.0	20.0	16.0	12.0
L_W(elbow)	23.9	23.0	22.9	23.0	22.3	21.3	20.3	19.3
L_W in elbow	55.1	51.9	50.6	50.3	47.0	43.2	40.0	34.3
$f_o D$, Hz-m	37.8	75	150	300	600	1,200	2,400	4,800
ΔL_W(elbow)	0.0	2.0	5.9	6.8	4.5	3.2	3.0	3.0
ΔL_W(reflect.)	6.8	3.4	0.8	0.0	0.0	0.0	0.0	0.0
ΔL_w(8 m duct)	4.2	3.8	3.0	1.8	1.3	1.3	1.3	1.3
Total attenuation	11.0	9.2	9.7	8.6	5.8	4.5	4.3	4.3
L_W(before grille)	44.1	42.7	41.9	41.7	41.2	38.7	35.8	30.0
L_W(grille)	55.6	55.6	55.6	55.6	55.6	55.6	55.6	55.6
CF_g	17	11	6	5	7	12	18	24
L_W(grille, octave band)	38.6	44.6	49.6	50.6	48.6	43.6	37.6	31.6
L_W to room	45.2	46.8	50.3	51.1	49.3	44.8	39.8	33.9

Noise Sources

The quantity D for the 900 mm main duct is, as follows:
$$D = (4S/\pi)^{1/2} = [(4)(0.810)/\pi]^{1/2} = 1.016 \, \text{m}$$

The velocity of the air before the elbow in the 900 mm main duct is found as follows:
$$u = Q_g/S = (5.040 \, \text{m}^3/\text{s})/(0.810 \, \text{m}^2) = 6.22 \, \text{m/s}$$

The Strouhal number for the 125 Hz octave band is calculated from its definition:
$$N_S = f_o D/u = (125)(1.016)/(6.22) = 20.4$$

The corresponding spectrum function is found in Table 5-19:
$$F_s = 33.9$$

The sound power generated by flow noise through the elbow is found from Eq. (5-64):
$$L_W(\text{elbow}) = 33.9 + 10\log_{10}(125) + 10\log_{10}(0.810)$$
$$+ 44.4 \log_{10}(6.22) - 54$$
$$L_W(\text{elbow}) = 33.9 + 21.0 + (-0.9) + 35.2 - 54 = 35.2 \, \text{dB}$$

The total sound power level at the elbow is found as follows:
$$L_W(b) = 10\log_{10}(10^{6.38} + 10^{3.52}) = 63.8 \, \text{dB}$$

For a square duct ($a = b$), the quantity $S^{1/2} = (ab)^{1/2} = a$, the side length of the duct cross section. For the 125 Hz octave band, the quantity $f_o S^{1/2} = f_o a = (125)(0.900) = 112.5$ Hz-m. The attenuation for the elbow is found from Table 5-18 to be $\Delta L_W = 4.4 \, \text{dB}$. The sound power level from the elbow to the tee is as follows:
$$L_W \,(\text{to tee}) = 63.8 - 4.4 = 59.4 \, \text{dB}$$

The spectrum function for the side-branch energy generation is found from Table 5-20 at a Strouhal number, $N_S = f_o D/u_1 = 20.4$, or $F_{sb} = 50.8$. The air velocity in the two branches may be determined as follows:
$$u_2 = Q_{g2}/S_2 = (5.040 - 1.440)/(0.810) = 4.44 \, \text{m/s}$$
$$u_3 = Q_{g3}/S_3 = (1.440)/(0.360) = 4.00 \, \text{m/s}$$

The branch functions given in Eqs. (5-68) and (5-69) may be determined:
$$L_2 = 46 \log_{10}(4.44) - 70 = -40.2 \, \text{dB}$$
$$L_3 = 23 \log_{10}(4.00) - 20 = -6.2 \, \text{dB}$$

These values may be combined according to Eq. (5-67):

$$L_{br} = 10\log_{10}(10^{-4.02} + 10^{-0.62}) = -6.2\,\text{dB}$$

The acoustic energy generated within the tee in the 125 Hz octave band is determined from Eq. (5-66):

$$L_W(\text{tee}) = 50.8 + 10\log_{10}(125) + 10\log_{10}(0.810) + (-6.2) - 1.5$$
$$= 59.7\,\text{dB}$$

The acoustic power in the tee is found by combining the energy before the tee and the energy generated within the tee:

$$L_W\text{ (in tee)} = 10\log_{10}(10^{5.94} + 10^{5.97}) = 62.6\,\text{dB}$$

The energy delivered to the 600 mm branch duct is found from Eq. (5-59):

$$L_W - L_W(\text{branch}) = 10\log_{10}\left(\frac{0.810 + 0.360}{0.360}\right) = 5.1\,\text{dB}$$

The sound power level delivered to the 600 mm side branch in the 250 Hz octave band is:

$$L_W(\text{branch}) = 62.6 - 5.1 = 57.5\,\text{dB}$$

The attenuation in the 12 m long run of the side branch may be calculated as follows. The attenuation per unit length is found from Table 5-15, with $D_e = 0.600$ m for the square cross section:

$$\Delta L_W(12\text{ m duct}) = (0.47\,\text{dB/m})(12\,\text{m}) = 5.6\,\text{dB}$$

The sound power level to the elbow in the 125 Hz octave band is as follows:

$$L_W\text{ (before elbow)} = 57.5 - 5.6 = 51.9\,\text{dB}$$

The energy generated in the elbow in the 600 mm duct may be determined by the same procedure as that for the previous elbow. The quantity D for the smaller duct is as follows:

$$D = (4S/\pi)^{1/2} = [(4)(0.360)/\pi]^{1/2} = 0.677\,\text{m}$$

The Strouhal number is calculated from its definition:

$$N_S = f_o D/u_3 = (125)(0.667)/(4.00) = 21.2$$

The spectrum function is found from Table 5-19, $F_s = 33.7$. The noise generated in the elbow is found from Eq. (5-64):

Noise Sources

$$L_W(\text{elbow}) = F_s + 10\log_{10} f_o + 10\log_{10} S + 44.4\log_{10} u_3 - 54$$

$$L_W(\text{elbow}) = 33.7 + 10\log_{10}(125) + 10\log_{10}(0.360) + 44.4\log_{10}(4.00) - 54$$

$$L_W(\text{elbow}) = 33.7 + 21 + (-4.4) + 26.7 - 54 = 23.0\,\text{dB}$$

The sound power level in the elbow in the 600 mm branch run is found by combining the energy generated and the energy to the elbow:

$$L_W\text{ (in elbow)} = 10\log_{10}(10^{5.19} + 10^{2.30}) = 51.9\,\text{dB}$$

It is noted that the noise generated in the smaller elbow is negligible in this example.

The attenuation in the elbow is found as follows. The quantity $f_o S^{1/2} = (125)(0.600) = 75$ Hz-m. The attenuation of the elbow in the 125 Hz octave band is found from Table 5-18, $\Delta L_W(\text{elbow}) = 2.0\,\text{dB}$.

The attenuation due to reflection at the open end of the 600 mm square duct is found from Table 5-16 at a value $f_o D = 75$ Hz-m for the 125 Hz octave band, $\Delta L_W(\text{reflection}) = 3.4\,\text{dB}$.

The attenuation in the 8 m long section of the branch is as follows:

$$\Delta L_W(8\,\text{m duct}) = (0.47\,\text{dB/m})(8\,\text{m}) = 3.8\,\text{dB}$$

The total attenuation from the elbow to the open end of the duct in the 125 Hz octave band is found by adding the three contributions (elbow, straight length, and reflection):

$$\Delta L_W(\text{total}) = 2.0 + 3.4 + 3.8 = 9.2\,\text{dB}$$

The sound power level before the grille is as follows:

$$L_W\text{ (before grille)} = 51.9 - 9.2 = 42.7\,\text{dB}$$

The grille pressure drop coefficient, from Table 5-21, is $C_D = 2.9$. The overall power level generated by the flow through the grille may be found from Eq. (5-70):

$$L_W(\text{grille}) = 10 + 10\log_{10} S + 30\log_{10} C_D + 60\log_{10} u$$

$$L_W(\text{grille}) = 10 + 10\log_{10}(0.360) + 30\log_{10}(2.9) + 60\log_{10}(4.00)$$

$$L_W(\text{grille}) = 10 + (-4.4) + 13.9 + 36.1 = 55.6\,\text{dB (overall)}$$

The peak frequency of the grille power level spectrum occurs at the following frequency, according to Eq. (5-72):

$$f_m = (150)(4.00) = 600\,\text{Hz}$$

This frequency falls in the 500 Hz octave band, so the frequency 125 Hz corresponds to $f_m/4$, and $CF_g = 11$, from Table 5-22. The sound power level generated by the flow through the grille for the 125 Hz octave band is found from Eq. (5-73):

$$L_W(\text{grille, octave band}) = 55.6 - 11 = 44.6\,\text{dB} \ (125\,\text{Hz octave band})$$

The sound power level delivered from the grille to the room may be determined by combining the sound power level before the grille and the power level generated in the grille:

$$L_W \text{ (to room)} = 10\log_{10}(10^{4.27} + 10^{4.46}) = 46.8\,\text{dB}$$

This calculation completes the first part of the problem, which is to determine the sound power level input to the room. Next, let us use this data to determine the steady-state sound pressure level in the room. We will continue to present a sample calculation for the 125 Hz octave band and present the results for the other octave bands in Table 5-25.

Values for the room constant in each octave band are given in Table 5-25. The dimensionless parameter needed to determine the directivity factor for the grille opening is as follows:

$$\frac{f_o S^{1/2}}{c} = \frac{(125)(0.360)^{1/2}}{(346.1)} = 0.217$$

The directivity factor is found from Table 5-23, $Q(\theta = 0°) = 3.7$.
The second term in Eq. (5-6) may be calculated:

$$10\log_{10}\left(\frac{4}{R} + \frac{Q}{4\pi r^2}\right) = 10\log_{10}\left[\frac{4}{7.1} + \frac{(3.7)}{(4\pi)(5.00)^2}\right] = -2.4\,\text{dB}$$

TABLE 5-25 Solution for Overall Sound Pressure Level in Example 5-7

Item	Octave band center frequency, Hz							
	63	125	250	500	1,000	2,000	4,000	8,000
L_w to room	45.2	46.8	50.3	51.1	49.3	44.8	39.8	33.9
R, m² (given)	4.2	7.1	8.6	11.8	15.1	17.8	19.1	22.0
$f_o S^{1/2}/c$	0.109	0.217	0.433	0.867	1.73	3.47	6.93	13.87
$Q(\theta = 0°)$	2.8	3.7	4.8	6.0	7.0	7.7	7.9	8.0
$10\log_{10}\left(\frac{4}{R} + \frac{Q}{4\pi r^2}\right)$	−0.2	−2.4	−3.2	−4.5	−5.4	−6.0	−6.3	−6.8
L_p, dB	45.1	44.5	47.2	46.7	44.0	38.9	33.6	27.2
Background L_p	55	50	45	40	35	30	28	28
Total L_p, dB	55.4	51.1	49.2	47.5	44.5	39.4	34.7	30.6

Noise Sources

The sound pressure level in the 125 Hz octave band may be found from Eq. (5-6) for sound propagated indoors:

$$L_p = 46.8 + (-2.4) + 0.1 = 44.5 \, \text{dB}$$

The background sound pressure level in the 125 Hz octave band is given as 50 dB in Table 5-25. The total octave band sound pressure level may be found by combining the noise from the grille and the background noise.

$$\text{Total } L_p(\text{octave band}) = 10 \log_{10}(10^{4.45} + 10^{5.00}) = 51.1 \, \text{dB}$$

The overall sound pressure level is found by combining the octave band sound pressure level values:

$$L_p(\text{overall}) = 10 \log_{10}(10^{5.54} + 10^{5.11} + \cdots + 10^{3.06}) = 58.2 \, \text{dB}$$

5.12 TRAFFIC NOISE

Since the early 1950s, the number of cars and trucks on the highways in the United States has increased greatly, as shown in Table 5-26. Increased environmental noise levels have accompanied the growth in the transportation sector. The U.S. Environmental Protection Agency (EPA) and the U.S. Department of Transportation (DoT) have recognized that significant noise problems associated with traffic flow exist. These agencies have proposed standards for allowable noise emission for interstate traffic and for acceptable noise levels for highway planning and siting.

Empirical relationships have been developed that can be used to predict the hourly energy-equivalent A-weighted sound level for freely flowing traffic (Transportation Research Board, 1976). For purposes of the correlation, the major traffic noise sources were classified as automobiles, medium

TABLE 5-26 Growth of Highway Vehicle Traffic in the United States

	Year					
Item	1950	1960	1970	1980	1990	2000
Population (millions)	151	181	204	229	257	289
Automobiles (millions)	40.4	61.7	87.0	122.6	172.4	242.5
Autos per person	0.268	0.341	0.426	0.535	0.671	0.839
Trucks/buses (millions)	8.8	12.2	19.3	28.4	43.9	67.2
Motorcycles (millions)	0.45	0.51	1.2	1.8	2.5	3.1

Source: EPA (1971b).

trucks (pick-up trucks, for example) and heavy trucks (18-wheelers, for example). It was found that the noise produced by all types of vehicles was proportional to the vehicle volume V, vehicles/hour, and inversely proportional to the equivalent distance from the highway D_E, meters, raised to the 1.5 power. For automobiles and medium trucks, the noise is directly proportional to the vehicle speed S, km/hour, raised to the 2.0 power. For heavy trucks, however, the noise was found to be inversely proportional to the truck speed.

The equivalent distance from the highway to the observer is the geometric average of the distance from the observer to the centerline of the nearest traffic lane, D_N, and the distance from the observer to the centerline of the farthest traffic lane, D_F, as illustrated in Fig. 5-6:

$$D_E = (D_N D_F)^{1/2} \tag{5-74}$$

The average width of a traffic lane is approximately 3.8 m (12.5 ft).

The correlations for the A-weighted equivalent sound level for each type of vehicle are given as follows:

(a) Automobiles:

$$L_e(A) = 10 \log_{10} V - 15 \log_{10} D_E + 20 \log_{10} S + 16 \tag{5-75}$$

(b) Medium trucks:

$$L_e(M) = 10 \log_{10} V - 15 \log_{10} D_E + 20 \log_{10} S + 26 \tag{5-76}$$

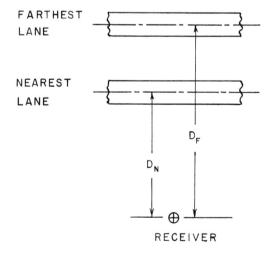

FIGURE 5-6 Illustration of the nearest and farthest lane distances for traffic noise.

Noise Sources

(c) Heavy trucks:

$$L_e(H) = 10\log_{10} V - 15\log_{10} D_E - 10\log_{10} S + 84 \qquad (5\text{-}77)$$

The total A-weighted sound level is found by combining the levels due to the three types of vehicle:

$$L_A = 10\log_{10}[10^{L_e(A)/10} + 10^{L_e(M)/10} + 10^{L_e(H)/10}] \qquad (5\text{-}78)$$

Certain adjustments must be made to the values obtained from the correlations to account for different highway conditions. If the road segment near the observer is not straight, then the roadway may be subdivided into several finite-length segments. The following factor Δ_1 must be added to each equivalent sound level value to correct for the fact that noise is generated from a finite length of roadway, instead of from a very long straight road. The geometry is illustrated in Fig. 5-7, where θ is the angle (expressed in degrees) subtended by the road segment:

$$\Delta_1 = 10\log_{10}(\theta/180°) \qquad (5\text{-}79)$$

The basic noise level is influenced by the condition of the road. The *normal* surface, from which the correlations were developed, consists of moderately rough asphalt or a concrete surface. For this surface, the adjustment $\Delta_2 = 0$. A *smooth* surface is one corresponding to seal-coated asphalt, for example. A very smooth surface road is not often encountered, because it has fairly low friction characteristics. For the smooth surface, the adjust-

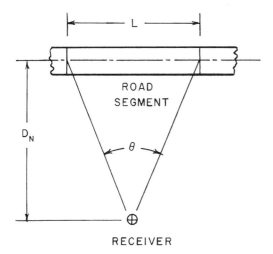

FIGURE 5-7 Road segment nomenclature.

ment $\Delta_2 = -5\,\text{dB}$. A *rough* surface is one corresponding to rough asphalt with large voids or for grooved concrete surfaces. For the rough surface, the adjustment $\Delta_2 = +5\,\text{dB}$. The adjustment should be applied only to automobile noise $L_e(A)$ and not to the truck noise. The noise generated by truck traffic is much less sensitive to the road condition than that generated by automobiles.

When trucks move upgrade, there is generally a need to change gears, with the resulting change in the noise emitted from the truck. For downgrade travel and for automobiles, no adjustment is needed ($\Delta_3 = 0$). The adjustment for trucks on an uphill grade (does not apply to cars) is given by the following expression:

$$\Delta_3 = \begin{cases} 0.6\,(\%G) & \text{for} \quad \%G < 8\% \text{ grade} \\ 5\,\text{dB} & \text{for} \quad \%G \geq 8\% \text{ grade} \end{cases} \quad (5\text{-}80)$$

Example 5-8. A four-lane highway has the following traffic data:

Automobiles: 1200 autos/hour at 90 km/hr (55.9 mph)
Medium trucks: 240 trucks/hour at 100 km/hr (62.1 mph)
Heavy trucks: 120 trucks/hour at 105 km/hr (65.2 mph)

The observer is located at a distance of 50 m (164 ft) from the centerline of the nearest lane. The distance from the observer to the centerline of the farthest lane is 65 m (213 ft). The road is straight and has an average surface. The grade is 3% uphill for trucks. Determine the equivalent A-weighted sound level due to traffic noise.

The equivalent distance between the observer and the highway is given by Eq. (5-74).

$$D_E = [(50)(65)]^{1/2} = 57.0\,\text{m} \;(187\,\text{ft})$$

The contribution of the automobiles to the noise is calculated from Eq. (5-75):

$$L_e(A) = 10\log_{10}(1200) - 15\log_{10}(57.0) + 20\log_{10}(90) + 16$$

$$L_e(A) = 30.8 - 26.3 + 39.1 + 16 = 59.6\,\text{dBA}$$

The medium truck contribution, without the grade correction, is calculated from Eq. (5-76).

$$L_e(M) = 10\log_{10}(240) - 15\log_{10}(57.0) + 20\log_{10}(100) + 26$$

$$L_e(M) = 23.8 - 26.3 + 40.0 + 26 = 63.5\,\text{dBA}$$

Noise Sources

The heavy truck contribution, without the grade correction, is found from Eq. (5-77):

$L_e(H) = 10 \log_{10}(120) - 15 \log_{10}(57.0) - 10 \log_{10}(105) + 84$

$L_e(H) = 20.8 - 26.3 - 20.2 + 84 = 58.3 \,\text{dBA}$

The road segment adjustment (straight road) and the surface adjustment (average surface) at zero. The grade adjustment for the trucks is found from Eq. (5-80):

$\Delta_3 = (0.6)(3.0) = 1.8 \,\text{dBA}$

The corrected values for the truck sound levels are as follows:

$L_e(M) = 63.5 + 1.8 = 65.3 \,\text{dBA}$

$L_e(H) = 58.3 + 1.8 = 60.1 \,\text{dBA}$

The overall sound level for the traffic noise is found by combining the three contributions according to Eq. (5-78):

$L_A = 10 \log_{10}(10^{5.96} + 10^{6.53} + 10^{6.01}) = 10 \log_{10}(5.324 \times 10^6)$

$L_A = 67.3 \,\text{dBA}$

5.13 TRAIN NOISE

The reaction of people to noise resulting from a train passing by differs from that produced by automobile and truck traffic. The noise due to the passage of a train has a definite beginning and ending and a finite duration. On the other hand, urban traffic noise is more or less continuous. There are fewer miles of train tracks than miles of highways, so train noise generally affects fewer people.

Railway noise in the community is often a short-term annoyance and not a threat for hearing damage. The ambient noise level is restored after the train has passed. Railway noise may produce a different psychological response than other noise sources. In fact, train sounds may be somewhat pleasant to retired railroad workers. As a result of these factors, train noise is often treated in terms of the community response to the noise of trains passing (Dept. of Transportation, 1978).

5.13.1 Railroad Car Noise

The standard railroad bed construction in the United States involves a tie-and-ballast construction. The ties are generally made of treated wood, and the ballast is a crushed rock aggregate placed between the ties and on drained and graded earth. The main function of the tie is to distribute the load from the steel rail section. This type of construction offers better sound attenuation than elevated structures or open concrete support structures. The sound generation correlations given in this section apply to the tie-and-ballast construction.

There are several contributions to railroad car noise generation, including (a) wheel/rail interaction, (b) car coupler interaction, and (c) vibration of structural components of the railroad car. When the railway and railroad car are properly maintained, these components are difficult to distinguish. As a general rule, the wheel/rail component is usually the main source of noise generated by a passing train (Vér, 1976).

The four main contributions to rail/wheel noise generation for railroad cars are (a) noise produced by rail roughness, (b) flat spots on the railroad car wheels, (c) gaps in the rail joints, and (d) rubbing of the wheel flange and the rail. The rubbing action of the wheel flange and the supporting rail can be significant for tracks with sharp curves. An increase in the noise level as much as 15 dBA has been reported (Cann et al., 1974). The high-frequency "squeal" and low-frequency "howling" sound of the railroad car going around a curve is usually not a major noise problem because the radius of most tracks is fairly large by design. The correlations in this section do not consider the effect of track curvature.

Impact noise occurs when a railroad car wheel with a flat spot rolls on the rail. The flat spots may result from non-uniform service wear or wear due to hard braking. When there are gaps between the joints of the rails, impact noise will occur when the railcar wheel moves across the joint. This noise is particularly noticeable if one rail is slightly higher at the joint than the adjoining rail.

There are several noise-control procedures that can be used to reduce the wheel/rail noise. The rails may be ground to provide a smoother and flatter rail surface, which reduces the noise by 3 to 6 dBA. The wheels may be turned or ground to eliminate flat spots. The wheel/rail noise may be reduced by as much as 8–10 dBA, depending on the severity of the wheel wear, by machining or grinding the wheels. Rail joints may be eliminated by using continuous welding of the rail joints. Noise reductions by as much as 8–10 dBA, depending on the degree of track unevenness, may be achieved by using continuous rails. Finally, some degree of noise reduction may be achieved by modifying the railcar support to include vibration damping in the suspension system (Lipscomb and Taylor, 1978).

The A-weighted sound level due to the passage of one train of cars, excluding the noise from the locomotive, is proportional to the time required for the train to pass, T_L/V, and proportional to the train speed, V, raised to the third power. The quantity T_L is the length of the train cars, not including the length of the locomotive, as shown in Fig. 5-8. The A-weighted noise

Noise Sources

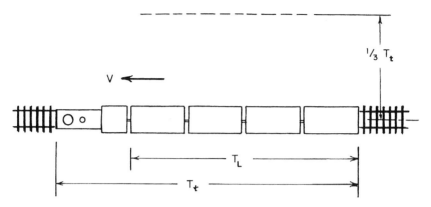

FIGURE 5-8 Train length for noise correlations.

level due to one train of cars passing at a distance $r_o = 30\,\text{m}$ (100 ft) is given by the following expression:

$$L_C = 10\log_{10}(T_L/V) + 30\log_{10}(V) + 43.5 \tag{5-81}$$

The total length of the railroad cars T_L is in units of meters, and the train speed V is in units of m/s. The average length of one railroad car is approximately 17.85 m (58.6 ft).

5.13.2 Locomotive Noise

Most of the locomotives in the United States are driven by diesel–electric systems. The diesel engines drive an onboard electric generator that, in turn, provides electrical energy to the drive-wheel electric motors.

The sources of noise from the diesel–electric locomotive include (a) diesel engine exhaust noise, (b) cooling fan noise, (c) engine structural vibration, and (d) traction motor blower. In addition, there is some noise generated due to wheel/rail interactions and vibration of the structural components of the locomotive body. The contribution of each of these noise sources is illustrated in Table 5-27 for a 3000 hp diesel–electric drive locomotive at full throttle. It is noted that diesel engine exhaust noise and cooling fan noise are predominant noise sources for the locomotive.

The exhaust system noise may be attenuated by about 6 dBA by using exhaust-driven turbochargers on the diesel engine. Exhaust silencers may also be used to reduce the exhaust noise. The installation of a silencer may present a difficult design problem, because of the limited space on board the locomotive. The noise from the locomotive under idle conditions is produced primarily by vibration of structural elements of the locomotive.

TABLE 5-27 Noise Contributions for a 3000 hp Diesel–Electric Driven Locomotive Under Full Throttle Conditions[a]

Noise source	L_A at 30 m, dBA (full throttle conditions)	Energy fraction, %
Engine exhaust	84	52
Cooling fan	83	41
Engine vibration	66.5	1
Traction motor blower	75	6
Overall sound level	87	100

[a]The noise levels are measured at a distance of 30 m (100 ft) from the locomotive.

Under idle conditions, the use of an exhaust silencer will not significantly influence the overall locomotive noise lvel.

The A-weighted sound level for a stationary locomotive at a distance of 30 m (100 ft) from the locomotive may be correlated by the following expression (Magrab, 1975):

$$L^\circ = 10 \log_{10}(\text{hp}) + 57.2 - \Delta_{tc} \qquad (5\text{-}82)$$

The quantity hp is the rating of the engine in horsepower; Δ_{tc} is 6 dB for a turbocharged engine and zero otherwise.

The A-weighted sound level due to the passage of N_L locomotives with a speed V at a distance r_o from the centerline of the tracks is given by the following expression:

$$L_L = L^\circ + 10 \log_{10}(\pi r_o/2V) + 10 \log_{10} N_L \qquad (5\text{-}83)$$

All locomotives are equipped with safety devices, such as horns, bells, or sirens. The sound from these devices can be 10–20 dB higher than the noise level of the train. The noise from these safety devices is usually considered as being necessary for the safe operation of the train, and is not considered when noise reduction procedures are proposed.

5.13.3 Complete Train Noise

The A-weighted sound level for one pass-by of the complete train, railroad cars plus locomotive, at a distance $r_o = 30$ m (100 ft) is found by combining the railroad car and locomotive noise levels:

$$L_1 = 10 \log_{10}(10^{L_C/10} + 10^{L_L/10}) \qquad (5\text{-}84)$$

Noise Sources

One of the purposes of predicting the train noise is to evaluate the noise impact on the areas surrounding the track. As discussed in Chapter 6, one parameter used as an indicator of community response to noise is the *day–night level*, L_{DN}. The day–night level is the energy-averaged A-weighted sound level with an extra (10 dB) emphasis on sound generated at night. The nighttime noise is usually more annoying than the same noise level occurring during the daytime. The day–night level due to pass-by of several trains at a distance $r_o = 30$ m is found from the following expression:

$$L_{DN}(r_o) = L_1 + 10 \log_{10} X - 49.37 \tag{5-85}$$

The quantity X is the effective number of pass-byes, with the nighttime traffic weighted 10 times as heavy as the daytime traffic:

$$X = N_d + 10 N_n \tag{5-86}$$

The quantity N_d is the number of pass-byes during the daytime, defined as the period between 7:00 a.m. and 10:00 p.m., and N_n is the number of pass-byes during the nighttime, defined as the period between 10:00 p.m. and 7:00 a.m.

The day–night sound level at any distance r from the centerline of the tracks depends on the distance. When the observer is within a distance equal to one-third of the total length of the train, T_t, the train radiates sound approximately as a line source:

$$L_{DN} = L_{DN}(r_o) - 10 \log_{10}(r/r_o) \quad \text{(for } r \leq T_t/3\text{)} \tag{5-87}$$

The reference distance is $r_o = 30$ m.

When the observer is located at a distance beyond one-third of the train length, the train appears more nearly as a point source, and the radiation approximates a spherical source. The day–night sound level in this case is given by the following expression:

$$L_{DN} = L_{DN}(r_o) - 10 \log_{10}(T_t/3r_o) - 20 \log_{10}(3r/T_t) \quad \text{(for } r > T_t/3\text{)} \tag{5-88}$$

The average length of one locomotive is 19.5 m (64 ft), and the range of locomotive lengths is from about 18 m (59 ft) to 21 m (68.9 ft). The average length of one railroad car is about 17.85 m (58.6 ft).

Example 5-9. A train is made up of two 2000 hp locomotives and 70 railroad cars. The train engine is not turbocharged. The train passes near the site of a proposed shopping center at a speed of 25 m/s (56 mph). The train passes four times during the day and two times during the night. The distance from the centerline of the tracks to the property line of the future shopping center is 240 m (787 ft). Determine the day–night sound level due to the pass-by of the trains. This information could be used in connection with a noise impact study of the shopping center site.

The total length of 70 railroad cars is found as follows, using the average car length:

$$T_L = (70 \text{ cars})(17.85 \text{ m/car}) = 1249.5 \text{ m } (4099 \text{ ft or } 0.776 \text{ miles})$$

The sound level due to one pass-by of the railroad cars is found from Eq. (5-81):

$$L_C = 10 \log_{10}(1249.5/25) + 30 \log_{10}(25) + 43.5$$
$$L_C = 17.0 + 41.9 + 43.5 = 102.4 \text{ dBA}$$

We note that the time required for the 70 railroad cars to pass by is $(1249.5/25) = 50$ sec.

The sound level generated by one stationary locomotive is found from Eq. (5-82):

$$L^\circ = 10 \log_{10}(2000) + 57.2 - 0 = 90.2 \text{ dBA}$$

The noise level due to two locomotives moving at 25 m/s is given by Eq. (5-83):

$$L_L = 90.2 + 10 \log_{10}[(\pi)(30)/(2)(25)] + 10 \log_{10}(2)$$
$$L_L = 90.2 + 2.8 + 3.0 = 96.0 \text{ dBA}$$

The combined sound level for the railroad cars and the locomotives at a distance of 30 m from the tracks is found from Eq. (5-84):

$$L_1 = 10 \log_{10}(10^{10.24} + 10^{9.60}) = 103.3 \text{ dBA}$$

The effective number of train pass-byes is found from Eq. (5-86):

$$X = N_d + 10 N_n = 4 + (10)(2) = 24$$

The day–night sound level at a distance of 30 m from the tracks is found from Eq. (5-85):

$$L_{DN}(r_o) = 103.3 + 10 \log_{10}(24) - 49.37 = 103.3 + 13.8 - 49.37$$
$$= 67.7 \text{ dBA}$$

The total length of the train is as follows:

$$T_t = (2 \text{ locomotives})(19.5 \text{ m/locomotives}) + 1249.5$$
$$= 1288.5 \text{ m } (4227 \text{ ft})$$

Then,

$$\tfrac{1}{3} T_t = (\tfrac{1}{3})(1288.5) = 429.5 \text{ m} > r = 240 \text{ m}$$

Noise Sources

The day–night level for the train noise at a distance of 240 m (787 ft) from the tracks is calculated from Eq. (5-87) in this case:

$$L_{DN} = 67.7 - 10\log_{10}(240/30) = 67.7 - 9.0 = 58.7\,\text{dBA}$$

This value is almost 4 dBA higher than the upper limit of 55 dBA recommended by the Environmental Protection Agency (EPA) for environmental noise.

PROBLEMS

5-1. A backward-curved-blade (BCB) centrifugal fan delivers $5.71\,\text{m}^3/\text{s}$ (12,100 cfm) of air against a static pressure of 218 Pa (0.875 in H_2O). The speed of the fan is 1468 rpm. The fan has 64 blades, and ducts are connected to both inlet and outlet of the fan. The fan is located in a room having a room constant of $4.00\,\text{m}^2$ at all frequencies, and the directivity factor for the fan is $Q = 2$. The fan housing transmission loss is given in Table 5-28. Determine the octave band sound pressure levels at a distance of 1.50 m (59.1 in) from the fan due to noise transmitted through the fan housing.

5-2. A propeller fan delivers $297.8\,\text{dm}^3/\text{s}$ (631 cfm) of air against a pressure head of 62.5 Pa (0.25 in H_2O). The fan has 4 blades and operates at 2400 rpm. The blade tone component of the noise from the fan is $B_T = 6\,\text{dB}$. Noise is radiated from the fan inlet to the outdoors, and there is negligible sound transmitted through the housing. The directivity factor for the fan is $Q = 2.455$. Determine the octave band sound pressure levels at a distance of 3.162 m (10.37 ft) from the fan inlet (outdoors). Attenuation by atmospheric air is negligible.

5-3. A radial-blade centrifugal fan used to move small pieces of scrap in air operates with a volumetric flow rate of $11.5\,\text{m}^3/\text{s}$ (24,370 cfm) against a static pressure head of 375 Pa (1.51 in H_2O). The fan speed is 1125 rpm, and the fan has 48 blades. The fan has both inlet and outlet ducts, so noise is transmitted to the outside through the fan housing only. The transmission loss for the housing is 12 dB for the

TABLE 5-28 Fan Housing Transmission Loss

	Octave band center frequency, Hz							
	63	125	250	500	1,000	2,000	4,000	8,000
Housing TL, dB	12	18	24	30	36	37	37	37

500 Hz octave band and 18 dB for the 1000 Hz octave band. The directivity index for the fan is DI = 3 dB. Determine the sound pressure level for the fan in the 500 Hz and in the 1000 Hz octave bands at a distance of 1.25 m (49.2 in) from the fan, if the fan is located outdoors.

5-4. A forward-curved-blade (FCB) centrifugal fan used in a residential air conditioning system handles a volumetric flow rate of 725 dm^3/s (1536 cfm) of air against a pressure rise of 375 Pa (1.51 in H$_2$O). The fan speed is 1482 rpm, and the fan has 42 blades. The fan has a duct connected on the outlet, but there is no duct on the fan inlet. Sound transmission through the fan housing is negligible. The fan is located outdoors, and the directivity factor for the fan is $Q = 4$. Determine the distance from the fan that an octave band sound pressure level of 55 dB is achieved in the 500 Hz octave band. Attenuation in the air around the fan is negligible.

5-5. A 50 hp (37.3 kW) drip-proof electric motor, operating at 1800 rpm, drives a centrifugal pump. The flow rate of water through the pump is 3.50 dm^3/s (7.416 cfm or 55.5 gpm), and the pressure rise through the pump is 3500 kPa (507.6 psi). The efficiency of the pump is 40%. The unit is located in a room that has a room constant of 2.50 m^2 at all frequencies. The directivity factor for the unit is $Q = 2$. Determine (a) the octave band sound pressure levels at a distance of 3.00 m (9.84 ft) from the unit and (b) the overall sound pressure level due to noise from the unit at a distance of 3 m from the unit. Note that the power (in units of hp) required to drive the pump is found from the following expression:

$$\text{hp} = \frac{Q_f \Delta P}{550 \varepsilon_p}$$

where Q_f is the volumetric flow rate (ft^3/sec), ΔP is the pressure rise (lb$_f$/ft^2), and ε_p is the pump efficiency. The numerical quantity in the denominator is a conversion factor for units, 550 ft-lb$_f$/hp-sec. If SI units are used (m^3/s and Pa), the conversion factor is 745.7 W/hp.

5-6. A 30 hp totally enclosed fan-cooled electric motor operates at 1800 rpm inside a room having a room cosntant of 20 m^2 (215 ft^2). The directivity factor for the motor is $Q = 4.00$. Determine (a) the A-weighted sound level and (b) the overall sound pressure level at a distance of 1.128 m (44.4 in) from the motor.

5-7. A 60 hp screw pump operates at 875 rpm in a room having a room constant of 200 m^2 (2153 ft^2). The pump is located near a corner of the room and the directivity factor for the pump is $Q = 7.30$. At a distance of 5.50 m (18.0 ft) from the pump, determine (a) the overall

Noise Sources

sound pressure level and (b) the sound pressure level in the 500 Hz octave band due to airborne noise from the pump.

5-8. A reciprocating pump operates at 1750 rpm and has a rated power requirement of 90 hp. The pump is located outdoors, and attenuation by the atmospheric air may be neglected. It is desired to locate the pump at a distance from a certain receiver location such that the overall sound pressure level caused by the pump does not exceed 95 dB. Determine the minimum distance between the pump and the receiver for $Q = 1$.

5-9. An axial flow gas compressor has a power rating of 2000 hp and operates at 3750 rpm. The number of blades in a single stage of the compressor is 32. The compressor is located outdoors, and the directivity factor for the compressor is $Q = 2$. At a distance of 60 m (197 ft) from the compressor, determine (a) the overall sound pressure level due to the compressor noise and (b) the octave band sound pressure level for the 1000 Hz octave band. Attenuation by atmospheric air may be neglected.

5-10. A 1000 kVA transformer is located in a room having a room constant of 2.20 m^2 (23.7 ft^2) at all frequencies. The transformer is forced-air cooled. The directivity factor for the transformer is $Q = 2$. Determine (a) the overall sound pressure level at a point 1.50 m (59.1 in) from the transformer and (b) the overall sound pressure level at a point 1.5 m from the transformer if the transformer is located outdoors. Attenuation by atmospheric air may be neglected.

5-11. An induced-draft cooling tower located outdoors has a propeller fan driven by a 25 hp motor. The directivity index for the cooling tower is $DI = 3 \text{ dB}$. Determine the distance from the cooling tower at which the overall sound pressure level is 70 dB. Determine the octave band sound pressure level in the 500 Hz and the 1000 Hz octave bands at this distance from the tower. Attenuation by atmospheric air may be neglected.

5-12. A natural-draft cooling tower has a water flow rate of 850 kg/s (1874 lb$_m$/sec) and a tower diameter of 9.50 m (31.2 ft). The total distance that the water falls in the tower is 11.4 m (37.4 ft). The distance between the bottom of the packing and the pond surface is 8.0 m (26.2 ft), and the packing material does not extend below the ring beam of the tower. Determine the A-weighted sound pressure level at a distance of 45 m (147.6 ft) from the tower.

5-13. Superheated steam at 1827 kPa (265 psia) and 500K (440°F) is vented to the atmosphere through a 4-in nominal SCH 40 pipe (inside diameter 102.3 mm = 4.028 in). The molecular weight for the steam is 18.016 g/mol, and the sonic velocity is 553 m/s (1814 fps). Determine the octave band sound pressure levels for the steam vent noise at a distance of 30 m (98.4 ft) from the vent and at an angle of 90° from the vent axis for the following octave bands: 250 Hz, 500 Hz, and 1000 Hz. Attenuation by the atmospheric air is negligible.

5-14. An air vent is located inside a room having a room constant of 125 m^2 (1345 ft^2). The air immediately upstream of the vent is at 810 kPa (117.5 psia) and 300K (80°F). It is desired to limit the overall sound pressure level to 85 dB at a distance of 3 m (9.84 ft) from the vent and at an angle of 120° from the vent axis. Determine the required diameter of the vent to achieve this condition.

5-15. The air vent from a compressed air source is a pipe having an inside diameter of 20.9 mm (0.824 in). The air in the line is at 305K (31.8°C or 89.3°F), for which the sonic velocity is 350 m/s (1148 fps). The vent line is located outdoors, and attenuation by atmospheric air is negligible. The measured octave band sound pressure level in the 500 Hz octave band 9 m (29.5 ft) from the vent at an angle of 30° from the vent axis is 55 dB. Determine (a) the pressure of the air in the vent line before the outlet of the pipe and (b) the overall sound pressure level.

5-16. A standard ball valve is located in a room with a room constant of 100 m^2 (1076 ft^2). The pressure of the air upstream of the valve is 950 kPa (137.8 psia) and the air temperature is 305K (89.3°F or 549°R). The flow-sizing coefficient for the valve is $C_g = 2100$ and the pressure drop across the valve is 500 kPa (72.5 psi). The valve is placed in an SCH 40 steel pipe. Determine (a) the A-weighted sound level at a distance of 800 mm (31.5 in) from the valve and (b) the volumetric flow rate through the valve.

5-17. A globe valve with standard trim ($C_1 = 30$; $C_g = 5850$) is located in a 4-in nominal (100-mm nominal) SCH 10 pipe (thickness, 3.05 mm = 0.120 in). The material of the pipe is stainless steel (density, 7920 kg/m^3 or 126.9 lb$_m$/ft^3). The gas flowing in the valve is oxygen (molecular weight, 32.0 g/mol), which enters the valve at 2500 kPa (362.6 psia) and 310K (558°R or 98.3°F). The pressure drop across the valve is 1500 kPa (217.6 psi). Determine the A-weighted sound level at a distance of 10 m (32.8 ft) from the valve, if the valve is located outdoors.

5-18. A globe valve with standard trim ($C_V = 110$) handles water at 25°C (77°F). The inlet pressure of the water is 500 kPa (72.5 psia), and the volumetric flow rate through the valve is 13.88 dm^3/s (220 gpm). The valve is located in a steel pipe of standard thickness (SCH 40). The vapor pressure of the water at 25°C is 3.17 kPa (0.460 psia). Determine the A-weighted sound level due to noise generated by

Noise Sources 221

flow through the valve at a distance of 3.65 m (12.0 ft) from the valve, if the valve is located outdoors.

5-19. Estimate the attenuation ΔL_W in the 500 Hz octave band and in the 4000 Hz octave band for a rectangular duct having dimensions 406 mm (16 in) by 457 mm (18 in). The length of the duct is 5.50 m (18.04 ft). The duct is internally lined with an acoustic material having an acoustic absorption coefficient $\alpha = 0.20$. Determine the attenuation for the same duct, if the duct is unlined.

5-20. A 90° elbow in a 406-mm (16-in) diameter duct has no turning vanes. The flow rate of air through the elbow is 1510 dm³/s (3200 cfm). For the 500 Hz octave band, determine the attenuation in the elbow and the flow-induced noise (dB) in the elbow.

5-21. The main duct before a 90° tee has a diameter of 406 mm (16 in) and an air flow rate of 1510 dm³/s (3200 cfm). The main duct after the tee has a diameter of 356 mm (14 in) and a flow rate of 910 dm³/s (1928 cfm). The branch duct has a diameter of 305 mm (12 in) and a flow rate of 600 dm³/s (1271 cfm). Determine the noise generated and transmitted into the side branch for the 500 Hz octave band. If the sound power level in the 500 Hz octave band before the tee is 65 dB, determine the sound power level after the tee in the branch duct in the 500 Hz octave band, including the effect of noise generation and sound transmitted from the main duct into the branch duct.

5-22. A circular ceiling diffuser has a diameter of 254 mm (10 in) and handles a flow rate of 250 dm³/s (530 cfm) of air at 280K (44°F). Determine the overall flow generated noise level for the grille, and the octave band noise generated (dB) in the 500 Hz octave band for the grille. If the density of the air is 1.260 kg/m³ (0.0787 lb$_m$/ft³), determine the pressure drop across the grille.

5-23. A commercial building is proposed for a site near an interstate highway, and an assessment of the anticipated traffic noise level at the site is required. From traffic records, the average traffic volume on the interstate highway near the site is 845 cars/hr; 275 medium trucks/hr; and 45 heavy trucks/hr. The average speeds of the vehicles are cars, 100 km/hr (62.1 mph); medium trucks, 115 km/hr (71.5 mph); and heavy trucks, 110 km/hr (68.4 mph). The distance from the centerline of the nearest lane to the property line is 120 m (394 ft), and the distance from the centerline of the farthest lane to the property line is 150 m (492 ft). At the proposed site, the interstate is straight and has a "normal" surface. The grade (affecting truck noise only) is 3% uphill. Determine the overall A-weighted sound level due to the traffic noise.

5-24. A company is planning to build an office building near a two-lane highway on which only automobile traffic is allowed. The average speed of the automobiles is 73.3 km/hr (45.5 mph), and the traffic volume is 100 cars/hr. The highway has a "normal" surface and is straight near the office site. The distance between the centerlines of the two lanes is $(D_F - D_N) = 4$ m (13.1 ft). The company requires that the A-weighted sound level due to the traffic noise be 45 dBA or less. Determine the minimum distance between the office site and the centerline of the nearest lane of the highway to achieve this condition.

5-25. The nearest lane of a highway is located 18.3 m (60.0 ft) from a hospital, and the farthest lane is 21.9 m (71.85 ft) from the hospital. Traffic is limited to automobiles only, and there are 90 automobiles/hr passing by the hospital. The road surface is "average," and the highway is straight at the hospital location. Determine the maximum automobile speed required to limit the traffic noise at the hospital location to 50 dBA.

5-26. Trains passing near a college campus have an average number of 40 railroad cars pulled by 2 locomotives at a speed of 15 m/s (33.6 mph). The railroad car length is 17.85 m (58.6 ft), and the locomotive length is 19.5 m (64.0 ft). The locomotives each have a rated horsepower of 1800 hp, and they have no turbochargers. There are 4 pass-byes during the daytime and 1 pass-by during the nighttime. Determine the day–night level due to the train pass-byes at a distance of (a) 40 m from the track and (b) 400 m from the track.

5-27. At a certain location near a railway line, 5 trains pass during the daytime and 2 trains pass during the nighttime. The trains have one 3000 hp locomotive 20 m (65.6 ft) long, with no turbocharger, and 40 railroad cars, each 17.85 m (58.6 ft) long. The average speed of the trains is 20 m/s (44.7 mph). At what distance from the tracks will the day–night level due to the train noise be 50 dBA?

REFERENCES

American Gas Association. 1969. Noise control for reciprocating and turbine engines driven by natural gas and liquid fuel. American Gas Association, New York.

ASHRAE. 1991. *ASHRAE Handbook. HVAC Applications.* American Society of Heating, Refrigerating and Air-Conditioning Engineers, Atlanta, GA.

Avallone, E. A. and Baumeister, III, T. 1987. *Marks' Standard Handbook for Mechanical Engineers,* 9th edn, pp. 14-48–14-56. McGraw-Hill, New York.

Noise Sources

Baumann, H. D. 1970. On the prediction of aerodynamically created sound pressure level of control valves, ASME Paper 70-WA/FE-28. American Society of Mechanical Engineers Winter Annual Meeting, New York.

Baumann, H. D. 1987. A method for predicting aerodynamic valve noise based on modified free jet theories, ASME Paper 87-WA/NCA-7. American Society of Mechanical Engineers Winter Annual Meeting, New York.

Beranek, L. L. and Vér, I. L. 1992. *Noise and Vibration Control Engineering*, p. 546. John Wiley and Sons, New York.

Bullock, C. E. 1970. Aerodynamic sound generation by duct elements. *American Society of Heating, Refrigerating and Air-Conditioning Engineers (ASHRAE) Trans*, Part II, 76: 97–108.

Burgess Industries. 1966. *Silencing Handbook*. Burgess Industries, Dallas, TX.

Cann, R. G., Fredberg, J. J., and Manning, J. E. 1974. Prediction and control of rail transit noise and vibration—a state of the art assessment. DoT Report PB 233 363. U.S. Department of Transportation, Washington, DC.

Dept. of Transportation. 1978. Transport of solid commodities via freight pipeline: noise impact assessment. Dept. of Transportation RSPA/DPB-50/78/35. U.S. Dept. of Transportation, Office of University Research, Washington, DC.

Diehl, G. H. 1972. Stationary and portable air compressors. Proceedings of the InterNoise 72 Conference, Tutorial Papers on Noise Control, pp. 154–158, Washington, DC.

Ellis, R. M. 1971. Cooling tower noise generation and radiation. *Sound and Vibration* 14: 171–182.

Environmental Protection Agency. 1971a. Median sound power levels for various types of equipment and operation, building equipment and home appliances. Report No. NTID 300.1. U.S. Environmental Protection Agency, Washington, DC.

Environmental Protection Agency. 1971b. Transportation noise and noise from equipment powered by internal combustion engines. Report No. NTID 300.13. U.S. Environmental Protection Agency, Washington, DC.

Faulkner, L. L. 1976. *Handbook of Industrial Noise Control*, pp. 473–477. Industrial Press, New York.

Graham, J. B. 1972. How to estimate fan noise. *Sound and Vibration* 6: 24–27.

Heitner, I. 1968. How to estimate plant noises. *Hydrocarbon Processing* 47: 67–74.

Hubert, M. 1970. Untersuchungen über Geräusche durchströmter Gitter, doctoral dissertation, Berlin Technical University, Berlin.

Lighthill, M. J. 1952. On sound generated aerodynamically. Proc. Roy. Soc. A211: 564.

Lipscomb, D. M. and Taylor, A. C. 1978. *Noise Control Handbook of Principles and Practices*, pp. 274–278. Van Nostrand Reinhold, New York.

McQuiston, F. C. and Parker, J. D. 1994. *Heating, Ventilating and Air Conditioning*, 4th edn, pp. 442–445. John Wiley and Sons, New York.

Magrab, E. A. 1975. *Environmental Noise Control*, pp. 161–162, John Wiley and Sons, New York.

Nakano, A. 1968. Characteristics of noise emitted by valves. Paper F-5-7, 6th International Congress of Acoustics, Tokyo, Japan.

Stiles, G. F. 1974. Identification, prediction, and attenuation of control valve noise. In: *Reduction of Machinery Noise*, pp. 286–297, Purdue University Press, Lafayette, IN.

Thumann, A. and Miller, R. K. 1986. *Fundamentals of Noise Control Engineering*, pp. 77–78. The Fairmont Press, Atlanta, GA.

Transportation Research Board. 1976. Highway noise generation and control. National Cooperative Highway Research Program (NCHRP) Report No. 173. Bolt, Beranek and Newman, Boston, MA.

Vér, I. L. 1976. Wheel/rail noise: impact noise generation by wheel and rail discontinuities. *J Sound Vibration*, 46: 25.

6
Acoustic Criteria

One of the first steps in the design of a system for noise reduction is to establish the acoustic criteria for the physical situation. This step is similar in principle to the determination of the failure mode in mechanical design. If rupture or breaking of the part constitutes failure, the ultimate strength of the material is used in the design of the part. The maximum stress to which the part is subjected is limited to a stress less than the ultimate strength of the material. There are different failure criteria for different cases. For example, if the part were subjected to dynamic loading (time-varying stress), the fatigue strength would be the material property that would be important in the design of the part. In some cases, excessive deflection may constitute failure, and stress is not involved in limiting the size of the part.

Similarly, we must determine the "failure criteria" for the specific acoustic design, so that we may design the system to prevent this "failure." In acoustic design, as well as in mechanical design, there are several different criteria for different applications. In some cases, the designer seeks to avoid permanent hearing loss for workers in an industrial area. In other cases, one may desire to avoid annoyance and unpleasant reactions from the community near a plant or other source of noise. Finally, the acoustic designer may wish to reduce the noise so that the noise does not interfere with the workers' communication or performance of their assigned tasks. We will examine some of these acoustic criteria and their applications in this chapter.

6.1 THE HUMAN EAR

To gain an appreciation of the damaging effects of sound on the human ear, one must understand the physical construction of the ear. The human ear is a remarkable acoustic system. The ear is capable of responding to sounds over a frequency range from about 16–20 Hz up to frequencies in the 16–20 kHz range. In addition, the ear can detect acoustic pressures as low as 20 µPa at a frequency of 1000 Hz and can withstand acoustic pressures as large as 2000 Pa for short times.

The acoustic particle velocity for sound in air at 20°C (68°F) for an acoustic pressure of 20 µPa may be calculated from Eq. (2-9):

$$u = \frac{p}{\rho c} = \frac{(20)(10^{-6})}{(413)} = 48.4 \times 10^{-9}\,\text{m/s} = 1.9 \times 10^{-6}\,\text{in./sec} = 0.165\,\text{in/day}$$

The corresponding particle displacement for a frequency of 1000 Hz may be found from Eq. (4-43):

$$\xi = \frac{u}{2\pi f} = \frac{(48.4)(10^{-9})}{(2\pi)(1000)} = 7.70 \times 10^{-12} = 7.70\,\text{pm} = 3 \times 10^{-10}\,\text{in}$$

The diameter of the nitrogen molecule is about 380×10^{-12} m or 380 pm (Reid and Sherwood, 1966). The human ear can detect particle displacements that are almost 1/50 of the diameter of a nitrogen molecule.

The human ear is one of the more intricate and complex mechanical structures in the body. As shown in Fig. 6-1, the ear consists of three main parts:

1. The outer ear, consisting of the *pinna* or visible ear, which acts as a horn to collect sound, and the *meatus* or auditory canal, which is terminated by the *tympanic membrane* or eardrum.
2. The middle ear, which involves three small bones: the *malleus* or "hammer", the *incus* or "anvil," and the *stapes* or "stirrup". These bones of the middle ear serve to transform the pressure variations in the air in the outer ear into mechanical motion. The *eustachian tube* in the middle ear serves to equalize the pressure between the outer and inner ear volumes.
3. The inner ear, which contains the *semicircular canals*, the fluid gyroscope associated with maintaining balance of the body, and the *cochlea*, which analyzes, converts, and transmits information about sound from the outer ear to the brain through the *auditory nerves*.

Acoustic Criteria

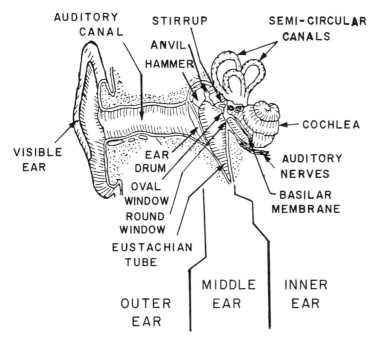

FIGURE 6-1 Cross-section of the human ear. (From *Engineering Principles of Acoustics*, D. D. Reynolds, 1981. By permission of Allyn and Bacon, Inc.)

The auditory canal acts as a resonator tube to increase the sound pressure level of the sound striking the visible ear by 10 dB to 20 dB at the eardrum, depending on the frequency of the sound. The resonant frequency of the auditory canal is on the order of 3 kHz, so the acoustic pressure increase is more pronounced in the 2–4 kHz octave bands. The approximate length of the auditory canal is 25–30 mm (1–1$\frac{1}{4}$ in).

The mechanical motion of the eardrum is transmitted and amplified by about 25 dB through the three-bone linkage in the middle ear. The hammer, attached at one end directly to the eardrum, is normally locked to the anvil. The anvil drives the stirrup, which is mounted into and sealed around the periphery of the oval window by a network of elastic fibers. When the ear is subjected to very intense sound, the contact between the hammer and the anvil is broken, so the three-bone set acts as a safety device to prevent damage to the oval window.

The main part of the inner ear is the cochlea, which is a bony tube about 34 mm (1.34 in) long, filled with liquid and coiled like a snail's shell. The cochlea makes about $2\frac{3}{4}$ turns around a central hollow passage that contains the nerve fibers going to the brain. The cochlea is illustrated in

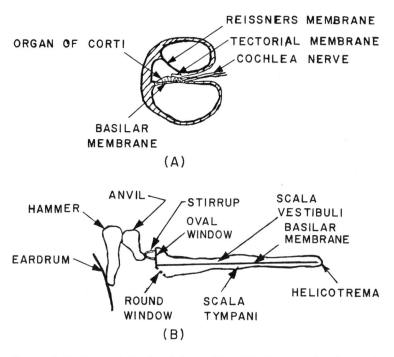

FIGURE 6-2 Internal details of the cochlea. (A) Cross-sectional view through one turn of the cochlea. (B) The cochlea is shown "rolled out" straight, instead of its actual coiled configuration. (From *Engineering Principles of Acoustics*, D. D. Reynolds, 1981. By permission of Allyn and Bacon, Inc.)

Fig. 6-2. There is a bony projection or shelf and a membrane called the *basilar membrane* that runs the length of the cochlea. The basilar membrane divides the cochlea into two chambers, the upper chamber or *scala vestibuli*, and the lower chamber or *scala tympani*. There is a small opening at the end of the cochlea, called the *helicotrema*, which provides a connecting passage between the upper and lower chambers. The basilar membrane varies in width from 0.2 mm (0.008 in) at the oval window to about 0.5 mm (0.020 in) at the end of the cochlea chamber.

The *organ of corti* is mounted about halfway along the spiral of the cochlea on the basilar membrane. The organ of corti is made up of about 30,000 hair cells, arranged in four rows, which are attached to the tectorial membrane in contact with the upper surface of the organ of corti. Any movement of the basilar membrane supporting the hair cells will cause the hair cells to bend. The bending of the small hairs produces the nerve impulses in the neurons that are transmitted to the brain. This is the component of the ear that can become permanently destroyed through long-

Acoustic Criteria

term exposure to loud noise. The hair cells become fatigued due to exposure to prolonged bending stress, and the cells die. When the hair cells die, they cannot be rejuvenated or resurrected. The person suffers permanent hearing loss when the hair cells die. No amount of "pre-conditioning" will strengthen the hairs to resist exposure to loud noise.

6.2 HEARING LOSS

Because of the acoustic characteristics of the outer ear and the mechanical characteristics of the middle ear, the human ear does not act as a linear transducer for sound pressure levels. The threshold of hearing as a function of frequency is given in Table 6-1. This table presents values of the sound pressure level for a pure tone that a person (below age 18 with no hearing loss) is just able to hear at the given frequency. It may be noted that the ear is most sensitive in the frequency range around 3000 Hz, which corresponds to the resonant frequency of the auditory canal. Because of poor acoustic impedance matching between the air outside the ear and the outer ear at frequencies below about 500 Hz, the ear can detect only sounds that have a sound pressure level greater than about 12 dB for frequencies of 250 Hz and lower.

For a sound pressure level of approximately 120 dB with a frequency between 500 Hz and 10 kHz, a person will experience a tickling sensation in the ears. This level represents the threshold of "feeling" or the beginning of discomfort due to noise. When the sound pressure level is increased above approximately 140 dB, the threshold of pain is reached. Continuous exposure to noise above 140 dB for a few minutes can result in permanent damage to elements of the ear.

TABLE 6-1 Threshold of Hearing[a]

Frequency, Hz	L_p(threshold), dB	Frequency, Hz	L_p(threshold), dB
31.5	59.2	2,000	2.4
63	36.0	3,000	−4.1
125	21.4	4,000	−3.6
250	12.1	5,000	0.2
500	6.5	6,000	5.2
1,000	3.6	8,000	17.3

[a] The table lists the sound pressure level of a pure tone that a person under age 18 with no hearing loss is just able to hear at the given frequency.
Source: ANSI (1967).

Hearing loss is defined as the change (increase) in the threshold of hearing at a given frequency. There is a naturally occurring loss of hearing that occurs with age, independent of occupational noise exposure. This hearing loss is called *presbycusis*. The shift in the hearing threshold with age is shown in Table 6-2 for men and women. The hearing loss that occurs with age is not included in the component of hearing loss associated with noise exposure.

There is a *temporary threshold shift* (TTS), in which a person loses some ability to detect weak sounds, but the ability is regained approximately 16 hours after the noise exposure is removed (Kryter, 1970). Noises having maximum energies in the low-frequency range (below about 250 Hz) will produce less TTS than noises having maximum energies in the high-frequency range (above about 2000 Hz). Exposure to a low-pitched "rumble" noise is less harmful to a person's hearing than exposure to a high pitched "screech." TTS cannot be relieved by medication, not even vitamin A: only getting away from the source of noise into a quieter region will promote recovery.

There is also a *noise-induced permanent threshold shift* (NIPTS), in which a person permanently loses the ability to detect weak sounds. The frequency range showing the most NIPTS is around 3 kHz, because the ear transmits sound at frequencies in the range from 1 kHz to 4 kHz most effectively. If a person is removed from the noisy environment, the

TABLE 6-2 Shift in the Average Threshold of Hearing with Age (Presbycusis) for Men and Women

Age, years	Frequency, Hz							
	500		1,000		2,000		4,000	
	Men	Women	Men	Women	Men	Women	Men	Women
25	0	0	0	0	1	0	4	0
30	0	1	1	1	2	1	8	2
35	1	3	2	3	3	4	12	4
40	2	4	4	4	6	5	17	7
45	4	6	5	6	8	8	23	10
50	5	8	7	8	12	10	28	13
55	7	10	9	11	16	13	35	18
60	8	12	11	13	20	15	31	22
65	10	14	13	15	24	18	47	26

Source: Beranek (1960).

Acoustic Criteria

NIPTS does not progress further, but the ear does not recover. The ear does not get "toughened" through exposure to noise. Excessive noise always causes hearing loss, even in teenagers who may feel they are bulletproof.

From work conducted at the Air Force Aerospace Medical Research Laboratory (Baughn, 1973), it was found that exposure to noise at levels less than 90 dBA during the normal 8-hour work day would result in 25 dB or more NIPTS for only 10% or less of the population. The complete results of the research are shown in Fig. 6-3.

6.3 INDUSTRIAL NOISE CRITERIA

Regulations for control and limitation of noise date back to early Roman times, when chariot races were prohibited on village streets at night because of the noise associated with the racing. On the other hand, noise was considered beneficial during the Middle Ages, when a town filled with noise

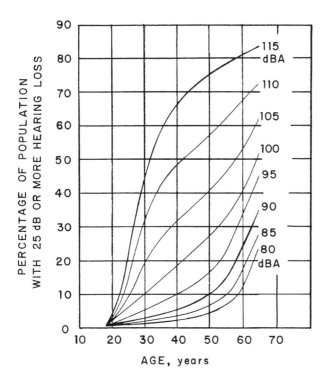

FIGURE 6-3 Percentage of workers developing a 25 dB or more hearing loss due to exposure to continuous A-weighted levels of noise. (From Baughn, 1973.)

indicated prosperity and health of the population. One of the main reasons for promotion of industrial noise control programs, however, is to prevent hearing impairment of workers due to occupational noise exposure.

Noise exposure criteria were developed in 1965 by the National Academy of Sciences and the National Research Council, Committee on Hearing, Bioacoustics and Biomechanics or CHABA (Kryter et al., 1965). The criteria for safe noise exposure levels were described in terms of pure tones and 1/3 octave and octave band data for noise. An acceptable noise level would produce NIPTS after 10 or more years of no more than 10 dB at 1 kHz and below, 15 dB at 2 kHz, and 20 dB at 3 kHz or higher. The CHABA criteria provide a good tool for hearing damage control; however, the criteria are not simple to use in industrial environments.

A safety regulation on industrial noise exposure was added to the Walsh–Healy Act in 1969. The Occupational Safety and Health Act of 1970 extended the scope of noise control legislation to all workers involved in interstate commerce activities. The Occupational Safety and Health Administration (OSHA) adopted a noise exposure limit of 90 dBA for an 8-hour working period. Higher noise exposures were allowed for shorter periods of time. For each 5 dBA increase of the noise exposure above 90 dBA, the allowed exposure time was halved. The daily noise exposure of 90 dBA corresponds to a noise level for which 10% or less of the population will experience a permanent hearing loss of 25 dB or less at 50 years of age, as indicated by Fig. 6-3.

The OSHA noise exposure criteria were verified in 1983 (OSHA, 1983). According to the OSHA criteria, continuous exposure to noise levels greater than 115 dBA are not permitted for any duration. The action level or level of noise exposure at which hearing conservation measures must be initiated was set at 85 dBA. The upper limit for impulsive noise exposure was set at 140 dBA. The OSHA regulation made two major provisions: (a) maximum levels of industrial noise exposure were set and an employee could not be exposed to noise levels exceeding these limits without hearing protection, and (b) required action by the employer was indicated if these noise levels were exceeded.

The permissible noise level for various times of exposure T is determined from the following expression:

$$L_A = 85 + \frac{5\log_{10}(16/T)}{\log_{10}(2)} \quad \text{(for } T \leq 16 \text{ hours)} \tag{6-1}$$

The permissible time for exposure to a continuous noise level L_A is determined from the following relationship:

Acoustic Criteria

$$T = \frac{16}{2^{(L_A - 85)/5}} \quad \text{(for 85 dBA} \leq L_A \leq 115 \, \text{dBA)} \quad (6\text{-}2)$$

The sound levels are measured on the A-scale of a standard sound level meter set on the "slow" response.

When the daily noise exposure is composed of two or more periods of noise exposure at different levels, the combined effect must be considered instead of the individual effect of each level. The following expression is used to determine the *noise exposure dosage* (NED) for situations in which the noise level varies during the working period:

$$\text{NED} = \frac{C_1}{T_1} + \frac{C_2}{T_2} + \frac{C_3}{T_3} + \cdots + \frac{C_n}{T_n} \quad (6\text{-}3)$$

The quantities C_1, C_2, etc., are the total times of exposure to the noise levels L_{A1}, L_{A2}, etc., in hours per day. The quantities T_1, T_2, etc., are the total permitted exposure times at the noise levels L_{A1}, L_{A2}, etc. For noise levels $L_A < 85 \, \text{dBA}$, $T = \infty$; for noise levels $L_A > 115 \, \text{dBA}$, $T = 0$.

If the noise exposure dosage exceeds 1.00, the employee noise exposure is in excess of the OSHA limits. Note that this procedure is similar in concept to the Palmgren–Minor linear damage rule for cumulative fatigue damage in mechanical parts (Collins, 1981).

If the noise level exceeds the allowable values according to the OSHA criteria, the employer should first conduct a noise survey to locate the areas in which the OSHA limits are exceeded and to locate the specific source of the noise. Next, engineering measures or controls should be implemented to attempt to reduce the worker nose exposure. Some examples of engineering controls measures include:

1. Substitution of quieter machinery, such as using larger slower machines, using belt drives instead of gear drives, or redesigning the equipment for lower noise emission.
2. Substitution of manufacturing processes, such as using welding instead of riveting.
3. Replacement of worn or loose parts.
4. Installation of vibration dampers and isolators.
5. Installation of flexible mountings and connectors.
6. Place the noise source within an enclosure or place acoustic barriers between the worker and the noise source.
7. Isolate the worker from the noise source by placing the worker and the machine controls in an acoustically treated room.

If engineering measures or controls are not feasible, then administrative controls should be examined. Some examples of administrative control measures include:

1. Arrange the work schedule such that the employee noise exposure is limited to acceptable values.
2. Increase the number of workers assigned to a specific task such that work in noisy areas can be completed in shorter times.
3. Perform occasional tasks involving work in high noise areas when a minimum number of employees will be exposed to the noise.

If neither engineering nor administrative control measures are practical, then the employee must be provided with personal hearing protection equipment and trained in the proper use of the equipment. It is the responsibility of the employer to enforce the proper use of hearing protection equipment. When noise levels of 85 dBA or higher are present, the employer must put in place an audiometric testing program. Audiometric testing must be conducted on each individual working in high-noise areas at regular intervals of time, often on an annual basis. Records of the audiometric tests and of daily worker noise exposure must be maintained.

Example 6-1. An employee works 1 hour where the sound level is 90 dBA. The worker inspects gauges and other items for 2 hours where the sound level is 92 dBA. A total of 3 hours is spent in an area around a compressor where the sound level is 94 dBA. The remaining 2 hours are spent in a relatively quiet office area where the sound level is 60 dBA. Is this employee's noise exposure in violation of the OSHA regulations?

The allowable time of exposure at each level is calculated from Eq. (6-2). For example, for a level of $L_{A_2} = 92 \, \text{dBA}$, we find the following allowable exposure time:

$$T_2 = \frac{16}{2^{(92-85)/5}} = 6.063 \text{ hours} = 6 \text{ hours } 3.8 \text{ minutes}$$

The other times are as follows:

$T_1 = 8$ hours at 90 dBA and $C_1 = 1$ hour
$T_2 = 6.063$ hours at 92 dBA and $C_2 = 2$ hours
$T_3 = 4.595$ hours at 94 dBA and $C_3 = 3$ hours
$T_4 = \infty$ at 60 dBA and $C_4 = 2$ hours

Acoustic Criteria

The noise exposure dosage is found from Eq. (6-3):

$$\text{NED} = \frac{1}{8} + \frac{2}{6.063} + \frac{3}{4.595} + 0 = 0.1250 + 0.3299 + 0.6529$$

$$\text{NED} = 1.1078 > 1$$

The noise exposure *does* exceed the OSHA limits.

What can we do about this noise exposure problem? First, we may try engineering measures or controls. The largest exposure (0.6529) occurs in the compressor room. The compressor could have acoustic treatment applied to reduce its noise generation. The required sound level in the compressor room may be found from Eq. (6-3):

$$\text{NED} = 1 = 0.1250 + 0.3299 + \frac{3}{T_3}$$

$$T_3 = \frac{3}{1 - 0.4549} = 5.504 \text{ hours}$$

The corresponding sound level is found from Eq. (6-1):

$$L_{A_3} = 85 + \frac{(5)\log_{10}(16/5.504)}{\log_{10}(2)} = 85 + 7.7 = 92.7 \text{ dBA}$$

We could apply acoustic material over the compressor surface or apply acoustic treatment to the walls and ceiling of the compressor room to reduce the sound level from 94 dBA to 92.7 dBA, or a reduction of only 1.7 dBA, which is probably feasible.

An alternative solution would be to use administrative measures or controls. For example, we could use two workers to complete the tasks in the compressor room in a shorter period of time. The maximum time for the workers in the compressor room is found from Eq. (6-3):

$$\text{NED} = 1 = 0.1250 + 0.3299 + \frac{C_3}{4.595}$$

$$C_3 = (4.595)(1 - 0.4549) = 2.50 \text{ hours} = 2 \text{ hours } 30 \text{ minutes}$$

The workers could spend the time difference $(3.00 - 2.50) = 0.50$ hour in the area where the sound level is less than 85 dBA, and compliance with OSHA requirements would be achieved.

6.4 SPEECH INTERFERENCE LEVEL

When exposed to two different sounds at the same time, the ear often perceives only the louder sound. This phenomenon is called *masking*. Masking results when the receptors in the cochlea are not available for processing the particular sound information because they are being stimulated by another signal. A noise signal that is spread out over a range of frequencies results in more masking near the center frequency than a pure tone at this frequency (Ehmer, 1959).

If the background noise level is excessive, a person may not be able to carry on a conversation or understand a telephone conversation. Because noise can interfere with speech intelligibility, this noise may disrupt work where communication is necessary.

The consonants contain much of the information conveyed in speech. The consonants are more easily masked by background noise than are the vowels, because the sounds of the consonants are generally weaker than those of the vowels. Nearly all of the information in speech is contained in the frequency range from about 200 Hz to 6000 Hz. The understanding of communication (speech intelligibility) is influenced by the type of communication (technical information is less readily transmitted than "small talk"), whether the two people "know" each other well or are relative strangers, and the length of the conversation.

One measure of the effect of background noise on speech intelligibility is the speech interference level (L_{SIL}). The SIL is defined as the arithmetic average of the sound pressure levels of the interfering noise in the four octave bands—500 Hz, 1000 Hz, 2000 Hz, and 4000 Hz—rounded to the nearest 1 dB value (ANSI, 1986). These octave bands contain the frequencies most important for communication. If an octave band analyzer is not available, the SIL can be estimated from the A-weighted sound level reading by the following expression:

$$L_{SIL} \approx L_A - 7\,dB \tag{6-4}$$

The SIL values resulting in various levels of vocal effort for face-to-face communication may be estimated from the following expression (Lazarus, 1987):

$$L_{SIL} = K - 20 \log_{10} r \quad (r \text{ is in meters}) \tag{6-5}$$

The constant K is given in Table 6-3. The data are based on the assumption that the information communicated is not familiar to the listener. For communication between women and men, the data for women should be used.

Generally, the voice level used by the speaker will change as the background noise level changes. In addition, the speaker may move closer to the listener as the background noise level increases. The expected voice levels for various SIL values of the background noise are given in Table 6-4.

The background noise SIL limits for telephone communication are given in Table 6-5 (Peterson and Gross, 1972). For speakerphones, SIL values of approximately 5 dB higher than those given in Table 6-5 may be

Acoustic Criteria

TABLE 6-3 Background Speech Interference Level[a] Limits for Face-to-Face Communication

Vocal Effort	K, dB Women	K, dB Men	Comment
Normal voice	50	54	Communication is satisfactory for the given vocal effort in this range
Raised voice	56	60	
Loud voice	62	66	
Very loud voice	67	71	
Shouting	72	76	Communication is difficult
Maximum shouting	75	79	Communication without amplification is impossible above this level
Limit for amplified speech	110	114	Vocal communication is impossible above this level

[a] $L_{SIL} = K - 20\log_{10} r$ (meters)
Source: Lazarus (1987).

tolerated, if the speakerphone is not located more than 1 m (39 in) from the person.

Example 6-2. In one area of an industrial plant, the octave band sound pressure level spectrum is given in Table 6-6. Determine the maximum distance between the speaker and listener (both males) for communication in a normal voice.

The speech interference level is found by averaging the sound pressure levels in the four octave bands, 500 Hz, 1000 Hz, 2000 Hz, and 4000 Hz:

$$L_{SIL} = \tfrac{1}{4}(73 + 69 + 65 + 59) = 66.5 \text{ dB} \rightarrow \text{Use 67 dB}$$

The distance between the people for conversation in a normal voice is given by Eq. (6-5):

$$L_{SIL} = 67 = 54 - 20\log_{10} r$$

$$r = 10^{-(13)/(20)} = 0.224 \text{ m} = 224 \text{ mm (8.8 in)}$$

Men would not typically carry on a "normal" conversation at a spacing of only about 225 mm or $8\tfrac{3}{4}$ in.

For conversation in a raised voice, the distance between the two people would be as follows:

$$r = 10^{(60-67)/20} = 0.447 \text{ m} = 447 \text{ mm (17.6 in)}$$

TABLE 6-4 Expected Voice Level for Face-to-Face Communication Corresponding to Various Background SIL Values

	Background speech interference level, L_{SIL}	
Vocal effort	Women	Men
Normal voice	45–49 dB	44–48 dB
Raised voice	53–60 dB	53–62 dB
Loud voice	61–71 dB	63–77 dB
Very loud voice	69–81 dB	71–91 dB
Shouting	77–91 dB	80–99 dB

Source: Lazarus (1987).

Even this distance is somewhat close for men to carry on a conversation comfortably in a raised voice. For conversation in a loud voice, the distance between the two people would be as follows:

$$r = 10^{(66-67)/20} = 0.891 \text{ m } (35.1 \text{ in})$$

This would represent a more comfortable distance between the people.

From Table 6-5, we observe that a telephone conversation would be difficult for the SIL of 67 dB.

6.5 NOISE CRITERIA FOR INTERIOR SPACES

We have all experienced the problem of attempting to work in an environment with a high background noise level. The noise may interfere with conversation with another worker, or the background noise may simply

TABLE 6-5 Speech Interference Level Limits for Telephone Communication

Speech interference level, L_{SIL}, dB	Telephone conversation category
< 63 dB	Satisfactory
63–78 dB	Difficult to understand conversation
78–83 dB	Unsatisfactory
> 83 dB	Impossible to understand conversation

Source: Peterson and Gross (1972).

Acoustic Criteria

TABLE 6-6 Sound Pressure Level Spectrum for Example 6-2

	Octave band center frequency, Hz							
	63	125	250	500	1,000	2,000	4,000	8,000
L_p(OB), dB	59	65	70	73	69	65	59	50

"get on our nerves" and interfere with effective concentration on the task at hand. The levels of noise may not be sufficiently high to produce damage to the person's hearing; however, work is degraded by the background noise.

Noise criteria (NC) curves were first introduced (Beranek, 1957) to evaluate existing noise problems in interior spaces such as offices, conference rooms, and homes. It was found that a background noise that fitted the original NC curves was not completely neutral. The noise had components that sounded both "hissy" and "rumbly." The original NC curves were also based on the "old" octave bands.

The NC curves were revised (Beranek, 1971) to produce a more nearly neutral background noise spectrum. These curves, called the *preferred noise criterion* (PNC) curves to distinguish them from the older NC curves, were also based on the present-day octave bands. Finally, the PNC curves were revised to make equal the perceived loudness for the octave bands that contain the same number of critical bands (Stevens, 1972). The rating number on the NCB curves is the average of the NCB values in the 500 Hz, 1000 Hz, 2000 Hz, and 4000 Hz octave bands, corresponding to the octave bands used in calculating the SIL.

The NCB curves specify the maximum noise levels in each octave band for a specified noise criterion rating. The NCB rating of a given noise spectrum is the highest penetration of the noise spectrum into the NCB curves. The numerical values for the NCB curves are given in Table 6-7 (Beranek, 1989).

The suggested NCB ratings for various activities and different interior spaces are shown in Table 6-8. The table values may be used to determine if an existing acoustic situation is satisfactory for its anticipated usage, and to determine the acoustic treatment required to make the background noise acceptable if the noise level is too high. The values given in Table 6-8 apply for background noise consisting of both equipment noise (air conditioning systems, machinery, etc.) and activity noise due to the activity of the people in the room.

The NCB curves may also be used to determine the acceptability of the space for speech communication and whether annoying "rumbles" or

TABLE 6-7 Octave Band Sound Pressure Levels Associated with the 1989 Balanced Noise Criterion (NCB) Curves

NCB, dB	Octave band center frequency, Hz							
	31.5	63	125	250	500	1,000	2,000	4,000	8,000
10	59	43	30	21	15	12	8	5	2
15	61	46	34	26	20	17	13	10	7
20	63	49	38	30	25	22	18	15	12
25	65	52	42	35	30	27	23	20	17
30	68	55	46	40	35	32	28	25	22
35	71	59	50	44	40	37	33	30	27
40	73	62	54	49	45	42	38	35	32
45	76	65	58	53	50	47	43	40	37
50	79	69	62	58	55	52	48	45	42
55	82	72	67	63	60	56	54	51	48
60	85	75	71	67	64	62	59	56	53
65	88	79	75	71	69	66	64	61	58
70	91	82	79	76	74	71	69	66	63
75	94	85	83	80	78	76	74	71	69

Source: Beranek (1989).

"hisses" are present in the background noise spectrum. These terms are more subjective than precise technical terms. If the L_{SIL} for the background noise is equal to or less than the NCB rating, the space generally will be acceptable for speech communication.

To determine whether there may be an annoying "rumble" sound, a value of

$$\text{NCB(rumble)} = L_{SIL} + 3\,\text{dB}$$

is calculated. The values for this NCB curve are compared with the octave band sound pressure levels for the background noise for the octave bands of 500 Hz or lower. If any octave band sound pressure level exceeds the NCB(rumble) curve, then there is a high probability that a "rumble" will be perceived in the background noise. Noise control procedures could be implemented to reduce the sound pressure level in the offending octave bands to acceptable values. This procedure is illustrated in the following example. If the octave band sound pressure level in the 63 Hz octave band exceeds about 75 dB, or if the octave band sound pressure level in the 31.5 Hz octave band exceeds about 70 dB, there will be a good chance that noticeable vibrations will occur in gypsum board structures, if any are present in the room. This condition should also be checked.

Acoustic Criteria

TABLE 6-8 Recommended Values of Noise Criteria (NCB) Ratings for Steady Background Noise in Various Indoor Spaces

Activity and type of space	NCB rating
Broadcast and recording studio:	
Distant microphone pickup used	10
Close microphone pickup used only	Not to exceed 25
Sleeping, resting, relaxing:	
Suburban and rural homes, apartments, hospitals	25–35
Urban homes, hotels, hospitals	30–40
Excellent listening conditions required:	
Concert halls, opera houses, recital halls	10–15
Very good listening conditions required:	
Large auditoriums, drama theaters, large churches	15–20
Small auditoriums, music rehearsal rooms, large conference rooms	25–30
Good listening conditions required:	
Private offices, school classrooms, small conference rooms libraries	30–40
Moderately good listening conditions required:	
Large offices, reception areas, retail stores, restaurants	35–45
Fair listening conditions required:	
Living rooms in dwellings (conversation and listening to television)	30–40
Lobbies, laboratory work spaces, general secretarial areas	40–50
Moderately fair listening conditions required:	
Light maintenance shops, industrial plant control rooms, kitchens, and laundries	45–55
Acceptable speech and telephone communication areas:	
Shops, garages	50–60
Speech communication not required:	
Factory and shop areas	55–70

Source: From L. L. Beranek and I. L. Vér, *Noise and Vibration Control Engineering*, 1992. By permission of John Wiley and Sons, Inc.

To determine if there may be an annoying "hiss" sound, an average of the NCB values for the 125 Hz, 250 Hz, and 500 Hz octave bands is calculated:

$$\text{NCB(hiss)} = [\text{NCB}(125\,\text{Hz}) + \text{NCB}(250\,\text{Hz}) + \text{NCB}(500\,\text{Hz})]/3$$

The values for this NCB curve are compared with the octave band sound pressure levels for the background noise for the octave bands of 1000 Hz or higher. If any octave band sound pressure level exceeds the NCB(hiss) curve, then there is a high probability that a "hiss" will be perceived in the background noise. Noise control procedures could be implemented to reduce the sound pressure level in the offending octave bands to acceptable values. This procedure is also illustrated in the following example.

Example 6-3. The sound pressure level spectrum for the air distribution system noise from Example 5-7 is given in Table 6-9. Suppose the room to which the air is distributed is a living room in a residence. Determine the NCB rating for the room.

The NCB values are found from Table 6-7 at the corresponding values of the octave band sound pressure level. The largest value of NCB is 43, which occurs for the 500 Hz and 1000 Hz octave bands. The noise criteria rating for the room is NCB-43.

It is noted from Table 6-8 for living rooms in dwellings (fair listening conditions required) that the recommended NCB rating is from NCB-30 to NCB-40. The calculated NCB rating exceeds this value by 3 dB. From Table 6-9, we see that sound pressure levels in the 500 Hz, 1000 Hz, and 2000 Hz octave bands produce a NCB rating above 40 dB. To reduce the NCB rating to NCB-40, we would need to reduce the octave band sound pressure levels to 45 dB (500 Hz), 42 dB (1000 Hz), and 38 dB (2000 Hz). This could be achieved by (a) increasing the room constant by adding acoustic treatment on the walls and ceiling or (b) decreasing the acoustic power input to the room by adding a plenum chamber after the fan outlet.

Let us check the speech interference level. Taking the average of the sound pressure levels in the octave bands from 500 Hz to 4000 Hz, we find the SIL:

$$L_{SIL} = \tfrac{1}{4}(48 + 45 + 39 + 35) = 41.75 \, \text{dB} \rightarrow \text{Use 42 dB}$$

In this case, $L_{SIL} <$ NCB-43, so background noise would allow satisfactory speech communication in the room.

TABLE 6-9 Solution for Example 6-3

	Octave band center frequency, Hz							
	31.5	63	125	250	500	1,000	2,000	4,000	8,000
L_p, dB	63	55	51	49	48	45	39	35	31
NCB	20	30	36	40	43[a]	43[a]	41	40	39
NCB-45	76	65	58	53	50				
NCB-40					45	42	38	35	32

[a] Largest values.

Acoustic Criteria 243

Next, let us check for annoyance due to "rumble" noise:

NCB(rumble) = 42 + 3 = 45 dB

The pertinent portion of the NCB-45 curve is given in Table 6-9. It is observed that all of the sound pressure level values in the range from 31.5 to 500 Hz are less than the NCB-45 values, so there will be no annoyance due to low-pitched rumble noise.

Finally, let us check for annoyance due to "hiss" noise. Averaging the NCB values for the 125 Hz, 250 Hz, and 500 Hz octave bands, we obtain the following value:

NCB(hiss) = (36 + 40 + 43)/3 = 39.7 dB → Use 40 dB

The pertinent portion of the NCB-40 curve is given in Table 6-9. It is observed that the sound pressure levels in the 1000 Hz and 2000 Hz octave bands exceed the NCB-40 values for those octave bands. The occupants of the room would probably experience some annoyance due to the perceived "hissing" noise of the air distribution system. This problem could be alleviated by reducing the sound pressure level in the 1000 Hz octave band from 45 dB to 42 dB, and by reducing the sound pressure level in the 2000 Hz octave band from 39 dB to 38 dB.

6.6 COMMUNITY REACTION TO ENVIRONMENTAL NOISE

Before a new plant is constructed or new equipment is installed, the effects of noise produced by the plant or equipment should be estimated. The response of the surrounding community to the additional environmental noise may be a factor in the site selection, the acoustic design of the plant, and public relations for the company. It would be much better to anticipate and correct noise problems before the plant is built or equipment is installed than to endure lawsuits because of noise produced by the plant or equipment.

There are several factors that influence the community tolerance for environmental noise. Some of these factors are listed as follows.

1. Discrete frequency sounds. A whistle tone is more annoying than a broadband hissing noise, even when the broadband noise level is somewhat higher than that of the pure tone noise.
2. Repetitiveness or fluctuation in sound level. An obvious change in sound level outdoors generally directs one's attention to the noise source. On the other hand, people tend to become acclimated to a steady marginal source of sound.

3. Sleep-disturbing noises. People usually become somewhat testy when their sleep is disturbed.
4. Ambient noise level. When the ambient or background noise level is very low, even the dripping of water will annoy some people. For locations in a busy urban area where traffic noise is present, an additional noise source may not produce a significant community reaction; whereas, the same additional noise source in a quiet rural setting would produce many complaints from the community.
5. Impulsive or startling noises. A sudden intrusion of noise may be unsettling for many people.
6. Visibility of the noise source. People tend to be more tolerant of marginal noise sources that are concealed from their view.
7. Noise that conveys unpleasant information. Most people prefer to hear pleasant music than the sound of shattering glass, for example.

Community noise rating curves for outdoor (environmental) noise have been developed (Stevens et al., 1955) based on principles similar to those used to establish the NCB curves for noise in interior spaces. These curves allow the estimation of the community response to the noise level outdoors. The base rating (N_o) curves are given in Table 6-10. The base

TABLE 6-10 Base Noise Rating Curves (N_o) for Environmental Noise Rating

Base noise rating, N_o	Octave band center frequency, Hz							
	63	125	250	500	1,000	2,000	4,000	8,000
25	55	43	36	29	25	22	18	16
30	59	47	40	34	30	27	24	22
35	63	52	45	39	35	32	30	27
40	67	56	49	44	40	37	35	33
45	71	61	54	49	45	42	40	38
50	74	65	58	53	50	47	45	44
55	78	70	63	58	55	53	50	48
60	82	74	68	63	60	58	55	54
65	86	78	72	68	65	63	60	59
70	90	83	77	73	70	68	66	65
75	94	87	82	78	75	73	71	70
80	97	91	86	83	80	78	76	75

Source: Thumann and Miller (1986).

Acoustic Criteria

noise rating is given by the highest penetration (largest value of N_o) of the measured octave band sound pressure level data into the noise rating curves.

A composite correction factor (CF) must be applied to the base noise rating value to allow for the various factors influencing a person's annoyance to the outdoor noise. The corrected composite noise rating (L_{CNR}) is determined from the following expression:

$$L_{CNR} = N_o + CF \tag{6-6}$$

Values for the component correction factors are given in Table 6-11. It is important to note that only one value of correction factor is used from each of the six categories of influencing factor. The overall correction factor is the sum of the values for the six individual effects.

The average community response to a noise with a given composite noise rating is summarized in Table 6-12. One must exercise some caution in using the data in Table 6-12, because it is somewhat subjective, and there is a range of responses in any population to the same environmental noise. Some people are intolerant of noise (and maybe they are naturally grouchy), whereas others have a high tolerance level for intrusive noise.

Example 6-4. An air vent located outdoors produces the noise spectrum shown in Table 6-13. The vent noise is broadband, and the sound is not impulsive. The vent operates about 30 times each hour, 1 minute duration, during the daytime and the evening, but not during the nighttime. The vent operates year-round. The vent is located in an area with light industry. Determine the environmental noise rating and the anticipated community reaction to the vent noise.

The base noise rating values from Table 6-10 are shown in Table 6-13. the largest value is 75 dB, which occurs for the 4000 Hz octave band. Thus, the base noise rating is $N_o = 75$ dB.

The correction factors may be found from Table 6-11:

Noise spectrum	0
Repetitiveness (10–60 times/hour)	−5
Time of day (during the evening)	−5
Season (year-round)	0
Area (light industry)	−10
Peak factor (non-impulsive)	0
Total	CF = −20 dB

The composite noise rating for the vent noise is found from Eq. (6-6):

$$L_{CNR} = N_o + CF = 75 + (-20) = 55 \text{ dB}$$

TABLE 6-11 Correction Factors (CF) for Various Influencing Factors for Community Noise Reaction[a]

Influencing factor	Possible condition	CF, dB
Noise spectrum	Noise with pure-tone components	+5
	Broadband noise	0
Repetitiveness	Continuous to 1/minute	0
	10–60 times/hour	−5
	1–10 times/hour	−10
	4–24 times/hour	−15
	1–4 times/day	−20
	1 time/day	−25
Time of day	Daytime only (7:00 a.m. to 6:00 p.m.)	−10
	Evening (6:00 p.m. to 10:00 p.m.)	−5
	Nighttime (10:00 p.m. to 7:00 a.m.)	0
Season of the year	Winter only	−5
	Summer (and winter)	0
Type of area	Rural	+10
	Suburban	+5
	Urban residential	0
	Residential with some business	−5
	Area with light industry	−10
	Area with heavy industry	−15
Peak factor	Impulsive sounds	+5
	Non-impulsive sounds	0

[a] Only one correction is applied from each of the six categories.
Source: From A. Thumann and R. K. Miller, *Fundamentals of Noise Control Engineering*. By permission of the Fairmont Press, Inc.

From Table 6-12, we see that the anticipated community response would be *widespread complaints*. If it is desired to reduce the community reaction to only mild annoyance, the composite noise rating would need to be reduced to $L_{\text{CNR}} = 45\,\text{dB}$, or the base noise rating would be $N_o = 65\,\text{dB}$. As noted from Table 6-13, a reduction in the octave sound pressure levels in the octave band from 1000 Hz to 8000 Hz would be required. The required reduction in octave band sound pressure level is greatest (11 dB and 10 dB, respectively) in the 4000 Hz and 8000 Hz octave bands.

Acoustic Criteria

TABLE 6-12 Average Community Reaction to Noise Based on the Composite Noise Rating L_{CNR}

Corrected composite noise rating, L_{CNR}, dB	Community response	Percent of population complaining
39 dB or less	No reaction	
40–45 dB	Mild annoyance	1
46–50 dB	Sporadic complaints	2
51–55 dB	Widespread complaints	7
56–69 dB	Threats of legal action	12
70 dB or greater	Vigorous legal action	22

6.7 THE DAY–NIGHT LEVEL

6.7.1 EPA Criteria

The U.S. Environmental Protection Agency (EPA) has investigated the effect of noise on people and the effect of the noise from the environment on the health and welfare of the affected people (EPA, 1974). The EPA concluded that the A-weighted sound level correlated as well with human response to noise as more complex measures. As a result, the A-weighted sound level was selected as the basis for environmental noise criteria.

Generally, the A-weighted sound level does not remain constant during any extended period at a particular location. It would be incorrect to average directly the decibel readings during the period. Instead, one should use the *energy-equivalent sound level*, L_{eq}, which is the sound level averaged on an energy basis:

$$L_{eq} = 10 \log_{10}[\Sigma t_j 10^{L_j/10}] \tag{6-7}$$

TABLE 6-13 Solution for Example 6-4

	Octave band center frequency, Hz							
	63	125	250	500	1,000	2,000	4,000	8,000
L_p(OB), dB	43	50	55	61	66	69	71	69
N_o, dB	—	33	46	58	66	71	75[a]	74
L_{CNR}-65	86	78	72	68	65	63	60	59
Reduction, dB	—	—	—	—	1	6	11	10

[a] Largest value.

The quantity t_j is the fraction of the time period that the noise has an A-weighted sound level of L_j, and L_j is the A-weighted sound level during the *j*th time interval. Many sound level meters have the feature that this quantity may be measured directly with the meter.

The EPA found (not surprisingly) that people were more sensitive to noise during the nighttime hours than during the daytime period. From our discussion of the environmental noise rating parameter in Sec. 6.6, we found that noise that occurred only during the daytime was about 10 dB less annoying than noise that occurred during the nighttime. The correction factor from Table 6-11 for noise during the daytime only is CF = −10 dB, whereas CF = 0 dB for noise during the nighttime.

Based on this observation, the EPA suggested a modification of the equivalent sound level, called the *day–night average sound level*, L_{DN}, to take into consideration the additional annoyance of noise at nighttime. The day–night level was developed originally to be used as an aid in land-use planning. For this parameter, the nighttime equivalent sound levels were increased by 10 dB for the time period from 10:00 p.m. to 7:00 a.m. This nighttime period involves 9 hours, or a fraction of 0.375 of the 24-hour day, and the daytime period involves 15 hours, or a fraction of 0.625 of the 24-hour day. The day–night level is, accordingly, defined by the following expression:

$$L_{DN} = 10 \log_{10}[(0.625)\, 10^{L_D/10} + (0.375)\, 10^{(L_N+10)/10}] \tag{6-8}$$

The quantity L_D is the equivalent sound level during the daytime hours, and L_N is the equivalent sound level during the nighttime.

According to the EPA studies, the effects given in Table 6-14 would be observed if the day–night level of 55 dBA is present. For outdoor activities that should be free of speech interference and produce no significant annoyance, the EPA recommends the criterion that $L_{DN} \leq 55$ dBA. Similarly, for indoor activities, the EPA recommendation is that $L_{DN} \leq 45$ dBA. For minimum hearing loss (no more than 5 dB noise-induced permanent threshold shift for 96% of the population) over a period of 40 years, the EPA recommends that the noise exposure during the 24-hour day be limited by $L_{DN} \leq 70$ dBA.

Example 6-5. During a 1-hour period, the A-weighted sound level is 70 dBA for 30 minutes, 75 dBA for 20 minutes, and 80 dBA for 10 minutes. Determine the energy equivalent sound level.

The fractions for each interval are $(30/60) = 0.5000$ for 70 dBA, 0.3333 for 75 dBA, and 0.1667 for 80 dBA. Using Eq. (6-7), we find the energy-equivalent sound level:

Acoustic Criteria

TABLE 6-14 Effects Corresponding to a Day–Night Level of 55 dBA

Condition	Magnitude of the effect
Speech indoors	100% sentence intelligibility with a 5 dB margin of safety
Speech outdoors	99% sentence intelligibility at 1 m (3.3 ft) spacing; 95% sentence intelligibility at 3.5 m (11.5 ft) spacing
Average community reaction	No evident reaction; 7 dB below the beginning of threats of legal action
Complaints	About 1% may complain, depending on the person's attitude and other non-noise-related factors
Annoyance	About 17% may be somewhat annoyed
Attitude toward area	Noise is essentially one of the least important factors influencing the person's attitude toward the area

Source: EPA (1974).

$$L_{eq} = 10\log_{10}[(0.5000)\,10^{7.0} + (0.3333)\,10^{7.5} + (0.1667)\,10^{8.0}]$$

$$L_{eq} = 10\log_{10}(3.2208 \times 10^7) = 75.1\,\text{dBA}$$

Example 6-6. The hourly equivalent sound levels measured outdoors at a particular location are given in Table 6-15. Determine the day–night level for this data.

During the daytime, the fraction of time for each sound level is calculated as follows:

50 dBA: 3 hours, or $t = 3/15 = 0.2000$

60 dBA: 10 hours, or $t = 10/15 = 0.6667$

70 dBA: 2 hours, or $t = 2/15 = 0.1333$

The equivalent sound level during the daytime is found from Eq. (6-7):

$$L_D = 10\log_{10}[(0.2000)\,10^{5.0} + (0.6667)\,10^{6.0} + (0.13333)\,10^{7.0}]$$

$$L_D = 63.1\,\text{dBA}$$

For the 9 hours during the nighttime, the fraction of time for each sound level is calculated as follows:

30 dBA: 5 hours, or $t = 5/9 = 0.5556$

40 dBA: 4 hours, or $t = 4/9 = 0.4444$

TABLE 6-15 Data for Example 6-6

Daytime				Nighttime	
Time[a]	L_A, dBA	Time[a]	L_A, dBA	Time[a]	L_A, dBA
7:00 a.m.	50	3:00 p.m.	60	10:00 p.m.	40
8:00 a.m.	60	4:00 p.m.	60	11:00 p.m.	40
9:00 a.m.	70	5:00 p.m.	70	12:00 mid	40
10:00 a.m.	60	6:00 p.m.	60	1:00 a.m.	30
11:00 a.m.	60	7:00 p.m.	60	2:00 a.m.	30
12:00 noon	60	8:00 p.m.	50	3:00 a.m.	30
1:00 p.m.	60	9:00 p.m.	50	4:00 a.m.	30
2:00 p.m.	60			5:00 a.m.	30
				6:00 a.m.	40

[a] "Time" refers to the hour beginning with the time given in the table.

The equivalent sound level during the nighttime is found from Eq. (6-7) also:

$$L_N = 10 \log_{10}[(0.5556) \, 10^{3.0} + (0.4444) \, 10^{4.0}] = 37.0 \, \text{dBA}$$

The day–night level is found from Eq. (6-8):

$$L_{DN} = 10 \log_{10}[(0.625) \, 10^{6.31} + (0.375) \, 10^{(37.0+10)/10}] = 61.1 \, \text{dBA(DN)}$$

We note that this value is greater than the EPA recommended value of 55 dBA for outdoor activity. To reduce the day–night level to 55 dBA, one could install barriers, for example, to reduce the noise during the daytime hours only. The required reduction value of the daytime sound level could be calculated as follows:

$$10^{5.50} = (0.625) \, 10^{L_D/10} + (0.375) \, 10^{4.70}$$

$$L_D = 56.8 \, \text{dBA}$$

If the sound level during the daytime could be reduced by $(63.1 - 56.8) = 6.3$ dBA, the day–night level would be reduced to 55 dBA.

6.7.2 Estimation of Community Reaction

If noise spectrum data are not available, the day–night level of the background noise, with suitable correctors, may be used to estimate the anticipated community response to the environmental noise:

$$L_{DN}(\text{corrected}) = L_{DN}(\text{measured}) + \text{CF}_{DN} \qquad (6\text{-}9)$$

Acoustic Criteria

The day–night sound level contains explicitly the effect of annoyance due to noise during the nighttime, so the other effects—such as location, time of the year, etc.—are accounted for with the correctors given in Table 6-16.

The anticipated community response to environmental noise in terms of the day–night level of the noise is given in Table 6-17. This data may be used in a manner similar to that for the environmental noise rating to design for satisfactory community response to planned introduction of a noise source outdoors.

Example 6-7. The noise levels in a normal suburban area are given in Table 6-18. The area has had some prior experience with intrusive noises. There are no pure tone components of the noise, and it is not impulsive. The

TABLE 6-16 Correctors to be Added to the Measured Day–Night Level for Various Influencing Factors[a]

Influencing factor	Description of condition	CF_{DN}, dBA
Noise spectrum	Pure tones or impulsive noise present	+5
	No pure tone or impulsive sounds	0
Type of location	Quiet suburban or rural community	+10
	Normal suburban community	+5
	Urban residential community	0
	Noisy urban residential community	−5
	Very noisy urban community	−10
Time of year	Summer or year-round	0
	Winter only or windows always closed	−5
Previous noise exposure	No prior experience with the intruding noise	+5
	Some prior experience with the noise or where the community is aware that good-faith efforts are being made to control noise	0
	Considerable experience with the noise and the group associated with the source of noise has good community relations	−5
	Aware that the noise source is necessary, of limited duration, and/or an emergency situation	−10

[a] Only one correction factor should be used from each category.

TABLE 6-17 Average Community Reaction to Noise Based on the Day–Night Level, L_{DN}

Corrected day–night level, L_{DN}(corrected)	Expected community response
< 62 dBA(DN)	No reaction
62–67 dBA(DN)	Complaints
67–72 dBA(DN)	Threats of community action
> 72 dBA(DN)	Vigorous community action

noise source will be present year-round. Determine the anticipated community response to the noise source.

The equivalent sound level for the daytime hours is found from Eq. (6-7):

$$L_D = 10 \log_{10}[(0.2667)\, 10^{6.0} + (0.4000)\, 10^{5.5} + (0.3333)\, 10^{5.0}]$$

$$L_D = 56.3\, dBA$$

The equivalent sound level during the nighttime is found from Eq. (6-7) also:

$$L_N = 10 \log_{10}[(0.2222)\, 10^{4.5} + (0.7778)\, 10^{4.0}] = 41.7\, dBA$$

The day–night level is found from Eq. (6-8):

$$L_{DN} = 10 \log_{10}[(0.625)\, 10^{5.63} + (0.375)\, 10^{(41.7+10)/10}] = 55.1\, dBA(DN)$$

TABLE 6-18 Data for Example 6-7

Duration	A-weighted level	Fraction
Daytime:		
4 hours	60 dBA	$t = 4/15 = 0.2667$
6 hours	55 dBA	$t = 6/15 = 0.4000$
5 hours	50 dBA	$t = 5/15 = 0.3333$
Nighttime:		
2 hours	45 dBA	$t = 2/9 = 0.2222$
7 hours	40 dBA	$t = 1/9 = 0.7778$

Acoustic Criteria

The correction factors for other influences are found in Table 6-16 as follows:

Noise spectrum	−0
Type of location	+5
Time of year (year-round)	−0
Previous noise exposure	−0
Total	$CF_{DN} = +5\,dBA$

The corrected day–night level is as follows:

$$L_{DN}(\text{corrected}) = 55.1 + 5 = 60.1\,dBA(DN)$$

The anticipated community reaction from Table 6-17 is *no reaction*.

6.8 HUD CRITERIA

The U.S. Department of Housing and Urban Development (HUD) was charged with developing guides for zoning modifications or for siting of dwellings where the noise and zoning regulations were already in existence. There was evidence that annoyance for a specific noise exposure depended on both the average level of the noise and on the variability of the source of noise (Griffiths and Langdon, 1968). The *noise pollution level*, L_{NP}, was developed to recognize this phenomenon (Schultz, 1972).

The noise pollution level is defined by the following expression:

$$L_{NP} = L_{eq} + 2.56\sigma \qquad (6\text{-}10)$$

The quantity L_{eq} is the energy-weighted equivalent A-weighted sound level, and σ is the standard deviation of the A-weighted levels. The coefficient 2.56 was selected as the best fit to data from studies of subjective response to variable noise levels. The standard deviation of the noise levels may be determined from the following equation:

$$\sigma^2 = \Sigma t_j L_j^2 - (\Sigma t_j L_j)^2 \qquad (6\text{-}11)$$

The quantities t_j are the fractions of time that the noise level L_j occurs, usually over a 24-hour period. If the probability distribution of the noise over time is approximately gaussian, the noise pollution level may be estimated by the following expression:

$$L_{NP} = L_{50} + \delta + \delta^2/60 \qquad (6\text{-}12)$$

The quantity $\delta = (L_{10} - L_{90})$, and L_{10}, L_{50}, and L_{90} are the A-weighted sound levels that are exceeded 10%, 50%, and 90% of the time, respectively.

The first term (L_{eq}) in Eq. (6-10) represents the equivalent continuous noise level, and the second term (2.56σ) represents the effect of fluctuations in the noise level, or the difference between the steady background noise and the intruding noise.

The HUD criteria for allowable noise pollution levels for new residential construction are given in Table 6-19 (Schultz, 1970). A builder may not be able to obtain federal funding or federally guaranteed loans unless the noise level at the proposed site is acceptable. The *clearly acceptable* category means that no special acoustic treatment is required for the construction because of environmental noise. The normally acceptable category means that the housing site may be acceptable (discretionary) if special acoustic treatment, such as an acoustic barrier, is used to reduce the noise level at the site. Also, the builder would need to assure the authorities that the indoor noise level would not exceed 55 dBA for more than 60 minutes or not exceed 45 dBA for more than 8 hours per day.

Example 6-8. The noise levels on the A-scale in a normal suburban area are given in Table 6-18. Determine the noise pollution level for the environmental noise and the HUD category for the noise.

Ordinarily, date would be gathered over shorter time periods than 1 hour for a noise impact study, and the data set would be much more extensive than that given in Table 6-18. This problem, however, is designed to illustrate the principles involved, without getting mired in mountains of data.

TABLE 6-19 Environmental Noise Level Criteria (HUD) for New Residential Construction

Category	Description	L_{NP}, dBA(NP)
Clearly acceptable	Noise does not exceed 45 dBA for 30 minutes in 24 hours	< 62
Normally acceptable	Noise does not exceed 65 dBA for 8 hours in 24 hours	62–74
Normally unacceptable	Noise exceeds 65 dBA for 8 hours in 24 hours, or loud repetitive sounds are present	74–88
Clearly unacceptable	Noise exceeds 80 dBA for 6 minutes in 24 hours or exceeds 75 dBA for 8 hours in 24 hours	> 88

Source: Schultz (1970).

Acoustic Criteria

The fractions of the entire day at which each sound level occurs may be calculated as follows:

At 60 dBA: $t = 4/24 = 0.1667$
At 55 dBA: $t = 6/24 = 0.2500$
At 50 dBA: $t = 5/24 = 0.2083$
At 45 dBA: $t = 2/24 = 0.0833$
At 40 dBA: $t = 7/24 = 0.2917$

The equivalent sound level is found from Eq. (6-7):

$$L_{eq} = 10 \log_{10}[(0.1667)\,10^{6.0} + (0.2500)\,10^{5.5} + (0.2083)\,10^{5.0} + \cdots]$$

$$L_{eq} = 10 \log_{10}(272{,}109) = 54.3\,\text{dBA}$$

The quantities needed in calculating the standard deviation are determined as follows:

$$\Sigma t_j(L_j)^2 = (0.1667)(60)^2 + (0.2500)(55)^2 + (0.2083)(50)2 + \cdots = 2512.5$$

$$\Sigma t_j L_j = (0.1667)(60) + (0.2500)(55) + (0.2083)(50) + \cdots = 49.583$$

The standard deviation for the data may be determined from Eq. (6-11):

$$\sigma^2 = 2512.5 - (49.583)^2 = 53.993$$

$$\sigma = (53.993)^{1/2} = 7.348\,\text{dBA}$$

The noise pollution level is found from Eq. (6-10):

$$L_{NP} = 54.3 + (2.56)(7.348) = 54.3 + 18.8 = 73.1\,\text{dBA(NP)}$$

It is noted from Table 6-19 that this value of noise pollution level falls into the *normally acceptable* category. It is also noted from the data in Table 6-18 that the noise does not exceed 65 dBA at all during the 24-hour period.

6.9 AIRCRAFT NOISE CRITERIA

The noise of automobile and truck traffic along a freeway is relatively uniform, with a few highs and lows in the noise level. On the other hand, the noise due to the fly-over of aircraft from an airport is neither constant nor uniform, even near the airport. Some large airports have takeoffs and landings as frequently as one every minute or two, whereas smaller airports may have only one or two takeoffs and/or landings each hour. The intrusive noise of the aircraft may be as much as 30–40 dBA higher than the ambient noise, in contrast to ground traffic noise. The intrusive noise for aircraft is often more annoying than the more steady ground traffic noise, so different methods have been examined to determine the effect of aircraft noise.

6.9.1 Perceived Noise Level

The *perceived noise* level, L_{PN}, was developed as a single-number rating of annoyance to noise, or aircraft noise, in particular (Kryter, 1959). The perceived noise level is developed from contours of equal perceived "noisiness" (Kryter and Pearsons, 1963). The units for the equal "noisiness" contours is called *noys*, in analogy with an older procedure of determining contours for equal "loudness" in *sones* (Fletcher and Munson, 1933; Stevens, 1972). The numerical values of the noys are selected such that a noise of 4 noys is perceived as four times as "noisy" as a noise of 1 noy. The conversion data needed for converting from octave band sound pressure level measurements to an effective noy value are given in Table 6-20 (Pinto, 1962).

The procedure to obtain an *effective perceived noise level*, L_{EPN}, dB(PN), from experimental octave band sound pressure level data is as follows. First, the octave band data is converted to noys using Table 6-20. Next, the effective noy value N_e is calculated from the following expression:

$$N_e = 0.3 \Sigma N_j + 0.7 N_{max} \tag{6-13}$$

TABLE 6-20 Contours of Equal Noisiness N, in noys

N, noys	Octave band center frequency, Hz							
	63	125	250	500	1,000	2,000	4,000	8,000
1	60	51	44	40	40	32	29	37
2	67	59	53	50	50	42	39	44
3	72	64	59	56	56	50	44	49
5	77	71	66	63	63	55	52	57
10	85	79	75	73	73	65	62	67
15	90	85	81	79	79	71	68	73
20	94	89	85	83	83	75	72	77
30	100	95	91	89	89	81	78	83
40	104	99	95	93	93	87	82	87
50	108	103	99	96	96	89	86	91
60	110	105	101	99	99	91	88	93
80	114	109	105	103	103	96	93	98
100	117.5	112.5	108.5	106.5	106.5	99	96	101
150	123	118	114	112	112	105	102	107
200	127.4	122.4	119.4	116.4	116.4	109	106	111
300	133.3	128.3	124.3	122.3	122.3	115	112	117

Source: Pinto (1962).

Acoustic Criteria

The first term is the sum of the noy values for the octave bands, and N_{max} is the largest noy value in all of the octave bands. Finally, the effective perceived noise level, L_{EPN}, is determined from the following relationship:

$$L_{EPN} = 40 + \frac{10 \log_{10} N_e}{\log_{10}(2)} \quad (6\text{-}14)$$

It is noted that an increase of 10 dB(PN) is equivalent to a doubling of the noisiness in noys.

Example 6-9. The sound level spectrum for a single fly-over at a particular location around an airport is given in Table 6-21. Determine the effective perceived noise level for the noise.

The noisiness values corresponding to each octave band sound pressure level are determined from Table 6-20 and listed in Table 6-21. The largest noy value or N_{max} value is 74 noy; this occurs in the 250 Hz octave band.

The effective noisiness is calculated from Eq. (6-13):

$$N_e = (0.3)(24 + 47 + 74 + 56 + \cdots) + (0.7)(74) = (0.3)(327) + 51.8$$

$$N_e = 149.9 \text{ noy}$$

The effective perceived noise level is found from Eq. (6-14):

$$L_{EPN} = 40 + \frac{10 \log_{10}(149.9)}{\log_{10}(2)} = 40 + 72.2 = 112.2 \text{ dB(PN)}$$

6.9.2 Noise Exposure Forecast

The *noise exposure forecast* (NEF) was developed to determine the land-use compatibility around a commercial (not military) airport with the noise generated by the aircraft using the airport (Galloway and Bishop, 1970). The aircraft noise is usually much more significant than any other noise source for those people living in the vicinity of the airport.

TABLE 6-21 Data for Example 6-9

	Octave band center frequency, Hz						
	63	125	250	500	1,000	2,000	4,000	8,000
L_p(OB), dB	97	102	104	98	90	85	80	76
N_j, noy	24	47	74[a]	56	32	40	35	19

[a] N_{max} value.

For a specific class of aircraft (i) on one of the flight paths (j) of the airport, the noise exposure forecast, NEF(i,j), is related to the effective perceived noise level, $L_{\text{EPN}}(i,j)$, and the number of daytime and nighttime flights, N_D and N_N. A landing is considered as one "flight," and a takeoff is considered as another "flight" in determining the NEF:

$$\text{NEF}(i,j) = L_{\text{EPN}}(i,j) + 10 \log_{10}[N_D(i,j) + 16.67 N_N(i,j)] - 88 \quad (6\text{-}15)$$

The value of $N_D(i,j)$ is the number of flights between 7:00 a.m. and 10:00 p.m. (daytime), and $N_N(i,j)$ is the number of flights between 10:00 p.m. and 7:00 a.m. (nighttime), for a particular class of aircraft (i) and a specific flight path (j). The factor 16.67 arises from the fact that the ratio of daytime hours to nighttime hours is $(15/9) = 1.667$, and the noise during the nighttime hours is weighted as 10 times as important as the daytime noise. The constant 88 is arbitrary; however, it was introduced to avoid confusion between the NEF and the composite noise rating, used previously for rating aircraft noise effects.

Values for the effective perceived noise level for various types of aircraft and the detailed method for calculation of this parameter from measurements are outlined in the literature (Pearsons and Bennett, 1974).

The NEF value at a specific location adjacent to the airport is found by adding (energywise) the NEF(i,j) for each class of aircraft along each flight path:

$$\text{NEF} = 10 \log_{10}\left[\sum_i \sum_j 10^{\text{NEF}(i,j)/10}\right] \quad (6\text{-}16)$$

For a rough approximation (within ± 3 dBA), the NEF can be calculated from measured values of the day–night level (EPA, 1974):

$$\text{NEF} \approx L_{\text{DN}} - 35 \quad (6\text{-}17)$$

Noise exposure forecast contours have been used by HUD in evaluating prospective land use around airports. The report on the determination of the NEF contours is usually included in environmental impact studies dealing with the noise from the aircraft operations around an airport area. Representative land-use compatibility recommendations are given in Table 6-22. The criteria of the U.S. Department of Housing and Urban Development (HUD) are shown in Table 6-23 (HUD, 1971).

The detailed construction of the NEF contours for a specific airport is a time-consuming process. An approximate procedure has been developed to estimate the location of the NEF-30 and NEF-40 contours, based on the flight schedules for the airport (HUD, 1971). The dimensions of the approximate NEF-30 and NEF-40 contours are illustrated in Fig. 6-4. The numer-

Acoustic Criteria

TABLE 6-22 Land-Use Compatibility as a Function of the Noise Exposure Forecast (NEF)

Land use	NEF			
	<24	24–30	30–40	>40
Residential	S	S	Q^a	U
Commercial; industrial	S	S	S	U
Hotels, offices, public buildings	S	S	Q^b	U
Schools, hospitals, churches	S	Q^b	U	U
Theaters, auditoriums	Q^c	Q^b	U	U

S = satisfactory; Q = questionable; U = unsatisfactory.
[a] Individuals may complain, and some may complain vigorously. New single-family dwelling construction should be avoided. Noise control features must be included in the building design for apartment buildings.
[b] Construction should be avoided unless a detailed analysis of noise control requirements is made and the building design contains the required noise control features.
[c] A detailed noise analysis is required for any auditorium where music is to be played.
Source: HUD (1971).

TABLE 6-23 HUD Site Acceptability Categories as Related to Airport Noise

Category	Location of the site from the runway center point
Clearly acceptable	Outside the NEF-30 controur, at a distance equal to or greater than the distance between the NEF-30 and NEF-40 contours
Normally acceptable	Outside the NEF-30 contour, at a distance less than the distance between the NEF-30 and NEF-40 contours
Normally acceptable	Between the NEF-30 and NEF-40 contours
Clearly unacceptable	Within the NEF-40 contour

Source: HUD (1971).

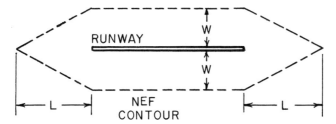

FIGURE 6-4 Description of dimensions used to determine the approximate location of the NEF-30 and NEF-40 contours. (From HUD, 1971.)

ical values of the dimensions L and W are given in Table 6-24. The number of "effective" flights considers the weighting of noise due to the nighttime flights:

$$N_{EF} = N_D + 16.67 N_N \qquad (6\text{-}18)$$

The application of this approximate procedure is illustrated in the following example.

It has been observed (Beranek, 1971, p. 583) that the simplified or approximate procedure is conservative; i.e., the NEF contours according to the approximate method are larger than those determined through a more detailed analysis. For this reason, the simplified approach should be

TABLE 6-24 Dimensions for NEF Contour Approximation

Effective number of flights, N_{EF} [a]	NEF-30		NEF-40	
	W	L	W	L
0–50	1000 ft (305 m)	1 mile (1.6 km)	0	0
50–500	0.5 mile (0.8 km)	3 miles (4.8 km)	1000 ft (305 m)	1 mile (1.6 km)
500–1,300	1.5 mile (2.4 km)	6 miles (9.7 km)	2000 ft (610 m)	2.5 miles (4.0 km)
> 1,300	2 miles (3.2 km)	10 miles (16.1 km)	3000 ft (915 m)	4 miles (6.4 km)

[a] The effective number of flights $N_{EF} = N_D + 16.67 N_N$
Source: HUD (1971).

Acoustic Criteria 261

used only for preliminary planning, and the detailed analysis should be utilized for more detailed planning purposes (Schultz, 1970).

Example 6-10. A small airport has two main runways: (a) an east–west runway, which is 3 miles (4.83 km) long, and (b) a north–south runway, which is 2 miles (3.22 km) long. The runways cross at a distance of 0.75 miles (1.21 km) from the west end of the east–west runway and 0.75 miles (1.21 km) from the south end of the north–south runway, as shown in Fig. 6-5. There are 90 daytime flights (takeoffs plus landings) and 12 nighttime flights during each 24-hour period. Determine the location of the NEF-30

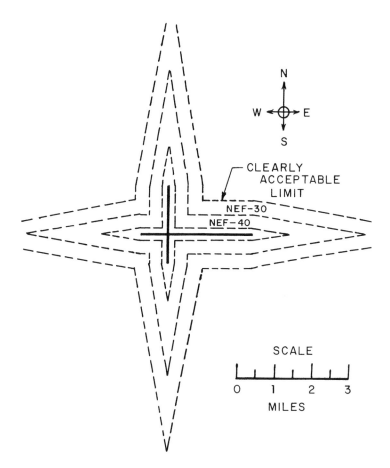

FIGURE 6-5 Diagram for Example 6-10.

and NEF-40 contours, using the approximate method. Also, determine the limit for the HUD clearly acceptable category for land use.

The equivalent number of flights is found from Eq. (6-18):

$$N_{EF} = 90 + (16.67)(12) = 90 + 200 = 290 \text{ flights}$$

The dimensions for the NEF contours are found from Table 6-24:

NEF-40: $L = 1$ mile (1.6 km) and $W = 1000$ ft (305 m)

NEF-30: $L = 3$ miles (4.8 km) and $W = 0.5$ mile (0.8 km)

These contours are shown in Fig. 6-5.

The limiting line for the HUD "clearly acceptable" category is found. The difference between the NEF-30 and NEF-40 contours is as follows:

$$\Delta L = 3 - 1 = 2 \text{ miles } (3.2 \text{ km})$$

$$\Delta W = 2640 - 1000 = 1640 \text{ ft} = 0.311 \text{ mile } (0.500 \text{ km})$$

The dimensions of the limiting line (measured from the runway) are as follows:

$$L = 3 + 2 = 5 \text{ miles } (8.0 \text{ km})$$

$$W = 0.50 + 0.311 = 0.811 \text{ miles} = 4280 \text{ ft } (1.30 \text{ km})$$

PROBLEMS

6-1. In a certain work area, 25 people of age 30 years have been exposed to a sound level of 105 dBA during the 8-hour working day over the previous 12 years. How many of the workers would be expected to suffer a noise-induced permanent threshold shift of 25 dB or more? For the males in the group, what would the permanent threshold shift be at 1000 Hz due to aging alone?

6-2. During a typical working day (8 hours), a worker in a shop must spend 2 hours operating a punch press, where the sound level is 97 dBA. The worker spends 4 hours preparing stock for the punch press in a space where the sound level is 92 dBA. The remainder of the day is spent in other work activities in an area where the sound level is 75 dBA. (a) Is this noise exposure in compliance with the OSHA regulations? What is the noise exposure dosage? (b) If the exposure is not in compliance, the worker may be allowed to spend more time on other work activities in the 75 dBA area and less time around the punch press. What is the maximum time that the worker can be allowed to work around the punch press in order to comply with OSHA regulations? If the exposure is in compliance, the worker may be allowed to spend more time around the press. If this is the case, how much time can the worker spend around the press and still be in compliance with the OSHA regulations?

6-3. In a textile mill, workers spend 2 hours preparing stock in an area where the sound level is 92 dBA, and 2 hours 40 minutes around the machine area. The remainder of the 8-hour day is spent in an area where the sound level is 70 dBA. Determine the maximum sound level around the machine area that is allowable according to the OSHA criteria.

6-4. A worker is exposed to a sound level of 100 dBA for 30 minutes in a metal forming area; then the worker spends 3 hours in a stock preparation room. The remainder of the 8-hour day is spent in the stockroom, where the noise level is 65 dBA. Determine the maximum allowable sound level in the stock preparation room for the worker to be in compliance with the OSHA criteria.

6-5. The data given in Table 6-25 were obtained in a shop room around a Summit punch press. Determine (a) the speech interference level and

TABLE 6-25 Data for Problems

	\multicolumn{8}{c}{Octave band center frequency, Hz}
	63	125	250	500	1,000	2,000	4,000	8,000
Data for Problems 6-5 and 6-8—shop room:								
L_p(OB), dB	46	48	59	63	61	58	58	56
Data for Problems 6-6 and 6-9—office:								
L_p(OB), dB	46	43	41	39	36	36	33	32
Data for Problem 6-7—library:								
L_p(OB), dB	60	57	54	47	40	35	26	20
Data for Problem 6-10—transformer:								
L_p(OB), dB	47	51	45	41	41	35	32	24
Data for Problem 6-11—cooling tower:								
L_p(OB), dB	70	71	69	67	60	55	47	41
Data for Problem 6-12—incinerator:								
L_p(OB), dB	71	82	75	73	63	58	50	48
Data for Problem 6-13—aeration system:								
L_p(OB), dB	70	60	54	51	50	53	49	44
Data for Problem 6-19—machine area:								
L_p(OB), dB	67	71	75	67	63	58	52	49

(b) the distance between speaker and listener at which conversation in a raised voice could easily be carried out. Would the given background noise be satisfactory for carrying out a telephone conversation?

6-6. The measured octave band sound pressure levels in an office are given in Table 6-25. Determine (a) the speech interference level and (b) the distance between female workers at which a conversation in a normal voice could be carried out.

6-7. The measured octave band sound pressure levels inside a library reading room are given in Table 6-25. The noise is due to a centrifugal ventilating fan. Determine the speech interference level and balanced noise criterion rating for the library room. Determine whether the acoustic environment would be satisfactory for "good listening" conditions for a library.

6-8. Determine the balanced noise criterion rating for the background noise in the shop room given in Table 6-25. Is this value for NCB suitable for a shop area?

6-9. Determine the balanced noise criterion rating for the background noise in the office given in Table 6-25. Is the background noise level satisfactory for large offices or reception areas? Would there be any problem with "hissy" sounds or "rumbly" sounds in the background noise?

6-10. An electric utility company is planning to construct a transformer station in a rural area. The expected noise spectrum due to the transformer noise at the property line is given in Table 6-25. The transformer noise involves no pure-tone components, and the transformer will be operating day and night year-round. There are no impulsive noises involved. Determine the composite noise rating and the probable community reaction to this noise source, based on the environmental noise rating.

6-11. A cooling tower has the sound level spectrum given in Table 6-25 at the property line of a plant. The cooling tower noise involves no pure-tone components, and the noise is continuous from the tower when it is operating. The tower is operated only during the daytime and the evening, and the tower is turned off at night. The cooling tower is operated year-round, and the sound is non-impulsive. The tower is located in a residential area where there are some businesses present. Determine the composite noise rating for the tower noise and the anticipated community reaction to the cooling tower noise.

6-12. An outdoor incinerator produces the sound pressure level spectrum given in Table 6-25 at the property line of a nearby residence. The incinerator noise has no pure-tone components and is non-impulsive. The unit is in operation all day from 7:00 a.m. to 10:00 p.m. year-round, and the noise is emitted five times per hour, on the average. The incinerator is located in a residential area that does contain some business in the area. Determine the composite noise rating and the anticipated community reaction to the incinerator noise.

6-13. Your company has installed an aeration system at a waste treatment plant outdoors. Some residents in the neighboring suburban area have initiated a lawsuit, charging that the noise is excessively annoying. The noise does involve pure-tone noise, but there is no impulsive noise emitted. The noise is continuous in nature, and is present year-round, all day and all night. The measured octave band sound pressure levels at the nearest resident's property line are given in Table 6-25. Determine the community composite noise rating. Based on the anticipated reaction of the "average person" to the noise, would the lawsuit be definitely justified?

6-14. At a certain location, the measured A-weighted sound levels are as follows:

$$7:00 \text{ a.m. to } 1:00 \text{ p.m.}, \quad L_A = 50 \text{ dBA}$$
$$1:00 \text{ p.m. to } 5:00 \text{ p.m.}, \quad L_A = 55 \text{ dBA}$$
$$5:00 \text{ p.m. to } 7:00 \text{ p.m.}, \quad L_A = 60 \text{ dBA}$$
$$7:00 \text{ p.m. to } 10:00 \text{ p.m.}, \quad L_A = 45 \text{ dBA}$$
$$10:00 \text{ p.m. to } 7:00 \text{ a.m.}, \quad L_A = 40 \text{ dBA}$$

Determine the day–night level at the given location.

6-15. At a proposed building site, the measured A-weighted sound levels are as follows:

$$7:00 \text{ a.m. to } 4:00 \text{ p.m.}, \quad L_A = 60 \text{ dBA}$$
$$4:00 \text{ p.m. to } 10:00 \text{ p.m.}, \quad L_A = 50 \text{ dBA}$$
$$10:00 \text{ p.m. to } 7:00 \text{ a.m.}, \quad L_A = 40 \text{ dBA}$$

Determine the day–night level at the proposed building site.

6-16. A noise impact study was made at the site of a proposed housing complex. The source of noise is the traffic from a nearby interstate highway. The noise involves no pure-tone components or impulsive sounds. The housing complex is located in an urban residential area, and the people do have some prior experience or exposure to the traffic noise. The traffic noise will be present year-round. The measured A-weighted sound levels for the site are given in Table 6-26. Determine the day–night level at the site location. Based on the day–night level, determine the anticipated community response to the noise.

TABLE 6-26 Data[a] for Problem 6-16

Daytime				Nighttime	
Time	L_A, dBA	Time	L_A, dBA	Time	L_A, dBA
7:30 a.m.	56	1:30 p.m.	59	10:30 p.m.	53
8:30 a.m.	58	2:30 p.m.	59	11:30 p.m.	52
9:30 a.m.	58	3:30 p.m.	59	12:30 a.m.	50
10:30 a.m.	59	4:30 p.m.	59	1:30 a.m.	48
11:00 a.m.	60	5:30 p.m.	59	2:30 a.m.	48
12:30 p.m.	59	6:30 p.m.	57	3:30 a.m.	48
		7:30 p.m.	55	4:30 a.m.	48
		8:30 p.m.	53	5:30 a.m.	50
		9:30 p.m.	53	6:30 a.m.	54

[a] The data are the energy-averaged values for the hour, i.e., the value for 7:30 a.m. corresponds to the energy-averaged noise level from 7:00 a.m. to 8:00 a.m.

6-17. Determine the noise pollution level for the noise given in Problem 6-14. In which HUD land-use category would this environmental noise fall?

6-18. For the housing noise impact study given in Problem 6-16, determine the noise pollution level. In which category does the environmental noise fall, in terms of the HUD criteria?

6-19. The octave band sound pressure levels measured around a machine area are given in Table 6-25. For this noise spectrum, determine the effective "noisiness" in noys and the effective perceived noise level dB(PN).

6-20. A small airport has two runways, as shown in Fig. 6-6. The east–west runway has a length of 11,880 ft (2.25 miles or 3.62 km). The second runway makes an angle of 60° with the east–west runway, and its length is 7920 ft (1.50 miles or 2.41 km). There are 40 daytime flights and 6 nighttime flights during each 24-hour period for the airport. Plot to scale (1 in = 1 mile) the NEF-30 and NEF-40 contours for the airport. A proposed building site is located 3960 ft (0.75 miles or 1.21 km) due north of the east end of the east–west runway. Determine the HUD site acceptability criterion for the location of the proposed building site.

Acoustic Criteria

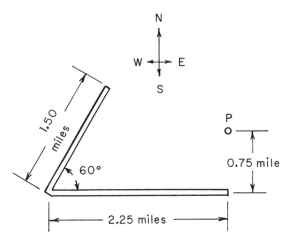

FIGURE 6-6 Diagram for Problem 6-2.

REFERENCES

ANSI. 1986. Rating noise with respect to speech interference, ANSI S3.14-1977 (R-1986). American National Standards Institute, New York.
ANSI. 1967. Normal equal-loudness contours of pure tones, ISO/R226. American National Standards Institute Inc., New York.
Baughn, W. L. 1973. Relation between daily noise exposure and hearing loss based on the evaluation of 6835 industrial noise exposure cases, Report No. AMRL-TR-73-53. Wright-Patterson AFB, U.S. Air Force.
Beranek, L. L. 1957. Revised criteria for noise in buildings. *Noise Control* 3: 19–27.
Beranek, L. L. 1960. *Noise Reduction*, pp. 496–501. McGraw-Hill, New York.
Beranek, L. L. 1971. *Noise and Vibration Control*, pp. 564–568. McGraw-Hill, New York.
Beranek, L. L. 1989. Balanced noise criterion (NCB) curves. *J. Acoust. Soc. Am.* 86: 650–664.
Collins, J. A. 1981. *Failure of Materials in Mechanical Design*, pp. 241–243. Wiley-Interscience, New York.
Ehmer, R. H. 1959. Masking of tones vs. noise bands. *J. Acoust. Soc. Am.* 31: 1253.
EPA. 1974. Information on levels of environmental noise requisite to protect public health and welfare with an adequate margin of safety, EPA 550/9-74-004. U.S. Environmental Protection Agency, Washington, DC.
Fletcher, H. and Munson, W. A. 1933. Loudness, its definition, measurement and calculation. *J. Acoust. Soc. Am.* 5(2): 82–105.
Galloway, W. J. and Bishop, D. E. 1970. Noise exposure forecast: evolution, evaluation, extensions and land use interpretations, FAA Report 70-9. Federal Aviation Administration, Washington, DC.

Griffiths, I. D. and Langdon, F. J. 1968. Subjective response to road traffic noise. *J. Sound and Vibration* 8: 16–32.
HUD. 1971. Noise abatement guidelines, Circular Policy No. 1390.2. U.S. Department of Housing and Urban Development, Washington, D.C.
Kryter, K. D. 1959. Scaling human reactions to the sound from aircraft. *J. Acoust. Soc. Am.* 31: 1415.
Kryter, K. D. 1970. *The Effects of Noise on Man*, p. 140. Academic Press, New York.
Kryter, K. D. and Pearsons, K. S. 1963. Some effects of spectral content and duration on perceived noise level. *J. Acoust. Soc. Am.* 35: 866–83.
Kryter, K. D., Ward, W. D., Miller, J. D., and Eldredge, D. H. 1965. Hazardous exposure to intermittent and steady-state noise. NAS-NRC Committee on Hearing, Bioacoustics and Biomechanics (CHABA), WG 46, Washington, DC. See also: *J. Acoust. Soc. Am.* 39: 451–64, 1966.
Lazarus, H. 1987. Prediction of verbal communication in noise – a development of generalized SIL curves and the quality of communication. *Applied Acoustics* 20:245–261, 21:325.
Lord, H., Gatley, W. S., and Evenson, H. A. 1980. *Noise Control for Engineers*, pp. 191–195. McGraw-Hill, New York.
OSHA. 1983. Occupational noise exposure: Hearing Conservation Amendment. Occupational Safety and Health Administration. *Federal Register* 48(46): 9738–9785.
Pearsons, K. S. and Bennett, R. L. 1974. Handbook of Noise Ratings, NASA CR-2376. National Aeronautics and Space Administration, Washington, DC.
Peterson, A. P. G. and Gross, E. E. 1972. *Handbook of Noise Measurement*, 7th edn, p. 38. General Radio, Concord, MA.
Pinto, R. M. N. 1962. Sex and acoustic trauma: audiologic study of 199 Brazilian airline stewards and stewardesses (VARIG). *Rev. Brazil. Med. (Rio)* 19: 326–327.
Reid, R. C. and Sherwood, T. K. 1966. *The Properties of Gases and Liquids*, 2nd edn, p. 633. McGraw-Hill, New York.
Schultz, T. J. 1970. Technical background for noise abatement in HUD's operating programs, Report 2005. Bolt, Beranek and Newman, Cambridge, MA.
Schultz, T. J. 1972. *Community Noise Ratings*, pp. 56–61. Applied Science Publishers, London.
Stevens, S. S. 1972. Perceived level of noise by Mark VII and decibels (E). *J. Acoust. Soc. Am.* 51: 575–600.
Stevens, K. N., Rosenblith, W. A., and Bolt, R. H. 1955. A community's reaction to noise. *Noise Control* 1: 63–71.
Thumann, A. and Miller, R. K. 1986. *Fundamentals of Noise Control Engineering*, pp. 56–61. The Fairmont Press, Atlanta, GA.

7
Room Acoustics

Room acoustics has been of interest since man first began to gather in auditoriums and churches. There was little information about the technical design of interior spaces for effective acoustic behavior until the beginning of the 20th century when Wallace Clement Sabine (Sabine, 1922) made extensive experimental studies of the acoustical properties of rooms, such as the Boston Symphony Hall. He developed empirical relationships to allow the designer to determine the amount of acoustic treatment required to achieve the desired acoustic behavior of the room.

In this chapter, we will examine the techniques that may be used to limit reverberation of sound in a room or to control the steady-state sound level in a room. The design of acoustic enclosures and acoustic barriers will also be considered.

7.1 SURFACE ABSORPTION COEFFICIENTS

7.1.1 Values for Surface Absorption Coefficients

Sound-absorbing materials are used to reduce the sound levels in a room or to reduce the reverberation, if either of these quantities are excessive. As will be discussed in Sec. 7.3, surface absorption materials do not change the sound coming directly from the source to the receiver. Instead, the surface treatment affects sound that has been reflected at least one time from the surfaces of the room. This sound is associated with the *reverberant sound field*.

The *surface absorption coefficient* α is defined as the ratio of the acoustic energy absorbed by the surface to the acoustic energy striking the surface:

$$\alpha = \frac{W_{abs}}{W_{in}} \qquad (7\text{-}1)$$

The energy absorbed at the surface may be transmitted through the material or may be dissipated within the material.

The surface absorption coefficient is generally a function of the frequency of the incident sound wave. Some representative values of the surface absorption coefficient for various interior surfaces are given in Appendix D.

The effect of people and furniture in a space is given by the product of the surface absorption coefficient and the area (αS), because the surface area of a person or chair is not always easy to determine. Values for this quantity are also given in Appendix D.

7.1.2 Noise Reduction Coefficient

In some cases, it is desirable to have a single-number rating to use in comparing the acoustic absorbing qualities of different materials. It is generally better to have available the spectrum of values (surface absorption coefficient as a function of frequency) for design purposes, however. The *noise reduction coefficient* (NRC) may be used for rough comparison of acoustic absorbing characteristics.

The noise reduction coefficient is defined as the average of the surface absorption coefficients in the 250 Hz, 500 Hz, 1000 Hz, and 2000 Hz octave bands. By convention, the NRC values are always rounded off to the nearest 0.05. For example, let us determine the NRC value for 1-inch thick fiberglass formboard. Using the values from Appendix D, we find the following:

$$\text{NRC} = (1/4)(0.34 + 0.79 + 0.99 + 0.93) = 0.7625$$

After rounding this value off to the nearest 0.05, we find the noise reduction coefficient for the formboard:

$$\text{NRC} = 0.75$$

The value of NRC-0.75 gives a rough measure of the effectiveness of the insulation in abosrbing sound; however, there is no indication that the material is actually more effective for high-frequency sound than for low-frequency sound.

7.1.3 Mechanism of Acoustic Absorption

When selecting acoustic materials for noise reduction, it is important for the designer to be aware of the mechanisms involved in absorption of the acoustic energy in the material. In addition, it is important to know the frequency distribution of the sound in the room, in order that the appropriate sound absorbing material may be matched with the acoustic field. A different material would be selected to absorb low-frequency sound than would be chosen to absorb high-frequency sound, generally. Let us consider three cases.

Porous felt-like sound absorbing materials are commercially available as mats, boards, or preformed components. Materials of manufacture include glass fibers, mineral or organic fibers, textiles or open cell foams, usually polyurethane foams. Representative curves for a porous felt-like material are shown in Fig. 7-1. It is noted that the absorption coefficient is smaller at low frequencies (250 Hz or lower), but is near unity at high

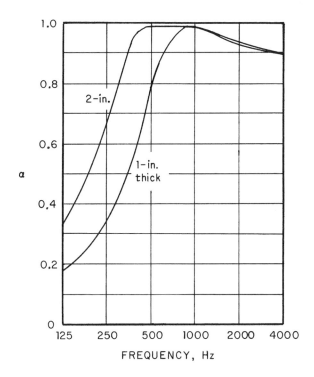

Figure 7-1 Surface absorption coefficient α for formboard, a porous felt-like material.

frequencies (1000 Hz or higher). The porous felt-like material is more effective in absorbing sound at higher frequencies than at lower frequencies.

The mechanism for absorption of acoustic energy for the porous materials is the fluid frictional energy dissipation between the air and the solid fibers. At high frequencies, the energy dissipation is larger because the particle velocity is larger than at low frequencies. The expansion and contraction of the air within the irregular spaces of the material also result in momentum losses for the air. The data presented in Fig. 7-1 also illustrate that the absorption coefficients are larger for the thicker material, which has more surface area for energy dissipation.

Unperforated panel absorbers involve one or more layers of metal or plywood with an air space behind the panel. Representative curves for a plywood panel are shown in Fig. 7-2. It is observed that the absorption coefficient is larger at the lower frequencies (500 Hz and below). The absorption coefficient is also larger for the thinner panel. The absorption coefficient may be increased by placing a porous acoustic absorbing material in the air space behind the panel.

The absorption for the unperforated panel is related to the transmission loss for the panel. As discussed in Chapter 4, the transmission loss for a

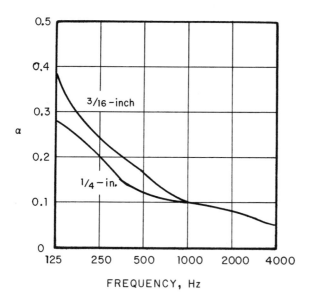

FIGURE 7-2 Surface absorption coefficient α for plywood panel with a 2-inch air space behind the panel.

Room Acoustics

panel in the stiffness-controlled region of acoustic behavior, encountered at low frequencies, decreases as the frequency is increased.

The absorption characteristics of perforated panels backed by an air space are illustrated in Fig. 7-3. The graph is shown for 1/2-inch (12.7 mm) thick perforated plywood panels with 3/16-inch (4.8 mm) diameter holes. The panel is backed with a 2.25 inch (57 mm) air space filled with a porous acoustic material. The absorption coefficient is largest in the mid-range of frequencies (250–500 Hz).

When the thickness of the panel is small, the absorption is primarily due to dissipation of the acoustic energy within the acoustic material behind the panel. For larger panel thickness, the perforated panel acts as a resonant cavity-type absorber (Helmholtz resonator). In this case, the absorption is greatest around the resonant frequency for the cavity. For example, the resonant frequency for the perforations in the panel with 11% open area is approximately 300 Hz.

In summary, we see that if we wish to absorb noise mainly in the high-frequency range (above about 1000 Hz), we should use a porous acoustic material on the surface. If we wish to absorb sound mainly in the low-frequency range (below about 250 Hz), we could use an unperforated panel with an air space behind the panel. For absorption of sound mainly

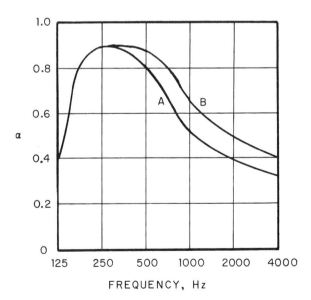

FIGURE 7-3 Surface absorption coefficient α for perforated plywood panel backed by a porous absorbent material. Curve A, 11% open area; curve B, 16.5% open area.

in the intermediate frequency range (250–1000 Hz), we would select a perforated panel with an absorbent material behind the panel. We would need to use combinations of these materials for effective absorption of sound over all frequencies.

7.1.4 Average Absorption Coefficient

In general, the various surfaces in a room will not have the same value of surface absorption coefficient at the same frequency. In this case, we need to determine an average value for the absorption coefficient for use in acoustic design. The appropriate average surface absorption coefficient is the surface-weighted average:

$$\bar{\alpha} = \frac{\Sigma \alpha_j S_j}{S_o} \tag{7-2}$$

The quantity S_o is the total surface area of the room. When people or furniture are present in the room, the absorption will be increased. To take this effect into account, the total absorption capacity (αS) for the people and furniture is added in the numerator of Eq. (7-2), but the surface area of the people and furniture is *not* included in the total surface area of the room S_o in the denominator.

The subjective acoustic characteristics of a room may be estimated from the value of the average surface absorption coefficient. The subjective perceptions are listed in Table 7-1 (Beranek, 1954).

7.2 STEADY-STATE SOUND LEVEL IN A ROOM

When a source of sound is turned on in a room, some of the energy emitted from the source is absorbed at the surfaces of the room and some energy is reflected back into the room. Steady-state conditions are usually achieved

TABLE 7-1 Subjective Acoustic Characteristics of Rooms

Room characteristic	Average absorption coefficient
Dead room	0.40
Medium-dead room	0.25
Average room	0.15
Medium-live room	0.10
Live room	0.05

Room Acoustics

after at time on the order of 0.25 seconds for most rooms. For a room having floor dimensions of 8 m × 8 m (26.2 ft × 26.2 ft), a sound wave will cross the room in (8 m)/(347 m/s) = 0.023 seconds. The sound wave will cross the room (0.25)/(0.023) = 10+ times in 0.25 seconds. When steady-state conditions have been achieved, the energy supplied by the sound source is equal to the energy absorbed by the room surfaces.

For steady-state conditions, we may divide the acoustic field in the room into two parts, as shown in Fig. 7-4:

1. The *direct sound field*, which consists of the acoustic energy associated with sound waves that have not struck surfaces in the room.
2. The *reverberant sound field*, which consists of the remainder of the energy.

The reverberant sound field is associated with all of the sound waves that have been reflected one or more times from the various surfaces in the room.

The acoustic energy density associated with the direct sound field is given by Eqs (2-21) and (2-27):

$$D_D = \frac{I_D}{c} = \frac{QW}{4\pi r^2 c} \tag{7-3}$$

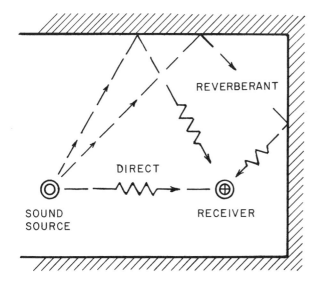

FIGURE 7-4 The reverberant and the direct sound fields.

The quantity Q is the directivity factor for the sound source, W is the acoustic power radiated by the sound source into the room, r is the distance between the source of sound and the receiver, and c is the speed of sound in the air in the room.

To calculate the reverberant sound field, let us consider the element shown in Fig. 7-5. The acoustic energy contained in the small elemental volume is equal to the product of the energy per unit volume (acoustic energy density) and the element volume:

$$dE = D_R \, dr \, dS \qquad (7\text{-}4)$$

The sound power radiated from the small area dS toward the surface element ΔS is equal to the change in energy of the element per unit time, where $c = dr/dt$.

$$dW = \frac{dE}{dt} = D_R \, c \, dS \qquad (7\text{-}5)$$

For a uniform reverberant sound field, the intensity of the acoustic energy radiated from the small area dS is the sound power per unit area:

$$dI = \frac{dW}{4\pi r^2} = \frac{D_R \, c \, dS}{4\pi r^2} \qquad (7\text{-}6)$$

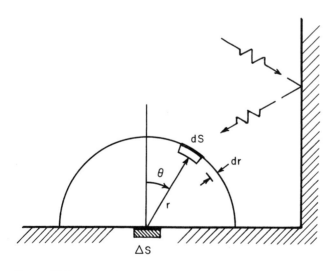

FIGURE 7-5 Reverberant sound field incident on a small element of the surface of the room ΔS.

Room Acoustics

The acoustic power incident on the room surface area increment ΔS from the small elemental area dS is the product of the energy per unit area and the projected area in the direction of the element dS:

$$dW_{in} = dI(\Delta S \cos\theta) \tag{7-7}$$

The differential area on the spherical surface is given by the following expression in spherical coordinates:

$$dS = r^2 \sin\theta \, d\theta \, d\varphi \tag{7-8}$$

If we make the substitutions from Eqs (7-6), (7-7), and (7-8) and integrate, we obtain the total reverberant acoustic power incident on the small surface ΔS:

$$dW_{in} = \frac{D_R c \, \Delta S}{4\pi r^2} \int_0^{2\pi} \int_0^{\pi/2} r^2 \sin\theta \cos\theta \, d\theta \, d\varphi$$

$$dW_{in} = \tfrac{1}{4} D_R c \, \Delta S \tag{7-9}$$

Assuming a uniform reverberant sound field, we may integrate Eq. (7-9) over the surface area of the room S_o to obtain the total reverberant acoustic power incident on the room surface:

$$W_{in} = \tfrac{1}{4} D_R c S_o \tag{7-10}$$

The acoustic power absorbed by the room surface is given by $(\bar{\alpha} W_{in})$, according to the definition of the surface absorption coefficient in Eq. (7-1). In steady state, the acoustic power absorbed is equal to the power that is supplied by the reverberant sound field:

$$W(1 - \bar{\alpha}) = \bar{\alpha} W_{in} = \tfrac{1}{4} D_R c \bar{\alpha} S_o \tag{7-11}$$

We may solve for the reverberant acoustic energy density from Eq. (7-11):

$$D_R = \frac{4W(1 - \bar{\alpha})}{c \bar{\alpha} S_o} = \frac{4W}{cR} \tag{7-12}$$

The quantity R is the *room constant*, defined by the following expression, for negligible energy attenuation in the room air:

$$R = \frac{\bar{\alpha} S_o}{1 - \bar{\alpha}} \tag{7-13}$$

The total acoustic energy density in the room in steady state is the sum of the contributions due to the direct sound field and the reverberant sound field:

$$D = D_R + D_D = \frac{4W}{cR} + \frac{QW}{4\pi r^2 c} = \frac{W}{c}\left(\frac{4}{R} + \frac{Q}{4\pi r^2}\right) = \frac{p^2}{\rho_o c^2} \tag{7-14}$$

The steady-state sound pressure may be found from Eq. (7-14):

$$p^2 = \rho_0 c W \left(\frac{4}{R} + \frac{Q}{4\pi r^2} \right) \quad (7\text{-}15)$$

Equation (7-15) may be written in an alternative form by introducing the acoustic reference quantities:

$$\frac{p^2}{p_{ref}^2} = \frac{W}{W_{ref}} \left(\frac{4}{R} + \frac{Q}{4\pi r^2} \right) \frac{\rho_0 c W_{ref}}{p_{ref}^2} \quad (7\text{-}16)$$

This expression may be converted to "level" form by taking $\log_{10}$ of both sides and multiplying through by 10:

$$L_p = L_W + 10 \log_{10} \left(\frac{4}{R} + \frac{Q}{4\pi r^2} \right) + 10 \log_{10} \left(\frac{\rho_0 c W_{ref}}{p_{ref}^2} \right) \quad (7\text{-}17)$$

$$\underset{\begin{bmatrix}\text{reverberant}\\ \text{sound field}\end{bmatrix}}{\uparrow} \quad \underset{\begin{bmatrix}\text{direct}\\ \text{sound field}\end{bmatrix}}{\uparrow}$$

As discussed in Sec. 5.1, the value of the last term in Eq. (7-17) for air at 101.3 kPa and 300K is 0.1 dB. The final form for the expression for the steady-state sound pressure level in a room may be written as follows:

$$L_p = L_W + 10 \log_{10} \left(\frac{4}{R} + \frac{Q}{4\pi r^2} \right) + 0.1 \quad (7\text{-}18)$$

The room constant R must be expressed in m² units, and the distance between the source and receiver r must be expressed in m units in Eq. (7-18).

An important observation may be made from Eq. (7-18). The term $(4/R)$ is associated with the reverberant sound field, which is dependent on the absorption characteristics of the room. The term $(Q/4\pi r^2)$ is associated with the sound coming directly from the sound source or the direct sound field, which is independent of the properties of the room. If the second term predominates, or if the direct sould field is much larger than the reverberant sound field, very little reduction in the sound pressure level can be achieved by adding more acoustic absorptive material to the room surfaces. In fact, it would be a waste of money and effort to buy and install acoustic material on the wall or ceiling, for this case. Other noise control techniques, such as using an acoustic barrier or enclosure, would be required in cases where the direct sound field is predominating. On the other hand, if the first term predominates, or if the reverberant sound field is much larger than the direct sound field, the steady-state sound pressure level can be reduced by adding acoustic material on the surfaces of the room.

Room Acoustics

Example 7-1. A room has dimensions of 6.20 m (20.3 ft) × 6.00 m (19.7 ft) × 3.10 m (10.2 ft) high, as shown in Fig. 7-6. The room has one solid wood door (1.20 m × 2.20 m or 3.94 ft × 7.22 ft) in the 6-m wall. The walls are plaster on lath. The ceiling is $\frac{1}{2}$-in acoustic tile on hard backing, and the floor is covered with a 3/8-in carpet on concrete floor, with no pad. A machine (sound power levels given in Table 7-2) is located in the room, and the directivity factor for the machine is $Q = 4$ for all frequencies. There are six adults standing in the room. Determine the octave band steady-state sound pressure level in the room at a distance of 6.00 m (19.7 ft) from the machine.

Let us carry out the calculations for the 500 Hz octave band in detail. The results for the other octave bands are given in Table 7-2. The surface absorption coefficients and surface areas are found as follows:

1. *Walls:*

$$\alpha_1 = 0.06$$

$$S_1 = (2)(6.20 + 6.00)(3.10) - (1.20)(2.20) = 75.64 - 2.64$$
$$= 73.00 \, \text{m}^2$$

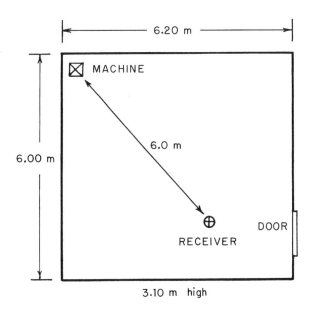

FIGURE 7-6 Diagram for Example 7-1.

TABLE 7-2 Solution for Example 7-1

	Octave band center frequency, Hz					
Item	125	250	500	1,000	2,000	4,000
L_W, dB (given)	52	57	60	56	50	43
$\bar{\alpha}$	0.1709	0.1784	0.2361	0.4034	0.4057	0.4113
R, m²	30.93	32.58	46.37	101.5	102.4	104.8
$10\log_{10}\left(\dfrac{4}{R}+\dfrac{Q}{4\pi r^2}\right)$	−8.6	−8.8	−10.2	−13.2	−13.2	−13.3
L_p(OB), dB	43.5	48.3	49.9	42.9	36.9	29.8

2. *Door:*

$$\alpha_2 = 0.05$$

$$S_2 = (1.20)(2.20) = 2.64\,\text{m}^2$$

3. *Ceiling:*

$$\alpha_3 = 0.55$$

$$S_3 = (6.20)(6.00) = 37.20\,\text{m}^2$$

4. *Floor:*

$$\alpha_4 = 0.21$$

$$S_4 = 37.20\,\text{m}^2$$

5. *People:*

$$(\alpha S) = (6\text{ people})(0.44) = 2.64\,\text{m}^2$$

The total surface area of the room is:

$$S_o = 75.64 + (2)(37.20) = 150.04\,\text{m}^2\ (1615\,\text{ft}^2)$$

The average surface absorption coefficient at 500 Hz may be found from Eq. (7-2):

$$\bar{\alpha} = \frac{(0.06)(73.00) + (0.05)(2.64) + (0.55)(37.20) + (0.21)(37.20) + 2.64}{(150.04)}$$

$$\bar{\alpha} = \frac{4.38 + 0.132 + 20.46 + 7.812 + 2.64}{150.04} = \frac{35.424}{150.04} = 0.2361$$

Room Acoustics

The room constant at 500 Hz is calculated from Eq. (7-13):

$$R = \frac{(150.04)(0.2361)}{(1 - 0.2361)} = 46.372 \, \text{m}^2$$

The steady-state sound pressure level in the 500 Hz octave band is found from Eq. (7-18):

$$L_p = 60 + 10\log_{10}\left[\frac{4}{46.372} + \frac{(4.0)}{(4\pi)(6.00)^2}\right] + 0.1$$

$$L_p = 60 + 10\log_{10}(0.08626 + 0.00884) + 0.1 = 60 + (-10.2) + 0.1$$

$$L_p = 49.9 \, \text{dB}$$

It is noted that the sound pressure level could be decreased by adding sound absorption material on the walls, because the reverberant sound field contribution (0.08626) is much larger (almost 10 times larger) than the direct sound field contribution (0.00884).

If we extrapolate the sound pressure spectrum to estimate $L_p(63 \, \text{Hz}) = 37 \, \text{dB}$, the overall sound pressure level due to the source is found as follows:

$$L_p = 10\log_{10}(10^{3.70} + 10^{4.35} + 10^{4.83} + \cdots + 10^{2.98}) = 53.4 \, \text{dB}$$

7.3 REVERBERATION TIME

When the source of sound in a room is suddenly turned off, a certain period of time is required before the sound energy is practically all absorbed by the surfaces in the room. For many applications, the duration of this time period is important for effective use of the space.

Let us consider the room shown in Fig. 7-7. The acoustic energy density associated with the sound after one reflection is given by the following expression:

$$D_1 = D_o(1 - \bar{\alpha}) \tag{7-19}$$

The quantity D_o is the original acoustic energy density before the sound strikes any walls. The acoustic energy density after the second reflection is found in a similar manner:

$$D_2 = D_1(1 - \bar{\alpha}) = D_o(1 - \bar{\alpha})^2 \tag{7-20}$$

By extension, we see that the acoustic energy density in the room after n reflections is given by the following expression:

$$D_n = D_o(1 - \bar{\alpha})^n \tag{7-21}$$

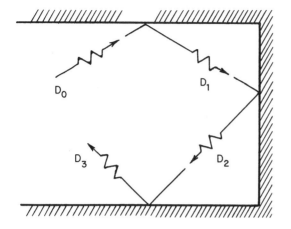

FIGURE 7-7 Reverberant sound field after the sound source has been turned off for various numbers of reflections.

The time between one reflection and the next reflection is related to the speed of sound c:

$$t_1 = d/c \tag{7-22}$$

The quantity d is the *mean free path* of the sound wave, or the average distance that the sound travels before being reflected. The mean free path is given by the following expression for a room (Pierce, 1981):

$$d = 4V/S_o \tag{7-23}$$

The quantity V is the total volume of the room. Making the substitution from Eq. (7-23) for the mean free path into Eq. (7-22), we obtain the following relationship for the time between reflections:

$$t_1 = \frac{4V}{S_o c} \tag{7-24}$$

The total time t after n reflections is given by the following:

$$t = nt_1 = \frac{4Vn}{cS_o} \tag{7-25}$$

The number of reflections n during the time t is found from Eq. (7-25):

$$n = \frac{cS_o t}{4V} \tag{7-26}$$

Room Acoustics

The expression for the acoustic energy density in the room as a function of the time after the source of sound has been turned off may be found by combining Eqs (7-21) and (7-26):

$$D = D_o(1 - \bar{\alpha})^{(S_o c/4V)t} \tag{7-27}$$

We may write the term involving the average surface absorption coefficient in the following form:

$$(1 - \bar{\alpha}) = \exp[\ln(1 - \bar{\alpha})] = \exp\left[-\ln\left(\frac{1}{1-\bar{\alpha}}\right)\right] \tag{7-28}$$

The acoustic energy density in Eq. (7-27) may be expressed in the following alternative form:

$$D = D_o \exp(-cat/4V) \tag{7-29}$$

The quantity a is called the *number of absorption units* and is defined by the following expression:

$$a = S_o \ln\left(\frac{1}{1-\bar{\alpha}}\right) \tag{7-30}$$

In architectural work, if the units of the surface area are expressed in ft^2, then the units for the absorption are called sabins, in honor of W. C. Sabine, who conducted the initial work in laying a foundation for the rational design for architectural acoustics. On the other hand, if the room surface area is expressed in m^2, the absorption units are sometimes called mks sabins. To avoid any confusion, we will express the absorption units directly in units of area.

The acoustic pressure and acoustic energy density are related by Eq. (2-20):

$$D = \frac{p^2}{\rho_o c} \quad \text{and} \quad D_o = \frac{p_o^2}{\rho_o c} \tag{7-31}$$

The quantity p_o is the original acoustic pressure before the sound source is turned off. Equation (7-29) may be written in terms of the acoustic pressures:

$$\frac{p^2}{p_o^2} = \exp(-cat/4V) = \frac{(p/p_{\text{ref}})^2}{(p_o/p_{\text{ref}})^2} \tag{7-32}$$

Taking $\log_{10}$ of both sides of Eq. (7-32) and multiplying by 10, we obtain the following result:

$$L_{p,o} - L_p = [10\log_{10}(e)](cat/4V) = 1.0857 cat/V \tag{7-33}$$

The *reverberation time* T_r is defined as the time required for the sound pressure level to decrease by 60 dB. Setting $(L_{p,o} - L_p) = 60\,\text{dB}$ in Eq. (7-33), we may solve for the reverberation time, $t = T_r$:

$$T_r = \frac{55.26 V}{ca} \tag{7-34}$$

This expression is called the *Norris–Eyring reverberation time* (Eyring, 1930), which is a modification of the expression originally developed by Sabine, who used $a = S_o \bar{\alpha}$.

If the walls of a room have significantly different surface absorption coefficients from those of the floor–ceiling combination, sound waves traveling between the walls will decay at a different rate from those traveling in the vertical direction. To take this phenomenon into consideration, it was proposed that the number of absorption units be expressed as three terms: for reflections between the side walls, the end walls, and the floor–ceiling combination (Fitzroy, 1959). The Fitzroy expression for the number of absorption units is given by the following expression:

$$\frac{1}{a} = -\frac{1}{S_o}\left[\frac{(S_x/S_o)}{\ln(1-\bar{\alpha}_x)} + \frac{(S_y/S_o)}{\ln(1-\bar{\alpha}_y)} + \frac{(S_z/S_o)}{\ln(1-\bar{\alpha}_z)}\right] \tag{7-35}$$

The quantity S_x is the side-wall surface area, S_y is the end-wall surface area, S_z is the floor–ceiling area, and S_o is the total surface area. The quantities $\bar{\alpha}_x$, $\bar{\alpha}_y$, and $\bar{\alpha}_z$ are the surface absorption coefficients for the side-wall surfaces, the end-wall surfaces, and the ceiling–floor area, respectively. If any of the three average surface absorption coefficients exceed 0.60, the corresponding term, $\ln(1-\bar{\alpha})$, is replaced by $(-\bar{\alpha})$ for that surface combination.

The effect of people and furniture in the room may be introduced by adding the absorption capacity of the empty room a_o, given by Eq. (7-35), to the absorption of the people and furniture:

$$a = a_o + \Sigma(\alpha S)_{\text{people}} \tag{7-36}$$

It has been shown (Fitzroy, 1959) that the reverberation time calculated from the Fitzroy equation, Eq. (7-35), is in better agreement with experimental data for rooms with nonuniform absorption than is Eq. (7-30). For a room in which the average floor–ceiling absorption coefficient (α_z) was about 13 times that of the other two combinations, the reverberation time calculated from Eq. (7-30) was found to be about 25% of the measured reverberation time, whereas Eq. (7-35) yielded values for the reverberation time within 3% of the measured values. If the average absorption coefficients of the surfaces are approximately the same, either relationship will predict values of reverberation time near that measured.

Room Acoustics

A value for the optimum reverberation time may be needed for design purposes. The optimum reverberation time depends on the usage of the space (church, music hall, conference room, etc.). An empirical relationship for the measured reverberation time in acoustically good concert halls and opera houses (Beranek, 1962) has been developed:

$$T_r = \frac{1}{0.1 + 5.4(S_F/V)} \qquad (7\text{-}37)$$

The quantity S_F is the total floor area (m^2) of the audience, orchestra, and chorus areas, and V is the volume of the space (m^3).

For design purposes, the optimum reverberation time for the 500 Hz octave band may be estimated from the following empirical expression (Beranek, 1954, p. 425):

$$T_{r,\text{opt}} = a + \frac{b \log_{10} V}{\log_{10} e} = a + 2.3026\, b \log_{10} V \qquad (7\text{-}38)$$

The quantity V is the volume of the space (m^3). The numerical values for the constants a and b are given in Table 7-3A. The reverberation time ratio from Table 7-3B may be used to estimate the optimum reverberation time for other octave bands for a room used for music. For rooms involving speech only, it is recommended that the value for reverberation time at 500 Hz be used for other frequencies.

Example 7-2. Determine the reverberation time for the 500 Hz octave band for the room given in Example 7-1. The sonic velocity for the air in the room (at 21°C or 70°F) is 343.8 m/s (1128 fps).

The average surface absorption coefficient and surface area were calculated previously:

$$\bar{\alpha} = 0.2361 \quad \text{and} \quad S_o = 150.04\, \text{m}^2$$

The number of absorption units calculated from Eq. (7-30) is as follows:

$$a = (150.04) \ln\left(\frac{1}{1 - 0.2361}\right) = 40.41\, \text{m}^2 \quad \text{(or, mks sabins)}$$

The volume of the room is as follows:

$$V = (6.20)(6.00)(3.10) = 115.32\, \text{m}^3\ (4072\, \text{ft}^3)$$

TABLE 7-3A Values for the Constants in Eq. (7-38) for the Optimum Reverberation Time at 500 Hz

Type of space	a	b
Catholic church; organ music	+0.098	1/5
Protestant church; synagogue; concert hall	−0.162	1/5
Music studio; opera house	−0.352	1/5
Conference room; movie theater	−0.101	2/15
Broadcast room for speech	−0.192	1/9

TABLE 7-3B Reverberation Time Ratio $[T_{t,opt}(f)/T_{r,opt}(500\,Hz)]$ for Music Rooms

Frequency, Hz	Ratio	Frequency, Hz	Ratio
31.5	2.40	1,000	1.05
63	1.93	2,000	1.05
125	1.46	4,000	1.05
250	1.13	8,000	1.05
500	1.00		

Source: Beranek (1954, p. 426).

Using the Norris–Eyring expression, Eq. (7-34), we may estimate the reverberation time:

$$T_r = \frac{(55.26)(115.32)}{(343.8)(40.41)} = 0.459\,s$$

Because the floor and ceiling in this example have surface absorption coefficients that are much different from that of the walls, the Fitzroy relationship, Eq. (7-35), would probably yield more accurate results for the reverberation time. The parameters for the side wall (the long wall) are as follows:

$$S_x = (2)(6.20)(3.10) = 38.44\,m^2$$

$$\bar{\alpha}_x = 0.06$$

$$S_x/S_o = (38.44)/(150.04) = 0.2562$$

Room Acoustics

The values for the end wall (the shorter wall) are as follows:

$$S_y = (2)(6.00)(3.10) = 37.20 \text{ m}^2$$

$$\bar{\alpha}_y = \frac{(0.06)(37.20 - 2.64) + (0.05)(2.64)}{(37.20)} = \frac{2.206}{37.20} = 0.0593$$

$$S_y/S_o = 0.2479$$

Finally, the parameters for the floor–ceiling combination are as follows:

$$S_z = (2)(6.20)(6.00) = 74.40 \text{ m}^2$$

$$\bar{\alpha}_z = \frac{(0.21)(37.20) + (0.55)(37.20)}{(74.40)} = \frac{28.27}{74.40} = 0.3800$$

$$S_z/S_o = 0.4959$$

The number of absorption units, without the effect of the people and furniture, is given by Eq. (7-35):

$$\frac{1}{a_o} = -\frac{1}{(150.04)}\left[\frac{(0.2562)}{\ln(1-0.06)} + \frac{(0.2479)}{\ln(1-0.0593)} + \frac{(0.4959)}{\ln(1-0.3800)}\right]$$

$$\frac{1}{a_o} = \frac{4.1406 + 4.0552 + 1.0373}{150.04} = \frac{9.2332}{150.04} = 0.06154 \text{ m}^{-2}$$

$$a_o = 16.250 \text{ m}^2$$

The effect of the people and furniture may be included by using Eq. (7-36):

$$a = 16.250 + 2.64 = 18.89 \text{ m}^2$$

The reverberation time for the 500 Hz octave band, using the Fitzroy relationship, is calculated from Eq. (7-34):

$$T_r = \frac{(55.26)(115.32)}{(343.8)(18.89)} = 0.981 \text{ s}$$

This value for the reverberation time would probably be more nearly in agreement with measured values for the room, because there will be sound waves moving between the walls that are not damped out as rapidly as the sound waves moving between the floor and ceiling.

Example 7-3. Suppose the room in Example 7-2 is a conference room. Determine the amount of 1-inch fiberglass formboard that must be added to the walls to reduce the reverberation time at 500 Hz, according to the Fitzroy relationship, to the optimum value.

The optimum reverberation time may be determined from Eq. (7-38):

$$T_{t,\text{opt}} = -0.101 + (2/15)(2.3026)\log_{10}(115.32) = 0.532 \text{ s}$$

The number of absorption units required to achieve this reverberation time is found from Eq. (7-34):

$$a = \frac{55.26V}{cT_r} = \frac{(55.26)(115.32)}{(343.8)(0.532)} = 34.842 \, \text{m}^2$$

The number of absorption units, excluding the people and furniture, is found from Eq. (7-36):

$$a_o = 34.842 - 2.64 = 32.202 \, \text{m}^2$$

There are several approaches that could be used in this case. Generally, it is better to distribute the sound absorption material than to concentrate the material on one surface. Let us determine the amount of formboard required to make the average surface absorption coefficients for the side walls and the end walls equal. Using the Fitzroy expression, Eq. (7-35), we obtain the following values:

$$\frac{1}{a_o} = \frac{1}{32.202} = -\frac{1}{(150.04)}\left[\frac{(0.2562 + 0.2479)}{\ln(1-\alpha)} + \frac{(0.4959)}{\ln(1-0.3800)}\right]$$

$$4.6593 = -\frac{0.5041}{\ln(1-\alpha)} + 1.0373$$

$$1 - \alpha = 0.8701 \quad \text{or,} \quad \bar{\alpha}_x = \bar{\alpha}_y = 0.1299$$

At 500 Hz, the surface absorption coefficient for fiberglass formboard is $\alpha_5 = 0.79$, from Appendix D. The required surface area S_5 covered by the formboard on the side walls may be found from the following expression:

$$\bar{\alpha}_x S_x = (0.1299)(38.44) = (0.06)(38.44 - S_5) + 0.79 S_5$$

$$S_5 = \frac{4.9934 - 2.3064}{(0.79 - 0.06)} = 3.680 \, \text{m}^2 \, (39.61 \, \text{ft}^2)$$

If the formboard is distributed evenly on both walls, the required surface area per wall is as follows:

$$\tfrac{1}{2} S_5 = 1.840 \, \text{m}^2 \, (19.81 \, \text{ft}^2)$$

This is a little less than 10% of each side-wall surface area, so it would be practical (and possible) to add the formboard to the side walls.

Similarly, the required area of formboard S_6 on the end walls to achieve the desired average surface absorption coefficient is found as follows:

$$\bar{\alpha}_y S_y = (0.1299)(37.20) = (0.06)(34.56 - S_6) + (0.05)(2.64) + 0.79 S_6$$

$$S_6 = \frac{4.8323 - 2.0736 - 0.1320}{(0.79 - 0.06)} = 3.598 \, \text{m}^2 \, (38.7 \, \text{ft}^2)$$

Room Acoustics

If half of the area is placed on each end wall, the amount of one end wall covered by the formboard is as follows:

$$\tfrac{1}{2}S_6 = 1.799 \text{ m}^2 \; (19.4 \text{ ft}^2)$$

This area is also about 10% of each end-wall surface area, so it would be possible to add the acoustic material on the end wall.

7.4 EFFECT OF ENERGY ABSORPTION IN THE AIR

As was discussed in Sec. 4.12, the effect of dissipation of acoustic energy in the air through which the sound wave is moving is generally important only for high-frequency sound and for sound transmitted over large distances. The same behavior is observed for sound transmitted in rooms. In this section, we will consider the effect of dissipation of energy in the air in a room on the steady-state sound level and on the reverberation time.

7.4.1 Steady-State Sound Level with Absorption in the Air

In addition to the energy absorbed at the walls, there is an attenuation of the sound due to absorption in the volume of air in the room. This effect also modifies the energy from the reverberant field. Equation (7-11) may be written in the following form to include these effects:

$$W(1-\bar{\alpha})\,e^{-md} = \bar{\alpha}W_{\text{in}} + (1 - e^{-md})W_{\text{in}} = \tfrac{1}{4}D_R c S_o (1 + \bar{\alpha} - e^{-md}) \tag{7-39}$$

The quantity d is the mean free path of the sound wave, given by Eq. (7-23), and m is the energy attenuation coefficient. The first term on the left side of Eq. (7-39) represents the energy delivered to the reverberant field or the energy that has not been absorbed after striking the walls or passing through the air before striking the walls. We may solve for the reverberant acoustic energy density, including the effect of energy absorption in the air:

$$D_R = \frac{4W(1-\bar{\alpha})\,e^{-md}}{cS_o(1+\bar{\alpha}-e^{-md})} = \frac{4W}{cR} \tag{7-40}$$

The room constant, including the effects of air attenuation, is given by the following expression from Eq. (7-40):

$$R = \frac{S_o(1+\bar{\alpha}-e^{-md})}{(1-\bar{\alpha})\,e^{-md}} \tag{7-41}$$

In general, the term md is usually small for rooms. As an example, the energy attenuation coefficient for air at 25°C (77°F) and 60% relative humidity at 2000 Hz is $m = 2.16\,\text{km}^{-1} = 0.00216\,\text{m}^{-1}$, from Table 4-8. Let us consider a room having dimensions $30\,\text{m} \times 15\,\text{m} \times 5\,\text{m}$ high (98.4 ft × 49.2 ft × 16.4 ft). The volume is $V = 2250\,\text{m}^3$, and the surface area is $S_\text{o} = 1350\,\text{m}^2$, so the mean free path is $d = 4V/S_\text{o} = 6.667\,\text{m}$. The quantity $md = (0.00216)(6.667) = 0.0144$. For small vlaues of md, we may expand the exponential expression and approximate the exponential by the first two terms in the series:

$$\text{e}^{-md} = 1 - md - \cdots \approx 1 - 4mV/S_\text{o} \qquad (7\text{-}42)$$

If we make the substitution from Eq. (7-42) for small md (or $md \leq 0.20$) into Eq. (7-41), we obtain the following expression for the room constant:

$$R = \frac{S_\text{o}(\bar{\alpha} + 4mV/S_\text{o})}{1 - \bar{\alpha} - (4mV/S_\text{o})} \qquad (7\text{-}43)$$

The direct sound field is also affected by the attenuation in the air in the room. The modification of Eq. (7-3) to include energy attenuation effects is as follows:

$$D_\text{D} = \frac{QW\,\text{e}^{-mr}}{4\pi c r^2} \qquad (7\text{-}44)$$

If the value of the parameter mr is less than about 0.10, the exponential in Eq. (7-44) is approximately unity and may be neglected. In the previous example, this would correspond to a distance between the sound source and receiver of $r = 0.10/0.00216 = 46.3\,\text{m}$ (152 ft).

The total acoustic energy density is the sum of the reverberant field, Eq. (7-40), and the direct field, Eq. (7-44):

$$D = D_\text{R} + D_\text{D} = \frac{W}{c}\left[\frac{4}{R} + \frac{Q\,\text{e}^{-mr}}{4\pi r^2}\right] = \frac{p^2}{\rho_\text{o} c^2} \qquad (7\text{-}45)$$

The corresponding sound pressure level may be found by introducing the reference pressure and reference power, then taking $\log_{10}$ of both sides of the resulting expression and multiplying by 10:

$$L_\text{p} = L_\text{W} + 10\log_{10}\left(\frac{4}{R} + \frac{Q\,\text{e}^{-mr}}{4\pi r^2}\right) + 0.1 \qquad (7\text{-}46)$$

Equation (7-43) must be used to evaluate the room constant, if air attenuation effects are to be included.

Example 7-4. Determine the octave band sound pressure levels for the 500 Hz and 4000 Hz octave bands for the room given in Example 7-1, if

Room Acoustics

air attenuation is considered. The air in the room is at 21°C (70°F) and 50% relative humidity.

From Table 4-8, we find the following values for the energy attenuation coefficient:

$$m = 0.39 \text{ km}^{-1} \text{ at } 500 \text{ Hz and } m = 6.11 \text{ km}^{-1} \text{ at } 4000 \text{ Hz}.$$

For a frequency of 500 Hz, the dimensionless parameter is as follows:

$$4mV/S_o = (4)(0.39)(10^{-3})(115.32)/(150.04) = 1.20 \times 10^{-3}$$

The room constant is found from Eq (7.43):

$$R = \frac{(150.04)(0.2361 + 0.0012)}{1 - 0.2361 - 0.0012} = 46.68 \text{ m}^2$$

The exponential factor in the direct sound field expression is as follows:

$$e^{-mr} = \exp[-(0.39)(10^{-3})(6)] = \exp(-0.00234) = 0.9977$$

The octave band sound pressure level for the 500 Hz octave band is as follows:

$$L_p = 60 + 10 \log_{10}\left[\frac{4}{46.68} + \frac{(4)(0.9977)}{(4\pi)(6)^2}\right] + 0.1$$

$$L_p(500 \text{ Hz}) = 60 + (-10.2) + 0.1 = 49.9 \text{ dB}$$

If we repeat the calculations for the 4000 Hz octave band, we obtain the following values:

$$4mV/S_o = 0.01878$$
$$R = 113.23 \text{ m}^2$$
$$e^{-mr} = 0.9640$$
$$L_p(4000 \text{ Hz}) = 43 + (-13.6) + 0.1 = 29.5 \text{ dB}$$

It is observed that the effect of air attenuation is negligible (less than 0.1 dB) in the 500 Hz octave band and is essentially negligible (about 0.3 dB difference) for the 4000 Hz octave band, in this example. If all the room dimensions were increased by a factor of 10, then the air attenuation would be more significant, particularly in the 4 kHz octave band.

7.4.2 Reverberation Time with Absorption in the Air

The general effect of air attenuation is to decrease the reverberation time, since there is an additional mechanism (air attenuation) present to remove energy from the acoustic field. If we include the effect of air attenuation, Eq.

(7-19) for the acoustic energy density after the first reflection would be modified as follows:

$$D_1 = D_o(1 - \bar{\alpha})e^{-md} \tag{7-47}$$

The acoustic energy density after the second reflection is found similarly:

$$D_2 = D_1(1 - \bar{\alpha})e^{-md} = D_o(1 - \bar{\alpha})^2 e^{-2md} \tag{7-48}$$

The acoustic energy density after n reflections is given by the following expression:

$$D_n = D_o(1 - \bar{\alpha})^n e^{-nmd} \tag{7-49}$$

If we substitute the number of reflections n from Eq. (7-26) in Eq. (7-49), the following relationship is obtained for the time dependence of the acoustic energy density, considering air attenuation:

$$D = D_o(1 - \bar{\alpha})^{(S_o c/4V)t} e^{-mct} \tag{7-50}$$

This expression may be written in the following alternative form:

$$D = D_o \exp\left[-\frac{cS_o t}{4V} \ln\left(\frac{1}{1-\bar{\alpha}}\right) - mct\right] \tag{7-51}$$

If we introduce the number of absorption units a from Eq. (7-30), Eq. (7-51) may be written in the following form:

$$D = D_o \exp\left[-\frac{ct}{4V}(a + 4mV)\right] \tag{7-52}$$

The acoustic energy density in terms of the acoustic pressure is given by Eq. (7-31). If we set the difference between the original sound pressure level and the sound pressure level after a time T_r (the reverberation time) equal to 60 dB, the following expression is obtained for the reverberation time, including the effect of attenuation in the air:

$$T_r = \frac{55.26V}{c(a + 4mV)} \tag{7-53}$$

According to Eq. (7-53), the effect of air attenuation increases the absorption from a to $(a + 4mV)$. The effect of air attenuation is more pronounced for large rooms (large volumes) than for small rooms.

Example 7-5. Determine the reverberation time in Problem 7-2 if the effect of air attenuation were to be considered. The air in the room is at 21°C (70°F) and 50% relative humidity. The energy attenuation coefficient at 500 Hz is $m = 0.39 \text{ km}^{-1}$ and the number of absorption units is $a = 18.89 \text{ m}^2$, according to the Fitzroy relationship.

Room Acoustics

The value of the parameter associated with air attenuation effects is as follows:

$$4mV = (4)(0.39)(10^{-3})(115.32) = 0.18\,\text{m}^2$$

The total absorption may be calculated:

$$a + 4mV = 18.89 + 0.18 = 19.07\,\text{m}^2$$

The reverberation time, including air attenuation, is found from Eq. (7-53):

$$T_r = \frac{(55.26)(115.32)}{(343.8)(19.07)} = 0.972\,\text{s}$$

The reverberation time calculated in Example 7-2, neglecting air attenuation, was 0.981 s, so the effect of air attenuation in this example is to decrease the reverberation time by about 1% or about 0.010 s.

7.5 NOISE FROM AN ADJACENT ROOM

In the previous material, we have considered the sound field produced by a source of sound within the room. There are situations where the noise source may be located in an adjacent room, such as the case of a conference room adjoining a machinery room. In this case, the acoustic properties of the adjacent room will influence the steady-state sound pressure level in the room to be considered. We would like to examine the case of a noise source in an adjacent room in this section.

7.5.1 Sound Source Covering One Wall

For the case of sound transmitted through a wall into a room, we may consider the wall itself as a source of sound, as far as the interior of the room is concerned. The reverberant sound field is independent of the location of the sound source, if the sound field may be considered to be diffuse. The acoustic energy density associated with the reverberant sound field is given by Eq. (7-12):

$$D_R = \frac{4W}{cR} \tag{7-54}$$

where quantity W is the acoustic power radiated from the wall, as shown in Fig. 7-8, R is the room constant, and c is the sonic velocity in the air.

For small distances from the wall, or for $r < (S_w/2\pi)^{1/2}$, the sound waves leaving the wall are practically plane waves, where S_w is the surface area of the wall. For plane sound waves, the acoustic energy density associated with the direct sound field is given by the following expression:

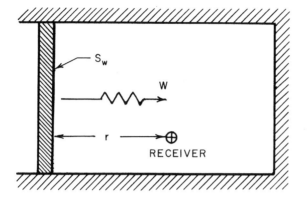

FIGURE 7-8 A wall of area S_w acting as a source of sound.

$$D_D = \frac{I_D}{c} = \frac{W}{S_w c} \tag{7-55}$$

The acoustic intensity I_D is the acoustic power per unit area, W/S_w.

The total steady-state acoustic energy density in the room is the sum of the reverberant and direct contributions:

$$D = D_D + D_R = \frac{W}{c}\left(\frac{4}{R} + \frac{1}{S_w}\right) = \frac{p^2}{\rho_o c^2} \tag{7-56}$$

We may solve for the steady-state sound pressure (squared) from Eq. (7-56) for the case of the receiver located near the wall source:

$$p^2 = \rho_o c W \left(\frac{4}{R} + \frac{1}{S_w}\right) \quad [\text{for } r < (S_w/2\pi)^{1/2}] \tag{7-57}$$

On the other hand, when the receiver is located a large distance from the wall, or for $r > (S_w/2\pi)^{1/2}$, the source of sound (the wall) acts as a source with a directivity factor $Q = 2$. In this case, the acoustic energy associated with the direct sound field is given by the following expression:

$$D_D = \frac{I_D}{c} = \frac{QW}{4\pi r^2 c} = \frac{W}{2\pi r^2 c} \tag{7-58}$$

The total acoustic energy density for this case is given as follows:

$$D = \frac{W}{c}\left(\frac{4}{R} + \frac{1}{2\pi r^2}\right) = \frac{p^2}{\rho_o c} \quad [\text{for } r > (S_w/2\pi)^{1/2}] \tag{7-59}$$

Room Acoustics

The corresponding steady-state sound pressure (squared) may be found from Eq. (7-59):

$$p^2 = \rho_0 c W \left(\frac{4}{R} + \frac{1}{2\pi r^2} \right) \quad [\text{for } r > (S_w/2\pi)^{1/2}] \tag{7-60}$$

Both Eqs (7-57) and (7-60) may be converted to "level" form by introducing the reference pressure and reference power, taking $\log_{10}$ of both sides, and multiplying by 10. The final result is as follows:

For $r < (S_w/2\pi)^{1/2}$:

$$L_p = L_W + 10 \log_{10} \left(\frac{4}{R} + \frac{1}{S_w} \right) + 0.1 \tag{7-61}$$

For $r > (S_w/2\pi)^{1/2}$:

$$L_p = L_W + 10 \log_{10} \left(\frac{4}{R} + \frac{1}{2\pi r^2} \right) + 0.1 \tag{7-62}$$

In these equations, we have taken the value of the constant term as follows:

$$10 \log_{10} \left(\frac{\rho_0 c W_{\text{ref}}}{p_{\text{ref}}^2} \right) = 0.1 \, \text{dB}$$

We note that the values for the term associated with the direct sound field are equal when $r = (S_w/2\pi)^{1/2} = r^*$.

7.5.2 Sound Transmission from an Adjacent Room

The result obtained in the previous section may be applied to the case of sound being generated in one room (by noisy equipment, for example) and transmitted through a wall of area S_w into an adjacent room, as shown in Fig. 7-9. Let us denote the room in which the source of sound is located as Room (1), and the room in which the receiver (people, for example) is located as Room (2). The steady-state sound pressure (squared) in the room with the source of sound may be determined from Eq. (7-15):

$$p_1^2 = \rho_0 c W \left(\frac{4}{R_1} + \frac{Q}{4\pi r_1^2} \right) \tag{7-63}$$

Similarly, we have shown that the steady-state sound pressure (squared) in the other room due to sound transmitted through the wall is given by Eq. (7-57) or Eq. (7-60):

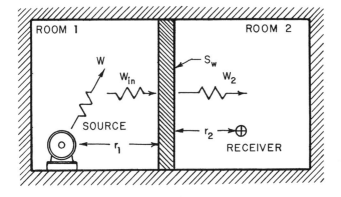

FIGURE 7-9 Sound transmitted from an adjacent room through an interior wall.

$$p_2^2 = \rho_0 c W_2 \left(\frac{4}{R_2} + \frac{1}{S_w} \right) \quad [\text{for } r_2 < (S_w/2\pi)^{1/2}] \quad (7\text{-}64)$$

$$p_2^2 = \rho_0 c W_2 \left(\frac{4}{R_2} + \frac{1}{2\pi r_2^2} \right) \quad [\text{for } r_2 > (S_w/2\pi)^{1/2}] \quad (7\text{-}65)$$

The sound power transmission coefficient a_t for the wall is defined by the following expression:

$$a_t = \frac{W_2}{W_{in}} = \frac{W_2}{I_{in} S_w} = \frac{W_2 R_1}{S_w W} \quad (7\text{-}66)$$

The acoustic power transmitted into the second room through the wall between the rooms may be found from Eq. (7-66):

$$W_2 = a_t S_w W / R_1 \quad (7\text{-}67)$$

If we substitute this result from Eq. (7-67) into Eqs (7-64) and (7-65), we obtain the following expressions for the sound pressure (squared):

$$p_2^2 = \frac{\rho_c c W}{R_1} \left(\frac{4 S_w}{R_2} + 1 \right) a_t \quad [\text{for } r_2 < (S_w/2\pi)^{1/2}] \quad (7\text{-}68)$$

$$p_2^2 = \frac{\rho_0 c W}{R_1} \left(\frac{4 S_w}{R_2} + \frac{S_w}{2\pi r_2^2} \right) a_t \quad [\text{for } r_2 > (S_w/2\pi)^{1/2}] \quad (7\text{-}69)$$

Note that the transmission loss for the wall is defined by Eq. (4-90):

$$\text{TL} = 10 \log_{10}(1/a_t) \quad (7\text{-}70)$$

Room Acoustics

Equations (7-68) and (7-69) may be converted to "level" form as follows:

$$L_{p2} = L_W - 10\log_{10}(R_1) + 10\log_{10}\left(\frac{4S_w}{R_2} + 1\right) - TL + 0.1$$
$$[\text{for } r_2 < (S_w/2\pi)^{1/2}] \tag{7-71}$$

$$L_{p2} = L_W - 10\log_{10}(R_1) + 10\log_{10}\left(\frac{4S_w}{R_2} + \frac{S_w}{2\pi r_2^2}\right) - TL + 0.1$$
$$[\text{for } r_2 > (S_w/2\pi)^{1/2}] \tag{7-72}$$

The sound pressure level in the first room, the room with the noise source, is given by Eq. (7-18):

$$L_{p1} = L_W + 10\log_{10}\left(\frac{4}{R_1} + \frac{Q}{4\pi r_1^2}\right) + 0.1 \tag{7-73}$$

The designer has several choices to produce a noise reduction in the room adjacent to the room with the noise source, including:

1. Increase the room constant R_1 for the room containing the sound source. This approach reduces noise in the adjacent room (Room 2) by reducing the reverberant noise before the sound is transmitted through the wall. If all other factors remain unchanged, increasing the average surface absorption coefficient, for example, from 0.10 to 0.20 in Room 1, will increase the room constant by a factor of 2.25 and decrease the sound pressure level in Room 2 by about 3.5 dB.
2. Increase the room constant R_2 for the adjacent room by adding acoustic treatment to the room surfaces, for example. If all other factors remain unchanged, increasing the average surface absorption coefficient from 0.10 to 0.20 in Room 2 will increase the room constant by a factor of 2.25 and decrease the sound pressure level in Room 2 by about 1.5 dB near the separating wall and up to about 3.5 dB far from the wall.
3. Increase the transmission loss for the wall between the two rooms. In many situations, this approach results in the most significant noise reduction. For example, if the wall thickness is doubled, the transmission loss will be increased by 6 dB and the sound pressure level in Room 2 will be decreased by 6 dB.
4. The sound pressure level in Room 2 can also be reduced by reducing the sound power level for the sound source in Room 1. This approach may not always be feasible.

Example 7-6. A Jordan refiner used in a paper mill is located in a room having a total surface area of 900 m² (9688 ft²) and an average absorption coefficient of 0.05. The refiner has a sound power level of 105 dB and a directivity factor of 2.0. The refiner is located 4 m (13.1 ft) from the wall of the operator's room. The operator's room has a total surface area of 100 m² (1076 ft²) and an average surface absorption coefficient of 0.35. The transmission loss for the wall between the refiner room and the operator's room is 30 dB. The surface area of the wall between the two rooms is 16 m² (172 ft²). The operator is located 1.5 m (4.9 ft) from the wall. Determine the steady-state sound pressure level in the refiner room and in the operator's room, neglecting the effect of air absorption.

The room constant for the refiner room may be calculated from Eq. (7-13):

$$R_1 = \frac{\bar{\alpha} S_1}{1 - \bar{\alpha}_1} = \frac{(0.05)(900)}{1 - 0.05} = 47.37 \, \text{m}^2$$

The steady-state sound pressure level in the refiner room is found from Eq. (7-73):

$$L_{p1} = 105 + 10 \log_{10} \left[\frac{4}{47.37} + \frac{2.0}{(4\pi)(4.0)^2} \right] + 0.1$$

$$L_{p1} = 105 + 10 \log_{10}(0.08444 + 0.00995) + 0.1 = 105 + (-10.3) + 0.1$$

$$L_{p1} = 94.8 \, \text{dB}$$

The room constant for the operator's room is as follows:

$$R_2 = \frac{\bar{\alpha}_2 S_2}{1 - \bar{\alpha}_2} = \frac{(0.35)(100)}{1 - 0.35} = 53.85 \, \text{m}^2$$

Let us calculate the following quantity:

$$r^* = (S_w / 2\pi)^{1/2} = (16/2\pi)^{1/2} = 1.596 \, \text{m} \, (5.24 \, \text{ft})$$

The distance between the operator (receiver) and the wall is as follows:

$$r_2 = 1.50 \, \text{m} < 1.596 \, \text{m} = r^*$$

The sound pressure level in the operator's room is found from Eq. (7-71):

$$L_{p2} = 105 - 10 \log_{10}(47.37) + 10 \log_{10} \left[\frac{(4)(16)}{(53.85)} + 1 \right] - 30 + 0.1$$

$$L_{p2} = 105 - 16.8 + 3.4 - 30 + 0.1 = 61.7 \, \text{dB}$$

Room Acoustics

7.6 ACOUSTIC ENCLOSURES

In a room containing a source of noise, such as a piece of machinery, acoustic treatment of the walls of the room may not reduce the sound level in the room sufficiently. This is especially true when the direct field predominates over the reverberant field at the receiver location. When a reduction in the sound level of more than about 10 dB is required, an enclosure for the noise source is often the most practical solution to control noise of an existing machine. Reductions in the noise levels by 20–30 dB are common with complete or full machine enclosures. Noise reductions as high as 50 dB may be achieved with special isolation treatment for the enclosure.

Generally, the only inherent disadvantage, if accessibility to the machine is not required, is the initial cost of the enclosure. If accessibility to the machine is required (to feed in material, to make adjustments, etc.), then a partial enclosure must be used, and careful attention must be directed to the design of the openings in the enclosure.

Examples of enclosures are shown in Fig. 7-10 (enclosure for an automatic press) and Fig. 7-11 (enclosure for a saw).

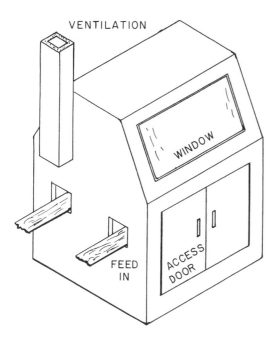

FIGURE 7-10   Enclosure for an automatic press.

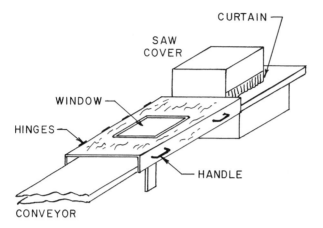

FIGURE 7-11 Enclosure for a gang rip saw for wood. The cover is 3-inch thick plywood lined with 1-inch thick polyurethane foam. (From *Handbook of Acoustical Enclosures and Barriers*, R. K. Miller and W. V. Motone, 1978. Used by permission of the Fairmont Press, Inc.)

7.6.1 Small Acoustic Enclosures

An enclosure is considered to be "small' if the bending wavelength of the enclosure wall is large compared with the largest panel dimension and if the wavelength of the sound inside the enclosure is large compared with the largest interior dimension of the enclosure. The wavelength of the bending wave is a function of the frequency:

$$\lambda_b = \left[\frac{\pi c_L h}{\sqrt{3} f}\right]^{1/2} \quad (7\text{-}74)$$

The quantity c_L is the speed of longitudinal waves in the enclosure wall material, see Eq. (4-156), and h is the thickness of the enclosure wall. For practical purposes, the enclosure may be considered to be "small" if the following condition is met:

$$\frac{L_{max}}{\lambda} = \frac{f L_{max}}{c} \leq 0.1 \quad (7\text{-}75)$$

The quantity L_{max} is the largest interior dimension of the enclosure, and c is the speed of sound for the air in the enclosure.

The enclosure acts to reduce the acoustic power radiated from the system, as shown in Fig. 7-12. If the acoustic power radiated from the

Room Acoustics

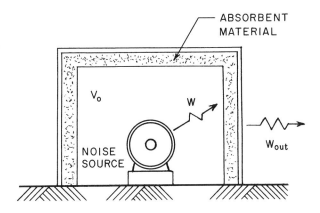

FIGURE 7-12 Nomenclature for an enclosure.

enclosure is denoted by W_{out}, the *insertion loss* (IL) for the system is defined by the following expression:

$$\text{IL} = 10 \log_{10}(W/W_{\text{out}}) = L_W - L_{W,\text{out}} \tag{7-76}$$

For a small enclosure, the air space and the enclosure walls are acoustically coupled. In this case, the surface absorption and wall transmission loss have little effect on the performance of the enclosure. The most important factor is the stiffness of the enclosure walls. The power ratio for a small sealed (no openings) enclosure may be calculated from the following expression (Vér, 1973):

$$\frac{W}{W_{\text{out}}} = \left[1 + \frac{V_o}{\rho_o c^2 \Sigma C_{wj}}\right]^2 \tag{7-77}$$

The quantity V_o is the volume of air in the enclosure, ρ_o is the density of the air in the enclosure, c is the speed of sound in the air in the enclosure, and C_{wj} are the volume compliances of each of the walls.

The volume compliance of a homogeneous panel with fixed (clamped) edges is found from the following expression (Timoshenko and Woinowsky-Krieger, 1959):

$$C_w = \frac{S_w^3 F(\beta)}{\beta^2 B} \tag{7-78}$$

The quantity S_w is the surface area of the panel; $\beta = a/b \geq 1$ is the aspect ratio for the panel, where a is the larger edge dimension of the panel, and b is the smaller edge dimension of the panel. The quantity B is the flexural rigidity for the panel. For a homogeneous panel of thickness h, with Young's modulus E and Poisson's ratio σ, the flexural rigidity is given by the following expression:

$$B = \frac{Eh^3}{12(1-\sigma^2)} \qquad (7\text{-}79)$$

The flexural rigidity for a two-layer panel is given by Eq. (4-184), and the flexural rigidity of a rib-stiffened panel is given by Eq. (4-190).

The function $f(\beta)$ in Eq. (7-78) for a panel with clamped edges may be estimated from the following expression:

$$F(\beta) = (3.50/\pi^8)\{1 + 1.033 \tanh[(\pi/2)(\beta - 1)]\} \qquad (7\text{-}80)$$

If all of the panels of the enclosure are made of the same material and have the same thickness h, the summation for all the walls of the enclosure in Eq. (7-78) may be written as follows:

$$\Sigma C_{wj} = \frac{12(1-\sigma^2)}{Eh^3} \Sigma(S_{wj}^3/\beta_j^2)F(\beta_j) \qquad (7\text{-}81)$$

It may be observed from Eq. (7-77) that the insertion loss for a small enclosure is increased if the enclosure walls are made stiffer, since an increase in the flexural rigidity results in a decrease in the panel volume complaince. A very small volume compliance of the enclosure panels results in a large value of the power ratio, W/W_{out}, or a small value of the power radiated from the enclosure, W_{out}, relative to the power radiated by the noise source, W.

A geometry that exhibits higher stiffness than the rectangular box geometry is a cylindrical body with two hemispherical end caps (Beranek and Vér, 1992). Insertion losses of as high as 50 dB have been achieved with this geometry using no absorbing material on the interior of the enclosure.

Another approach to achieving small enclosure wall compliance is to use a composite material consisting of a honeycomb core between two plates (Fuchs, et al., 1989). In the research reported by these authors, the plate facing the noise source had circular openings over each honeycomb cavity, which produced a resonator element to absorb energy. The plate containing the holes was covered with a thin membrane to prevent contaminants from entering the cavities and to provide additional energy dissipation. Insertion loss values on the order of 20 dB were obtained with a 100-mm thick wall for the frequency range from 31.5 Hz to 8000 Hz.

Example 7-7. A small motor is to be enclosed in a rectangular enclosure having dimensions 200 mm long × 100 mm wide × 100 mm high (7.87 in ×

Room Acoustics

3.94 in × 3.94 in). There is no sound transmitted through the floor of the enclosure, so only the two end walls and three side walls are considered. The enclosure is constructed of 14-gauge steel sheet, 1.9 mm (0.0747 in) thick. The air volume within the enclosure is 1.50 dm³ (91.5 in³), and the air is at 30°C (86°F), for which $\rho_o = 0.859 \text{ kg/m}^3$ (0.0536 lb$_m$/ft³) and $c = 349$ m/s (1145 fps). Determine the insertion loss for the enclosure for a frequency of 125 Hz.

From Appendix C, we find the following property values for steel:

Young's modulus, $E = 200$ GPa (29×10^6 psi)
Poisson's ratio, $\sigma = 0.27$
Speed of longitudinal waves, $c_L = 5110$ m/s (16,770 fps)

Let us check the condition given by Eq. (7-75) to determine the applicability of the "small-enclosure" analysis:

$$\frac{fL_{max}}{c} = \frac{(125)(0.200)}{(349)} = 0.0716 < 0.10$$

Next, let us check the condition given by Eq. (7-74) to determine the wavelength of bending waves in the panels of the enclosure:

$$\lambda_b = \left[\frac{(\pi)(5100)(0.0019)}{\sqrt{3}(125)}\right]^{1/2} = 0.375 \text{ m} = 375 \text{ mm} > 200 \text{ mm}$$

The panel dimensions do meet the conditions for the "small-enclosure" analysis.

The compliance for each of the enclosure walls may be calculated. For the end walls, the aspect ratio is $\beta_1 = 100/100 = 1$. The value of the function defined by Eq. (7-80) is as follows:

$$F(\beta_1) = (3.50)/(\pi^8) = 3.689 \times 10^{-3}$$

There are two end walls, so the first term in the summation in Eq. (7-81) has the following value:

$$\frac{2S_{w1}^3 F(\beta_1)}{\beta_1^2} = \frac{(2)(0.010)^3 (3.689)(10^{-3})}{(1.00)^2} = 7.378 \times 10^{-9} \text{ m}^6$$

The aspect ratio for the side walls is $\beta_2 = 200/100 = 2$:

$$F(\beta_2) = (3.50/\pi^8)\{1 + (1.033)\tanh[(\pi/2)(2.00-1)]\} = 7.183 \times 10^{-3}$$

There are three side walls, so the second term in the summation in Eq. (7-81) has the following values:

$$\frac{3S_{w2}^3 F(\beta_2)}{\beta_2^2} = \frac{(3)(0.020)^3(7.183)(10^{-3})}{(2.00)^2} = 4.310 \times 10^{-9}\,\mathrm{m}^6$$

The summation of the compliances for the walls of the enclosure may be calculated from Eq. (7-81):

$$\Sigma C_{wj} = \frac{(12)(1 - 27^2)(7.378 + 4.310)(10^{-9})}{(200)(10^9)(0.0019)^3} = 94.79 \times 10^{-12}\,\mathrm{m}^5/\mathrm{N}$$

(or m^3/Pa)

The sound power ratio may be found from Eq. (7-77):

$$\frac{W}{W_{\mathrm{out}}} = \left[1 + \frac{(1.50)(10^{-3})}{(0.859)(349)^2(94.79)(10^{-12})}\right]^2 = (1 + 151.2)^2 = 23{,}179$$

The insertion loss is found from its definition, Eq. (7-76):

$$\mathrm{IL} = 10\log_{10}(23{,}179) = 43.7\,\mathrm{dB}$$

7.6.2 Large Acoustic Enclosures

An enclosure may be considered to be "large" when the enclosure volume exhibits a large number of resonant modes of vibration. A large enclosure usually meets the following condition:

$$\frac{fV_o^{1/3}}{c} \geq 1 \qquad (7\text{-}82)$$

The quantity V_o is the volume of air in the enclosure, f is the frequency of the sound in the enclosure, and c is the sonic velocity in the air in the enclosure.

There are several paths along which sound may be transmitted from the noise source within the enclosure to the space outside the enclosure, including (a) through the enclosure walls, (b) through openings in the enclosure walls, and (c) through solid structural supports. The magnitude of the sound transmitted through the walls of the enclosure is a function of the sound power transmission coefficient a_t of the walls, as discussed in Chapter 4. The magnitude of the sound leaking through openings in the enclosure can also be expressed in terms of an equivalent sound power transmission coefficient (Mechel, 1986). By using proper vibration isolation, the transmission of sound through solid supports should be reduced to a negligible contribution for the enclosure to be effective in noise control.

Room Acoustics

It is important that a large fraction of the acoustic energy radiated from the noise source inside the enclosure be dissipated within the enclosure. But, it is equally important to block the transmission of sound through the enclosure walls. To achieve this condition, the enclosure walls are usually constructed of a composite material, with the inside layer having a large surface absorption coefficient and the other layer or layers having a large transmission loss or small sound power transmission coefficient.

The acoustic power radiated by the noise source W is equal to the sound power absorbed or dissipated at the wall surface plus the energy transmitted through the walls:

$$W = W_{tr} + W_{abs} = (\Sigma S_j a_{tj}/S_o)W_{inc} + (\Sigma S_j \alpha_j/S_o)W_{inc} \tag{7-83}$$

The quantity W_{inc} is the acoustic power incident on the enclosure walls from the noise source, S_j is the surface area of the jth component of the enclosure walls, and S_o is the total surface area of the enclosure. The power radiated from the surface of the enclosure to the surrounding space is the power that has been transmitted through the walls:

$$W_{out} = (\Sigma S_j a_{tj}/S_o)W_{inc} \tag{7-84}$$

The sound power ratio for the enclosure may be found by dividing the power from Eq. (7-83) by the power from Eq. (7-84):

$$\frac{W}{W_{out}} = 1 + \frac{\Sigma S_j \alpha_j}{\Sigma S_j a_{tj}} \tag{7-85}$$

The insertion loss is defined by Eq. (7-76).

The acoustic pressure within the enclosure may be determined from Eq. (7-10):

$$W_{inc} = \tfrac{1}{4} D_R c S_o = \frac{p^2 c S_o}{4\rho_o c^2} = \frac{p^2 S_o}{4\rho_o c} \tag{7-86}$$

The incident power W_{inc} may be found from Eq. (7-83) and combined with Eq. (7-86):

$$\frac{p^2}{\rho_o c} = \frac{4W}{\Sigma S_j a_{tj} + \Sigma S_j \alpha_j} \tag{7-87}$$

For partial enclosures or enclosures with openings, the absorptivities and sound power transmission coefficients for the openings are required. Some typical property values for materials used for covers of openings are given in Table 7-4.

The design value for the "absorptivity" of a simple opening is $\alpha = 1.00$. Although the sound power transmission coefficient for an opening

TABLE 7-4 Acoustic Properties for Some Materials Used for Covers of Openings in Enclosures

Material	Octave band center frequency, Hz					
	125	250	500	1,000	2,000	4,000
	Surface absorption coefficient, α					
Glass	0.18	0.06	0.04	0.03	0.03	0.02
Polyvinyl chloride (Plexiglas™)	0.20	0.07	0.05	0.04	0.04	0.03
Leaded vinyl curtain	0.33	0.88	0.79	0.69	0.53	0.26
	Sound power transmission coefficient, a_t					
Glass, $\frac{1}{4}$ in thick	0.020	0.0050	0.0032	0.0020	0.0016	0.0013
Double glass, $\frac{1}{4}$ in × $\frac{1}{2}$ in × $\frac{1}{4}$ in	0.005	0.004	0.004	0.002	0.0016	0.0010
Polyvinyl chloride film:						
0.0015 in thick, 1 layer	0.95	0.90	0.63	0.170	0.043	0.013
0.0015 in thick, 2 layers	0.90	0.70	0.17	0.043	0.013	0.013
Polyvinyl sheet (Plexiglas™), $\frac{1}{4}$ in thick	0.025	0.020	0.0063	0.0016	0.0005	0.000063
Polyvinyl sheet (Plexiglas™), $\frac{1}{2}$ in thick	0.0079	0.0050	0.0025	0.00063	0.00063	0.00020
Leaded vinyl curtain, 0.064 in thick	0.050	0.025	0.010	0.0025	0.0008	0.0003
Leaded vinyl curtain, 2 in thick	0.063	0.025	0.0050	0.0005	0.00016	0.00013
Polycarbonate film (Lexan™), 0.25 in thick	0.063	0.016	0.0040	0.0020	0.0020	0.0020
PC film, 0.50 in thick	0.016	0.0040	0.0020	0.0020	0.0020	0.0020
PC film, 2 layers, each 0.25 in thick with 2-in space	0.013	0.0020	0.00063	0.00025	0.00010	0.00006

would also be equal to 1, there is a directional effect for the opening, as far as the operator is concerned. In this case, the transmission coefficient should be modified for the directivity and diffraction effects on the opening. The following values are recommended for the effective transmission coefficient for simple openings (no cover) in an enclosure (Faulkner, 1976), assuming that the operator is located in front of the enclosure:

(a) Front opening, $a_t = 1$
(b) Side or top opening:
   no reflective surfaces nearby, $a_t = 1/3$
   with reflective surfaces nearby, $a_t = 2/3$

Room Acoustics 307

(c) Back opening:
 no reflective surfaces nearby, $a_t = 1/6$
 with reflective surfaces nearby, $a_t = 1/3$

If there are ventilating ducts or "sound traps" connected to the enclosure, an equivalent sound power transmission coefficient may be related to the decrease in the sound power level from the duct inlet to outlet, ΔL_w:

$$a_{t,\text{eff}} = 10^{-\Delta L_w/10} \tag{7-88}$$

The change in the sound power level in the ventilation duct may be estimated according to the material presented in Sec. 5.11.

Example 7-8. A production machine has a sound power level spectrum given in Table 7-5. The machine is operated in a room having dimensions of 20 m × 20 m × 4 m high (65.6 ft ×65.6 ft × 13.1 ft high). The room has an average surface absorption coefficient as given in Table 7-5. The directivity factor for the machine is unity, and the operator is located 3 m (9.8 ft) from the machine. It is desired to reduce the noise from the machine by placing the machine inside an enclosure having a width of 1.80m (5.91 ft), a length of 1.20 m (3.94 ft), and a height of 1.00 m (3.28 ft). To allow for material flow into and out of the enclosure, there are two openings (one in each side) 300 mm × 200 mm (11.8 in × 7.87 in), as shown in Fig. 7-13. The enclosure is constructed of 25 mm (1 in) thick plywood, covered with a 1-in layer of acoustic absorbent material on the inside. The transmission loss and absorption coefficients for the materials are given in Table 7-5. Determine the sound pressure level at the operator's location both with and without the enclosure in place.

Let us first determine the sound pressure level if the enclosure were not in place. The calculations will be made in detail for the 500 Hz octave band, and the results for the other octave bands are given in Table 7-5. The surface area of the room is found as follows:

$$S_o = (2)(20 + 20)(4.00) + (2)(20)(20) = 1120 \, \text{m}^2 \, (12{,}060 \, \text{ft}^2)$$

The room constant (at 500 Hz) is found from Eq. (7-13):

$$R = \frac{\bar{\alpha} S_o}{1 - \bar{\alpha}} = \frac{(0.051)(1120)}{(1 - 0.051)} = 60.19 \, \text{m}^2$$

The sound pressure level at the operator's location, without the enclosure in place, is calculated from Eq. (7-18):

TABLE 7-5 Solution for Example 7-8

	Octave band center frequency, Hz					
Item	125	250	500	1,000	2,000	4,000
Given data:						
Room average absorption coefficient, $\bar{\alpha}$	0.035	0.044	0.051	0.070	0.043	0.056
Sound power level, L_W, dB	103	109	114	117	113	107
Room constant, R, m²	40.62	51.55	60.19	84.30	47.88	66.44
$10\log_{10}\left[\dfrac{4}{R} + \dfrac{1}{2\pi r^2}\right]$, dB	−9.7	−10.6	−11.2	−12.5	−10.3	−11.6
Enclosure TL, dB	18.4	19.0	19.0	19.0	19.0	25.0
Enclosure α_1	0.16	0.27	0.63	0.97	0.99	0.96
Without the enclosure in place:						
$L_p^o(\text{OB})$, dB	93.4	98.5	102.9	104.6	102.8	95.5
With the enclosure:						
α_{t1}	0.0145	0.01259	0.01259	0.01259	0.01259	0.00316
$S_1 a_{t1}$, m²	0.1162	0.1012	0.1012	0.1012	0.1012	0.0254
$S_2 a_{t2}$, m²	0.0040	0.0040	0.0040	0.0040	0.0040	0.0040
$\Sigma S_j a_{tj}$, m²	0.1202	0.1052	0.1052	0.1052	0.1052	0.0294
$S_1 \alpha_1$	1.286	2.171	5.065	7.780	7.960	7.718
$S_2 \alpha_2$	0.120	0.120	0.120	0.120	0.120	0.120
$\Sigma S_j \alpha_j$, m²	1.406	2.291	5.185	7.900	8.080	7.838
W/W_{out}	12.70	22.78	50.29	76.10	77.81	267.6
IL, dB	11.0	13.6	17.0	18.8	18.9	24.3
$L_{W,\text{out}}$, dB	92.0	95.4	97.0	98.2	94.2	82.7
$L_p(\text{OB})$, dB	82.4	84.9	85.9	85.8	83.9	71.2

Room Acoustics

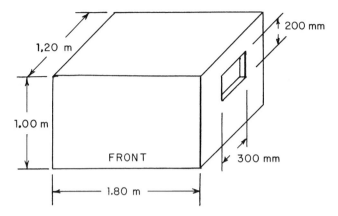

FIGURE 7-13 Diagram for Example 7-8.

$$L_p^o = 114 + 10\log_{10}\left[\frac{4}{60.19} + \frac{1}{(4\pi)(3.00)^2}\right] + 0.1$$

$$L_p^o = 114 + (-11.2) + 0.1 = 102.9\,\text{dB}$$

The calculations for the other octave bands are given in Table 7-5. If the octave bands are combined to obtain the A-weighted sound level, we obtain the following value, without the enclosure:

$$L_A^o = 108.4\,\text{dBA} \quad \text{(without the enclosure)}$$

According to Eq. (6-2), the maximum time per day that the worker could be exposed to this noise level and be in compliance with the OSHA criteria is as follows:

$$T = \frac{16}{2^{(108.4-85)/5}} = 0.624\,\text{hours} = 37.4\,\text{min}$$

The volume of the enclosure, not considering the volume occupied by the machine, is as follows:

$$V_o = (1.80)(1.20)(1.00) = 2.160\,\text{m}^3\ (76.28\,\text{ft}^3)$$

If we take the air temperature as 30°C (86°F), for which the sonic velocity is 349 m/s (1145 fps), the left side of Eq. (7-82) may be calculated for a frequency of 500 Hz:

$$\frac{fV_o^{1/3}}{c} = \frac{(500)(2.160)^{1/3}}{(349)} = 1.85 > 1$$

The enclosure meets the "large enclosure" criterion for frequencies of 500 Hz and higher, and is fairly close to meeting the criterion for the 250 Hz octave band. The insertion loss prediction, using the large enclosure relationships, for the 125 Hz octave band would be questionable, because ($fV_o^{1/3}/c = 0.46 < 1$) for this octave band. Generally, the 125 Hz octave band does not contribute as much for the A-weighted levels as does the higher octave bands, so the error (for the A-weighted sound level) is probably not significant for this problem.

Suppose the machine base covers the floor of the enclosure, such that the floor area is not involved in the calculations. The surface area of the enclosure, excluding the floor area and area of the openings, is as follows:

$$S_1 = (2)(1.20 + 1.80)(1.00) + (1.20)(1.80) - (2)(0.31)(0.20) = 8.04 \, \text{m}^2$$

The area of the two openings is as follows:

$$S_2 = (2)(0.30)(0.20) = 0.120 \, \text{m}^2$$

The sound power transmission coefficient for the walls may be found from the definition of transmission loss given by Eq. (4-90):

$$a_{t1} = 10^{-TL/10} = 10^{-1.9} = 0.01259$$

The openings are in the sides of the enclosure, and there are no reflective surfaces nearby, so the effective transmission coefficient for the openings is $a_{t2} = 1/3$. The summation of the transmission coefficients is as follows:

$$\Sigma S_j a_{tj} = (8.04)(0.01259) + (0.120)(1/3) = 0.1052 \, \text{m}^2$$

The summation of the surface absorption coefficients for the 500 Hz octave band is as follows, using $\alpha_2 = 1$ for the openings:

$$\Sigma S_j \alpha_j = (8.04)(0.63) + (0.12)(1.00) = 5.185 \, \text{m}^2$$

The sound power ratio for the enclosure is found from Eq. (7-85):

$$\frac{W}{W_{out}} = 1 + \frac{\Sigma S_j \alpha_j}{\Sigma S_j a_{tj}} = 1 + \frac{5.185}{0.1052} = 1 + 49.29 = 50.29$$

The insertion loss is found from Eq. (7-76):

$$IL = L_W - L_{W,out} = 10 \log_{10}(50.29) = 17.0 \, \text{dB}$$

The sound power level for the sound radiated from the surface of the enclosure in the 500 Hz octave band is as follows:

$$L_{W,out} = 114 - 17.0 = 97.0 \, \text{dB}$$

Room Acoustics 311

The corresponding sound pressure level for the 500 Hz octave band is found from Eq. (7-18):

$$L_p = 97.0 - 11.2 + 0.1 = 85.9 \, dB$$

The sound pressure levels for the other octave bands are given in Table 7-5. If these values are combined, the following values for the A-weighted sound level is obtained:

$$L_A = 89.8 \, dBA \quad \text{(with the enclosure)}$$

This noise level is in compliance for 8-hour per day exposure, according to OSHA criteria.

7.6.3 Design Practice for Enclosures

For satisfactory performance, there are several design guidelines for enclosures that have been developed by manufacturers (Miller and Montone, 1978).

If the *enclosure walls* are constructed of a composite material, such as a steel sheet and acoustic foam, the backing sheet should be at least 18-gauge galvanized steel (0.0478 in or 1.2 mm thick). The perforated face sheet should be at least 22-gauge galvanized steel, 0.0299 in (0.76 mm) thick. The perforations should be about 5/64 in (2 mm) diameter on 5/32 in (4 mm) staggered centers. The thickness of the acoustic material between the facings should be at least 2 in or 50 mm thick for best performance.

All *access doors* should be provided with double seals and double-action latches. Some typical door seal configurations are shown in Fig. 7-14.

The *enclosure window*, if access through the window is not required, should be constructed of safety glass or plastic (such as a polycarbonate) at least 1/2 in (12 mm) thick. It is important to provide gaskets for all windows to prevent leakage of noise around the windows. If access is required through the opening, then the openings can be covered with an acoustic curtain, such as a couple of leaded vinyl sheets about 1/8 in (3 mm) thick with staggered slits for easier access.

In many cases, a fan system is required to provide ventilation (at least one air change per minute), cooling, and particle removal from the interior of the enclosure. Lined ducts or "sound traps" should be provided for both the fresh air intake and the fan exhaust ducts. A typical design for a sound trap is shown in Fig. 7-15.

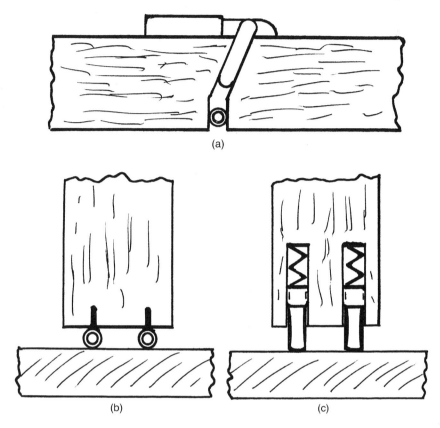

FIGURE 7-14 Example of door seals for an enclosure: (a) compression seal, (b) drop seal, and (c) spring-loaded seal.

7.7 ACOUSTIC BARRIERS

Acoustic barriers are commonly used for the control of noise in outdoor applications, such as reduction of highway noise to the surrounding areas, reduction of noise from transformer stations, and reduction of noise from construction equipment. Barriers are also used to reduce noise in indoor applications, such as in open-plan offices and schools and for machines that cannot be totally enclosed. For indoor applications, barriers are effective only for those cases in which the direct sound field is predominant at the receiver location, although the absorptive surface of the barrier will also reduce the reverberant field somewhat. In general, barriers are more effective in reducing high-frequency noise than for low-frequency noise.

Room Acoustics

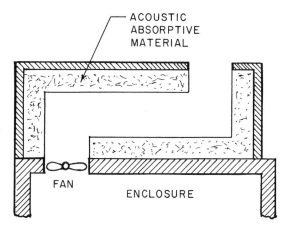

FIGURE 7-15 Typical sound trap for ventilating fans in an enclosure.

Sound interacts with a barrier in three ways: (a) reflection from the barrier surface, (b) direct transmission through the barrier, and (c) diffraction over the top of the barrier. The barrier should have a high transmission loss to be effective in blocking the sound. Also, the barrier should have an absorptive covering for indoor applications.

7.7.1 Barriers Located Outdoors

For transmission of sound across a barrier located outdoors, the following expression has been developed for the sound pressure level L_p at the receiver position due to a point noise source having a sound power level L_W on the opposite side of the barrier (Maekawa, 1968):

$$L_p = L_W + \text{DI} - 20\log_{10}(A + B) - 10\log_{10}\left(\frac{1}{a_b + a_t}\right) - 10.9 \quad (7\text{-}89)$$

The quantities A and B are distances from the noise source to the top of the barrier and from the top of the barrier to the receiver, respectively, as illustrated in Fig. 7-16. The quantity a_b is the *barrier coefficient*, and a_t is the sound power transmission coefficient for the barrier wall.

For a point source, the barrier coefficient may be found from the following expression:

$$a_b = \frac{\tanh^2[(2\pi N)^{1/2}]}{2\pi^2 N} \quad \text{(for } N < 12.7\text{)} \quad (7\text{-}90a)$$

$$a_b = 0.0040 \quad \text{(for } N \geq 12.7\text{)} \quad (7\text{-}90b)$$

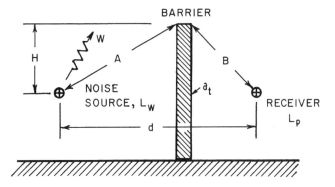

FIGURE 7-16 Barrier dimensions.

The quantity N is the *Fresnel number*, which is the ratio of the difference between the direct path length d and the path length over the barrier to one-half of the wavelength of the sound λ:

$$N = \frac{2}{\lambda}(A + B - d) = \frac{2f}{c}(A + B - d) \tag{7-91}$$

The quantity f is the frequency of the sound wave and c is the sonic velocity in the air around the barrier.

For traffic noise from a roadway, the difference between the sound pressure level at the source with the barrier L_p and without the barrier L_p^o is given by the following expression (Barry and Reagan, 1978):

$$L_p^o - L_p = 15 \log_{10}\left(\frac{A+B}{d}\right) + 10 \log_{10}\left(\frac{1}{a_b^{3/4} + a_t}\right) \tag{7-92}$$

The barrier coefficient a_b is given by Eq. (7-90).

For effective noise control, the barrier should be sufficiently massive that the sound transmitted directly through the barrier is negligible compared with the sound transmitted over the wall. For design purposes, the sound power transmission coefficient a_t should be less than about 1/8 of the barrier coefficient:

$$a_t < a_b/8 \quad \text{(design condition)} \tag{7-93}$$

Example 7-9. A concrete barrier 100 mm (4 in) thick is to be built around a transformer station located outdoors. The top of the barrier is 2.50 m (8.2 ft) high above the transformer and is located 10 m (32.8 ft) from the transformer. The property line (receiver) is located 30 m (98.4 ft) from the transfor-

Room Acoustics

mer. The directivity factor for the transformer may be taken as $Q = 1$. The sound power level spectrum and the transmission loss for the barrier as a function of frequency are given in Table 7-6. Determine the sound pressure level spectrum without the barrier in place and with the barrier.

Let us work out the calculations for the 500 Hz octave band in detail. Without the barrier, the sound pressure level may be determined from Eq. (5-5), neglecting attenuation in the atmospheric air:

$$L_p^o = L_W + \mathrm{DI} - 20\log_{10} d - 10.9 = 106 + 0 - 20\log_{10}(30) - 10.9$$
$$L_p^o = 106 - 29.5 - 10.9 = 65.5\,\mathrm{dB}$$

The results for the other octave bands are shown in Table 7-6. If we use these values to determine the A-weighted sound level, we find the following value:

$$L_A^o = 69.6\,\mathrm{dBA}$$

If the transformer noise were continuous (both day and night), the day–night level may be found from Eq. (6-8):

$$L_{DN} = 10\log_{10}[(0.625)\,10^{6.96} + (0.375)\,10^{7.96}] = 76.0\,\mathrm{dBA}$$

If the transformer were located in an urban residential area (no corrections), the anticipated community response, from Table 6-17, would involve vigorous community action.

The barrier could be placed around the transformer station to alleviate this negative community response. The dimensions for the barrier, assuming the source (transformer) and the receiver are at the same elevation, are as follows:

$$A = (10^2 + 2.50^2)^{1/2} = 10.308\,\mathrm{m} \quad (33.82\,\mathrm{ft})$$
$$B = (20^2 + 2.50^2)^{1/2} = 20.156\,\mathrm{m} \quad (66.13\,\mathrm{ft})$$
$$(A + B - d) = 10.308 + 20.156 - 30.0 = 0.463\,\mathrm{m} \quad (1.520\,\mathrm{ft})$$

The Fresnel number at 500 Hz, for a sonic velocity of $c = 347\,\mathrm{m/s}$, corresponding to an air temperature of 300K (80°F), is calculated from Eq. (7-91):

$$N = \frac{(20)(500)(0.463)}{(347)} = 1.335 < 12.7$$
$$2\pi N = (2\pi)(1.335) = 8.391$$

The barrier coefficient is found from Eq. (7-90a):

$$a_b = \frac{\tanh^2(\sqrt{8.391})}{(\pi)(8.391)} = 0.03748$$

TABLE 7-6 Solution for Example 7-9

Item	Octave band center frequency, Hz							
	63	125	250	500	1,000	2,000	4,000	8,000
Given data:								
L_W, dB	112	116	110	106	106	100	95	89
TL, dB	36	38	38	38	38	44	50	56
Without the barrier in place:								
L_p°, dB	71.6	75.6	69.6	65.6	65.6	59.6	54.6	48.6
With the barrier in place:								
N	0.167	0.334	0.667	1.335	2.671	5.342	10.68	21.4
a_b	0.1807	0.1216	0.0710	0.0375	0.0189	0.0095	0.0047	0.0040
a_t	0.0003	0.0002	0.0002	0.0002	0.0002	4×10^{-5}	1×10^{-5}	3×10^{-6}
$-10\log_{10}(a_b + a_t)$	7.4	9.1	11.5	14.2	17.2	20.2	23.2	24.0
L_p, dB	64.0	66.3	57.9	51.2	48.2	39.2	31.2	24.4

The sound power transmission coefficient at 500 Hz is as follows:

$$a_t = 10^{-TL/10} = 10^{-3.8} = 0.00016 < (0.03748)/(8) = 0.00469$$

$$a_b + a_t = 0.03748 + 0.00016 = 0.03764$$

The sound pressure level in the 500 Hz octave band with the barrier in place is found from Eq. (7-89):

$$L_p = 106 + 0 - 20\log_{10}(10.308 + 20.156) - 10\log_{10}(1/0.03764) - 10.9$$

$$L_p = 106 - 29.7 - 14.2 - 10.9 = 51.2 \, \text{dB}$$

The results for the other octave bands are given in Table 7-6. The corresponding A-weighted sound level is as follows:

$$L_A = 55.3 \, \text{dBA} \quad \text{(with the barrier in place)}$$

The day–night level, for continuous noise day and night, is found as in the previous calculation:

$$L_{DN} = 61.7 \, \text{dBA}$$

If the transformer were located in an urban residential area (no corrections), the anticipated community response, from Table 6-17, would be "no reaction."

We may make the following observations relative to this example. First, we see that the barrier is most effective in reducing high-frequency noise. For the 63 Hz octave band, the barrier reduces the sound pressure level by $(71.6 - 64.0) = 7.6$ dB. On the other hand, the barrier reduces the sound pressure level by $(48.6 - 24.4) = 24.2$ dB in the 8000 Hz octave band. The reduction in the 8000 Hz octave band is more than 3 times that for the 63 Hz octave band.

Secondly, we note that the A-weighted sound level is reduced by $(69.6 - 55.3) = 14.3$ dBA. The use of a barrier changes the anticipated community response from "vigorous pursuit of legal action" to "no reaction," without the need for a complete enclosure of the transformer station.

7.7.2 Barriers Located Indoors

When a barrier is located indoors, it will primarily affect the direct sound field; however, there will also be an effect on the reverberant sound field due to surface absorption of the barrier wall, as illustrated in Fig. 7-17.

The acoustic energy density associated with the direct sound field, if attenuation in the room air is negligible, is given by the following expression:

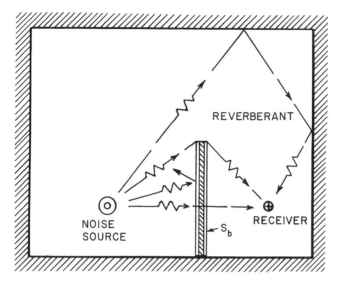

FIGURE 7-17 Barrier located indoors.

$$D_D = \frac{WQ(a_b + a_t)}{4\pi(A+B)^2 c} \tag{7-94}$$

The quantity W is the acoustic power for the sound source and Q is the directivity factor for the source.

The room constant is changed by the insertion of the barrier, because additional surface is exposed to the reverberant field. The room constant with the barrier in place is given by the following expression:

$$R_b = \frac{\bar{\alpha} S_o + S_b(\alpha_1 + \alpha_2)}{1 - \bar{\alpha} - (S_b/S_o)(\alpha_1 + \alpha_2)} \tag{7-95}$$

The quanity $\bar{\alpha}$ is the average surface absorption coefficient for the room and S_o is the surface area of the room (excluding the surface area of the barrier). The quantities α_1 and α_2 are the surface absorption coefficients for the front and back sides of the barrier and S_b is the surface area of *one side* of the barrier.

The sound pressure level, with the barrier in place, is found from the following expression, for an indoor application:

$$L_p = L_W + 10\log_{10}\left[\frac{4}{R_b} + \frac{Q(a_b + a_t)}{4\pi(A+B)^2}\right] + 0.1 \tag{7-96}$$

Room Acoustics

For indoor applications, the barrier is most effective in reducing noise for the situations in which the term associated with the direct sound field, $Q/4\pi r^2$, is much larger (say, more than 6 times larger) than the term associated with the reverberant sound field, $4/R$, in Eq. (7-17).

Example 7-10. A machine has an octave band sound power level of 109 dB for the 1000 Hz octave band. The machine is located in a room having dimensions of 30 m × 30 m × 5 m high (98.4 ft × 98.4 ft × 16.4 ft high), with an average surface absorption coefficient of 0.35. The machine has a directivity factor of 2.0, and the operator is located at a distance of 3.00 m (9.84 ft) from the machine. It is desired to reduce the noise received by the operator by placing a barrier 5.00 m long × 3.00 m high (16.40 ft × 9.84 ft) at a distance of 1.00 m (3.28 ft) from the machine, as shown in Fig. 7-18. The operator's ear and the machine center are both 1.50 m (59 in) above the floor. The transmission loss for the barrier is 31 dB in the 1000 Hz octave band. The surface absorption coefficient for one side (facing the operator) of the barrier is 0.90, and the absorption coefficient for the other side is 0.20. Determine the sound pressure level in the 1000 Hz octave band at the operator's ear without the barrier and with the barrier in place.

The surface area of the room is as follows:

$$S_o = (2)(30 + 30)(5) + (2)(30)(30) = 2400 \text{ m}^2 \quad (25{,}830 \text{ ft}^2)$$

The room constant without the barrier is found from Eq. (7-13):

$$R = \frac{\bar{\alpha} S_o}{1 - \bar{\alpha}} = \frac{(0.35)(2400)}{1 - 0.35} = 1292 \text{ m}^2$$

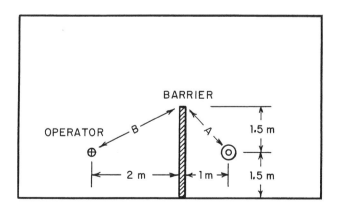

FIGURE 7-18 Diagram for Example 7-10.

The sound pressure level, without the barrier in place, is given by Eq. (7-18):

$$L_p^o = 109 + 10\log_{10}\left[\frac{4}{1292} + \frac{2.0}{(4\pi)(3.00)^2}\right] + 0.1$$

$$L_p^o = 109 + 10\log_{10}(0.003095 + 0.017674) + 0.1 = 92.3 \text{ dB}$$

At this point, we may note that the barrier should be effective in reducing the noise to the operator, because the direct field is $(0.017675/0.003095) = 5.7$ or about 6 times as large as the reverberant field.

The room constant, with the barrier in place, is calculated from Eq. (7-95):

$$R_b = \frac{(0.35)(2400) + (5.00)(3.00)(0.90 + 0.20)}{1 - 0.35 - (15/2400)(0.90 + 0.20)} = 1332 \text{ m}^2$$

The presence of the barrier in the room increases the room constant by about 3%.

The distances for the barrier are found as follows:

$$A = (1.00^2 + 1.50^2)^{1/2} = 1.8028 \text{ m} \quad (5.915 \text{ ft})$$
$$B = (2.00^2 + 1.50^2)^{1/2} = 2.5000 \text{ m} \quad (8.2021 \text{ ft})$$
$$A + B - d = 1.8028 + 2.5000 - 3.00 = 1.3028 \text{ m} \quad (4.274 \text{ ft})$$

The Fresnel number from Eq. (7-91) is as follows:

$$N = \frac{2f(A + B - d)}{c} = \frac{(2)(1000)(1.3028)}{(347)} = 7.509 < 12.7$$

$$2\pi N = (2\pi)(7.509) = 47.18$$

The barrier coefficient is found from Eq. (7-90a):

$$a_b = \frac{\tanh^2[(47.18)^{1/2}]}{(\pi)(47.18)} = 0.006747$$

The sound power transmission coefficient for the barrier is as follows:

$$a_t = 10^{-TL/10} = 10^{-3.10} = 0.000794$$

The sound power level at the operator's location with the barrier in place is found from Eq. (7-96):

$$L_p = 109 + 10\log_{10}\left[\frac{4}{1332} + \frac{(2.0)(0.006747 + 0.000794)}{(4\pi)(1.8028 + 2.500)^2}\right] + 0.1$$

$$L_p = 109 + 10\log_{10}(0.003004 + 0.0000648) + 0.1 = 84.0 \text{ dB}$$

Room Acoustics 321

The use of the barrier reduced the noise in the 1000 Hz octave band from 92.3 dB to 84.0 dB, or a reduction of 8.3 dB. The reverberant field contribution is reduced from 0.003095 to 0.003004, or a reduction of about 3%. On the other hand, the direct field contribution is reduced from 0.017674 to 0.0000648, or a reduction of more than 99%.

PROBLEMS

7-1. Determine the NRC rating for $1\frac{1}{2}$-in thick polyurethane foam.

7-2. It is desired to select a perforated ceiling tile thickness, if the tile is placed on a hard backing, such that the NRC rating is NRC-0.60 or higher. Determine a suitable ceiling tile thickness.

7-3. An auditorium has dimensions of 15 m × 12 m × 4 m high (49.2 ft × 39.4 ft × 13.1 ft high). The walls are 3/8-inch plywood paneling, the floor is covered with indoor–outdoor carpet on the concrete slab, and the ceiling is 3/4-inch perforated ceiling tile on a furring backing. There are 10 windows of ordinary glass, each 1.75 m × 1.00 m (68.9 in × 39.4 in), and the windows are closed. There are no draperies on the windows. When the auditorium is in use, there are 100 people seated in the wood theater chairs. A ventilating fan (directivity factor, $Q = 2$) located in the room has a sound power level of 72 dB in the 500 Hz octave band. Determine the steady-state sound pressure level in the 500 Hz octave band at a distance of 8.00 m (26.2 ft) from the fan for the following conditions: (A) for the empty room, without the chairs and people, and (B) for the room with the chairs and people present.

7-4. An office workroom has dimensions of 3.70 m × 3.40 m × 2.60 m high (12.1 ft × 11.2 ft × 8.5 ft high). The walls and ceiling are plaster on lath and the floor is linoleum tile on concrete. There is one door (surface absorption coefficient $\alpha = 0.050$ in the 500 Hz octave band), having dimensions of 2.20 m × 0.90 m (86.6 in × 35.4 in). An office machine in the room has a sound power level of 68 dB and a directivity factor of $Q = 2.8$ in the 500 Hz octave band. Determine the steady-state sound pressure level in the room in the 500 Hz octave band at a distance of 2.00 m (6.56 ft) from the machine. What fraction of the floor area would need to be covered with 44 oz pile carpet (uncoated backing) on a 40 oz pad to reduce the steady-state pressure level in the 500 Hz octave band by 4.0 dB?

7-5. A natatorium has dimensions of 125 m × 80 m × 10 m high (410 ft × 262 ft × 32.8 ft high). The swimming pool inside the natatorium has dimensions of 50 m × 100 m (164 ft × 328 ft). The floor and walls of the room are rubber tile on concrete, and the ceiling is painted concrete block. A water pump (directivity factor $Q = 1$) has a sound power level of 86.5 dB in the 500 Hz octave band. It is desired to cover a portion of the walls of the natatorium with 8-in painted sound-absorbing structural masonry blocks (three insulation-filled cavities) such that the sound pressure level in the 500 Hz octave band due to the pump noise is 60 dB at a distance of 20 m (65.6 ft) from the pump when there are no people in the natatorium. Determine the required surface area of the masonry blocks.

7-6. A room has dimensions of 10 m × 10 m × 4 m high (32.8 ft × 32.8 ft × 13.1 ft high). The surface absorption coefficients for the room surfaces in the 500 Hz octave band are as follows: walls, 0.20; floor, 0.25; and ceiling, 0.55. There are 12 people standing in the room. It is desired to place a loudspeaker in the room such that a steady-state sound pressure level of 70 dB is produced at a distance of 6 m (19.7 ft) from the loudspeaker. Determine the required acoustic power output, watts, for the loudspeaker, which has a directivity factor $Q = 2$.

7-7. A warehouse room has dimensions of 25 m × 12 m × 6 m high (82 ft × 39.4 ft × 19.7 ft high). The floor and walls are unpainted poured concrete, and the floor is wood (same properties as wood door). One wall has a metal door (same properties as unglazed brick), having dimensions of 4.50 m × 2.5 m high (14.8 ft × 8.2 ft high). A forklift truck in the room has the sound power level spectrum given in Table 7-7. The directivity factor for the lift truck is also given in Table 7-7. Determine the A-weighted sound level at a distance of 10 m (32.8 ft) from the forklift truck.

7-8. A room has dimensions of 3.70 m × 3.40 m × 2.60 m high (12.1 ft × 11.2 ft × 8.5 ft high). there are two windows (closed) of ordinary glass 1.20 m × 1.20 m (47.2 in × 47.2 in) in the long wall. The walls and ceiling are gypsum plaster on metal lath, and the floor is linoleum tile on concrete. For the 500 Hz octave band, deter-

TABLE 7-7 Data for Problem 7-7

	Octave band center frequency, Hz					
	63	125	250	500	1,000	2,000	4,000
Sound power level, L_W, dB	80	88	94	95	90	83	71
Directivity factor, Q	2.0	2.2	2.4	3.0	3.6	3.8	4.0

mine the following, using the Norris–Eyring relationship, Eq. (7-30): (A) the reverberation time for the empty room, and (B) the reverberation time if four upholstered chairs, with a person seated in each chair, are present in the room. The room air is at 23°C (73°F), for which the density is 1.193 kg/m³ (0.0745 lb$_m$/ft³) and the sonic velocity 345 m/s (1132 ft/sec).

7-9. A small lecture room has dimensions of 12 m × 7 m × 3.35 m high (39.4 ft × 23.0 ft × 11.00 ft high). The wall is covered with gypsum plaster on lath, the ceiling is painted poured concrete, and the floor is covered with linoleum tile on concrete. The area of the closed windows (ordinary glass) in the long wall is 9.0 m² (96.9 ft²), and the area of the solid wood door in the short wall is 1.9 m² (20.5 ft²). There are 12 high school students seated in desks in the room. The air temperature is 23°C (73°F), for which the density is 1.193 kg/m³ (0.0745 lb$_m$/ft³) and the sonic velocity 345 m/s (1132 ft/sec). It is desired that the reverberation time, using Eq. (7-30), be reduced to 1.40 s for the 500 Hz octave band by covering a portion of the ceiling with 3/4-inch perforated ceiling tile on furring backing. Determine the required area of the ceiling that must be covered with tile to achieve this condition.

7-10. Repeat Problem 7-9, using the Fitzroy relationship for the absorption units.

7-11. A room having dimensions of 18 m × 11 m × 3.8m (59 ft × 36 ft × 2.5 m high) is used as a band practice room. The floor of the room is rubber tile on concrete, and the ceiling is 1/2-inch acoustic tile. The upper half of the wall surface is covered with 1/2-inch acoustic tile and the other half is plaster with a smooth finish on lath. There is one solid wood door in the short wall, with an area of 1.80 m² (19.4 ft²). There are 50 people seated in wooden chairs (same properties as high school students seated in desks). Determine the reverberation time in the 500 Hz octave band for the room, using Eq. (7-30) for the absorption units. The air in the room is at 24°C (75°F), for which the density is 1.188 kg/m³ (0.0742 lb$_m$/ft³) and the sonic velocity is 346 m/s (1134 ft/sec).

7-12. A classroom has dimensions of 9.00 m × 7.00 m × 3.00 m high (29.5 ft × 23.0 ft × 9.84 m high). The sonic velocity of the air in the room is 344.3 m/s (1130 fps). The reverberation time for the empty room is 1.80 s for the 500 Hz octave band. If 30 high school students enter the room and stand around, determine the resulting reverberation time in this case for the 500 Hz octave band.

7-13. A room having dimensions of 25 m × 25 m × 5 m high (82 ft × 82 ft × 16.4 ft high) has walls of unpainted unglazed brick, a ceiling of fibrous plaster, and the floor is covered with 5/8-inch thick wool pile carpet with a pad. There are 20 people in wood theater seats in the room. There are two solid wood doors (area, 2.00 m² or 21.5 ft²) in the walls. The air in the room is at 25°C (77°F), for which $\rho_o = 1.184 \, \text{kg/m}^3$ (0.0739 lb$_m$/ft³) and $c = 346$ m/s (1136 ft/sec). Using the Fitzroy relationship, determine the reverberation time in the 500 Hz octave band for the room.

7-14. Determine the optimum reverberation time in the 500 Hz octave band for a movie theater having dimensions of 20 m × 42.5 m × 10 m high (65.6 ft × 139.4 ft × 32.8 m high).

7-15. An auditorium has dimensions of 45 m × 60 m × 15 m high (147.6 ft × 196.9 ft × 49.2 ft high). The surface absorption coefficient for the walls and ceiling is 0.10, and the absorption coefficient for the floor is 0.05, both in the 2000 Hz octave band. There are 1200 empty wood theater seats in the room, and the room air is at 25°C (77°F) and 40% relative humidity, for which $c = 346$ m/s (1136 ft/sec). The ventilation grille in one wall produces a noise source level of 78 dB in the 2000 Hz octave band, and has a directivity factor of $Q = 2$. Determine the steady-state sound pressure level in the 2000 Hz octave band in the room at a distance of 40 m (131.2 ft) from the noise source, including the effect of attenuation in the air.

7-16. A basketball gymnasium has dimensions 50 m × 40 m × 10 m high (164 ft × 131 ft × 32.8 ft high). The surface absorption coefficient for the walls and floor is 0.08, and the surface absorption coefficient for the ceiling is 0.10. There are 400 people in the auditorium with $(\alpha S) = 0.40 \, \text{m}^2$/person. These values are for the 4000 Hz octave band. The air in the gymnasium is at 30°C (86°F) and 40% relative humidity ($c = 349$ m/s or 1145 ft/sec). Determine the reverberation time for the 4000 Hz octave band, including the effect of attentuation in the air in the room.

7-17. An office room having dimensions of 3.70 m × 3.40 m × 2.50 m high (12.1 ft × 11.2 ft × 8.20 ft high) has an average surface absorption coefficient of 0.115 in the 500 Hz octave band. Air-handling equipment is located in an adjacent room across the shorter wall. The equipment room has dimensions of 3.40 m × 3.50 m × 2.50 m high (11.2 ft × 11.5 ft × 8.20 ft high) and has an average surface absorption coefficient of 0.040. The directivity factor for the air-handling equipment is $Q = 1.380$, and the equipment is located 1.50 m (4.92 ft) from the interior wall. The sound power level for the air-handling equipment in the 500 Hz octave band is 95 dB, and the transmission loss for the interior wall is 25 dB. The air temperature is 25°C (77°F) in both rooms, for which the sonic velocity is 346.1 m/s (1135 ft/sec).

Determine the steady-state sound pressure level in the 500 Hz octave band: (A) in the air-handling room near the interior wall and (B) in the office space at a distance of 2.0 m (6.56 ft) from the interior wall.

7-18. An apartment room has dimensions of 7.50 m × 6.00 m × 2.50 m high (24.6 ft × 19.7 ft × 8.20 ft high). The average surface absorption coefficient in the 500 Hz octave band for the room is 0.160. In a room adajcent to the shorter wall, an inconsiderate neighbor insists on playing his stereo so loud that he produces a sound power level of 98 dB in the 500 Hz octave band from the stereo set. The directivity factor is $Q = 1$ for the stereo in the 500 Hz octave band. The transmission loss for the wall between the rooms is 26 dB, and the room constant for the thoughtless neighbor's room is 5.00 m². Determine the sound pressure level in the 500 Hz octave band in the apartment room due to the neighbor's stereo at a distance of 3.0 m (9.84 ft) from the shorter wall. What is the sound pressure level in the stereo enthusiast's room at a distance of 2.0 m (6.56 ft) from the stereo?

7-19. A classroom having a room constant of 50 m² is located adjacent to an equipment room having a room constant of 5.00 m². A ventilation fan operates in the equipment room at a distance of 2.50 m (8.20 ft) from the separating wall. The fan has a directivity index $DI = 4.8$ dB and radiates acoustic power into the equipment room at a rate of 3.162 mW. The wall between the two rooms has dimensions of 5.00 m × 2.50 m high (16.40 ft × 8.20 ft high). A student is sitting at a distance of 1.20 m (3.94ft) from the separating wall. It is desired to limit the sound pressure level due to the transmitted noise to 45 dB at the student's location. Determine the required transmission loss for the wall between the rooms.

7-20. A small enclosure is used to reduce the noise from an office machine. The enclosure is constructed of 2024 aluminum and has the following dimensions: 400 mm × 350 mm × 250 mm high (15.7 in × 13.8 in × 9.84 in high). The thickness of the panels is 3.26 mm (0.128 in). The internal volume of the enclosure, excluding the volume of the machine, is 18 dm³ (0.636 ft³). The air inside the enclosure is at 25°C (77°F), for which the density is 1.184 kg/m³ (0.0739 lb$_m$/ft³) and the sonic velocity is 345 m/s (1132 ft/sec). Using the "small enclosure" analysis, determine the insertion loss for the enclosure for a frequency of 125 Hz. The bottom of the enclosure is covered, so it does not enter in the acoustic calculations.

7-21. An office has a room constant of 3.72 m² for the 500 Hz octave band. There is a noisy office machine (sound power level, 65 dB; directivity factor, $Q = 1$) in the room. It is desired to enclose the machine with an enclosure having dimensions of 900 mm × 700 mm × 600 mm high (35.4 in × 27.6 in × 23.6 in high), with an opening in the back of the enclosure, 100 mm × 200 mm (3.94 in × 7.87 in). The opening faces a reflective surface (the wall). The uncovered floor area within the enclosure is 0.430 m^2 (4.63 ft^2), and the surface absorption coefficient for the floor is 0.20. The other surfaces of the enclosure are covered with acoustic foam having a surface absorption coefficient of 0.60. The transmission loss for the enclosure walls and top is 30 dB. Negligible sound is transmitted through the floor ($a_t \approx 0$ for the floor). Determine the sound pressure level in the 500 Hz octave band at a location of 1.50 m (4.92 ft) from the office machine (or the enclosure): (A) with no enclosure and (B) with the enclosure in place.

7-22. An enclosure is built around a natural gas compressor located outdoors. The enclosure has dimensions of 4.00 m × 4.00 m × 3.00 m high (13.1 ft × 13.1 ft × 9.84 ft high). The compressor foundation completely covers the floor, so that sound is not transmitted through the floor of the enclosure. The compressor has a sound power level of 120 dB and a directivity factor of $Q = 1$. The enclosure walls and ceiling are covered with an acoustic material having a surface absorption coefficient of 0.50. The transmission loss for the walls and ceiling is 30 dB. A square opening 305 mm × 305 mm (12.0 in × 12.0 in) is located in the front of the enclosure, and there is no covering over the opening. Determine the sound pressure level at a distance of 32 m (105.0 ft) in front of the enclosure. Attenuation in the air may be neglected.

7-23. A barrier having a total height of 7.60 m (24.9 ft) is to be installed between a natural gas compressor unit and a rural residence. The sound power level for the compressor is 105 dB in the 1000 Hz octave band, and the compressor directivity factor is $Q = 1.80$. The compressor center is 1.60 m (5.25 ft) above the ground. The distance from the barrier to the compressor unit is 4.00 m (13.1 ft), and the distance between the barrier and the residence is 8.00 m (26.2 ft). The ear of the receiver at the residence may be considered to be 1.60 m (5.25 ft) above the ground. The transmission loss for the barrier is 30 dB. The air around the compressor is at 27°C (81°F), for which the sonic velocity is 347.3 m/s (1139 ft/sec). Determine the sound pressure level in the 1000 Hz octave band at the residence: (A) without the barrier and (B) with the barrier in place.

7-24. A person is riding a motorcycle at a distance of 12 m (39.4 ft) from a concrete wall having a transmission loss of 35 dB. The sound power level of the motorcycle in the 1000 Hz octave band is 115 dB. The height of the wall above the level of the motorcycle is 4.00 m (13.1 ft). On the other side of the wall at a distance of 3.00 m (9.84 ft) from the wall, a person is relaxing in a hammock. The air temperature is 27°C (81°F), for which the sonic velocity is 347.3 m/s (1139 ft/sec). Determine the sound pressure level in the 1000 Hz octave band, due to the motorcycle noise, at the location of the person in the hammock.

7-25. The octave band sound pressure levels, $L_p^o(OB)$, at a distance of 57 m (187 ft) from a highway with no barrier, are given in Table 7-8. It is proposed to build a barrier having a height of 8.00 m (26.2 ft) above the traffic level. The barrier is to be located 20 m (65.6 ft) from the traffic and 37 m (121.4 ft) from the receiver, who is also at the traffic level. Ambient air temperature is 20°C (68°F), for which the sonic velocity is 343.2 m/s (1126 ft/sec). The transmission loss (TL) for the proposed barrier is given in Table 7-8. Determine the following at the receiver location: (A) the A-weighted sound level without the barrier, (B) the octave band sound pressure levels with the barrier in place, and (C) the A-weighted sound level with the barrier in place. Neglect air attenuation.

7-26. A barrier 3.00 m high × 15 m wide (9.84 ft × 49.2 ft) is installed in a room to protect the workers from excessive noise generated by a punch press, which would produce a sound pressure level of 100 dB in the 1000 Hz octave band at the worker's station if the barrier were not present. The punch press and the worker's ear are both 1.50 m (4.92 ft) from the floor, and the barrier extends to the floor. The dimensions of the room are 20 m × 20 m × 5 m high (65.6 ft × 65.6 ft × 16.4 ft high), and the average surface absorption coefficient for the room is 0.750 at 1000 Hz. The transmission loss for the barrier is 35 dB, and the surface absorption coefficient for each side of the barrier is 0.950. The directivity factor for the punch press is $Q = 2.00$. The punch press is located 1.50 m (4.92 ft) from the barrier, and the worker is also located 1.50 m (4.92 ft) from the barrier on the other side. The air in the room is at 27°C (81°F), for which the sonic velocity is 347.3 m/s (1139 ft/sec). Determine

TABLE 7-8 Data for Problem 7-25

Item	Octave band center frequency, Hz							
	63	125	250	500	1,000	2,000	4,000	8,000
$L_p^o(OB)$, dB	68	69	67	65	64	60	52	44
TL, dB		14	20	26	32	32	32	42	50

the sound pressure level for a frequency of 1000 Hz at the operator's station with the barrier in place. Neglect the attenuation in the air in the room.

7-27. A shop has dimensions of 10 m × 15 m × 8 m high (32.8 ft × 49.2 ft × 26.2 m high). The average surface absorption coefficient for the shop without the barrier is 0.263. There is a machine along one wall which has a sound power level of 92 dB and directivity factor $Q = 4$ in the 500 Hz octave band. A barrier 4.0 m high × 6.0 m wide (13.1 ft × 19.7 ft) is placed 2.50 m (8.20 ft) from the machine. The operator is located at a distance of 1.50 m (4.92 ft) from the barrier on the other side. The side of the barrier facing the machine has a surface absorption coefficient of 0.12, and the side facing the operator has a surface absorption coefficient of 0.88 for the 500 Hz octave band. The machine and the operator's ear are both 1.50 m (4.92 ft) from the floor, and the barrier extends to the floor. The air in the shop is at 29°C (84°F), for which the sonic velocity is 348.5 m/s (1143 ft/sec). Determine the sound pressure level in the 500 Hz octave band at the operator's location. Neglect attenuation in the air.

REFERENCES

Barry, T. M. and Reagan, J. A. 1978. FHWA highway traffic noise prediction model, Report No. FHWA-RD-77-108. U.S. Federal Highway Administration, Washington, DC.

Beranek, L. L. 1954. *Acoustics*, p. 316. McGraw-Hill, New York.

Beranek, L. L. 1962. *Music, Acoustics, and Architecture*. John Wiley and Sons, New York.

Beranek, L. L. and Vér, I. L. 1992. *Noise and Vibration Control Engineering*, p. 490. John Wiley and Sons, New York.

Eyring, C. F. 1930. Reverberation time in "dead" rooms. *J. Acoust. Soc. Am.* 1: 217–241.

Faulkner, L. L. 1976. *Handbook of Industrial Noise Control*, p. 199. Industrial Press, New York.

Fitzroy, D. 1959. Reverberation formula which seems to be more accurate with non-uniform distribution of absorption. *J. Acoust. Soc. Am.* 31: 893.

Fuchs, H. M., Ackermann, U., and Frommhold, W. 1989. Development of membrane absorbers for industrial noise abatement. *Bauphysik* 11(H.1): 28–36.

Maekawa, Z. 1968. Noise reduction by screens. *Appl. Acoust.* 1: 157–173.

Mechel, F. P. 1986. The acoustic sealing of holes and slits in walls. *J. Sound Vibr.* 3(2): 297–336.

Miller, R. K. and Montone, W. V. 1978. *Handbook of Acoustical Enclosures and Barriers*, pp. 89–122. The Fairmont Press, Atlanta, GA.

Pierce, A. D. 1981. *Acoustics: An Introduction to Its Physical Principles and Applications*, pp. 260–262. McGraw-Hill, New York.
Sabine, W. C. 1922. *Collected Papers on Acoustics*. Peninsula Publishing, Los Altos, CA.
Timoshenko, S. and Woinowsky-Kriger, S. 1959. *Theory of Plates and Shells*, 2nd edn, pp. 197–202. McGraw-Hill, New York.
Vér, I. L. 1973. Reduction of noise by acoustic enclosures. In: *Isolation of Mechanical Vibration, Impact and Noise*, Vol. 1, pp. 192–220. American Society of Mechanical Engineering, New York.

8
Silencer Design

A *silencer* is an important noise control element for reduction of machinery exhaust noise, fan noise, and other noise sources involving flow of a gas. In general, a silencer may be defined as an element in the flow duct that acts to reduce the sound transmitted along the duct while allowing free flow of the gas through the flow passage. A silencer may be *passive*, in which the sound is attenuated by reflection and absorption of the acoustic energy within the element. An *active* silencer is one in which the noise is canceled by electronic feedforward and feedback techniques. In this chapter, we will examine several types of passive silencers, also called *mufflers*. The detailed design procedures for mufflers are available in the literature (Munjal, 1987).

Passive silencers may be of the reactive or dissipative type. In this chapter, we will consider two types of *reactive mufflers*—the *side branch muffler* or resonator chamber muffler and the *expansion chamber muffler*—in which the main mechanism for attenuation of sound passing through the muffler is reflection of the acoustic energy back to the source. The other passive muffler considered in this chapter is the dissipative muffler, in which the primary mechanism for acoustic energy attenuation in the muffler is absorption of acoustic energy within the lining of the muffler.

8.1 SILENCER DESIGN REQUIREMENTS

The optimum design of a silencer involves several requirements, some of which may be in conflict with others; consequently, the muffler design will

Silencer Design

involve consideration of the interactions of the various design criteria. The design requirements are now considered.

1. *Acoustic requirements.* The minimum reduction of the noise by the muffler is usually specified as a function of frequency, either in octave bands or in 1/3 octave bands. The most frequently used acoustic performance parameters include (a) the *insertion loss*, IL, which is the difference in sound pressure level for the surroundings due to the insertion of the silencer into the system; (b) the *noise reduction*, NR, which is the difference in sound pressure level between the point immediately upstream and the point immediately downstream of the muffler; and (c) the *transmission loss*, TL, which is the change in sound power level across the muffler, if there were no energy reflected back to the muffler in the tail pipe. The insertion loss and the noise reduction usually depend on the characteristics of the tail pipe, in addition to the muffler parameters. The transmission loss usually depends only on the characteristics of the muffler.
2. *Aerodynamic requirements.* The maximum allowable pressure drop of the gas flowing through the muffler is usually determined by the application. The pressure drop for air-handling systems is usually limited to a few inches of water: 4 in H_2O or 1 kPa or less, for example (McQuiston and Parker, 1994). On the other hand, internal combustion engines may operate with back pressures as high as 25–30 kPa (3.6–4.4 psi) (Heywood, 1988).
3. *Geometric requirements.* In many cases, such as in automotive applications, there are limitations on the physical size and the shape of the muffler. This requirement often interacts with the acoustic requirements.
4. *Mechanical and material requirements.* Although the internal gauge pressure within most mufflers is relatively small, the mechanical design of the muffler must be considered. In applications involving high-temperature gases or corrosive gases, the materials selected for the muffler must be compatible with the fluid handled. If there are suspended particles (soot, for example) in the gases, the mechanical design must be such that these particles are not easily trapped within the muffler. This requirement may interact with the geometric requirements.
5. *Economic requirements.* Although this requirement is listed last, it is often the most critical one. The muffler must be designed for minimum cost, subject to the constraints of the other require-

ments. Economic considerations include both the initial (purchase) cost and the operating (maintenance) costs.

8.2 LUMPED-PARAMETER ANALYSIS

The *lumped-parameter model* is utilized in many areas of physical analysis. The advantage of the lumped-parameter model is that the governing equations are either ordinary differential equations or algebraic equations. The properties of the lumped-parameter elements (coefficients) usually have a physical interpretation, and the acoustic lumped-parameter model has analogues in terms of corresponding mechanical or electric systems. Lumped-parameter models in acoustic analysis are generally valid for situations in which $ka < 1$ or $fa/c < 1/2\pi \approx 0.16$, where a is a characteristic dimension of the physical system and f is the frequency.

8.2.1 Acoustic Mass

For a mechanical system, Newton's second law of motion may be written in terms of the change in velocity with respect to time:

$$F_{net} = m \frac{dv}{dt} \tag{8-1}$$

The quantity m is the mass being accelerated and v is the velocity of the mass. This expression is identical in mathematical form to the relationship for the voltage change across an inductive element in an electrical circuit:

$$\Delta e = L_E \frac{di}{dt} \tag{8-2}$$

The quantity L_E is the mutual inductance (units: henry), and i is the electrical current (amperes). The electric voltage is analogous to the net mechanical force, and the electric current is analogous to the velocity of the mass.

The relationships given by Eqs (8-1) and (8-2) suggest that an analogous relationship may be developed for acoustic systems in which a portion of the system is accelerated. In the acoustic systems, however, it is more convenient to use acoustic pressure instead of the force and the acoustic volume velocity ($U = Su$) instead of the particle velocity. Newton's second law of motion for an accelerated mass may be written in the following form:

$$F_{net} = S \Delta p = m \frac{d(U/S)}{dt} \tag{8-3}$$

$$\Delta p = \frac{m}{S^2} \frac{dU}{dt} = M_A \frac{dU}{dt} \tag{8-4}$$

Silencer Design

The quantity M_A is the acoustic mass (units: kg/m^4). The physical systems analogous to mechanical mass are shown in Fig. 8-1. The acoustic pressure is analogous to the electric voltage (or mechanical force), and the volumetric flow rate is analogous to the electric current (or velocity of a mass).

The expression for the acoustic mass for a tube, as shown in Fig. 8-2, may be developed. The mass of gas within the tube is given by:

$$m = \rho_o(\pi a^2 L) \tag{8-5}$$

The quantity a is the radius of the tube and L is the tube length. Making the substitution for the mass into Eq. (8-4), we obtain the following expression:

$$M_A = \frac{\pi a^2 L \rho_o}{(\pi a^2)^2} = \frac{\rho_o L}{\pi a^2} \quad \text{(long tube)} \tag{8-6}$$

Actually, there is an additional mass of gas at each end of the tube that is also accelerated. This additional mass must be added to the mass within

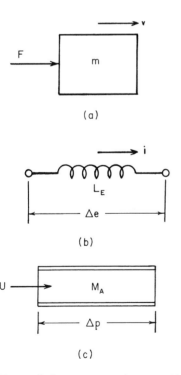

FIGURE 8-1 Inertance elements: (a) mechanical mass, (b) electrical inductance, and (c) acoustic mass.

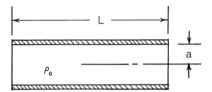

FIGURE 8-2 Mass fluid with a density ρ_o within a circular tube of radius a and length L.

the tube to determine the correct acoustic mass. There are two cases that may be considered for the end of the tube, as shown in Fig. 8-3:

(a) Flanged end (Pierce, 1981):

$$\Delta m = (8/3)a^3 \rho_o = \rho_o \pi a^2 \, \Delta L_1 \tag{8-7}$$

The additional equivalent length to account for the mass of gas at the end that is accelerated is given by the following expression:

$$\Delta L_1 = \frac{8a}{3\pi} \quad \text{(flanged end)} \tag{8-8}$$

(b) Free end:

$$\Delta m = 0.613 \pi a^3 \rho_o = \rho_o \pi a^2 \, \Delta L_2 \tag{8-9}$$

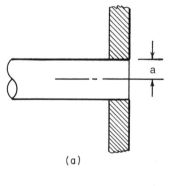

FIGURE 8-3 End conditions for a tube: (a) flanged end and (b) free end.

Silencer Design

The additional equivalent length to account for the mass of gas at the end, for a tube with a free end, that is accelerated is given by the following expression:

$$\Delta L_2 = 0.613a \quad \text{(free end)} \tag{8-10}$$

The total acoustic mass for a tube may be written in the form:

$$M_A = \frac{\rho_o L_e}{\pi a^2} \tag{8-11}$$

The equivalent mass for the tube is given by the following:

$$L_e = L + \Delta L_a + \Delta L_b \tag{8-12}$$

The quantities ΔL_a and ΔL_b are the additional equivalent end corrections for each end of the tube, depending on the type of end termination.

If the instantaneous volumetric flow rate is sinusoidal, we may write the following:

$$U(t) = U_m e^{j\omega t} \tag{8-13}$$

If we make this substitution into Eq. (8-4), we find the following relationship between the instantaneous acoustic pressure difference and the instantaneous volumetric flow rate:

$$\Delta p(t) = j\omega M_A U_m e^{j\omega t} = \Delta p_m e^{j\omega t} \tag{8-14}$$

The acoustic reactance for a mass element may be defined, as follows:

$$X_A = \frac{\Delta p}{U} = j\omega M_A \tag{8-15}$$

The quantity $\omega = 2\pi f$ is the circular frequency for the sound wave.

8.2.2 Acoustic Compliance

The change in force acting on a mechanical spring element (an energy storage element) is proportional to the displacement of the ends of the spring, or the velocity integrated over the time during which the displacement occurs:

$$\Delta F_s = K_S \int u\,dt = \frac{1}{C_M} \int u\,dt \tag{8-16}$$

The quantity K_S is the spring constant, and $C_M = 1/K_S$ is the mechanical compliance (units: m/N). The corresponding relationship for an electrical capacitor is as follows:

$$\Delta e = \frac{1}{C_E} \int i \, dt \qquad (8\text{-}17)$$

The quantity C_E is the electrical capacitance (units: farad).

The corresponding capacitance element for the acoustic system is a volume of gas that is compressed and expanded by the gas entering the volume at a volumetric flow rate U, as shown in Fig. 8-4. For a fixed volume V, we may write the conservation of mass principle for the gas as follows:

$$V \frac{d\rho}{dt} = \rho_o U \qquad (8\text{-}18)$$

The acoustic compression/expansion process is thermodynamically reversible and adiabatic (isentropic) for small amplitudes. The pressure–density relationship for such a process is as follows:

$$p/\rho^\gamma = \text{constant} = P_o/\rho_o^\gamma \qquad (8\text{-}19)$$

$$dp = (P_o/\rho_o^\gamma)\gamma\rho^{\gamma-1} \, d\rho \qquad (8\text{-}20)$$

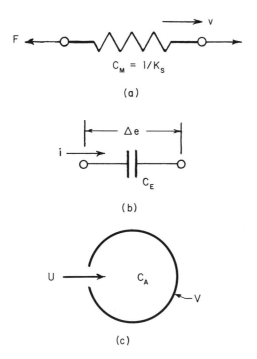

FIGURE 8-4 Compliance elements: (a) mechanical spring, (b) electrical capacitor, and (c) acoustic compliance.

Silencer Design

The quantity P_o is atmospheric pressure and γ is the specific heat ratio for the gas. The time rate of change of the density of the gas within the volume may be found from Eq. (8-20):

$$\frac{d\rho}{dt} = \frac{\rho_o^\gamma}{\gamma_o P_o \rho^{\gamma-1}} \frac{dp}{dt} \tag{8-21}$$

Combining Eqs (8-18) and (8-21), we obtain the following expression for the volumetric flow rate of gas into the volume:

$$U = \frac{V(\rho_o/\rho)^{\gamma-1}}{\gamma P_o} \frac{dp}{dt} \approx \frac{V}{\gamma P_o} \frac{dp}{dt} \tag{8-22}$$

The expression for the speed of sound for an ideal gas ($P_o = \rho_o RT$) is given by Eq. (2-1):

$$c^2 = \gamma RT = \gamma P_o / \rho_o \tag{8-23}$$

Making this substitution into Eq. (8-22), we obtain the following relationship for the volumetric flow rate:

$$U = \frac{V}{\rho_o c^2} \frac{dp}{dt} \tag{8-24}$$

If we separate variables and integrate Eq. (8-24), we obtain the analogous relationship for an acoustic compliance element:

$$\Delta p = \frac{\rho_o c^2}{V} \int U \, dt = \frac{1}{C_A} \int U \, dt \tag{8-25}$$

The quantity C_A is the acoustic compliance (units: m³/Pa or m⁵/N):

$$C_A = \frac{V}{\rho_o c^2} \tag{8-26}$$

If we make the substitution for the volumetric flow rate from Eq. (8-14) into Eq. (8-25), we may obtain the following expression for the instantaneous acoustic pressure difference for a compliance element:

$$\Delta p(t) = \frac{U_m}{j\omega C_A} e^{j\omega t} = -\frac{jU_m}{\omega C_A} e^{j\omega t} = \Delta p_m \, e^{j\omega t} \tag{8-27}$$

The acoustic reactance for a compliance element may be defined as follows:

$$X_A = \frac{\Delta p}{U} = -\frac{j}{\omega C_A} \tag{8-28}$$

8.2.3 Acoustic Resistance

The mechanical resistance in a system is provided by a damper, often considered as a *viscous* or *linear* damper, in which the force on the damper is directly proportional to the velocity, as shown in Fig. 8-5:

$$F_d = R_M v \tag{8-29}$$

The quantity R_M is the mechanical resistance or damping coefficient (units: N-s/m).

The analogous electrical quantity is the electrical resistance R_E, defined by Ohm's law:

$$\Delta e = R_E i \tag{8-30}$$

The quantity R_E is the electrical resistance (units: ohm).

The analogous acoustic resistance is defined in a similar manner:

$$\Delta p = R_A U = R_A S u \tag{8-31}$$

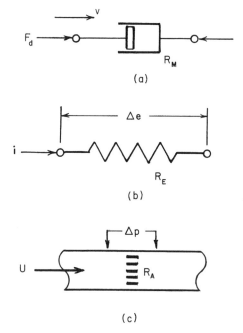

FIGURE 8-5 Resistance elements: (a) mechanical damper, (b) electrical resistor, and (c) acoustic resistance or constriction.

Silencer Design

The quantity R_A is the acoustic resistance (units: Pa-s/m^3 = N-s/m^5). The units for the acoustic resistance (Pa-s/m^3) are sometimes called acoustic ohms. In contrast to the electrical resistance, the acoustic resistance is often a function of the frequency of the sound wave. In acoustic systems, the resistance may be provided by restrictions, such as screens.

The acoustic resistance of tubes depends on the size of the boundary layer near the tube wall relative to the tube radius or diameter. This ratio is given by the following expression:

$$r_v = a(2\pi f \rho/\mu)^{1/2} \tag{8-32}$$

The quantity μ is the viscosity of the gas in the tube and a is the tube radius. The frictional resistance for the gas within a tube is given by the following expressions:
(a) small tube, $r_v < 4\sqrt{2} = 5.66$:

$$R_A = \frac{8\mu L}{\pi a^4} \tag{8-33}$$

(b) intermediate tube, $r_v > 5.66$:

$$R_A = \frac{(4\pi f \mu \rho)^{1/2}}{\pi a^2}\left(\frac{L}{a} + 2\right) \tag{8-34}$$

The acoustic resistance of an orifice of negligible thickness is given by the following expression:

$$R_A = \begin{cases} \dfrac{2\pi f^2 \rho}{c} & \text{(for } ka < \sqrt{2}\text{)} \\ \dfrac{\rho c}{\pi a^2} & \text{(for } ka > \sqrt{2}\text{)} \end{cases} \tag{8-35a} \tag{8-35b}$$

The quantity $k = 2\pi f/c$ is the wave number and a is the orifice radius.

8.2.4 Transfer Matrix

The transfer matrix approach to silencer design has been used extensively since large-capacity digital computers have become available. Let us consider the acoustic mass element shown schematically in Fig. 8-6. The input and output variables are the acoustic pressure and volumetric flow rate. The following set of equations may be written for steady-state operation:

$$p_2 = p_1 + j\omega M_A U_1 \tag{8-36a}$$

$$U_2 = U_1 \tag{8-36b}$$

340   **Chapter 8**

FIGURE 8-6 Transfer element schematic for an acoustic mass. The "outputs" are the acoustic pressure p_2 and the acoustic volumetric flow rate U_2; the "inputs" are the corresponding values of p_1 and U_1.

These two relationships may be written as a single matrix equation:

$$\begin{Bmatrix} p_2 \\ U_2 \end{Bmatrix} = \begin{bmatrix} 1 & j\omega M_A \\ 0 & 1 \end{bmatrix} \begin{Bmatrix} p_1 \\ U_1 \end{Bmatrix} = \begin{bmatrix} T_{11} & T_{12} \\ T_{21} & T_{22} \end{bmatrix} \begin{Bmatrix} p_1 \\ U_1 \end{Bmatrix} \qquad (8\text{-}37)$$

The T matrix is called the *transfer matrix*. For a mass element, the transfer matrix is obtained from Eq. (8-37):

$$[T]_{\text{mass}} = \begin{bmatrix} 1 & j\omega M_A \\ 0 & 1 \end{bmatrix} \qquad (8\text{-}38)$$

The transfer matrices for the compliance and resistance element may be written in a similar manner:

$$[T]_{\text{comp}} = \begin{bmatrix} 1 & -j/\omega C_A \\ 0 & 1 \end{bmatrix} \qquad (8\text{-}39)$$

$$[T]_{\text{rest}} = \begin{bmatrix} 1 & R_A \\ 0 & 1 \end{bmatrix} \qquad (8\text{-}40)$$

For two or more elements in series, as shown in Fig. 8-7, the output of element A is the input to element B:

$$\begin{Bmatrix} p_3 \\ U_3 \end{Bmatrix} = [T]_A \begin{Bmatrix} p_2 \\ U_2 \end{Bmatrix} = [T]_A [T]_B \begin{Bmatrix} p_1 \\ U_1 \end{Bmatrix} \qquad (8\text{-}41)$$

We observe from Eq. (8-41) that the overall transfer matrix for two or more elements in series is the matrix product of the individual transfer matrices:

$$[T]_o = [T]_A [T]_B \qquad (8\text{-}42)$$

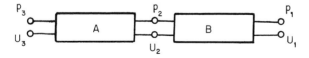

FIGURE 8-7 Acoustic elements in series.

Silencer Design

8.3 THE HELMHOLTZ RESONATOR

The primary element of the side-branch muffler is a resonator volume (*Helmholtz resonator*). We need to become familiar with the characteristics of this acoustic system before considering the application in silencer design. In addition, the analysis of the Helmholtz resonator illustrates the principles of the lumped-parameter analysis for acoustic systems (Howe, 1976).

8.3.1 Helmholtz Resonator System

The Helmholtz resonator and the analogous mechanical and electrical systems are shown in Fig. 8-8. The system consists of an acoustic resistance R_A, an acoustic mass M_A, and an acoustic compliance C_A. The elements are in "series," such that the volumetric flow rate is the same for each element. In the acoustic system, the pressure is analogous to the mechanical force or the electrical voltage, and the volumetric flow rate is analogous to the velocity of a mass or the electrical current.

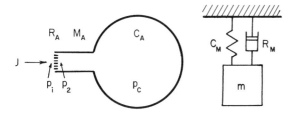

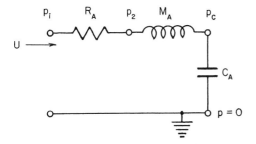

FIGURE 8-8 Equivalent circuit diagram for the Helmholtz resonator. The pressure at one terminal of the acoustic compliance is atmospheric pressure, or the acoustic pressure is zero. The acoustic pressure at the other end of the compliance terminal is p_C. The Helmholtz resonator is analogous to the mechanical system with a spring, damper, and mass.

8.3.2 Resonance for the Helmholtz Resonator

The resonant frequency for the Helmholtz resonator is similar in principle to the natural frequency for vibration of a spring–mass system. For an electrical system, resonance occurs at a frequency such that the electrical reactance is zero. The resonant frequency for the Helmholtz resonator is given by the following expression:

$$f_o = \frac{1}{2\pi(M_A C_A)^{1/2}} \tag{8-43}$$

This relationship may be used for design purposes to determine the resonator frequency for given resonator dimensions, or the required dimensions may be found to achieve a given resonant frequency.

Example 8-1. A Helmholtz resonator is to be constructed of a cylinder with the diameter equal to the length, $D = L_1$. The opening to the resonator is an orifice with a radius $a = 10$ mm (0.394 in) and a thickness $L = 1$ mm (0.039 in). It is desired to select the resonator dimensions such that the resonant frequency is 250 Hz. The gas in the resonator volume is air at 21°C (70°F) and 101.3 kPa (14.7 psia), for which the density is 1.200 kg/m³ (0.0749 lb$_m$/ft³) and the sonic velocity is 343.8 m/s (1128 ft/sec).

Both ends of the hole in the resonator wall are "flanged," so the equivalent length of the resonator inlet may be found from Eqs (8-8) and (8-12):

$$L_e = 1.00 + (2)(8/3\pi)(10.0) = 17.98 \text{ mm}$$

The acoustic mass may be calculated from Eq. (8-11):

$$M_A = \frac{\rho_o L_e}{\pi a^2} = \frac{(1.200)(0.01798)}{(\pi)(0.010)^2} = 68.68 \text{ kg/m}^4$$

The required acoustic compliance for the resonator may be found from Eq. (8-43):

$$C_A = \frac{1}{4\pi^2 f_o^2 M_A} = \frac{1}{(4\pi^2)(250)^2(68.68)} = 5.901 \times 10^{-9} \text{ m}^5/\text{N} = \frac{V}{\rho_o c^2}$$

The resonator volume required to achieve a resonant frequency of 250 Hz may be determined:

$$V = (5.901)(10^{-9})(1.200)(343.8)^2 = 0.837 \times 10^{-3} \text{ m}^3 = 0.837 \text{ dm}^3$$

The cylindrical volume (for $D = L_1$) is given by the following:

$$V = \tfrac{1}{4}\pi D^2 L_1 = \tfrac{1}{4}\pi D^3$$

Silencer Design

The resonator dimensions may be determined:

$$D = L_1 = [(4/\pi)(0.837)(10^{-3})]^{1/3} = 0.1021 \text{ m} = 102.1 \text{ mm} \quad (4.02 \text{ in})$$

The wavelength of the sound wave in the resonator is as follows:

$$\lambda = \frac{c}{f_o} = \frac{343.8}{250} = 1.375 \text{ m} \gg L_1 = 0.1021 \text{ m}$$

The wavelength is much larger than the resonator dimensions, so the lumped-parameter approach is valid.

8.3.3 Acoustic Impedance for the Helmholtz Resonator

For steady-state operation, the acoustic impedance for the Helmholtz resonator with all elements in series may be written in the following complex form:

$$Z_A = \frac{p_{in}}{U} = R_A + j\omega M_A - \frac{j}{\omega C_A} \quad (8\text{-}44)$$

$$Z_A = R_A + jM_A\left(\omega - \frac{1}{\omega M_A C_A}\right) = R_A + jM_A\left(\omega - \frac{\omega_o^2}{\omega}\right) \quad (8\text{-}45)$$

The quantity ω_o is the circular resonant frequency:

$$\omega_o = 2\pi f_o = \frac{1}{(M_A C_A)^{1/2}} \quad (8\text{-}46)$$

The acoustic impedance for the Helmholtz resonator may be written in the following alternative form:

$$Z_A = R_A\left[1 + j\frac{2\pi f_o M_A}{R_A}\left(\frac{f}{f_o} - \frac{f_o}{f}\right)\right] \quad (8\text{-}47)$$

Let us define the *acoustic quality factor* Q_A by the following relationship:

$$Q_A = \frac{2\pi f_o M_A}{R_A} \quad (8\text{-}48)$$

The acoustic quality factor gives a measure of the quality or sharpness of tuning for the Helmholtz resonator (Nilsson, 1983). A large value for the acoustic quality factor usually implies that there is a small acoustic resistance present. The quality factor in the acoustic system is analogous to the damping ratio (or, actually, the reciprocal of the damping ratio) for the mechanical vibratory system. The acoustic impedance for the Helmholtz resonator may be written in terms of the acoustic quality factor:

$$Z_A = R_A\left[1 + jQ_A\left(\frac{f}{f_o} - \frac{f_o}{f}\right)\right] \qquad (8\text{-}49)$$

The magnitude of the acoustic impedance may be found from Eq. (8-49):

$$|Z_A| = (\text{Re}^2 + \text{Im}^2)^{1/2} = R_A[1 + Q_A^2(f/f_o - f_o/f)^2]^{1/2} \qquad (8\text{-}50)$$

The phase angle ϕ between the incident acoustic pressure and the volume flow rate through the necktube of the resonator may also be determined from Eq. (8-49):

$$\tan\phi = \text{Im}/\text{Re} = Q_A[(f/f_o) - (f_o/f)] \qquad (8\text{-}51)$$

8.3.4 Half-Power Bandwidth

The acoustic quality factor gives a measure of how sharply the resonator "resonates." Another qualitative measure of the sharpness of tuning is the acoustic half-power bandwidth. The average acoustic power delivered to the resonator is equal to the power dissipated in the resistance element, because the capacitive and mass elements only momentarily store energy and do not dissipate energy:

$$W = U^2 R_A = (p_{\text{in}}/|Z_A|)^2 R_A \qquad (8\text{-}52)$$

The magnitude of the acoustic impedance may be found from Eq. (8-50):

$$W = \frac{(p_{\text{in}}^2/R_A)}{1 + Q_A^2[(f/f_o) - (f_o/f)]^2} \qquad (8\text{-}53)$$

The power delivered to the resonator at the resonant frequency, $f = f_o$, is given by the following:

$$W_o = \frac{p_{\text{in}}^2}{R_A} \qquad (8\text{-}54)$$

Equation (8-53) may be written in dimensionless form as follows:

$$\frac{W}{W_o} = \frac{1}{1 + Q_A^2[(f/f_o) - (f_o/f)]^2} \qquad (8\text{-}55)$$

The ratio of acoustic power delivered to the resonator may also be expressed in "level" form as follows:

$$\Delta L_W = L_{W_o} - L_W = 10\log_{10}\{1 + Q_A^2[(f/f_o) - (f_o/f)]^2\} \qquad (8\text{-}56)$$

Silencer Design

A plot of the acoustic power delivered to the resonator as a function of frequency is shown in Fig. 8-9. We may observe several facts from this figure. First, the acoustic power is maximum when the frequency is equal to the resonant frequency. Secondly, the curve is "spread out" for small values of the acoustic quality factor (large damping), whereas it has a more sharp peak for large values of the acoustic quality factor (small damping). For frequency ratios larger than about $(f/f_o) = 10$, the sound power level decreases at a rate of about 6 dB/octave.

A quantitative measure of the sharpness of the peak in the power curve for the Helmholtz resonator is given by the acoustic half-power bandwidth. This quantity is defined as the difference between the two frequencies f_1 and f_2 at which the acoustic power delivered to the resonator is one-half of the power delivered at resonance. This condition corresponds to the frequencies at which $(W/W_o) = \frac{1}{2}$ in Eq. (8-55):

$$W/W_o = \tfrac{1}{2} = \{1 + Q_A^2[(f_{1,2}/f_o) - (f_o/f_{1,2})]^2\}^{-1} \tag{8-57}$$

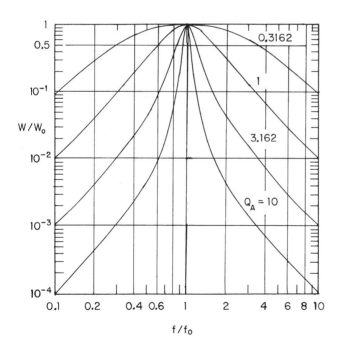

FIGURE 8-9 Graph of the ratio of the acoustic power delivered to a Helmholtz resonator at any frequency f to the power delivered at the resonant frequency f_o. The plot is made for various values of the acoustic quality factor Q_A.

The two solutions for Eq. (8-57) are as follows:

$$\frac{f_1}{f_o} = \frac{(1+4Q_A^2)^{1/2}-1}{2Q_A} \quad \text{(lower frequency)} \quad (8\text{-}58)$$

$$\frac{f_2}{f_o} = \frac{(1+4Q_A^2)^{1/2}+1}{2Q_A} \quad \text{(upper frequency)} \quad (8\text{-}59)$$

The acoustic half-power bandwidth is found by taking the difference of the two frequencies given by Eqs (8-58) and (8-59):

$$\frac{f_2-f_1}{f_o} = \frac{\Delta f}{f_o} = \frac{1}{Q_A} \quad (8\text{-}60)$$

If we multiply Eq. (8-58) by Eq. (8-59), we obtain the following:

$$\frac{f_1 f_2}{f_o^2} = \frac{(1+4Q_A^2)-1}{4Q_A^2} = 1$$

$$f_o = (f_1 f_2)^{1/2} \quad (8\text{-}61)$$

The resonant frequency is the geometric mean of the upper and lower frequencies of the acoustic half-power bandwidth.

Example 8-2. Consider the resonator in Example 8-1, for which the resonant frequency was 250 Hz. Determine the magnitude of the acoustic impedance at 250 Hz and at 125 Hz, the acoustic half-power bandwidth, and the acoustic power delivered at 250 Hz and at 125 Hz. The incident sound pressure level is 80 dB. The viscosity of air at 21°C is $\mu = 15.35\,\mu\text{Pa-s}$.

First, let us determine the acoustic resistance for the opening of the resonator. The parameter from Eq. (8-32) may be calculated for the resonant frequency:

$$r_v = a(2\pi f_o \rho_o/\mu)^{1/2} = (0.010)[(2\pi)(250)(1.200)/(15.35)(10^{-6})]^{1/2}$$

$$r_v = 110.8 > 5.66$$

For a frequency $f = 125\,\text{Hz}$, we find $r_v = 78.3 > 5.66$. Both cases fall in the "intermediate tube" range, so the acoustic resistance may be calculated from Eq. (8-34). The acoustic resistance at the resonant frequency, $f_o = 250\,\text{Hz}$, is as follows:

$$R_A = \frac{[(4\pi)(250)(1.200)(15.35)(10^{-6})]^{1/2}}{(\pi)(0.010)^2}\left(\frac{1}{10}+2\right) = 1608\,\text{Pa-s/m}^3$$

Similarly, we find the following value for the acoustic resistance at a frequency $f = 125\,\text{Hz}$:

Silencer Design

$$R_A = (1608)(125/250)^{1/2} = 1137 \, \text{Pa-s/m}^3$$

The acoustic quality factor, using the acoustic resistance at resonance, is found from Eq. (8-48):

$$Q_A = \frac{(2\pi)(250)(68.68)}{(1608)} = 67.09$$

Similarly, the acoustic quality factor, using the acoustic resistance at 125 Hz, is as follows:

$$Q_A = (67.09)(1608/1137) = 94.88$$

The acoustic impedance at resonance ($f = f_o$) is equal to the acoustic resistance, as seen from Eq. (8-50). At a frequency of 250 Hz, the acoustic impedance is as follows:

$$|Z_A| = R_A = 1608 \, \text{Pa-s/m}^3 = 1.608 \, \text{kPa-s/m}^3$$

The acoustic impedance at a frequency of 125 Hz may be calculated from Eq. (8-50), using the quality factor corresponding to 125 Hz:

$$|Z_A| = (1137)\{1 + (94.88)^2[(250/125) - (125/250)]^2\}^{1/2}$$

$$|Z_A| = (1137)(142.3) = 161.82 \times 10^3 \, \text{Pa-s/m}^3 = 161.82 \, \text{kPa-s/m}^3$$

The phase angle between the incident acoustic pressure and the volumetric flow rate is $\phi = 0$ at resonance. The phase angle at 125 Hz may be found from Eq. (8-51):

$$\tan \phi = (94.88)[(125/250) - (250/125)] = -142.3$$

$$\phi = -89.5° = -1.564 \, \text{rad}$$

The phase angle for 125 Hz is almost $-90°$ or $-\frac{1}{2}\pi$ rad.

The acoustic half-power bandwidth may be found from Eq. (8-60):

$$\Delta f = \frac{f_o}{Q_A} = \frac{250}{67.09} = 3.73 \, \text{Hz} = f_2 - f_1$$

The lower frequency f_1 is found from Eq. (8-58):

$$\frac{f_1}{f_o} = \frac{[1 + (4)(67.09)^2]^{1/2} - 1}{(2)(67.09)} = 0.9926$$

$$f_1 = (0.9926)(250) = 248.1 \, \text{Hz}$$

The upper frequency f_2 is found as follows:

$$f_2 = f_1 + \Delta f = 248.1 + 3.7 = 251.8 \, \text{Hz}$$

The incident sound pressure is found from the sound pressure level:
$$p_{in} = (20)(10^{-6})\,10^{80/20} = 0.200\,\text{Pa}$$

The acoustic power delivered to the resonator at the resonant frequency is found from Eq. (8-54):
$$W_o = \frac{p_{in}^2}{R_A} = \frac{(0.200)^2}{(1608)} = 24.86 \times 10^{-6}\,\text{W} = 24.86\,\mu\text{W}$$

The acoustic power delivered to the resonator at 125 Hz is found from Eq. (8-55):
$$\frac{W}{W_o} = \frac{1}{1+(94.88)^2[(125/250)-(250/125)]^2} = 0.04937 \times 10^{-3}$$
$$W = (24.86)(10^{-6})(0.04937)(10^{-3}) = 1.227 \times 10^{-9}\,\text{W} = 1.227\,\text{nW}$$

We observe that the resonator readily accepts energy around the resonant frequency; however, the energy delivered to the resonator rapidly decreases as the frequency is increased (or decreased) from the resonant value. This characteristic is important when considering the use of a side-branch muffler, which is based on the use of a Helmholtz resonator element.

8.3.5 Sound Pressure Level Gain

Another parameter of interest for the Helmholtz resonator is the difference between the incident sound pressure level and the sound pressure level within the resonator volume. The sound pressure level gain L_G for the resonator is defined as follows:

$$L_G = L_{p,c} - L_{p,in} = 20\log_{10}(p_c/p_{in}) \tag{8-62}$$

The quantity p_c is the acoustic pressure in the cavity or resonator volume and p_{in} is the acoustic pressure incident on the resonator system.

The magnitude of the acoustic pressure within the cavity may be found from Eq. (8-28) for the capacitive element:

$$|p_c| = \frac{U}{\omega C_A} \tag{8-63}$$

The volumetric flow rate and the incident pressure are related through the system impedance, given by Eq. (8-50):

$$|p_{in}| = |Z_A||U| \tag{8-64}$$

We may combine Eqs (8-63) and (8-64) to obtain the expression for the acoustic pressure ratio:

Silencer Design

$$\frac{|p_c|}{|p_{in}|} = \frac{1}{\omega C_A |Z_A|} = \frac{M_A/R_A}{\omega C_A M_A(|Z_A|/R_A)} \tag{8-65}$$

If we make the substitution for the resonant frequency from Eq. (8-43) and the acoustic impedance from Eq. (8-50), we may write Eq. (8-65) in the following form:

$$\frac{|p_c|}{|p_{in}|} = \frac{(f_o/f)Q_A}{[1 + Q_A^2(f/f_o - f_o/f)^2]^{1/2}} \tag{8-66}$$

This expression has the same mathematical form as that for the mechanical transmissibility, which we will examine in Chapter 9. It is noted that, at resonance $f/f_o = 1$, and the pressure ratio is equal to the acoustic quality factor Q_A. The sound pressure level at resonance is given by Eq. (8-62):

$$L_{G_o} = 20 \log_{10} Q_A \quad \text{(at resonance)} \tag{8-67}$$

Example 8-3. Determine the sound pressure level gain for the Helmholtz resonator given in Example 8-2 at resonance (250 Hz) and at 125 Hz.

The sound pressure level gain at resonance may be found from Eq. (8-67):

$$L_{G_o} = 20 \log_{10}(67.09) = 36.5 \, \text{dB}$$

We note that the acoustic pressure and acoustic pressure level in the resonator volume for resonant condition are as follows:

$$p_c = Q_A p_{in} = (67.09)(0.200) = 13.42 \, \text{Pa}$$

$$L_{p,c} = L_{p,in} + L_G = 80 + 36.5 = 116.5 \, \text{dB}$$

The pressure ratio for a frequency of 125 Hz may be found from Eq. (8-66):

$$\frac{|p_c|}{|p_{in}|} = \frac{(250/125)(94.88)}{\{1 + (94.88)^2[(125/250) - (125/250)]^2\}^{1/2}} = 0.1481 \times 10^{-3}$$

The sound pressure level gain at 125 Hz is as follows:

$$L_G = 20 \log_{10}(0.1481 \times 10^{-3}) = -76.6 \, \text{dB}$$

The sound pressure level in the resonator cavity for a frequency of 125 Hz is much less than the sound pressure level of the incident sound wave:

$$L_{p,c} = L_{p,in} + L_G = 80 + (-76.6) = 3.4 \, \text{dB}$$

We observe that the Helmholtz resonator in this example serves as a good amplifier of sound only in the vicinity of the resonant frequency. This characteristic may be utilized when Helmholtz resonators are used in connection with stereo systems to emphasize certain frequency ranges.

8.4 SIDE-BRANCH MUFFLERS

The *side-branch muffler* is one type of silencer used to reduce noise emission in a restricted frequency range from a mechanical system. The side-branch muffler consists of a Helmholtz resonator connected to the main tube through which the sound is transmitted. The silencer acts to reduce sound transmission primarily by reflecting acoustic energy back to the source, so it is classed as a reactive silencer; however, some energy is dissipated within the acoustic resistance element of the silencer.

Some typical configurations for the side-branch muffler are shown in Fig. 8-10.

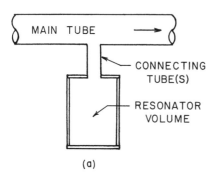

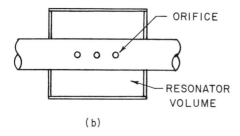

FIGURE 8-10 Configurations for side-branch mufflers: (a) resonator connected by a tube or tubes and (b) resonator connected through orifices.

Silencer Design 351

8.4.1 Transmission Loss for a Side-Branch Muffler

The acoustic impedance for the Helmholtz resonator in the side-branch muffler is given by Eq. (8-44):

$$Z_{Ab} = R_A + jX_A \qquad (8\text{-}68)$$

The acoustic reactance X_A of the resonator is given by the following expression:

$$X_A = 2\pi f M_A - \frac{1}{2\pi f C_A} \qquad (8\text{-}69)$$

The quantities M_A and C_A are the acoustic mass and acoustic compliance for the resonator.

The instantaneous acoustic pressure $p_1(t)$ in the main tube upstream of the junction or side branch may be written as follows:

$$p_1(t) = A_1 \, e^{j(\omega t - kx)} + B_1 \, e^{j(\omega t + kx)} \qquad (8\text{-}70)$$

The first term represents the incident sound wave at the junction, and the second term represents the sound wave reflected back toward the source. Similarly, the instantaneous acoustic pressure downstream of the junction or side branch may be written in the following form, assuming that there is negligible energy reflected beyond the junction:

$$p_2(t) = A_2 \, e^{j(\omega t - kx)} \qquad (8\text{-}71)$$

The acoustic pressure at the branch may be written in terms of the acoustic impedance of the resonator:

$$p_b(t) = U_b(t) Z_{Ab} = U_b(t)(R_A + jX_A) \qquad (8\text{-}72)$$

As indicated in Fig. 8-11, the acoustic pressure at the junction ($x = 0$) is the same for the three elements:

$$p_1(x=0) = p_2(x=0) = p_b \qquad (8\text{-}73)$$

$$A_1 + B_1 = A_2 = U_b(R_A + jX_A) \qquad (8\text{-}74)$$

The volumetric flow rate upstream of the junction or side branch may be written in terms of the acoustic velocity $u_1(t)$ and the cross-sectional area S of the main tube:

$$U_1(t) = S u_1(t) = (S/\rho_o c)[A_1 \, e^{j(\omega t - kx)} - B_1 \, e^{j(\omega t + kx)}] \qquad (8\text{-}75)$$

The volumetric flow rate downstream of the junction may be written in a similar fashion:

$$U_2(t) = S u_2(t) = (S/\rho_o c) A_2 \, e^{j(\omega t - kx)} \qquad (8\text{-}76)$$

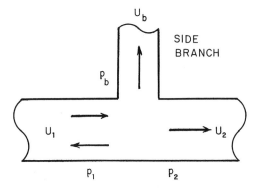

FIGURE 8-11 Physical conditions at the junction of a side branch. Subscript 1 denotes upstream conditions, subscript 2 denotes downstream conditions, and subscript b denotes conditions for the side branch.

At the junction ($x = 0$), we have the following condition for the volume flow rate:

$$U_1(x = 0) = U_2(x = 0) + U_b \tag{8-77}$$

$$A_1 - B_1 = A_2 + \frac{(\rho_o c/S)(A_1 + B_1)}{R_A + jX_A} \tag{8-78}$$

If we eliminate the term B_1 between Eqs (8-74) and (8-78), we obtain the following expression for the ratio of the coefficients A_1 (*incident sound wave*) and A_2 (*transmitted sound wave*):

$$\frac{A_1}{A_2} = \frac{(\rho_o c/2S) + R_A + jX_A}{R_A + jX_A} \tag{8-79}$$

The square of the magnitude of this ratio is as follows:

$$\left|\frac{A_1}{A_2}\right|^2 = \frac{[R_A + (\rho_o c/2S)]^2 + X_A^2}{R_A^2 + X_A^2} \tag{8-80}$$

The sound power transmission coefficient a_t for the side-branch muffler is defined as the ratio of the sound power transmitted to the sound power incident on the junction:

$$a_t = \frac{W_{tr}}{W_{in}} = \frac{|p_{tr}|^2}{|p_{in}|^2} = \left|\frac{A_2}{A_1}\right|^2 = \frac{R_A^2 + X_A^2}{[R_A + (\rho_o c/2S)]^2 + X_A^2} \tag{8-81}$$

The transmission loss TL for the muffler is related to the sound power transmission coefficient:

Silencer Design

$$\text{TL} = 10 \log_{10}(1/a_t) \tag{8-82}$$

If more than one tube is used to connect the resonator volume and the main tube, the total acoustic mass for N_t tubes having the same length and diameter is given by the following expression:

$$M_A = \frac{\rho_o(L + \Delta L_a + \Delta L_b)}{\pi a^2 N_t} = \frac{\rho_o L_e}{\pi a^2 N_t} \tag{8-83}$$

The quantity L is the length of each connecting tube, and ΔL_a and ΔL_b are the equivalent lengths to account for end effects at each end of the connecting tube. If the end of the connecting tube is flush with the main tube wall or the wall of the resonator volume (flanged end), use Eq. (8-8) for the end correction. If the connecting tube protrudes into the main tube or the resonator volume (free end), use Eq. (8-10) for the end correction.

The expression for the acoustic compliance C_A for the resonator is given by Eq. (8-26).

In many design cases, additional acoustic resistance, in the form of screens and other elements, must be added to achieve the required total acoustic resistance. The total acoustic resistance R_A is related to the specific acoustic resistance R_S (resistance for a unit area) by the following:

$$R_A = \frac{R_S}{\pi a^2 N_t} \tag{8-84}$$

The specific acoustic resistance for one screen layer R_{S1} is given in Table 8-1 for several screen sizes. The specific acoustic resistance for N_S layers of screens is $R_S = N_S R_{S1}$.

TABLE 8-1 Specific Acoustic Resistance for Wire Screen

Mesh size, wires/inch	Screen thickness		R_{S1}, rayl, N-s/m^3
	mm	inch	
30	0.66	0.026	5.67
50	0.44	0.0173	5.88
65	0.33	0.0129	6.40
100	0.23	0.0091	9.10
120	0.184	0.0072	13.5
200	0.114	0.0045	24.6
325	0.073	0.0029	49.1

The acoustic reactance for the resonator, using Eq. (8-69) with the resonant frequency given by Eq. (8-43), may be written in the following form:

$$X_A = 2\pi f M_A [1 - (f_o/f)^2] \qquad (8\text{-}85)$$

The acoustic quality factor may be introduced from Eq. (8-48):

$$X_A = R_A Q_A [(f/f_o) - (f_o/f)] \qquad (8\text{-}86)$$

The final expression for the sound power transmission coefficient (or the reciprocal) for the side-branch muffler is obtained by making the substitution from Eq. (8-86) into Eq. (8-81):

$$\frac{1}{a_t} = \frac{\beta^2 + Q_A^2[(f/f_o) - (f_o/f)]^2}{1 + Q_A^2[(f/f_o) - (f_o/f)]^2} \qquad (8\text{-}87)$$

The quantity β is defined as follows:

$$\beta = 1 + \frac{\rho_o c}{2SR_A} \qquad (8\text{-}88)$$

The quantity S is the cross-sectional area of the main tube from the source of sound, and ρ_o and c are the density and speed of sound, respectively, for the gas in the main tube.

At resonance ($f = f_o$), we see from Eq. (8-87) that the sound power transmission coefficient is a minimum (or the reciprocal is a maximum) and has the following value:

$$(1/a_t)_{\max} = \beta^2 \qquad (8\text{-}89)$$

The transmission loss for the muffler at resonance is a maximum:

$$TL_o = TL_{\max} = 10 \log_{10}(\beta^2) = 20 \log_{10} \beta \qquad (8\text{-}90)$$

The half-power bandwidth frequencies for the side-branch muffler may be determined by using Eqs (8-87) and (8-89):

$$\frac{1/a_t}{(1/a_t)_{\max}} = 0.500 = \frac{\beta^2 + Q_A^2[(f/f_o) - (f_o/f)]^2}{\beta^2\{1 + Q_A^2[(f/f_o) - (f_o/f)]^2\}} \qquad (8\text{-}91)$$

If we solve the quadratic equation, Eq. (8-91), for the two solutions, we obtain the lower frequency (f_1) and the upper frequency (f_2) of the half-power frequency band:

$$f_1/f_o = \gamma\{[1 + (1/\gamma^2)]^{1/2} - 1\} \qquad (8\text{-}92)$$

$$f_2/f_o = \gamma\{[1 + (1/\gamma^2)]^{1/2} + 1\} \qquad (8\text{-}93)$$

Silencer Design

The quantity γ is defined by the following expression:

$$\gamma = \frac{\beta}{2Q_A(\beta^2 - 2)^{1/2}} \tag{8-94}$$

We observe from Eq. (8-94) that the value of the quantity β must meet the condition $\beta > \sqrt{2}$ if the side-branch muffler is to be used.

The sound power transmission coefficients at the upper and lower frequencies of the half-power band are related to the transmission loss at resonance:

$$TL_1 = TL_2 = 10\log_{10}[0.500(1/a_t)_{max}] = TL_o - 3.0\,dB \tag{8-95}$$

The transmission loss at resonance is given by Eq. (8-90).

Example 8-4. A side-branch muffler has a resonator volume of $3.00\,dm^3$ ($0.1059\,ft^3$). Three side-branch tubes are used, and each tube has a diameter of 24 mm (0.945 in) and a length of 75 mm (2.95 in). The side-branch tubes are flush with the main tube surface at one end and with the resonator container wall at the other end. The acoustic resistance is provided by one layer of 100-mesh screen in each tube. The diameter of the main tube is 150 mm (5.91 in). The fluid flowing in the main tube is air at 325°C (617°F) and 110 kPa (15.96 psia), for which the sonic velocity $c = 490\,m/s$ (1608 ft/sec) and density $\rho_o = 0.641\,kg/m^3$ ($0.040\,lb_m/ft^3$). Determine the transmission loss for the muffler at a frequency of 125 Hz.

The equivalent length for the side-branch tubes (flanged ends) is as follows:

$$L_e = L + 2(8\pi/3)a = 75 + (2)(8\pi/3)(12) = 276\,mm$$

The acoustic mass for the side-branch tubes may be calculated from Eq. (8-83):

$$M_A = \frac{(0.641)(0.276)}{(\pi)(0.012)^2(3)} = 130.4\,kg/m^4$$

The acoustic compliance for the resonator may be found from Eq. (8-26):

$$C_A = \frac{(3.00)(10^{-3})}{(0.641)(490)^2} = 19.49 \times 10^{-9}\,Pa\text{-}s/m^3$$

The resonant frequency for the silencer is found from Eq. (8-43):

$$f_o = \frac{1}{(2\pi)[(130.4)(19.49)(10^{-9})]^{1/2}} = 99.8\,Hz$$

The specific acoustic resistance for one layer of screen is $R_{S1} = 9.10$ N-s/m^3 from Table 8-1. The acoustic resistance for the muffler may be calculated from Eq. (8-84):

$$R_A = \frac{(1)(9.10)}{(3)(\pi)(0.012)^2} = 6705 \text{ Pa-s/m}^3 = 6.705 \text{ kPa-s/m}^3$$

The acoustic quality factor for the muffler is given by Eq. (8-48):

$$Q_A = \frac{(2\pi)(99.8)(130.4)}{(6705)} = 12.20$$

The cross-sectional area of the main tube is as follows:

$$S = (\tfrac{1}{4}\pi)(0.150)^2 = 0.01767 \text{ m}^2$$

The β-parameter may be calculated from Eq. (8-88):

$$\beta = 1 + \frac{(0.641)(490)}{(2)(0.01767)(6705)} = 1 + 1.326 = 2.326$$

The sound power transmission coefficient may be found from Eq. (8-87) for a frequency of 125 Hz:

$$\frac{1}{a_t} = \frac{(2.326)^2 + (12.20)^2[(125/99.8) - (99.8/125)]^2}{1 + (12.20)^2[(125/99.8) - (99.8/125)]^2} = 1.139$$

The transmission loss at a frequency of 125 Hz is found from Eq. (8-82):

$$\text{TL} = 10\log_{10}(1.139) = 0.6 \text{ dB}$$

The transmission loss at the resonant frequency ($f_o = 99.8$ Hz) is found from Eq. (8-90).

$$\text{TL}_o = 20\log_{10}(2.326) = 7.3 \text{ dB}$$

We note that there is a considerable difference between the transmission loss at resonance and at the higher frequency of 125 Hz.

The parameter γ given by Eq. (8-94) may be calculated:

$$\gamma = \frac{(2.326)}{(2)(12.20)[(2.326)^2 - 2]^{1/2}} = 0.0516$$

The upper and lower frequencies for the half-power band may be calculated from Eqs (8-92) and (8-93):

$$f_1/f_o = (0.0516)\{[1 + (1/0.0516)^2]^{1/2} - 1\} = 0.9497$$

$$f_1 = (99.8)(0.9497) = 94.8 \text{ Hz}$$

Silencer Design

$$f_2/f_o = (0.0516)\{[1 + (1/0.0516)^2]^{1/2} + 1\} = 1.0529$$

$$f_2 = (99.8)(1.0529) = 105.1 \text{ Hz}$$

The half-power bandwidth is relatively small (about ±5%) for the muffler in this example.

8.4.2 Directed Design Procedure for Side-Branch Mufflers

In many silencer design situations, the following parameters are known or specified: (a) minimum acceptable transmission loss, TL_{min}; (b) primary range of frequencies for the silencer operation, f_1 and f_2; (c) cross-sectional area of the main tube, S; and (d) the type of gas in the silencer, so that the density ρ_o and speed of sound c are known. There are often more "unknowns" than design equations; therefore, there are several possible solutions for a particular set of design parameters. The design equations may be organized, however, in a form that allows a more directed design procedure. This directed design procedure is outlined in the following material for a side-branch muffler.

D1. The design resonant frequency for the resonator may be determined from the primary range of frequencies for the silencer and Eq. (8-61):

$$f_o = (f_1 f_2)^{1/2} \tag{8-96}$$

D2. The maximum transmission loss for the muffler or the transmission loss at resonance may be determined from the minimum acceptable transmission loss and Eq. (8-95):

$$TL_o = TL_{min} + 3 \text{ dB} \tag{8-97}$$

D3. The dimensionless parameter β may be determined from the value of the transmission loss at resonance and Eq. (8-90):

$$\beta = 10^{TL_o/20} \tag{8-98}$$

D4. The required acoustic resistance for the muffler may be determined from the value of the β-parameter and Eq. (8-88):

$$R_A = \frac{\rho_o c}{2(\beta - 1)S} = \frac{R_S}{\pi a^2 N_t} \tag{8-99}$$

The quantity a is the radius of a connecting tube, N_t is the number of connecting tubes, and R_S is the specific acoustic resistance for the connecting tubes.

D5. The acoustic quantity factor may be determined from the frequencies and the β-parameter with Eqs (8-92), (8-93), and (8-94). If we subtract

Eq. (8-92) from Eq. (8-93), we obtain the following relationship for the parameter γ:

$$\gamma = \frac{f_2 - f_1}{2f_o} \qquad (8\text{-}100)$$

Using Eq. (8-94) to eliminate the parameter γ, we obtain the following relationship for the acoustic quality factor:

$$Q_A = \frac{f_o \beta}{(f_2 - f_1)(\beta^2 - 2)^{1/2}} \qquad (8\text{-}101)$$

D6. The required acoustic mass for the muffler may be determined from the acoustic quality factor and Eq. (8-48):

$$M_A = \frac{Q_A R_A}{2\pi f_o} \qquad (8\text{-}102)$$

D7. The required acoustic compliance may be determined from the acoustic mass, the resonant frequency, and Eq. (8-43):

$$C_A = \frac{1}{4\pi^2 f_o^2 M_A} = \frac{V}{\rho_o c^2} \qquad (8\text{-}103)$$

D8. Practical choices must be made for two of the following quantities to completely specify the design: specific acoustic resistance, R_S; number of side-branch tubes, N_t; side-branch tube radius, a; resonator volume, V; and length of the side-branch tube, L. The remaining three quantities may be calculated from the previously determined values of the acoustic mass, M_A; the acoustic resistance, R_A; and the acoustic compliance, C_A.

This design procedure is illustrated in the following example.

Example 8-5. A side-branch muffler is to be constructed by placing a 300 mm (11.81 in) inside diameter cylinder around the 250 mm (9.84 in) inside diameter circular duct used in an air conditioning system, as shown in Fig. 8-12. The resonator volume is connected to the main duct through circular holes covered with wire screen. The thickness of the duct is 1.20 mm (0.047 in). The muffler is to have a minimum transmission loss of 6 dB over the frequency range from 177 Hz to 354 Hz (the 250 Hz octave band). the fluid flowing through the main duct is air at 10°C (50°F) and 105 kPa (15.23 psia), for which the density $\rho_o = 1.292 \text{ kg/m}^3$ (0.0807 lb$_m$/ft^3) and the sonic velocity $c = 337.3$ m/s (1107 ft/sec). Determine the muffler dimensions.

The design resonant frequency is found from Eq. (8-96), if we set the design frequency range as the half-power band range:

Silencer Design

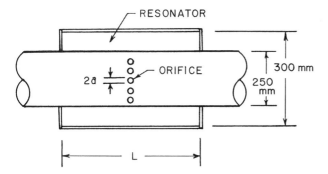

FIGURE 8-12 Figure for Example 8-5. The main tube diameter $D_1 = 250$ mm and the resonator volume OD is $D_2 = 300$ mm.

$$f_o = (f_1 f_2)^{1/2} = [(177)(354)]^{1/2} = 250 \text{ Hz}$$

The maximum transmission loss, which occurs at the resonant frequency, is as follows:

$$TL_o = TL_{min} + 3.0 \text{ dB} = 6.0 + 3.0 = 9.0 \text{ dB}$$

The damping parameter β is found from Eq. (8-98):

$$\beta = 10^{9.0/20} = 2.8216$$

The cross-sectional area for the main duct is as follows:

$$S = (\tfrac{1}{4}\pi)(0.250)^2 = 0.0491 \text{ m}^2$$

The required acoustic resistance may be calculated from Eq. (8-99):

$$R_A = \frac{(1.292)(337.3)}{(2)(2.8216 - 1)(0.0491)} = 2436 \text{ Pa-s/m}^3$$

The acoustic quality factor for the resonator may be found from Eq. (8-101):

$$Q_A = \frac{(250)(2.8216)}{(354 - 177)[(2.8216)^2 - 2]^{1/2}} = 1.6322$$

The required acoustic mass may be determined from Eq. (8-102):

$$M_A = \frac{(1.6322)(2436)}{(2\pi)(250)} = 2.531 \text{ kg/m}^4$$

The required acoustic compliance for the resonator is given by Eq. (8-103):

$$C_A = \frac{1}{(4\pi^2)(250)^2(2.531)} = 0.1601 \times 10^{-6} \text{ m}^5/\text{N}$$

If we use one layer of screen over each of N_t holes, the acoustic resistance and acoustic mass are given by Eqs (8-99), with $R_S = R_{S1}$, and (8-83) with flanged-end tubes:

$$R_A = \frac{R_S}{\pi a^2 N_t}$$

$$M_A = \frac{\rho_o[L + (16a/3\pi)]}{\pi a^2 N_t}$$

Solving for the ratio of the acoustic mass and acoustic resistance, we obtain the following expression:

$$\frac{M_A}{R_A} = \frac{\rho_o[L + (16a/3\pi)]}{R_S}$$

Let us try a 200-mesh wire screen, for which $R_{S1} = 24.6\,\text{N-s/m}^3$, for the acoustic resistance. If we solve for the required hole radius a, we obtain:

$$a = \frac{3\pi}{16}\left[\frac{M_A R_S}{R_A \rho_o} - L\right] = \frac{3\pi}{16}\left[\frac{(2.531)(24.6)}{(2436)(1.292)} - 0.0012\right]$$

$$a = 0.0109\,\text{m} = 10.9\,\text{mm}\ (0.429\,\text{in})$$

The hole diameter is:

$$2a = 21.8\,\text{mm}\ (0.858\,\text{in})$$

The required number of holes may be found as follows:

$$N_t = \frac{R_S}{\pi a^2 R_A} = \frac{(24.6)}{(\pi)(0.0109)^2(2436)} = 27.0\,\text{holes}$$

There are other possible solutions. For example, if we use 120-mesh screens, we would find that the hole radius would be $a = 5.7\,\text{mm}$ (0.224 in) and the number of holes would be $N_t = 55$ holes.

The required volume of the resonator is found from the acoustic compliance:

$$V = C_A \rho_o c^2 = (0.1601)(10^{-6})(1.292)(337.3)^2 = 0.02353\,\text{m}^3\ (0.831\,\text{ft}^3)$$

The required length of the resonator is as follows:

$$L_2 = \frac{(0.02353)}{(\tfrac{1}{4}\pi)\{(0.300)^2 - [0.250 + (2)(0.0012)]^2\}} = \frac{0.02353}{0.02065}$$

$$L_2 = 1.139\,\text{m}\quad (44.86\,\text{in})$$

The summary of the design for the side-branch muffler is as follows:

Muffler length, $L_2 = 1.139\,\text{m}$ (44.86 in)

Hole diameter, $2a = 21.8\,\text{mm}$ (0.858 in)
Resistance element, 1 layer of 200-mesh wire screen

The sound power transmission coefficient for a frequency $f = 63\,\text{Hz}$ may be found from Eq. (8-87):

$$\frac{1}{a_t} = \frac{(2.8216)^2 + (1.6322)^2[(63/250) - (250/63)]^2}{1 + (1.6322)^2[(63/250) - (250/63)]^2} = 1.1842$$

The transmission loss for the muffler at 63 Hz is as follows:

TL $(63\,\text{Hz}) = 10\log_{10}(1.1842) = 0.7\,\text{dB}$

8.4.3 Closed Tube as a Side-Branch Muffler

For a side-branch muffler with a resonator volume attached, the muffler operates quite well around the single resonant frequency of the resonator volume. A long tube, however, has several resonant frequencies (theoretically, an infinite number of resonant frequencies). The resonant frequencies for the closed tube shown in Fig. 8-13 are given by the following expression (Kinsler et al., 1982):

$$f_o = \frac{\left(n - \tfrac{1}{2}\right)c}{2L_e} \qquad (n = 1, 2, 3, \ldots) \tag{8-104}$$

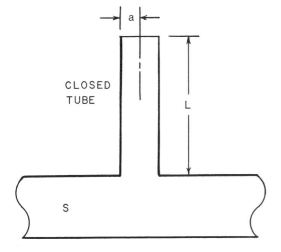

FIGURE 8-13   Closed tube of radius a and length L as a side branch.

The equivalent length of the closed tube is given by Eqs (8-8) and (8-12) for the open end:

$$L_e = L + (8/3\pi)a \tag{8-105}$$

The quantity L is the length of the tube and a is the radius of the tube.

The acoustic reactance for the closed tube is given by the following relationship, valid for $ka < 1$ (Reynolds, 1981):

$$X_A = -\frac{\rho_o c}{\pi a^2} \cot(kL_e) \tag{8-106}$$

The quantity k is the wave number:

$$k = \frac{2\pi f}{c} \tag{8-107}$$

The total acoustic resistance for the tube involves the energy dissipated due to fluid friction within the tube, as expressed by the attenuation coefficient σ, and the resistance provided by a screen or other element at the inlet of the tube, if such an element is used:

$$R_A = \frac{\rho_o c \sigma L + R_S}{\pi a^2} \tag{8-108}$$

The quantity R_S is the specific acoustic resistance for the screens or other elements. The attenuation coefficient may be found from the following expression:

$$\sigma = \frac{(\pi f \mu_e / \rho_o)^{1/2}}{ac} \tag{8-109}$$

The quantity μ_e is the effective viscosity for the gas, which includes the effect of heat conduction (Rayleigh, 1929):

$$\mu_e = \mu \left[1 + \frac{(\gamma - 1)}{(\gamma \mathrm{Pr})^{1/2}} \right] \tag{8-110}$$

The quantity μ is the viscosity of the gas, γ is the specific heat ratio, and Pr is the Prandtl number for the gas.

The sound power transmission coefficient for the closed-tube side-branch muffler may be obtained by substituting the expressions for the acoustic resistance and acoustic reactance given by Eqs (8-106) and (8-108) into Eq. (8-81):

$$\frac{1}{a_t} = \frac{[\sigma L + (R_S/\rho_o c) + (\pi a^2/2S)]^2 \tan^2(kL_e) + 1}{[\sigma L + (R_S/\rho_o c)]^2 \tan^2(kL_e) + 1} \tag{8-111}$$

Silencer Design

The sound power transmission coefficient reaches a minimum at the resonant frequency given by Eq. (8-104). The value (which is a maximum for $1/a_t$) is given by the following expression:

$$(1/a_t)_{max} = \left[1 + \frac{(\pi a^2/2S)}{\sigma L + (R_S/\rho_o c)}\right]^2 \qquad (8\text{-}112)$$

From Eq. (8-111), we note that for the frequencies for which

$$kL_e = 2\pi f L_e/c = n\pi \qquad (n = 1, 2, 3, \ldots) \qquad (8\text{-}113)$$

the term $\tan(kL_e) = 0$, and $(1/a_t) = 1$. The frequencies given by Eq. (8-113) correspond to frequencies at which sound is transmitted through the system without attenuation.

Example 8-6. A tube having an inside diameter of 54.8 mm (2.157 in) is attached to a pipe having an inside diameter of 161.5 mm (6.357 in) through which noise is transmitted. It is desired to select a closed-tube side-branch muffler that has a resonant frequency of 250 Hz. The gas in the tube is air at 21°C (70°F) and 101.3 kPa (14.7 psia), for which the sonic velocity $c = 343.8$ m/s (1128 ft/sec), density $\rho_o = 1.200$ kg/m³ (0.075 lb$_m$/ft³), viscosity $\mu = 18.17$ µPa-s, Prandtl number $Pr = 0.710$, and specific heat ratio $\gamma = 1.400$. Determine the required length of the attached tube and the transmission loss at 250 Hz and 177 Hz.

The equivalent length of the closed tube may be found from Eq. (8-104), using $n = 1$, because this value results in the shortest tube length:

$$L_e = \frac{c}{4f_o} = \frac{(343.8)}{(4)(250)} = 0.3438 \text{ m } (13.54 \text{ in})$$

The required length of the closed tube is calculated from Eq. (8-105):

$$L = L_e - (8/3\pi)a = 0.3438 - (8/3\pi)(0.0274) = 0.3205 \text{ m } (12.62 \text{ in})$$

The effective viscosity for the gas is calculated from Eq. (8-110):

$$\mu_e = (18.17)\left\{1 + \frac{(1.40 - 1)}{[(1.40)(0.710)]^{1/2}}\right\} = (18.17)(1.401) = 25.46 \text{ µPa-s}$$

The attenuation coefficient at 250 Hz is found from Eq. (8-109):

$$\sigma = \frac{[(\pi)(250)(25.46)(10^{-6})/(1.200)]^{1/2}}{(0.0274)(343.8)} = 0.01370 \text{ m}^{-1}$$

$$\sigma L = (0.01370)(0.3205) = 0.00439$$

The cross-sectional area of the main tube is as follows:
$$S = (\tfrac{1}{4}\pi)(0.1615)^2 = 0.02048 \text{ m}^2$$

The sound power transmission coefficient at the resonant frequency $f_o = 250$ Hz is found from Eq. (8-112):
$$(1/a_t) = \left[1 + \frac{(\pi)(0.0274)^2}{(2)(0.02048)(0.00439)} \right]^2 = (14.117)^2 = 199.28$$

The transmission loss at 250 Hz is as follows:
$$\text{TL}_o = 10 \log_{10}(199.28) = 23.0 \text{ dB}$$

The following values for the parameters for 177 Hz may be calculated:
$$kL_e = \frac{2\pi f L_e}{c} = \frac{(2\pi)(177)(0.3438)}{(343.8)} = 1.111 \text{ rad}$$

$$\sigma = \frac{[(\pi)(177)(25.46)(10^{-6})/(1.200)]^{1/2}}{(0.0274)(343.8)} = 0.01153 \text{ m}^{-1}$$

$$\sigma L = (0.01153)(0.3205) = 0.003695$$

$$\frac{\pi a^2}{2S} = \frac{(\pi)(0.0274)^2}{(2)(0.02048)} = 0.05758$$

The sound power transmission coefficient at 177 Hz is calculated from Eq. (8-111):
$$\frac{1}{a_t} = \frac{(0.003695 + 0.05758)^2 \tan^2(1.111) + 1}{(0.003695)^2 \tan^2(1.111) + 1} = \frac{1.0153}{1.0000557} = 1.0153$$

The transmission loss at 177 Hz is as follows:
$$\text{TL}(177 \text{ Hz}) = 10 \log_{10}(1.0153) = 0.066 \text{ dB} \approx 0.1 \text{ dB}$$

Suppose one layer of 200-mesh screen ($R_S = 24.6$ Pa-s/m^3) is placed over the open end of the tube:
$$\frac{R_S}{\rho_o c} = \frac{(24.6)}{(1.20)(343.8)} = 0.05963$$

The sound power transmission coefficient at resonance (250 Hz) with the screen is as follows:
$$(1/a_t)_{\max} = \left[1 + \frac{(0.0578)}{(0.00439 + 0.05963)} \right]^2 = (1.899)^2 = 3.608$$

Silencer Design

The addition of the screen results in a lower value of the transmission loss at resonance:

$$TL_o = 10 \log_{10}(3.608) = 5.6 \, \text{dB}$$

The sound power transmission coefficient at 177 Hz may be calculated as follows:

$$1/a_t = \frac{(0.003655 + 0.5963 + 0.05758)^2 \tan^2(1.111) + 1}{(0.003695 + 0.05963)^2 \tan^2(1.111) + 1} = \frac{1.0596}{1.01635}$$

$$1/a_t = 1.0426$$

The transmission loss at 177 Hz is slightly higher with the addition of the screen:

$$TL(177 \, \text{Hz}) = 10 \log_{10}(1.0426) = 0.18 \, \text{dB} \approx 0.2 \, \text{dB}$$

8.4.4 Open Tube (Orifice) as a Side Branch

Let us consider the case in which an open tube or orifice (a short, open tube) is used as the side branch. The reactance for a tube open to the surroundings ($C_A = \infty$) is equal to the acoustic mass for the tube:

$$X_A = 2\pi f M_A = \frac{2f \rho_o L_e}{a^2} = \frac{\rho_o c k L_e}{\pi a^2} \tag{8-114}$$

The quantity $k = 2\pi f/c$ is the wave number. The lumped-parameter analysis is valid only for small values of ka, so the acoustic resistance for the open end (valid for $ka < 1.4$) is given by Eq. (8.35a):

$$R_A = \frac{\rho_o c k^2}{2\pi} = \frac{2\pi f^2 \rho_o}{c} \tag{8-115}$$

The sound power transmission coefficient (or the reciprocal) for the open-tube side-branch muffler may be found by making the substitutions from Eqs (8-114) and (8-115) into Eq. (8-81):

$$\frac{1}{a_t} = \frac{\left(\frac{\rho_o c k^2}{2\pi} + \frac{\rho_o c}{2S}\right)^2 + \left(\frac{\rho_o c k L_e}{\pi a^2}\right)^2}{\left(\frac{\rho_o c k^2}{2\pi}\right)^2 + \left(\frac{\rho_o c k L_e}{\pi a^2}\right)^2} \tag{8-116}$$

This relationship may be simplified by canceling the term ($\rho_o c$) and dividing through by the last term in the numerator and denominator:

$$\frac{1}{a_t} = \frac{[(ka^2/2L_e) + (\pi a^2/2SkL_e)]^2 + 1}{(ka^2/2L_e)^2 + 1} \tag{8-117}$$

If we substitute for the wave number k in terms of the frequency f, we obtain the alternative expression for the sound power transmission coefficient:

$$\frac{1}{a_t} = \frac{\{\pi(fL_e/c)(a/L_e)^2 + [(a^2/4S)/(fL_e/c)]\}^2 + 1}{[\pi(fL_e/c)(a/L_e)^2]^2 + 1} \qquad (8\text{-}118)$$

For small frequencies (or, for $f < 0.1c/S^{1/2}$), Eq. (8-118) approaches the limiting expression, as follows:

$$\frac{1}{a_t} \rightarrow \left[\frac{(a^2/4S)}{(fL_e/c)}\right]^2 + 1 \qquad (8\text{-}119)$$

On the other hand, for large frequencies (or, for $f > c/S^{1/2}$), the sound power transmission coefficient given by Eq. (8-118) approaches a limiting value of unity, or the transmission loss approaches zero at large frequencies.

If an additional acoustic resistance, such as a screen, is placed over one end of the short tube, the sound power transmission coefficient expression is modified as follows:

$$\frac{1}{a_t} = \frac{\{\pi(fL_e/c)(a/L_e)^2 + [(R_S/2\pi\rho_o c)/(fL_e/c)] + [(a^2/4S)/(fL_e/c)]\}^2 + 1}{\{[\pi(fL_e/c)(a/L_e)^2] + [(R_S/2\pi\rho_o c)/(fL_e/c)]\}^2 + 1} \qquad (8\text{-}120)$$

The quantity R_S is the specific acoustic resistance of the screens. The low-frequency limiting value of the sound power transmission coefficient with the use of screen resistances is as follows:

$$\frac{1}{a_t} \rightarrow \left[\frac{(R_S/2\pi\rho_o c) + (a^2/4S)}{(fL_e/c)}\right]^2 + 1 \qquad (8\text{-}121)$$

The use of a screen resistance increases the low-frequency transmission loss. At high frequencies, the transmission loss approaches zero, whether screens are used or not used.

Example 8-7. A hole having an inside diameter of 54.8 mm (2.157 in) is placed in a pipe having an inside diameter of 161.5 mm (6.357 in) and wall thickness of 3.40 mm (0.134 in) through which noise is transmitted. The gas in the tube is air at 21°C (70°F) and 101.3 kPa (14.7 psia), for which the sonic velocity $c = 343.8$ m/s (1128 ft/sec), and density $\rho_o = 1.200$ kg/m^3 (0.075 lb$_m$/ft^3). Determine the transmission loss at 250 Hz and 125 Hz.

First, let us check the validity of the lumped-parameter model:

$$ka = \frac{2\pi f a}{c} = \frac{(2\pi)(250)(0.0274)}{(343.8)} = 0.125 < 1$$

Silencer Design

The lumped-parameter model does apply for this example. Let us also calculate the following parameter:

$$\frac{c}{S^{1/2}} = \frac{(343.8)}{(0.02048)^{1/2}} = 2402 \text{ Hz}$$

The frequency $f = 250$ Hz is near $(0.1)(2402) = 240.2$ Hz, so the low-frequency approximation, Eq. (8-119), does apply; however, let us use the general expression for this frequency.

The equivalent length of the hole, which has "flanged ends" at both inlet and outlet, is given by Eqs (8-8) and (8-12):

$$L_e = L + 2(8/3\pi)a = 3.40 + (2)(8/3\pi)(27.4) = 49.92 \text{ mm}$$

Let us calculate the following dimensionless parameters for a frequency of $f = 250$ Hz:

$$\frac{fL_e}{c} = \frac{(250)(0.04992)}{(343.8)} = 0.0363$$

$$\frac{a^2}{4S} = \frac{(0.0274)^2}{(4)(0.02048)} = 0.00916$$

The sound power transmission coefficient for a frequency of 250 Hz may be found from Eq. (8-118):

$$\frac{1}{a_t} = \frac{[(\pi)(0.0363)(27.4/49.92)^2 + (0.00916/0.0363)]^2 + 1}{[(\pi)(0.0363)(274/49.92)^2]^2 + 1}$$

$$\frac{1}{a_t} = \frac{(0.0344 + 0.2525)^2 + 1}{(0.0344)^2 + 1} = \frac{1.0823}{1.0012} = 1.081$$

The transmission loss is as follows:

$$\text{TL} = 10 \log_{10}(1.081) = 0.3 \text{ dB} \qquad (\text{for } f = 250 \text{ Hz})$$

The frequency $f = 125$ Hz falls in the range for which the low-frequency approximation applies, because 125 Hz $< (0.1)(2402) = 240.2$ Hz. For 125 Hz, the frequency parameter has the following value:

$$\frac{fL_e}{c} = \frac{(125)(0.04992)}{(343.8)} = 0.01815$$

The sound power transmission coefficient may be evaluated from Eq. (8-119):

$$\frac{1}{a_t} = 1 + \left(\frac{0.00916}{0.01815}\right)^2 = 1 + (0.505)^2 = 1.255$$

The transmission loss is as follows:

$$\text{TL} = 10\log_{10}(1.255) = 1.0\,\text{dB} \quad (\text{for } f = 125\,\text{Hz})$$

The transmission loss for the orifice side-branch muffler is considerably smaller than that for the closed-tube side-branch muffler at 250 Hz. On the other hand, the orifice has somewhat better performance at lower frequencies, although the transmission loss is not very large at a frequency of 125 Hz.

8.5 EXPANSION CHAMBER MUFFLERS

The *expansion chamber muffler* is a reactive-type muffler, because the reduction of noise transmission through the muffler is achieved by reflecting back to the source a portion of the energy entering the muffler. There is generally a negligible amount of energy dissipation within the muffler. The expansion chamber muffler consists of one or more chambers or expansion volumes which act as resonators to provide an acoustic mismatch for the acoustic energy being transmitted along the main tube. Some typical configurations for the expansion chamber muffler are shown in Fig. 8-14.

8.5.1 Transmission Loss for an Expansion Chamber Muffler

At the junction of the inlet tube and the expansion chamber, the instantaneous acoustic pressure in the inlet tube and in the expansion chamber are equal; and, similarly, the instantaneous acoustic pressures are equal at the junction of the expansion chamber and the outlet tube for the expansion chamber muffler. The instantaneous volumetric flow rates, $U(t) = Su(t)$, are equal on each side of the inlet and outlet junction. These conditions are the same as those used for analysis of transmission of sound from medium 1 through medium 2 into medium 3, as discussed in Sec. 4.7. Instead of the characteristic impedance $Z_o = \rho_o c$, the acoustic impedance $Z_A = \rho_o c/S = Z_o/S$ appears in the final expression for the sound power transmission coefficient. The gas is the same in all sections of the muffler, so that the impedance ratios become area ratios:

$$Z_1/Z_3 = Z_{A1}/Z_{A3} = S_3/S_1 \equiv \nu \qquad (8\text{-}122)$$

$$Z_1/Z_2 = Z_{A1}/Z_{A2} = S_2/S_1 \equiv m \qquad (8\text{-}123)$$

$$\frac{Z_2}{Z_3} = \frac{Z_{A2}}{Z_{A3}} = \frac{S_3}{S_2} = \frac{S_3/S_1}{S_2/S_1} = \frac{\nu}{m} \qquad (8\text{-}124)$$

Silencer Design

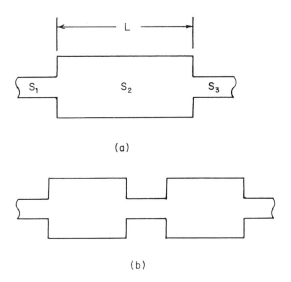

FIGURE 8-14 Configurations for expansion chamber mufflers: (a) single expansion volume and (b) double expansion volume.

If these substitutions are made into Eq. (4-123), the following expression is obtained for the sound power transmission coefficient (or the reciprocal) for an expansion chamber muffler:

$$\frac{1}{a_t} = \frac{(1+v)^2 \cos^2(kL) + (m+v/m)^2 \sin^2(kL)}{(4v)} \tag{8-125}$$

The quantity $k = 2\pi fL/c$ is the wave number and L is the length of the expansion chamber.

The expression for the sound power transmission coefficient, Eq. (8-125), may be written in a different form by substituting the trigonometric identity:

$$\cos^2(kL) = 1 - \sin^2(kL) \tag{8-126}$$

$$\frac{1}{a_t} = \frac{(1+v)^2 + [(m+v/m)^2 - (1+v)^2]\sin^2(kL)}{4v} \tag{8-127}$$

The second term in the numerator of Eq. (8-127) may be rearranged as follows:

$$\frac{1}{a_t} = \frac{(1+v)^2 + [(m-v/m)^2 - (1-v)^2]\sin^2(kL)}{4v} \tag{8-128}$$

For the special case in which the inlet tube and the outlet tube have the same cross-sectional area, $v = S_3/S_1 = 1$, Eq. (8-128) reduces to the following expression:

$$1/a_t = 1 + \tfrac{1}{4}(m - 1/m)^2 \sin^2(kL) \tag{8-129}$$

The transmission loss obtained from Eq. (8-129) is plotted in Fig. 8-15.

It is noted from Eq. (8-128) that the transmission loss is a maximum for the frequencies corresponding to the following condition:

$$kL = \frac{2\pi f_o L}{c} = \left(n - \tfrac{1}{2}\right)\pi \quad (n = 1, 2, 3, \ldots) \tag{8-130}$$

$$f_o = \frac{\left(n - \tfrac{1}{2}\right)c}{2L} \quad (n = 1, 2, 3, \ldots) \tag{8-131}$$

The maximum transmission loss (which occurs at the frequency f_o) may be found by combining Eqs (8-128) and (8-130):

$$\left(\frac{1}{a_t}\right)_{max} = \frac{(1 + v)^2 + (m - v/m)^2 - (1 - v)^2}{(4v)} \tag{8-132}$$

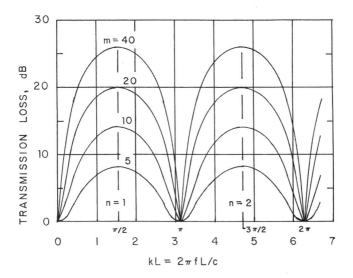

FIGURE 8-15 Plot of the transmission loss as a function of frequency for a single-chamber expansion chamber muffler.

Silencer Design

For the special case of $v = 1$, the maximum transmission loss expression is as follows:

$$(1/a_t)_{\max} = 1 + \tfrac{1}{4}(m - 1/m)^2 \tag{8-133}$$

We also observe that the transmission loss is a minimum for the frequencies corresponding to the following condition:

$$kL = n\pi \quad (n = 1, 2, 3, \ldots) \tag{8-134}$$

$$f_p = \frac{nc}{2L} \quad (n = 1, 2, 3, \ldots) \tag{8-135}$$

The minimum transmission loss expression is as follows:

$$\left(\frac{1}{a_t}\right)_{\min} = \frac{(1+v)^2}{4v} \tag{8-136}$$

For the special case of $v = 1$, $(1/a_t)_{\min} = 1$ and $\mathrm{TL}_{\min} = 0\,\mathrm{dB}$.

8.5.2 Design Procedure for Single-Expansion Chamber Mufflers

If the same parameters are given or known as in the case of the side-branch muffler design given in Sec. 8.4.2, we may develop a similar design procedure for the expansion chamber muffler with a single expansion volume.

D1. The resonant frequency (or center frequency) for the optimum expansion chamber muffler is the arithmetic average of the low and high operational frequencies, f_1 and f_2, of the muffler:

$$f_o = \tfrac{1}{2}(f_1 + f_2) \tag{8-137}$$

D2. The optimum length of the expansion chamber corresponds to the situation in which the transmission loss is maximized at the center or resonant frequency, as given by Eq. (8-131):

$$L = \frac{(n - \tfrac{1}{2})c}{2f_o} \quad (n = 1, 2, 3, \ldots) \tag{8-138}$$

D3. The value of the integer n may be estimated by satisfying the following condition:

$$k_2 L - k_1 L \approx \tfrac{1}{2}\pi \tag{8-139}$$

If we substitute for the expansion chamber length from Eq. (8-138), the following expression is obtained for the approximate value of the integer n:

$$n \approx \tfrac{1}{2} + \frac{f_o}{2(f_2 - f_1)} \tag{8-140}$$

The quantities f_1 and f_2 are the lower and upper frequencies for the main operational range of the muffler. Note that n must be an integer ($n = 1$ or 2 or 3, etc.), so that Eq. (8-140) may not be satisfied exactly. As illustrated in the following example, the transmission loss should be checked at the end frequencies.

D4. The expansion ratio required to achieve the minimum allowable transmission loss (which occurs at the end frequencies) may be found from Eq. (8-128), using $k_1 L = 2\pi f_1 L/c$. If practical dimensions cannot be achieved in this design step, the use of multiple expansion chambers may be required.

Example 8-8. A cylindrical air conditioning duct has a diameter of 250 mm (9.84 in). The fluid flowing in the duct is air at 10°C (50°F) and 105 kPa (15.23 psia), for which the sonic velocity $c = 337.3$ m/s (1107 ft/sec) and the density $\rho_o = 1.292$ kg/m³ (0.0807 lb$_m$/ft³). It is desired to design a single expansion chamber muffler that has a minimum transmission loss of 6 dB for the frequency range between 177 Hz and 354 Hz.

The center frequency for the muffler may be determined from Eq. (8-137):

$$f_o = \tfrac{1}{2}(177 + 354) = 265.5 \,\text{Hz}$$

The estimated value of the integer n may be found from Eq. (8-140):

$$n \approx \tfrac{1}{2} + \frac{(265.5)}{(2)(354 - 177)} = 1.25$$

We could possibly use either $n = 1$ or $n = 2$.

Let us check the results when using $n = 2$ first. The length of the muffler is found from Eq. (8-138):

$$L = \frac{(2 - \tfrac{1}{2})(337.3)}{(2)(265.5)} = 0.953 \,\text{m} \;(37.5 \,\text{in})$$

At the lower frequency (177 Hz) in the operational range for the muffler, we find the following value for $k_1 L = 2\pi f_1 L/c$:

$$k_1 L = \frac{(2\pi)(177)(0.953)}{(337.3)} = 3.1416 \,\text{rad} = \pi \,\text{rad}$$

As shown by Eq. (8-134), the transmission loss for $v = 1$ and $kL = \pi$ is TL $= 0$, so the value of $n = 2$ cannot be used.

Let us try the other possibility, $n = 1$. The required muffler length is found from Eq. (8-138):

Silencer Design

$$L = \frac{(1-\frac{1}{2})(337.3)}{(2)(265.5)} = 0.318\,\text{m} = 318\,\text{mm}\ (12.50\,\text{in})$$

For the lower frequency, f_1, we find the following value for the parameter $k_1 L$:

$$k_1 L = \frac{(2\pi)(177)(0.318)}{(337.3)} = 1.0472\,\text{rad} = \tfrac{1}{3}\pi\,\text{rad} = 60°$$

The sound power transmission coefficient (or the reciprocal) at the lower frequency corresponds to the minimum design transmission loss:

$$\text{TL}_{\min} = 6\,\text{dB} = 10\log_{10}(1/a_t)$$

$$1/a_t = 10^{0.60} = 3.981$$

The required size of the expansion chamber may be found from Eq. (8-129):

$$1/a_t = 3.981 = 1 + \tfrac{1}{4}(m - 1/m)^2 \sin^2(\pi/3)$$

$$m - 1/m = \frac{(2)(\sqrt{2.981})}{\sin(60°)} = 3.9874$$

$$m^2 - 3.9874m - 1 = 0$$

If we solve for the area ratio, we obtain the following value:

$$m = 1.9937 + [(1.9937)^2 + 1]^{1/2} = 4.224 = S_2/S_1 = (D_2/D_1)^2$$

The required expansion chamber diameter is as follows:

$$D_2 = (250)(4.224)^{1/2} = 514\,\text{mm}\ (20.2\,\text{in})$$

The maximum transmission loss for the muffler occurs at the center frequency, $f_o = 265.5$ Hz. The value is determined from Eq. (8-133):

$$(1/a_t)_{\max} = 1 + \tfrac{1}{4}[4.224 - (1/4.224)^2] = 4.975$$

$$\text{TL}_{\max} = 10\log_{10}(4.975) = 7.0\,\text{dB}$$

8.5.3 Double-Chamber Mufflers

For mufflers with two or more expansion chambers, the analysis is most conveniently carried out using the transfer matrix approach (Beranek and Vér, 1992). The results for the transmission loss for the double-chamber muffler with an external connecting tube and expansion chambers having equal lengths, as shown in Fig. 8-16(a), is given by the following expression (Davis et al., 1954):

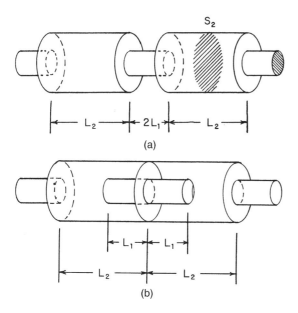

FIGURE 8-16 Nomenclature for double-chamber expansion chamber mufflers: (a) external connecting tube and (b) internal connecting tube.

$$\frac{1}{a_t} = \frac{F_1^2 + F_2^2}{16m^2} \qquad (8\text{-}141)$$

The quantities F_1 and F_2 are defined by the following expressions:

$$F_1 = (m+1)^2 \cos[2k(L_1 + L_2)] - (m-1)^2 \cos[2k(L_2 - L_1)] \qquad (8\text{-}142)$$

$$F_2 = \tfrac{1}{2}(m + 1/m)\{(m+1)^2 \sin[2k(L_1 + L_2)]$$
$$- (m-1)^2 \sin[2k(L_2 - L_1)]\} - (m - 1/m)(m^2 - 1)\sin(2kL_1)$$
$$(8\text{-}143)$$

The quantity L_1 is the half-length of the connecting tube (total length, $2L_1$), L_2 is the length of one expansion chamber, $m = S_2/S_1 =$ cross-sectional area ratio for chamber and inlet tube (assumed to have the same diameter as the connecting tube), and $k = 2\pi f/c =$ wave number.

The transmission loss for the double-chamber muffler is generally larger than that for a single-chamber muffler. There is a low-frequency region present, however, in which the TL is relatively small. The low-frequency pass band appears in the double-chamber muffler as a result of resonance between the connecting tube and the expansion chambers. When

Silencer Design

the connecting tube is made longer, the low-frequency pass band frequency width is made smaller. The upper frequency (called the *cut-off frequency*) in this pass band may be estimated from the following relationship:

$$f_c = \frac{c}{2\pi[mL_1L_2 + L_2(L_2 - L_1)/3]^{1/2}} \qquad (8\text{-}144)$$

The connecting tube length should be selected such that the primary operating frequency range of the muffler lies above the cut-off frequency f_c. The maximum transmission loss in the first band above the pass band is increased as the length of the connecting tube is increased; however, at higher frequencies, there are regions of low TL with frequency width on the order of 50 Hz or more for long connecting tube lengths. These pass bands would be objectionable if there were a significant fraction of the sound energy incident on the muffler in this frequency range.

For the double-chamber muffler with an internal connecting tube, as shown in Fig. 8-16(b), the following expression has been obtained for the sound power transmission coefficient (or the reciprocal):

$$1/a_t = G_1^2 + G_2^2 \qquad (8\text{-}145)$$

The quantities G_1 and G_2 are defined by the following expressions:

$$G_1 = \cos(2kL_2) - (m-1)\sin(2kL_2)\tan(kL_1) \qquad (8\text{-}146)$$

$$G_2 = \tfrac{1}{2}(m-1)\tan(kL_1)[(m+1/m)\cos(2kL_2) - (m-1/m)] \\ + \tfrac{1}{2}(m+1/m)\sin(2kL_2) \qquad (8\text{-}147)$$

The quantity L_1 is the half-length of the connecting tube (total length, $2L_1$), L_2 is the length of one expansion chamber, $m = S_2/S_1$ = cross-sectional area ratio for chamber and inlet tube (assumed to have the same diameter as the connecting tube), and $k = 2\pi f/c$ = wave number.

A low-pass band region is also present at frequencies below f_c, given by Eq. (8-144), as in the case of the muffler with the external connecting tube. There is a significant difference between the performance of the two mufflers, however, when the connecting tube length $2L_1$ is equal to the expansion chamber length L_2. The frequency band over which the muffler has a high TL for the internal tube muffler is about twice that of the external tube muffler, for the same dimensions. The largest attenuation occurs when $kL_1 = \tfrac{1}{2}\pi$, since the term $\tan(kL_1)$ in Eqs (8-146) and (8-147) becomes infinitely large under this condition, or the sound power transmission coefficient a_t approaches zero when $kL_1 = \tfrac{1}{2}\pi$.

Example 8-9. An expansion chamber muffler has two expansion chambers, each having a length of 300 mm (11.81 in) and a diameter of 200 mm

(7.874 in). The connecting tube between the two chambers is an internal tube having a diameter of 100 mm (3.937 in) and a length of 200 mm (7.874 in). The inlet and outlet tubes for the muffler also have diameters of 100 mm (3.937 in). The gas flowing through the muffler is air at 10°C (50°F) and 105 kPa (15.23 psia), for which the sonic velocity $c = 337.3$ m/s (1107 ft/sec) and the density $\rho_o = 1.292$ kg/m^3 (0.0807 lb$_m$/ft^3). Determine the transmission loss for the muffler at a frequency of 250 Hz.

The area ratio for the muffler is as follows:

$$m = S_2/S_1 = (D_2/D_1)^2 = (200/100)^2 = 4.00$$

Let us calculate the following dimensionless parameters:

$$2kL_2 = \frac{(2)(2\pi)(250)(0.300)}{(337.3)} = 2.794 \, \text{rad} = 160.1°$$

$$kL_1 = \frac{(2\pi)(250)(0.100)}{(337.3)} = 0.4657 \, \text{rad} = 26.7°$$

The parameters from Eqs (8-146) and (8-147) may now be calculated:

$$G_1 = \cos(2.794) - (4-1)\sin(2.794)\tan(0.4657) = -0.9403 - 0.5133$$

$$G_1 = -1.4536$$

$$G_2 = \tfrac{1}{2}(4-1)\tan(0.4657)[(4+1/4)\cos(2.794) - (4-1/4)]$$
$$+ \tfrac{1}{2}(4+1/4)\sin(2.794) = (0.7538)(-7.7461) + 0.7235$$

$$G_2 = -5.1159$$

The reciprocal of the sound power transmission coefficient for the double-chamber muffler with an internal connecting tube may be found from Eq. (8-145):

$$1/a_t = G_1^2 + G_2^2 = (-1.4536)^2 + (-5.1159)^2 = 28.285$$

The transmission loss for the muffler at 250 Hz is as follows:

TL $= 10 \log_{10}(28.285) = 14.5$ dB

Let us check the cut-off frequency given by Eq. (8-144):

$$f_c = \frac{(337.3)}{(2\pi)[(4.00)(0.300)(0.100) + (0.300)(0.300 - 0.100)]^{1/2}} = 126.5 \, \text{Hz}$$

This frequency is well below the frequency (250 Hz) in the previous calculation.

8.6 DISSIPATIVE MUFFLERS

Dissipative mufflers or silencers differ from reactive mufflers in that noise reduction in a dissipative muffler is achieved primarily by attenuation of the acoustic energy within the lining or other elements within the muffler. The dissipative muffler may also reflect some of the acoustic energy back to the source, similar to the action of the reactive muffler. Some configurations of dissipative mufflers are shown in Fig. 8-17.

Dissipative mufflers usually have wide-band noise reduction characteristics. The sharp peaks and valleys in the transmission loss curves that are found for reactive mufflers are usually not present for dissipative mufflers. As a result of this characteristic, dissipative mufflers are useful for solving noise control problems involving continuous noise spectra, such as fan noise, intake and exhaust noise from gas turbines, and noise through access openings in acoustic enclosures.

For the dissipative muffler shown in Fig. 8-18, the instantaneous acoustic pressure and instantaneous volume flow, including the effect of acoustic energy attenuation, may be written as follows:

$$p(x, t) = A \, e^{-\sigma x} \, e^{j(\omega t - kx)} + B \, e^{\sigma x} \, e^{j(\omega t + kx)} \tag{8-148}$$

$$U(x, t) = (SA/\rho_o c) \, e^{-\sigma x} e^{j(\omega t - kx)} - (SB/\rho_o c) \, e^{\sigma x} e^{j(\omega t + kx)} \tag{8-149}$$

The quantity σ is the *attenuation coefficient* for the muffler lining, and the quantities A and B are constants to be determined from the boundary conditions. The first term on the right side of Eqs (8-148) and (8-149) represents the sound wave traveling in the $+x$ direction, and the second term represents the sound wave traveling in the $-x$ direction.

Suppose we take the origin ($x = 0$) at the interface of the inlet tube and the muffler. At this point, the acoustic pressure on the inlet tube side $p_1(0, t)$ and on the muffler side $p_2(0, t)$ must be equal. Also, the volumetric flow quantities $U_1(0, t)$ and $U_2(0, t)$ must be the same at the interface. These two conditions yield the following relationships between the constants:

$$A_1 + B_1 = A_2 + B_2 \tag{8-150}$$

$$A_1 - B_1 = m(A_2 + B_2) \tag{8-151}$$

The quantity m is the area ratio, $m = S_2/S_1$. It is assumed that the inlet and outlet tubes have the same dimensions, $S_1 = S_3$.

At the other end of the muffler ($x = L$), we may equate the instantaneous acoustic pressures $p_2(L, t)$ and $p_3(L, t)$. The volume flow quantities $U_2(L, t)$ and $U_3(L, t)$ are also equal to the exit interface of the muffler. In calculating the transmission loss for the muffler, it is assumed that there are

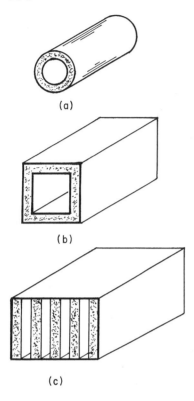

FIGURE 8-17 Configurations for dissipative mufflers: (a) circular lined chamber, (b) rectangular lined chamber, and (c) baffle-type muffler.

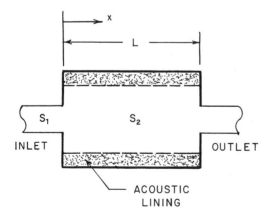

FIGURE 8-18 Nomenclature for dissipative muffler.

Silencer Design

no reflected sound waves in the exit tube, $B_3 = 0$. The application of these conditions yields two additional relationships between the constants:

$$A_2 e^{-\sigma L} e^{-jkL} + B_2 e^{\sigma L} e^{jkL} = A_3 \tag{8-152}$$

$$m(A_2 e^{-\sigma L} e^{-jkL} - B_2 e^{\sigma L} e^{jkL}) = A_3 \tag{8-153}$$

The sound power transmission coefficient is the ratio of the power transmitted to the power incident on the muffler:

$$a_t = \frac{W_{tr}}{W_{in}} = \frac{S_3 I_{tr}}{S_1 I_{in}} = \frac{p_{tr}^2}{p_{in}^2} = \frac{|A_3|^2}{|A_1|^2} \tag{8-154}$$

If we solve Eqs (8-150) through (8-153) simultaneously, we obtain the following expression for the reciprocal of the sound power transmission coefficient for the dissipative muffler:

$$1/a_t = C_1^2 \cos^2(kL) + C_2^2 \sin^2(kL) \tag{8-155}$$

The constants are given by the following expressions:

$$C_1 = \cosh(\sigma L) + \tfrac{1}{2}(m + 1/m) \sinh(\sigma L) \tag{8-156}$$

$$C_2 = \sinh(\sigma L) + \tfrac{1}{2}(m + 1/m) \cosh(\sigma L) \tag{8-157}$$

$$k = \frac{2\pi f c}{c} = \text{wave number} \tag{8-158}$$

The effect of the absorptive lining may be investigated by considering two limiting cases for Eq. (8-155). First, for small attenuation ($\sigma L \ll 1$), or for $\sigma L \leq 0.20$ for practical purposes, the hyperbolic functions approach the following limiting values within 2%:

$$\cosh(\sigma L) \to 1 \quad \text{and} \quad \sinh(\sigma L) \to \sigma L \quad (\text{for } \sigma L < 0.2) \tag{8-159}$$

If we make the substitutions from Eq. (8-159) into Eq. (8-155), we obtain the following expression for the small-attenuation limit:

$$1/a_t = [1 + \tfrac{1}{2}(m + 1/m)\sigma L]^2 + \tfrac{1}{4}(m - 1/m)^2 \sin^2(kL) \tag{8-160}$$

Representative values for the transmission loss for small attenuation are listed in Table 8-2.

At the other limit of very large attenuation ($\sigma L \gg 1$), or for $\sigma L \geq 5$ for practical purposes, the hyperbolic functions approach the following limiting values within 1%:

$$\sinh(\sigma L) \approx \cosh(\sigma L) \to \tfrac{1}{2} e^{\sigma L} \quad (\text{for } \sigma L \geq 5) \tag{8-161}$$

TABLE 8-2 Transmission Loss for a Dissipative Silencer with Small Values of Attenuation ($\sigma L \leq 0.2$)[a]

	$kL = \tfrac{1}{2}\pi$		$kL = \pi$	
Attenuation	$\sigma L = 0$	$\sigma L = 0.20$	$\sigma L = 0$	$\sigma L = 0.20$
$1/a_t$	4.516	5.546	1.000	2.031
TL, dB	6.5	7.4	0.0	3.1

[a]The area ratio for data in this table is $m = 4$. The quantity $k = 2\pi f/c =$ wave number.

If we make the substitutions from Eq. (8-161) into Eq. (8-155), we obtain the following expression for the large attenuation limit:

$$1/a_t = \tfrac{1}{4} e^{2\sigma L}[1 + \tfrac{1}{2}(m + 1/m)]^2 \tag{8-162}$$

Taking $\log_{10}$ of both sides of Eq. (8-162) and multiplying by 10, we obtain the transmission loss expression for the large-attenuation limit:

$$\text{TL} = 8.6859\sigma L + 20\log_{10}[\tfrac{1}{2} + \tfrac{1}{4}(m + 1/m)] \tag{8-163}$$

The first term in Eq. (8-163) represents the attenuation provided by the lining, and the second term represents the effect of reflection of acoustic energy back to the source as a result of the change in cross-sectional area of the flow passage.

Example 8-10. A dissipative muffler has a length of 825 mm (32.48 in) and a diameter of 446 mm (17.55 in). The diameter of the inlet and outlet tubes is 152 mm (6.00 in). The attenuation coefficient is 1.25 Np/m or $(8.6859\sigma) = 10.86$ dB/m. The gas flowing through the muffler is air at 600K (620°F) and 110 kPa (15.96 psia), for which the density is 0.639 kg/m³ (0.0399 lb$_m$/ft³) and the sonic velocity is 491.0 m/s (1611 ft/sec). The frequency of the sound being transmitted is 250 Hz. Determine the transmission loss for the muffler.

Let us first calculate the pertinent dimensionless parameters:

$$\sigma L = (1.25)(0.825) = 1.0313$$

$$kL = \frac{2\pi f L}{c} = \frac{(2\pi)(250)(0.825)}{(491.0)} = 2.6393 \text{ rad}$$

$$m = S_2/S_1 = (D_2/D_1)^2 = (446/152)^2 = 8.6096$$

Silencer Design

The coefficients defined by Eqs (8-156) and (8-157) may be determined:

$$C_1 = \cosh(1.0313) + \tfrac{1}{2}(8.6096 + 1/8.6096)\sinh(1.0313)$$

$$C_1 = 1.5805 + 5.3402 = 6.9207$$

$$C_2 = \sinh(1.0313) + \tfrac{1}{2}(8.6096 + 1/8.6096)\cosh(1.0313)$$

$$C_2 = 1.2240 + 6.8958 = 8.1198$$

The reciprocal of the sound power transmission coefficient may be found from Eq. (8-155):

$$1/a_t = (6.9207)^2 \cos^2(2.6393) + (8.1198)^2 \sin^2(2.6393)$$

$$1/a_t = 36.796 + 15.280 = 52.076$$

The transmission loss is as follows:

$$\text{TL} = 10\log_{10}(52.076) = 17.2\,\text{dB}$$

The reciprocal of the sound power transmission coefficient for an unlined expansion chamber muffler having the same area ratio as the muffler in this example may be found from Eq. (8-129):

$$1/a_t = 1 + \tfrac{1}{4}(8.6096 - 1/8.6096)^2 \sin^2(2.6393) = 5.180$$

The transmission loss for the unlined expansion chamber muffler is as follows:

$$\text{TL} = 10\log_{10}(5.180) = 7.1\,\text{dB}$$

The attenuation by the liner increases the transmission loss for the muffler by 10.1 dB in this case.

8.7 EVALUATION OF THE ATTENUATION COEFFICIENT

The best source of information for the attenuation coefficient is experimental data on the lining material. In some cases, this information is not available from manufacturer's data. For muffler design, we may need to estimate the attenuation coefficient before the muffler is built. After the unit has been constructed, the prototype should be tested to verify the design calculations.

8.7.1 Estimation of the Attenuation Coefficient

The attenuation coefficient (uncorrected for random incidence end effects) may be estimated from the following expression (Beranek, 1960):

$$\sigma = \frac{\pi f P_w}{2 S_2} \left[\frac{\rho_e Y}{2\kappa} \right]^{1/2} [(1+\psi^2)^{1/2} - 1]^{1/2} \qquad (8\text{-}164)$$

The quantities appearing in Eq. (8-164), which will be discussed in more detail in the following section, are as follows:

P_w = perimeter of the flow passage in the muffler
S_2 = open cross-sectional area of the muffler
ρ_e = effective density for the gas within the lining material
Y = porosity of the lining
κ = effective elasticity coefficient for the gas within the lining material

The quantity ψ is a dimensionless parameter defined by:

$$\psi = \frac{R_e}{2\pi f \rho_e} \qquad (8\text{-}165)$$

The quantity R_e is the effective flow resistance per unit thickness for the lining material.

Two general classes of acoustic lining materials are considered: (a) a semirigid material and (b) a soft blanket material. A semirigid material is one in which the solid portion of the material is relatively rigid, e.g., an acoustic ceiling tile. A soft blanket material is one in which the solid portion of the material is relatively flexible, e.g., a panel of glass fiber acoustic material.

An alternative method, based on an empirical curve fit of attenuation data, has also been developed (Beranek and Vér, 1992, p. 214). The attenuation coefficients for mineral wool and for glass fibrous material have been correlated by an expression in the following form, valid in the range $0.3 < \psi_o < 120$,

$$\sigma = \frac{\pi f P_w}{2 S c_o} a \psi_o^b \qquad (8\text{-}166)$$

The quantity c_o is the speed of sound at ambient temperature (T_o) for the gas in the lining, and the quantity ψ_o is defined by the following expression:

$$\psi_o = \frac{R_1 (T/T_o)^{-1.65}}{2\pi f \rho_o} \qquad (8\text{-}167)$$

The quantity R_1 is the specific flow resistance per unit thickness for the acoustic material, T is the absolute temperature of the gas within the acoustic material, and T_o is ambient temperature (300K). The temperature ratio term corrects for density and viscosity variation with

Silencer Design

temperature. The values for the regression constants a and b are given in Table 8-3.

8.7.2 Effective Density

The inertia effect of the fibers in a semirigid material is negligible, because the fibers do not move appreciably with the gas when a sound wave passes through the material. For the semirigid acoustic material, the effective density is given by the following relationship:

$$\rho_e = \rho_o \phi_s \qquad (8\text{-}168)$$

The quantity ϕ_s is called the *structure factor* (Zwikker and Kosten, 1949), which takes into account the effect of cavities and pores that are perpendicular to the direction of sound propagation in the material. The structure factor may be approximated by the following relationship:

$$\phi_s = 1 + 4.583(1 - Y) \qquad (6\text{-}169)$$

The quantity Y is the *porosity* of the material, defined as the ratio of the void volume or volume occupied by the gas within the material to the total volume of the lining material.

The inertia effect of the fibers in a soft blanket material at low frequencies is small, because the fibers move along with the motion of the gas within the material. At the limit of low frequencies, the effective density of a soft blanket material is equal to the bulk density of the blanket material plus the weighted density of the gas within the material. For high frequencies, the inertia effect of the fibers is large and the fibers are not able to follow the motion of the gas within the material. For the limit of high frequencies, the effective density of soft blanket materials becomes the same as that of rigid materials. The effective density for the soft blanket materials may be calculated from the following relationship:

TABLE 8-3 Values for the Regression Constants in Eq. (8-166)

Material	Range[a]	a	b
Mineral wool	$\psi_o < 6.4$	0.605	0.663
	$\psi_o \geq 6.4$	0.810	0.502
Glass fiber	$\psi_o < 6.4$	0.618	0.674
	$\psi_o \geq 6.4$	0.919	0.458

[a]The quantity ψ_o is defined by Eq. (8-167).

$$\rho_e = \frac{\rho_o \phi_s \{1 + [Y + (\rho_m/\rho_o\phi_s)]\psi_1^2\}}{1 + \psi_1^2} \qquad (8\text{-}170)$$

The quantity ρ_o is the density of the gas within the material, ρ_m is the bulk density of the lining material, Y is the porosity for the lining material, and ϕ_s is the structure factor. The quantity ψ_1 is defined by the following expression:

$$\psi_1 = \frac{R_1}{2\pi f \rho_m} \qquad (8\text{-}171)$$

We note that for small frequencies ($\psi_1 \to \infty$), the effective density from Eq. (8-170) approaches the following limiting value:

$$\rho_e \to \rho_o \phi_s Y + \rho_m \quad \text{(small frequencies)} \qquad (8\text{-}172)$$

For large frequencies ($\psi_1 \to 0$), the effective density from Eq. (8-170) approaches the other limiting value:

$$\rho_e \to \rho_o \phi_s \quad \text{(large frequencies)} \qquad (8\text{-}173)$$

8.7.3 Effective Elasticity Coefficient

The quantity κ in Eq. (8-164) is the *effective elasticity coefficient* for the gas within the lining material. The term κ/Y is actually the effective bulk modulus or compressibility coefficient for the gas. When the frequency of the sound is less than about 100 Hz, there is sufficient time for energy to be exchanged between the gas and the fibers of the lining material, such that the compression and expansion of the gas takes place almost at constant temperature (isothermally). For an ideal gas ($P = \rho R T$), the isothermal bulk modulus may be evaluated (Van Wylen et al., 1994):

$$\kappa_T = \rho \left(\frac{\partial P}{\partial \rho}\right)_T = \rho_o R T_o = P_o \qquad (8\text{-}174)$$

The quantity P_o is the absolute pressure of the gas within the lining material.

On the other hand, when the frequency of the sound wave is greater than about 1000 Hz, there is not sufficient time between cycles for significant energy transfer to take place between the fibers and the gas, so the compression and expansion process is practically adiabatic. For an ideal gas, the adiabatic bulk modulus is given by the following relationship:

$$\kappa_s = \gamma P_o \qquad (8\text{-}175)$$

The quantity γ is the specific heat ratio for the gas within the liner. For air and most diatomic gases, $\gamma = 1.40$.

Silencer Design

The effective elasticity coefficient for the gas within the liner material may be determined from the following relationship:

$$\kappa = \begin{cases} P_o & \text{(for } f \leq 100\,\text{Hz)} \\ [(3 - 2\gamma) + (\gamma - 1)\log_{10}(f)]P_o & \text{(for } 100\,\text{Hz} < f < 1000\,\text{Hz)} \\ \gamma P_o & \text{(for } f \geq 1000\,\text{Hz)} \end{cases}$$

(8-176)

8.7.4 Effective Specific Flow Resistance

For a semirigid material, the fibers of the material do not move appreciably when a gas moves through the material. The velocity of the fibers is practically zero, so the effective specific acoustic resistance for the semirigid material is equal to the actual specific acoustic resistance of the lining (rayl/m):

$$R_e = R_1 \quad \text{(semirigid material)} \tag{8-177}$$

For the soft blanket materials at low frequencies, the fibers move along with the air, such that the effective specific acoustic resistance approaches zero at low frequencies. For high frequencies, the fibers cannot follow the motion of the air and remain more nearly stationary. In this case, the effective specific resistance approaches the actual specific acoustic resistance. The general expression for the effective acoustic resistance for a soft blanket material is the following:

$$R_e = \frac{R_1}{1 + \psi_1^2} \quad \text{(soft blanket material)} \tag{8-178}$$

The quantity ψ_1 is defined by Eq. (8-171).

The specific acoustic resistance per unit thickness R_1 must be determined experimentally, in general, by measuring the pressure drop Δp through a material sample of known thickness Δx and surface area S, with a measured volume flow rate U through the material. The specific acoustic resistance may be calculated from the following expression:

$$R_1 = \frac{S\,\Delta p}{U\,\Delta x} \tag{8-179}$$

Correlations have been developed for some commonly used acoustic lining materials. The specific acoustic resistance of bulk glass-fiber materials has been correlated with the bulk density ρ_m of the material and the diameter of the fibers δ_f (Nichols, 1947):

$$R_1 = 3180(\rho_m)^{1.53}(\delta_f)^{-2} \tag{8-180}$$

Equation (8-180) is not a dimensionless correlation. The constant 3180 applies only when the bulk density is in kg/m³ and the fiber diameter is in μm. The units for the specific acoustic resistance are rayl/m = (Pa-s/m)/m = Pa-s/m².

Another empirical correlation has been developed for several commercial lining materials:

$$R_1 = C_m(\rho_m/\rho_{ref})^n \qquad (8\text{-}181)$$

The quantity ρ_m is the bulk density of the material and the reference density $\rho_{ref} = 16.018 \text{ kg/m}^3 = 1.00 \text{ lb}_m/\text{ft}^3$. Values for the regression constants C_m and n for several materials are given in Table 8-4.

TABLE 8-4 Values of the Regression Coefficients C_m and n in Eq. (8-181) for Various Commercial Acoustic Materials[a]

Material	C_m	n
Thermoflex 300[b]	27,500	1.45
Thermoflex 400[b]	20,300	1.45
Thermoflex 600[b]	46,300	1.45
Thermoflex 800[b]	16,100	1.45
Spincoustic[b]	1,850	1.82
Spintex 305-3.5[b]	1,874	1.70
Spintex 305-4.5[b]	1,500	1.50
Aluminum wool, 30–50 μm diameter	42.2	3.123
Basalt wool, 4–7 μm diameter	1,418	1.55
Kaoline wool, 1–3 μm diameter	15,910	1.36
Kaowool blanket	1,800	1.48
Glass fiber, 15–20 μm diameter	267	1.83
Glass fiber, 3–7 μm diameter	5,890	1.40
Glass fiber, 2–4 μm diameter	9,340	1.56

[a] $R_1 = C_m(\rho_m/\rho_{ref})^n$ where ρ_m is the bulk density of the material, and the reference density $\rho_{ref} = 16.018 \text{ kg/m}^3 = 1.00 \text{ lb}_m/\text{ft}^3$. The units for R_1 are rayl/m.
[b] Material originally manufactured by Johns Manville Corporation.

Silencer Design

8.7.5 Correction for Random Incidence End Effects

The attenuation coefficient, calculated according to the previous correlations, must be corrected for random incidence end effects, because the sound wave is not always a plane wave within the muffler. The corrected value of the attenuation coefficient is found by adding the random incidence correction $\Delta(\sigma L)$:

$$(\sigma L)_{\text{corr}} = \sigma L + \Delta(\sigma L) \tag{8-182}$$

The end correction may be evaluated from the following expressions (Beranek, 1960, p. 449):

(a) For $(S_2)^{1/2}/\lambda = f(S_2)^{1/2}/c \leq 0.09$:

$$\Delta(\sigma L) = 0 \tag{8-183a}$$

(b) For $0.09 < (S_2)^{1/2}/\lambda < 1.00$:

$$\Delta(\sigma L) = 0.5756\{1 + [1 + 1.912 \log_{10}(\sqrt{S_2}/\lambda)]^{1/3}\} \tag{8-183b}$$

(c) For $(S_2)^{1/2}/\lambda \geq 1.00$:

$$\Delta(\sigma L) = 1.1513 \tag{9-183c}$$

The quantity S_2 is the open cross-sectional area of the muffler chamber and λ is the wavelength of the sound wave in the muffler.

Example 8-11. A dissipative muffler having a length of 4.500 m (14.76 ft) and a diameter of 1.500 m (4.92 ft) is lined with a Kaowool blanket material, which is a soft blanket-type material. The unit specific acoustic resistance for the material is $R_1 = 180{,}000$ rayl/m, the mean density of the material is 100 kg/m^3 (6.24 lb$_m$/ft^3), and the porosity of the material is 0.960. The fluid flowing through the muffler is air at 450K (177°C or 350°F) and 140 kPa (20.3 psia), for which the density is 1.084 kg/m^3 (0.0677 lb$_m$/ft^3), the sonic velocity is 425 m/s (1394 ft/sec), and the specific heat ratio $\gamma = 1.40$. The frequency of the sound wave in the muffler is 2 kHz. Determine the attenuation coefficient.

First, we may calculate the attenuation coefficient, uncorrected for end effects. The structure factor for the lining material may be estimated from Eq. (8-169):

$$\phi_s = 1 + (4.583)(1 - 0.960) = 1.1833$$

The dimensionless parameter involving the specific acoustic resistance is found from Eq. (8-171):

$$\psi_1 = \frac{R_1}{2\pi f \rho_m} = \frac{(180{,}000)}{(2\pi)(2000)(100)} = 0.1432$$

The effective density for the gas in the soft blanket material is found from Eq. (8-170):

$$\rho_e = \frac{(1.084)(1.1833)\{1 + [0.96 + (100/1.2827)](0.1432)^2\}}{1 + (0.1432)^2}$$

$$\rho_e = (1.2827)(2.5666) = 3.292 \text{ kg/m}^3 \ (0.2055 \text{ lb}_m/\text{ft}^3)$$

The frequency of 2 kHz is greater than 1000 Hz, so the effective elastic constant for the gas in the lining material is found from Eq. (8-176) for the adiabatic case:

$$\kappa = \gamma P_o = (1.40)(140) = 196 \text{ kPa}$$

The effective specific acoustic resistance for the soft blanket material is found from Eq. (8-178):

$$R_e = \frac{R_1}{1 + \psi_1^2} = \frac{(180{,}000)}{1 + (0.1432)^2} = 176{,}380 \text{ rayl/m}$$

The dimensionless parameter defined by Eq. (8-165) may be calculated:

$$\psi = \frac{R_e}{2\pi f \rho_e} = \frac{(180{,}000)}{(2\pi)(2000)(3.292)} = 4.351$$

The cross-sectional area and perimeter of the muffler chamber may be determined:

$$\frac{S_2}{P_w} = \frac{\tfrac{1}{4}\pi D_2^2}{\pi D_2} = \tfrac{1}{4} D_2 = (\tfrac{1}{4})(1.500) = 0.375 \text{ m}$$

The attenuation coefficient, uncorrected for end effects, may now be found from Eq. (8-164):

$$\sigma = \frac{(\pi)(2000)}{(2)(0.375)} \left[\frac{(3.292)(0.960)}{(2)(196)(10^3)}\right]^{1/2} [(1 + 4.351^2)^{1/2} - 1]^{1/2}$$

$$\sigma = (8378)(0.002839)(1.8613) = 44.275 \text{ Np/m}$$

The wavelength of the sound wave in the muffler is found as follows:

$$\lambda = c/f = (425)/(2000) = 0.2125 \text{ m} \quad (8.37 \text{ in})$$

Let us calculate the dimensionless parameter involved in the end correction:

$$(S_2)^{1/2}/\lambda = [(\tfrac{1}{4}\pi)(1.500)^2]^{1/2}/(0.2125) = (1.329)/(0.2125) = 6.26 > 1$$

Silencer Design

The end correction may be evaluated from Eq. (8-183c):

$$\Delta(\sigma L) = 1.1513$$

The attenuation coefficient, corrected for end effects, may be evaluated from Eq. (8-182):

$$\sigma = 44.275 + (1.1513)/(4.50) = 44.275 + 0.256 = 44.531 \text{ Np/m}$$

The attenuation coefficient may also be expressed in decibel units:

$$8.6859\sigma = 386.8 \text{ dB/m}$$

Let us evaluate the attenuation coefficient from Eq. (8-166) for comparison. The density and sonic velocity for air at 300K and 101.3 kPa are $\rho_o = 1.177 \text{ kg/m}^3$ and $c_o = 347.2 \text{ m/s}$. The dimensionless parameter defined by Eq. (8-167) may be calculated:

$$\psi_o = \frac{(180,000)(425/300)^{-1.65}}{(2\pi)(2000)(1.177)} = 6.850$$

If we take the constants from Table 8-3 for mineral wool, we may evaluate the following term, for $\psi_o > 6.4$:

$$a\psi_o^b = (0.810)(6.850)^{0.502} = 2.128$$

The attenuation coefficient may now be calculated from Eq. (8-166):

$$\sigma = \frac{(\pi)(2000)(2.128)}{(2)(0.375)(347.2)} = 53.29 \text{ Np/m}$$

This value is about 20% larger than the value calculated through the more exact procedure (44.53 Np/m). This difference is probably reasonable in view of the scatter of experimental data on the attenuation coefficient parameters.

8.8 COMMERCIAL SILENCERS

Commercial silencers generally use designs that combine features from basic side-branch, expansion chamber, and dissipative silencers. One example of an industrial silencer is shown in Fig. 8-19. The silencer involves two expansion chambers connected by a double-tube arrangement. Absorbent material is provided to attenuate the noise in the higher frequency range.

Some automotive silencers use a two-pass internal tube arrangement, in which the inlet and outlet tubes are extended into the expansion volume, and several holes are placed in the inlet and outlet tubes, as shown in Fig.

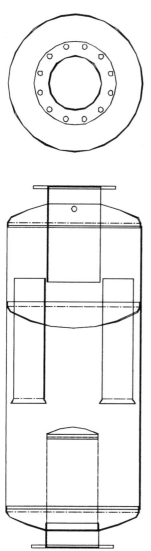

FIGURE 8-19 Industrial silencer. (By permission on MAXIM Silencer Engineering, Beaird Industries, Shreveport, LA.)

Silencer Design

8-20. This design involves both the side-branch principle and the wave-reflection principle discussed in previous sections.

Some geometries for commercial panel-type silencers are shown in Fig. 8-21. These units are primarily dissipative-type silencers. Water-filled tubes have been included within the open spaces of the silencer to provide "heat recovery" capabilities for the silencer, as well as attenuation of the noise from a gas turbine located upstream of the silencer.

8.9 PLENUM CHAMBERS

A *plenum chamber* is similar to a dissipative silencer, but it is also similar to a small room with a noise source in one wall. The dimensions of the plenum chamber are usually larger than the wavelength of the sound being attenuated. In addition, the inlet and outlet of the plenum chamber are usually not placed on the same axis, which is often the case for dissipative silencers.

One of the applications for plenum chambers is to smooth out flow fluctuations and poor velocity distribution after a fan or blower in an air distribution system, in addition to attenuation of the noise generated by the fan or blower.

The results of experimental measurements of plenum chamber acoustic performance are available in the literature (Wells, 1958; Benade, 1968). The following analysis yields expressions that predict acoustic performance in good agreement with experimental data for wavelengths smaller than the dimensions of the chamber. At low frequencies (wavelengths much larger than the chamber dimensions), the calculated transmission loss may be 5–10 dB smaller than experimental data, i.e., the calculations are "conservative" as far as the noise control situation is concerned. When the "room

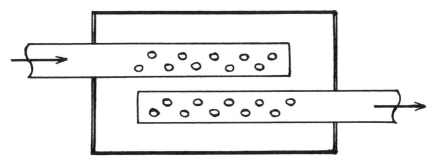

FIGURE 8-20 Schematic of a silencer for automotive applications.

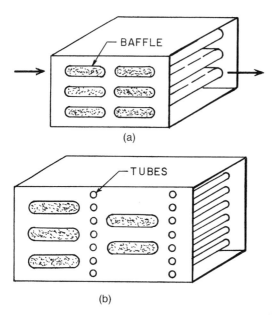

FIGURE 8-21 Baffle-type or panel-type silencers: (a) in-line panels and (b) staggered panels with heat-recovery water-filled tubes.

constant" for the plenum is very large (or the surface absorption coefficient for the lining approaches unity), the transmission loss for the plenum is determined primarily by the direct acoustic energy from the inlet to the outlet.

Let us consider the plenum chamber shown in Fig. 8-22. The acoustic intensity associated with the inlet energy W_{in} that is radiated directly from the inlet (area S_o) to the outlet is given by the following expression:

$$I_D = \frac{QW_{in}}{4\pi d^2} \tag{8-184}$$

The directivity factor $Q = 2$, if the inlet opening is located near the center of the inlet side of the plenum chamber. If the inlet opening is located near the top or bottom edge of the inlet side of the plenum chamber, $Q = 4$. The quantity d is the slant distance between the center of the inlet and center of the outlet opening of the plenum chamber. For the plenum chamber shown in Fig. 8-22, the distance d is given by the following expression:

$$d = [(L-h)^2 + H^2]^{1/2} \tag{8-185}$$

Silencer Design

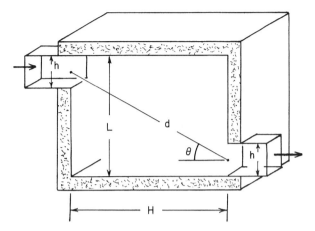

FIGURE 8-22 Configuration for a plenum chamber.

The quantity L is the vertical dimension of the plenum chamber (if the openings are located in the vertical sides of the chamber), H is the horizontal distance between the inlet and outlet openings, and h is the height (or diameter, if circular) of the openings.

An alternative arrangement for the inlet and outlet openings for a plenum chamber is shown in Fig. 8-23. For this case, the angle θ is given by the following expression:

$$\cos\theta = L_1/d \tag{8-186}$$

The dimension L_1 is the distance between the center of the inlet opening and the chamber side in which the outlet opening is located, and d is the slant distance between the two openings.

The acoustic energy $W_{\text{out,D}}$ associated with the direct sound field at the outlet is found by multiplying the intensity by the area of the outlet opening projected in the direction of the inlet opening:

$$W_{\text{out,D}} = \frac{QW_{\text{in}}S_o\cos\theta}{4\pi d^2} \tag{8-187}$$

The angle θ is defined in Fig. 8-22, and is given by the following expression:

$$\cos\theta = H/d \tag{8-188}$$

The acoustic energy density associated with the reverberant sound field in the plenum chamber may be determined from Eq. (7-12):

$$D_R = \frac{4W_{\text{in}}}{cR} \tag{8-189}$$

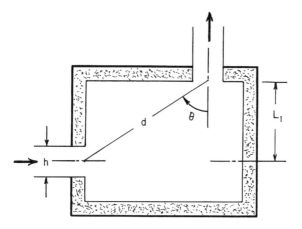

FIGURE 8-23 Plenum chamber with an alternative location for the outlet.

The quantity R is the room constant for the plenum chamber:

$$R = \frac{S\bar{\alpha}}{1-\bar{\alpha}} \tag{8-190}$$

The surface area S is the total surface area of the chamber, including the lined surface area S_L and the area of each opening S_o:

$$S = S_L + 2S_o \tag{8-191}$$

The average surface absorption coefficient $\bar{\alpha}$ may be determined from the following expression, assuming that the absorption coefficient for the openings is unity:

$$\bar{\alpha} = \frac{\bar{\alpha}_L S_L + 2S_o}{S} \tag{8-192}$$

The sound power at the outlet opening associated with the reverberant sound field is given by Eq. (7-10):

$$W_{\text{out},R} = \tfrac{1}{4} D_R c S_o = W_{\text{in}} S_o / R \tag{8-193}$$

The total energy leaving the plenum chamber is the sum of the direct and reverberant components from Eqs (8-187) and (8-193):

$$W_{\text{out}} = W_{\text{in}} S_o \left(\frac{Q \cos \theta}{4 \pi d^2} + \frac{1}{R} \right) \tag{8-194}$$

Silencer Design

The sound pressure transmission coefficient may be determined from Eq. (8-194):

$$a_t = \frac{W_{out}}{W_{in}} = \frac{S_o Q \cos \theta}{4\pi d^2} + \frac{S_o}{R} \tag{8-195}$$

The transmission loss for the chamber is given by the following expression:

$$\text{TL} = 10 \log_{10}(1/a_t) \tag{8-196}$$

The first term in Eq. (8-194) represents the energy transmitted directly from the inlet to the outlet opening of the plenum chamber. The second term represents the energy from the reverberant field within the chamber.

A double-chamber plenum, as shown in Fig. 8-24, may be used to reduce the effect of the direct sound transmission. If we denote the first or inlet section by subscript 1 and the second or outlet section by subscript 2, the sound power transmission coefficient for the double-chamber plenum may be estimated from Eq. (7-69) with the transmission coefficient of the opening between the two chambers taken as unity and $S_w = S_{o1}$:

$$a_t = \frac{W_{out}}{W_{in}} = \frac{S_{o1}}{R_1}\left(\frac{S_{o2} Q_2 \cos \theta_2}{4\pi d_2^2} + \frac{4 S_{o2}}{R_2}\right) \tag{8-197}$$

The quantities S_{o1} and S_{o2} are the outlet areas for each chamber and R_1 and R_2 are the room constants for each chamber, given by Eq. (8-190).

Example 8-12. A plenum chamber has dimensions of 0.800 m (31.5 in) × 0.800 m × 1.600 m (63.0 in) long. The inlet and outlet ducts have a height of $h = 300$ mm (11.8 in) and a width of 400 mm (15.75 in). The inlet and outlet duct openings are located along the edge of the plenum ($Q = 4$), as shown in Fig. 8-22. The plenum chamber is lined with 1-in thick acoustic foam, the

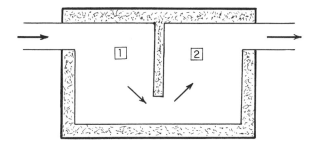

FIGURE 8-24 Double-chamber plenum.

properties of which are given in Appendix D. Determine the transmission loss for the plenum chamber.

The slant distance between the inlet and outlet openings is as follows:

$$d^2 = (L-h)^2 + H^2 = (0.800 - 0.300)^2 + 1.600^2 = 2.810 \, \text{m}^2$$

$$d = 1.676 \, \text{m} \quad (5.50 \, \text{ft})$$

The cosine of the angle between the slant distance and the normal to the outlet opening is as follows:

$$\cos\theta = \frac{H}{d} = \frac{1.600}{1.676} = 0.9545 \quad \text{or} \quad \theta = 17.35°$$

The areas of the openings and the lined portion of the chamber are as follows:

$$S_o = (0.300)(0.400) = 0.120 \, \text{m}^2 \quad (1.292 \, \text{ft}^2)$$

$$S_L = (2)(0.800 + 0.800)(1.600) + (2)(0.800)(0.800) - (2)(0.120)$$

$$S_L = 6.160 \, \text{m}^2 \quad (66.31 \, \text{ft}^2)$$

Let us work out the transmission loss for the octave band with a center frequency of 500 Hz. The surface absorption coefficient for the acoustic foam at 500 Hz is 0.51. The average surface absorption coefficient in the 500 Hz octave band for the plenum is found using Eq. (8-192):

$$\bar{\alpha} = \frac{(6.160)(0.51) + (2)(0.120)}{(6.400)} = 0.5284$$

The room constant for the plenum is found from Eq. (8-190):

$$R = \frac{(6.400)(0.5284)}{1 - 0.5284} = 7.170 \, \text{m}^2$$

The sound power transmission coefficient for the 500 Hz octave band is determined from Eq. (8-195):

$$a_t = \frac{(0.120)}{(7.170)} + \frac{(4)(0.9545)(0.120)}{(4\pi)(1.676)^2} = 0.01674 + 0.01297 = 0.02971$$

The transmission loss for the plenum in the 500 Hz octave band is as follows:

$$TL = 10\log_{10}(1/0.02971) = 15.3 \, \text{dB}$$

The calculations may be repeated for the other octave bands. The results are summarized in Table 8-5.

Silencer Design

TABLE 8-5 Data for Example 8-12

	Octave band center frequency, Hz					
	125	250	500	1,000	2,000	4,000
α_L	0.16	0.28	0.51	0.78	0.99	0.99
$\bar{\alpha}$, average	0.1915	0.3070	0.5284	0.7883	0.9904	0.9904
R, m²	1.516	2.835	7.170	23.824	658.54	658.54
a_t	0.0921	0.0553	0.02971	0.01801	0.01316	0.01316
TL, dB	10.4	12.6	15.3	17.4	18.8	18.8

PROBLEMS

8-1. A Helmholtz resonator is constructed of a thin-walled sphere, having an inside diameter of 101.6 mm (4.00 in). The effect of the thickness of the sphere on the acoustic mass may be neglected. What diameter hole should be drilled in the sphere if it is to resonate at a frequency of 320 Hz? The gas in the resonator is air at 300K (80°F) and 101.3 kPa (density, 1.177 kg/m³ = 0.0735 lb$_m$/ft³; sonic velocity, 347.2 m/s = 1139 ft/sec; specific heat ratio, 1.400). What would be the resonant frequency if a hole of twice the diameter as in the previous case were used, instead of the original hole?

8-2. An acoustic wall treatment consists of cinder blocks with a slot cut into the side of each block. The cavity within the block has dimensions of 101.6 mm (4.00 in) × 254 mm (10.00 in) × 152.4 mm (6.00 in). The slot dimensions are 9.53 mm (0.375 in) wide and 114.3 mm (4.50 in) long, and the slot thickness is 25.4 mm (1.00 in). The slot may be treated as a tube with an equivalent radius of $a = (5wt/6\pi)^{1/2}$, where w is the length of the slot and t is the width of the slot. The cavity is filled with air at 300K (80°F) and 101.3 kPa (14.7 psia), for which the density is 1.177 kg/m³ (0.0735 lb$_m$/ft³) and the sonic velocity is 347.2 m/s (1139 ft/sec). Treating the cavity as a Helmholtz resonator, determine the resonant frequency for the cavity.

8-3. The development of the expression for the acoustic compliance of a volume of gas, Eq. (8-26), involves the condition that the container is "rigid," or the container does not deform as the pressure within the container changes. The change of radius for a thin-walled spherical shell of initial radius b and wall thickness h is given by the following expression:

$$\frac{dr}{dt} = \frac{(1-\sigma)b^2}{2Eh}\frac{dp}{dt}$$

The quantity σ is Poisson's ratio for the shell material and E is Young's modulus. The change in internal volume of the shell is given by:

$$\frac{dV}{dt} = 4\pi b^2 \frac{dr}{dt}$$

Using these expressions in connection with Eq. (8-26) and the conservation of mass principle, show that the additional acoustic compliance due to the deformation of the spherical container of volume V is given by the following expression:

$$\Delta C_A = \frac{3(1-\sigma)bV}{2Eh}$$

The total acoustic compliance is $[C_A$ (Eq. 8-26) $+\Delta C_A]$. Determine the numerical value for the additional acoustic compliance ΔC_A, if the sphere has a radius of 150 mm (5.91 in) and a wall thickness of 0.51 mm (0.020 in). The sphere material is aluminum, for which Young's modulus is 73.1 GPa (10.6×10^6 psi) and Poisson's ratio is 0.33. The volume of the sphere is given by:

$$V = \tfrac{4}{3}\pi b^3$$

Compare this value of the additional acoustic compliance with the acoustic compliance given by Eq. (8-26), if the gas in the sphere is air at 300K and 101.3 kPa (density, 1.177 kg/m^3 = 0.0735 lb$_m$/ft^3; sonic velocity, 347.2 m/s = 1139 ft/sec). Is the effect of the flexibility of the container significant in this case?

8-4. An automobile with one window partially open may be treated as a Helmholtz resonator for relatively low frequencies. If the internal volume of the passenger compartment for an automobile is 1.60 m^3 (56.5 ft^3), and the window opening is 250 mm × 600 mm (9.84 in × 23.62 in) × 35 mm (1.38 in) thick, determine the resonant frequency for the automobile treated as a Helmholtz resonator. The air in the automobile is at 305K (31.8°C or 89.2°F) and 101.3 kPa (14.7 psia), for which the density is 1.158 kg/m^3 (0.0723 lb$_m$/ft^3) and the sonic velocity is 350 m/s (1148 ft/sec). The acoustic mass for the rectangular opening having a width w and a length t, with a thickness L, is given by the following expression:

$$M_A = \frac{\rho_o L_e}{wt}$$

Silencer Design

The equivalent length is $L_e = L + 2\Delta L$. For a rectangular opening, the additional length ΔL due to the end effects is given by the following expression:

$$\Delta L = \frac{8wt[1 + (w/t) + (t/w)]}{9\pi(w+t)}$$

Also, determine the acoustic quantity factor Q_A and the half-power bandwidth Δf for the automobile as a Helmholtz resonator around the resonant frequency. The acoustic resistance for the thin rectangular opening may be estimated from the following expression for $(k^2 wt) < 1$:

$$R_A = \frac{\pi^2 f^2 \rho_o [(w/t) + (t/w)]}{4c}$$

For $(k^2 wt) > 1$, $R_A = \rho_o c/wt$. The quantity k is the wave number.

8-5. A rigid-wall back-enclosed loudspeaker cabinet has inside dimensions of 300 mm (11.81 in) × 500 mm (19.69 in) × 400 mm (15.75 in). The front of the cabinet has a circular hole cut into it. The diameter of the hole is 50 mm (1.969 in), and the thickness of the panel is 60 mm (2.362 in). The gas in the cabinet is air at 22°C (295.2K or 72°F) and 101.3 kPa (14.7 psia), for which the density is 1.196 kg/m³ (0.0747 lb$_m$/ft³), and the sonic velocity is 344.4 m/s (1130 ft/sec). A single layer of 200-mesh screen having a specific acoustic resistance (resistance for a unit area) of 24.6 rayl (total acoustic resistance, 12,530 acoustic ohms) is placed over the opening in the cabinet. This screen provides the primary acoustic resistance (other fluid frictional resistances may be neglected). A sound wave having a sound pressure level of 80 dB is incident on the opening. Determine (a) the resonant frequency for the cabinet, considered as a Helmholtz resonator; (b) the acoustic quality factor Q_A for the cabinet; (c) the acoustic power delivered to the air in the cabinet at the resonant frequency; (d) the acoustic power delivered to the air in the cabinet at 63 Hz; and (e) the sound pressure level gain, dB, for the cabinet at the resonant frequency and at 63 Hz.

8-6. A ventilating duct has a square cross section with dimensions 300 mm × 300 mm (11.81 in), with a thickness of 1.65 mm (0.065 in). A side-branch muffler is constructed by drilling a hole 160 mm (6.30 in) in diameter in one wall of the duct and surrounding the duct with a closed chamber. The gas flowing through the duct is air at 40°C (313.2K or 104°F) and 102 kPa (14.79 psia), for which the density is 1.135 kg/m³ (0.0709 lb$_m$/ft³) and the sonic velocity is 354.7 m/s (1164 ft/sec). Determine the volume of the closed chamber such that the muffler will resonate at a frequency of 30 Hz, the transmission loss for the muffler at 30 Hz, and the transmission loss for the muffler at 60 Hz.

8-7. It is desired to design a side-branch muffler for a minimum transmission loss of 6 dB for the frequency range between 200 Hz and 400 Hz. The gas within the system is air at 600K (327°C or 620°F) and 101.3 kPa (14.7 psia), for which the density is 0.588 kg/m³ (0.0367 lb$_m$/ft³) and the sonic velocity is 491 m/s (1611 ft/sec). The main tube in which the air is flowing has a diameter of 50.8 mm (2.00 in). Two side-branch tubes are to be used, with each tube having a length equal to 4 times the tube radius ($L = 4a$). For the resonator, a cylindrical vessel is to be used, with its diameter equal to its length. Layers of 200-mesh wire screen are to be used as the acoustic resistance element. Determine the length and diameter of the resonator chamber, the length and diameter of each side-branch tube, and the number of layers of wire screen required in each side-branch tube.

8-8. A side–branch muffler is to be constructed of a hollow spherical container connected to the main tube by a tube having a length-to-diameter ratio ($L/2a$) of 0.1875. The inside diameter of the main tube is 30 mm (1.181 in). The significant acoustic resistance is provided by an acoustic material for which the specific acoustic resistance is given by $R_s = C_1 t$, where t is the material thickness and $C_1 = 30{,}000$ rayl/m. The side-branch muffler is to have a minimum transmission loss of 11 dB in the frequency range between 150 Hz and 416.7 Hz. The gas flowing through the system is air at 350K (77°C or 170°F) and 110.5 kPa (16.03 psia), for which the density is 1.100 kg/m³ (0.0687 lb$_m$/ft³) and the sonic velocity is 375 m/s (1230 ft/sec). Determine the diameter of the spherical resonator, the diameter of the side-branch tube, and the thickness of the acoustic material in the side-branch tube.

8-9. A side-branch muffler is attached to a duct with a square cross section 200 mm (7.874 in) × 200 mm. The resonator volume has a square cross section, 200 mm × 200 mm, and a length L_1. The connecting tube is a circular tube with a diameter of 32 mm (1.26 in) and a length of 32 mm. One end of the connecting tube is attached to the main duct (flanged end) and the other end protrudes into the resonator volume (free end), as shown in Fig. 8-25. Wire screens are placed over the connecting tube opening to provide an acoustic resistance of 555 Pa-s/m³. The resonant frequency for the system is 43.3 Hz. Determine the length of the resonator chamber and the transmission loss for the system at a frequency of 34.5 Hz, for the

Silencer Design

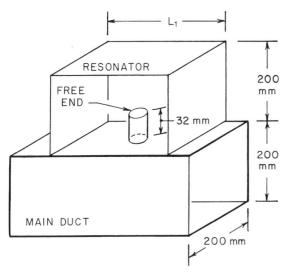

FIGURE 8-25 Sketch for Problem 8-9.

gas density of 1.143 kg/m^3 ($0.0714 \text{ lb}_m/\text{ft}^3$) and sonic velocity of 350 m/s (1148 ft/sec).

8-10. A side-branch muffler is attached to a pipe having an inside diameter of 102.3 mm (4.028 in) by using a tube having a length of 32.18 mm (1.267 in) and an inside diameter of 35 mm (1.378 in). The resonator volume is 1.50 dm^3, and the acoustic resistance in the connecting tube is 8422 Pa-s/m^3. The gas in the system is air at 302.8 K (29.6°C or 85.3°F) and 101.3 kPa (14.7 psia), for which the density is 1.166 kg/m^3 ($0.0728 \text{ lb}_m/\text{ft}^3$) and the sonic velocity is 348.8 m/s (1144 ft/sec). Determine the transmission loss for the system at a frequency of 200 Hz.

8-11. A water pipe has an inside diameter of 26.6 mm (1.047 in). A closed pipe having the same diameter is connected through a tee to the main pipe to act as a side-branch muffler. The water in the pipe is at 20°C (68°F), for which the density is 998.2 kg/m^3 ($62.32 \text{ lb}_m/\text{ft}^3$), the sonic velocity is 1483.2 m/s (4866 ft/sec), and the viscosity is 1.00 mPa-s ($2.42 \text{ lb}_m/\text{ft-hr} = 20.9 \times 10^{-6} \text{ lb}_f\text{-sec/ft}^2$). The only acoustic resistance is provided by the frictional energy dissipation in the side pipe. Determine the smallest length of the side pipe to achieve resonance at 125 Hz. What is the transmission loss for the side-branch muffler at 125 Hz and at 120 Hz?

8-12. A 21.3 mm (0.839 in) diameter hole is drilled in a pipe having an inside diameter of 40.9 mm (1.610 in) and a wall thickness of 3.7 mm (0.146 in). The gas in the pipe is air at 305K (31.8°C or 89.2°F) and 101.3 kPa (14.7 psia), for which the density is 1.158 kg/m^3 (0.0723 lb$_m$/ft^3) and the sonic velocity is 350 m/s (1148 ft/sec). The opening is open to atmospheric air. Determine the transmission loss for the hole as a side-branch muffler for a frequency of 500 Hz and for 1000 Hz.

8-13. Determine the length and diameter of the expansion chamber for a non-dissipative expansion chamber muffler for use in a marine propulsion system, in which the gas flowing through the muffler is air at 400K (126.8°C or 260.2°F) and 105 kPa (15.23 psia). The density of the gas is 0.915 kg/m^3 (0.0571 lb$_m$/ft^3), and the sonic velocity is 400.9 m/s (1315 ft/sec). The muffler must have a minimum transmission loss of 16.9 dB for the frequency range between 200 Hz and 400 Hz. The inlet and outlet tubes connecting the expansion chamber each have the same diameter of 54 mm (2.126 in).

8-14. A single-chamber expansion chamber muffler has an inside diameter of 108.2 mm (4.260 in) and a length of 1.524 m (5.00 ft). The inlet and outlet tubes each have an inside diameter of 22.5 mm (0.886 in). The gas flowing in the system is air at 80°C (353.2K or 176°F) and 101.4 kPa (14.71 psia), for which the density is 1.00 kg/m^3 (0.06243 lb$_m$/ft^3) and the sonic velocity is 376.7 m/s (1236 ft/sec). Determine the transmission loss for the muffler for a frequency of 62 Hz.

8-15. A single-chamber expansion chamber muffler has a length of 133.3 mm (5.248 in) and an expansion chamber diameter of 54.0 mm (2.127 in). The main tube in which the muffler is placed has an inside diameter of 18.0 mm (0.709 in). The gas in the system is air at 398.2K (125°C or 257°F) and 120 kPa (17.40 psia), for which the density is 1.050 kg/m^3 (0.0656 lb$_m$/ft^3). Determine the frequency at which the transmission loss is 12 dB for the muffler. Also, determine the frequency at which the transmission loss is zero for the muffler.

8-16. The inlet and outlet tubes for a double-chamber expansion chamber muffler have an inside diameter of 55 mm (2.165 in), and each expansion chamber has an inside diameter of 220 mm (8.661 in), and length of 580 mm (22.83 in). The connecting tube between the two expansion chambers has an inside diameter of 55 mm (2.165 in) and a length of 440 mm (17.32 in). The gas flowing through the muffler is air at 398.2K (125°C or 257°F) and 104.57 kPa (15.17 psia), for which the density is 0.915 kg/m^3 (0.0571 lb$_m$/ft^3) and the sonic velo-

Silencer Design 403

city is 400.0 m/s (1312 ft/sec). For a frequency of 125 Hz. Determine the transmission loss for (a) an internal-connecting tube muffler and (b) an external-connecting tube muffler.

8-17. A dissipative muffler used on an aircraft engine test stand has a length of 4.500 m (14.76 ft) and a diameter of 1.500 m (4.92 ft). The lining of the muffler has an attenuation coefficient of $\sigma = 0.50$ Np/m or $(8.6859\sigma) = 4.343$ dB/m. The gas flowing through the muffler is air at 896.0K (622.8°C or 1153°F) and 140 kPa (20.31 psia), for which the density is 0.544 kg/m^3 (0.0340 lb$_m$/ft^3) and the sonic velocity is 600.0 m/s (1969 ft/sec). The inside diameter of the inlet and outlet tubes is 500 mm (19.685 in). Determine the transmission loss for the muffler at a frequency of 3 kHz.

8-18. For a dissipative muffler, the muffler length is 760 mm (29.92 in) and the inside diameter of the muffler is 202.7 mm (7.980 in). The inlet and outlet pipes for the muffler also have inside diameters of 202.7 mm. The attenuation coefficient for the muffler lining material is $\sigma = 7.676$ Np/m or $(8.6859\sigma) = 66.67$ dB/m. The gas flowing through the muffler is air at 340.7K (67.5°C or 153.5°F) and 110.0 kPa (15.95 psia), for which the density is 1.125 kg/m^3 (0.0702 lb$_m$/ft^3) and the sonic velocity is 370.0 m/s (1214 ft/sec). Determine the transmission loss for the muffler at a frequency of 1000 Hz.

8-19. A dissipative muffler involves flat panel baffle elements with each panel having a length of $L = 3.000$ m (9.84 ft), a width of 1.224 m (4.02 ft), and a thickness of 102 mm (4.02 in). The spacing between the baffles is $b = 102$ mm (4.02 in). The cross section of the duct before and after the muffler is 1.224 m (4.02 ft) × 1.224 m. The attenuation coefficient, corrected for random incidence end effects, for the muffler dissipative panels is given by the following relationship:

$$\sigma = \frac{0.4636}{b}\left\{\frac{[1+3(f/f_m)^2]^{1/2}-1}{1+3(f/f_m)^2}\right\}^{1/2} + \frac{\Delta(\sigma L)}{L}$$

The quantity $f_m = 1000$ Hz, and f is the frequency. For the calculation of the random incidence end effects, use the flow area between a pair of panels, $S^* = (1.224)(0.102) = 0.1248$ m^2 (1.343 ft^2), or $(S^*)^{1/2} = 0.353$ m (1.159 ft). The total free-flow area within the muffler is $S_2 = 6S^* = 0.7488$ m^2 (8.060 ft^2). The gas flowing through the muffler is air at 398.2K (125°C or 257°F) and 105.1 kPa (15.24 psia), for which the density is 0.920 kg/m^3 (0.0574 lb$_m$/ft^3) and the sonic velocity is

400 m/s (1312 ft/sec). Determine the transmission loss for the muffler at a frequency of 2000 Hz.

8-20. A dissipative muffler is lined with a soft blanket-type acoustic material having a mean density of 128 kg/m^3 (7.99 lb$_m$/ft^3), a specific acoustic resistance of $R_1 = 200{,}000$ rayl/m, and a porosity of 0.950. The gas flowing through the lining material is air at 90°C (563.2K or 194°F) and 103.4 kPa (15.0 psia), for which the density is 0.640 kg/m^3 (0.0400 lb$_m$/ft^3), the sonic velocity is 475.7 m/s (1561 ft/sec), and the specific heat ratio is $\gamma = 1.400$. The length of the muffler lining is 4.500 m (14.76 ft), and the inside diameter of the lining is 1.500 m (4.92 ft). Determine the corrected attenuation coefficient for the material at a frequency of 500 Hz and a frequency of 3 kHz.

8-21. A dissipative muffler has internal cross-section dimensions of 800 mm (31.50 in) × 250 mm (9.84 in), with a length of 2.40 m (7.874 ft). The lining material is a semirigid material having a mean density of 175 kg/m^3 (10.93 lb$_m$/ft^3), a porosity of 0.900, and a specific acoustic resistance of 49,480 rayl/m. The gas flowing in the muffler is air at 304.9K (31.7°C or 89.1°F) and 105.0 kPa (15.23 psia), for which the density is 1.200 kg/m^3 (0.0749 lb$_m$/ft^3), the sonic velocity is 350 m/s (1148 ft/sec), and the specific heat ratio is 1.400. Determine the corrected attenuation coefficient for the lining material at a frequency of 2000 Hz and at 500 Hz.

8-22. A plenum chamber, similar to that shown in Fig. 8-22, has internal dimensions of 3.048 m (10.000 ft) high, 1.219 m (4.000 ft) wide, and 1.905 m (6.250 ft) long. The inlet and outlet openings for the plenum are rectangular, with dimensions of 610 mm (24.0 in) high × 915 mm (36.0 in) wide. The inlet is located at the top of the 3.048 m × 1.219 m side, and the outlet is located at the bottom of the opposite side of the plenum. The plenum chamber is lined with an acoustic material having a surface absorption coefficient of 0.780. Determine the transmission loss for the plenum chamber.

8-23. A plenum chamber, similar to that shown in Fig. 8-22, has a height of 1.351 m (4.432 ft), a width of 2.100 m (6.890 ft), and a length of 1.600 m (5.249 ft). The inlet is located at the upper edge of the 1.351 m × 2.100 m wall, and the outlet is located at the lower edge of the opposite wall. The dimensions of the inlet and outlet openings are 150 mm (5.906 in) high × 201 mm (7.913 in) wide. Determine the required surface absorption coefficient of the plenum acoustic lining material to achieve a transmission loss of 25.0 dB for the plenum chamber. What would the transmission loss be for the plenum cham-

Silencer Design

ber if the lining material had a surface absorption coefficient of unity?

8-24. A double-chamber plenum, similar to that shown in Fig. 8-24, has a height of 2.00 m (6.562 ft), a width of 1.225 m (4.019 ft), and each chamber has a length of 0.900 m (2.953 ft). Each of the three openings has a height of 200 mm (7.87 in) and a width of 250 mm (9.84 in). The inlet of the first chamber is located at the upper edge of the 2 m × 1.225 m wall, and the opening between the chambers is located at the lower edge of the opposite wall. The outlet of the second chamber is located at the upper edge of the wall opposite the opening between the chambers. The surface absorption coefficient for the plenum lining material is 0.622. Determine the transmission loss for the plenum chamber.

REFERENCES

Benade, A. H. 1968. On the propagation of sound waves in a cylindrical conduit. *J. Acoust. Soc. Am.* 44(2): 616–623.

Beranek, L. L. 1960. *Noise Reduction*, pp. 246–269. McGraw-Hill, New York.

Beranek, L. L. and Vér, I. L. 1992. *Noise and Vibration Control Engineering*, pp. 289–395. John Wiley and Sons, New York.

Davis, D. D., Stokes, G. M., Moore, D., and Stevens, G. L. 1954. Theoretical and experimental investigation of mufflers with comments on engine-exhaust muffler design. NACA Report 1192.

Heywood, J. B. 1988. *Internal Combustion Engine Fundamentals*, pp. 212–215. McGraw Hill, New York.

Howe, M. S. 1976. On the Helmholtz resonator. *J. Sound Vibr.* 13:427–440.

Kinsler, L. E., Frey, A. R., Coppens, A. B., and Sanders, J. V. 1982. *Fundamentals of Acoustics*, 3rd edn, p. 201. John Wiley and Sons, New York.

McQuiston, F. C. and Parker, J. D. 1994. *Heating, Ventilating, and Air Conditioning*, 4th edn, pp. 461–534. John Wiley and Sons, New York.

Munjal, M. 1987. *Acoustics of Ducts and Mufflers*. John Wiley and Sons, New York.

Nichols, R. H. 1947. Flow resistance characteristics of fibrous acoustical materials. *J. Acoust. Soc. Am.* 19: 866–871.

Nilsson, J. W. 1983. *Electric Circuits*. Addison-Wesley, Reading, MA.

Pierce, A. D. 1981. *Acoustics*, pp. 345–348, McGraw-Hill, New York.

Rayleigh, J. W. S. 1929. *The Theory of Sound*, Sect. 348–350. Macmillan, New York.

Reynolds, D. D. 1981. *Engineering Principles of Acoustics*, p. 341. Allyn and Bacon, Boston.

Wells, R. J. 1958. Acoustic plenum chambers. *Noise Control* 4(4): 9–15.

Van Wylen, G., Sonntag, R. and Borgnakke, C. 1994. *Fundamentals of Classical Thermodynamics*, 4th edn, p. 450. John Wiley and Sons, New York.

Zwikker, C. and Kosten, C. W. 1949. *Sound Absorbing Materials*. Elsevier Press, New York.

9
Vibration Isolation for Noise Control

One of the major sources of noise in mechanical equipment is the noise produced by energy radiated from vibrating solid surfaces in the machine. In addition, noise may be produced when vibratory motion or forces are transmitted from the machine to its support structure, through connecting piping, etc. Noise reduction may be achieved by isolating the vibrations of the machine from the connected elements. For the case of noise generated by a vibrating panel on the machine, noise reduction may be achieved by using damping materials on the panel to dissipate the mechanical energy, instead of radiating the energy into the surrounding air. In this chapter, we will consider some of the techniques for vibration isolation for machinery and examine some of the materials used in isolation of vibrations from equipment.

There are at least two types of vibration isolation problems that the engineer may be called upon to solve: (a) situations in which one seeks to reduce forces transmitted from the machine to the support structure and (b) situations in which one seeks to reduce the transmission of motion of the support to the machine. Some examples of the first case include reciprocating engines, fans, and gas turbines. An example of the second case is the mounting of electrical equipment in an aircraft or automobile such that motion of the vehicle is not "fed into" the equipment.

Vibration Isolation for Noise Control

9.1 UNDAMPED SINGLE-DEGREE-OF-FREEDOM (SDOF) SYSTEM

Many aspects of vibration isolation may be understood by examination of an SDOF system consisting of a mass and a linear spring, as shown in Fig. 9-1. For a more extensive treatment of mechanical vibrations, there are several references available in the literature (Rao, 1986; Tongue, 1996; Thomson and Dahleh, 1998). Let us denote the mechanical mass of the system by M and the spring constant (force per unit displacement) by K_S. Using Newton's second law of motion ($F_{net} = Ma$), the equation of motion for the system may be written in the following form:

$$M \frac{d^2 y}{dt^2} + K_S y = 0 \qquad (9\text{-}1)$$

The displacement of the mass from its equilibrium position is denoted by y and the symbol t denotes time.

Let us define the following parameter:

$$\omega_n^2 = \frac{K_S}{M} \qquad (9\text{-}2)$$

This quantity is called *the undamped natural frequency* of the system. As will be shown in this section, ω_n is the frequency at which the system will oscillate after being disturbed from its static equilibrium position by an initial displacement or an initial velocity.

Making the substitution from Eq. (9-2) into Eq. (9-1), we obtain the following result:

$$\frac{d^2 y}{dt^2} + \omega_n^2 y = 0 \qquad (9\text{-}3)$$

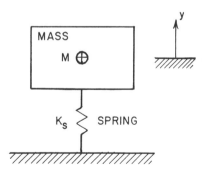

FIGURE 9-1   Undamped SDOF vibrating system.

The general solution of Eq. (9-3) may be written in either complex notation or in terms of trigonometric functions directly:

$$y(t) = A e^{j\omega_n t} + B e^{-j\omega_n t} = C_1 \cos(\omega_n t) + C_2 \sin(\omega_n t) \tag{9-4}$$

The constants of integration in Eq. (9-4) are determined by the initial conditions for the system. For example, suppose we have the following conditions at the initial time $t = 0$:

(a) initial displacement: $y(0) = y_o$
(b) initial velocity:

$$v(0) = \frac{dy(0)}{dt} = v_o$$

If we make these substitutions into Eq. (9-4), the following results are obtained:

$$C_1 = y_o \quad \text{and} \quad C_2 = v_o/\omega_n \tag{9-5}$$

$$A = \tfrac{1}{2}[y_o - j(v_o/\omega_n)] \quad \text{and} \quad B = \tfrac{1}{2}[y_o + j(v_o/\omega_n)] \tag{9-6}$$

The motion of a free undamped SDOF system may be found by substituting the expressions for the constants of integration into Eq. (9-4):

$$y(t) = y_o \cos(\omega_n t) + (v_o/\omega_n) \sin(\omega_n t) \tag{9-7}$$

$$y(t) = \tfrac{1}{2} y_o (e^{j\omega_n t} + e^{-j\omega_n t}) - \tfrac{1}{2} j(v_o/\omega_n)(e^{j\omega_n t} - e^{-j\omega_n t}) \tag{9-8}$$

We note that the two expressions are identical because of the following identities:

$$\cos(\omega_n t) = \tfrac{1}{2}(e^{j\omega_n t} + e^{-j\omega_n t}) \quad \text{and} \quad \sin(\omega_n t) = -\tfrac{1}{2} j(e^{j\omega_n t} - e^{-j\omega_n t})$$

We may write the expression for the motion of the undamped SDOF system in an alternative form:

$$y(t) = C \cos(\omega_n t + \phi) \tag{9-9}$$

Using the trigonometric identity for the cosine of the sum of two angles, Eq. (9-9) may be written as follows:

$$y(t) = C[\cos \phi \cos(\omega_n t) - \sin \phi \sin(\omega_n t)] \tag{9-10}$$

By comparing Eqs (9-10) and (9-7), we note that the following relations exist:

$$C \cos \phi = y_o \quad \text{and} \quad C \sin \phi = -v_o/\omega_n \tag{9-11}$$

Vibration Isolation for Noise Control

The values for the constants C and ϕ may be obtained as follows:

$$C^2(\cos^2\phi + \sin^2\phi) = C^2 = y_0^2 + (v_0/\omega_n)^2 \tag{9-12}$$

$$\tan\phi = \frac{\sin\phi}{\cos\phi} = -\frac{v_0}{\omega_n y_0} \tag{9-13}$$

The final expression for the displacement of the system is as follows:

$$y(t) = [y_0^2 + (v_0/\omega_n)^2]^{1/2} \cos(\omega_n t + \phi) \tag{9-14}$$

We observe from Eq. (9-14) that the motion is sinusoidal or simple harmonic with a frequency (in radians/second, for example) of ω_n. The undamped natural frequency f_n may also be expressed in Hz units:

$$f_n = \frac{\omega_n}{2\pi} = \frac{(K_S/M)^{1/2}}{2\pi} \tag{9-15}$$

The *static deflection* of the system (denoted by d) is the deflection of the spring due to the weight of the attached mass:

$$Mg = K_S d \quad \text{or} \quad d = Mg/K_S \tag{9-16}$$

The quantity g is the acceleration due to gravity. At standard conditions, $g = 9.806 \text{ m/s}^2$ (32.174 ft/sec^2 or 386.1 in/sec^2). By combining Eqs (9-15) and (9-16), we find the relationship between the static deflection and the undamped natural frequency for the system:

$$f_n = \frac{(g/d)^{1/2}}{2\pi} \tag{9-17}$$

In practice, the static deflection may be easily measured, and the undamped natural frequency may then be determined from experimental measurements of the static deflection.

Example 9-1. A machine has a mass of 50 kg (110.2 lb$_m$). It is desired to design the support system such that the undamped natural frequency is 5 Hz. Determine the required spring constant for the support system and the corresponding value of the static deflection.

The undamped natural frequency is given by Eq. (9-15). The required spring constant is found as follows:

$$K_S = M(2\pi f_n)^2 = (50)[(2\pi)(5)]^2$$

$$K_S = 49{,}350 \text{ N/m} = 49.35 \text{ kN/m} \quad (281.8 \text{ lb}_m/\text{in})$$

The static deflection is found from Eq. (9-16):

$$d = \frac{Mg}{K_S} = \frac{(50)(9.806)}{(49.35)(10^3)} = 9.93 \times 10^{-3}\,\text{m} = 9.93\,\text{mm} \quad (0.391\,\text{in})$$

9.2 DAMPED SINGLE-DEGREE-OF-FREEDOM (SDOF) SYSTEM

All mechanical systems have some amount of damping or energy dissipation associated with the motion of the system, so we need to examine the effects of damping on the system vibration. It should be noted that a large amount of damping is not always a good thing to have, especially if one wishes to reduce the force transmitted at frequencies much above the undamped natural frequency for the system.

Let us consider the system shown in Fig. 9-2, in which a mass M is connected to a support through a spring (spring constant K_S) and a viscous damper. The force produced by the viscous damper is proportional to the velocity difference across the damper:

$$F_d(t) = R_M v(t) = R_M \frac{dy}{dt} \quad (9\text{-}18)$$

The quantity R_M is the coefficient of viscous damping or the mechanical resistance, which has units of N-s/m. This combination of units is sometimes called a mechanical ohm, in analogy with the electrical system, i.e., 1 mech ohm = 1 N-s/m.

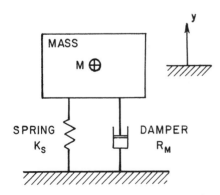

FIGURE 9-2 Damped SDOF vibrating system.

Vibration Isolation for Noise Control

We obtain the following result by applying Newton's second law of motion to the mass shown in Fig. 9-2:

$$M \frac{d^2 y}{dt^2} + R_M \frac{dy}{dt} + K_S y = 0 \qquad (9\text{-}19)$$

If we divide through by the mass M and introduce the undamped natural frequency from Eq. (9-2), we obtain the following result:

$$\frac{d^2 y}{dt^2} + \frac{R_M}{M} \frac{dy}{dt} + \omega_n^2 y = 0 \qquad (9\text{-}20)$$

The general solution for Eq. (9-20) is as follows:

$$y(t) = A e^{s_1 t} + B e^{s_2 t} \qquad (9\text{-}21)$$

The quantities s_1 and s_2 are given by:

$$s_1 = -(R_M/2M) + [(R_M/2M)^2 - \omega_n^2]^{1/2} \qquad (9\text{-}22)$$

$$s_2 = -(R_M/2M) - [(R_M/2M)^2 - \omega_n^2]^{1/2} \qquad (9\text{-}23)$$

There are three difference cases as far as the vibratory motion of the mass with damping is concerned. These cases depend on the nature of the second term in Eqs (9-22) and (9-23).

9.2.1 Critically Damped System, $(R_M/2M) = \omega_n$

The value of the damping coefficient in this case is called the *critical damping coefficient* $R_{M,cr}$:

$$R_{M,cr} = 2M\omega_n = 2M(K_S/M)^{1/2} = 2(K_S M)^{1/2} \qquad (9\text{-}24)$$

For any condition, the *damping ratio* ζ is defined by the following ratio:

$$\zeta = \frac{R_M}{R_{M,cr}} = \frac{R_M}{2(K_S M)^{1/2}} \qquad (9\text{-}25)$$

The factors from Eqs (9-22) and (9-23) may be expressed in terms of the damping ratio:

$$s_1 = -[\zeta - (\zeta^2 - 1)^{1/2}]\omega_n \qquad (9\text{-}26)$$

$$s_2 = -[\zeta + (\zeta^2 - 1)^{1/2}]\omega_n \qquad (9\text{-}27)$$

For the case of critical damping, $\zeta = 1$, we obtain repeated solutions for the differential equation, Eq. (9-20), or $s_1 = s_2 = -\omega_n$. For this situation, the general solution has the following form:

$$y(t) = (C_1 + C_2 t) e^{-\omega_n t} \qquad (9\text{-}28)$$

In particular, if the initial displacement is $y(0) = y_o$ and the initial velocity is $v(0) = v_o$, we may evaluate the constants of integration:

$$C_1 = y_o \quad \text{and} \quad C_2 = v_o + y_o \omega_n \qquad (9\text{-}29)$$

The motion of the mass is described by the following relationship:

$$y(t) = [y_o + (v_o + y_o \omega_n)t] e^{-\omega_n t} \qquad (9\text{-}30)$$

Equation (9-30) shows that, for the case of a free critically-damped system, there is no oscillatory motion. The system simply moves somewhat slowly back to its static equilibrium position.

9.2.2 Over-Damped System, $\zeta > 1$

In this case, the second term in Eqs (9-22) and (9-23) is real and negative, and not imaginary, so there is no free oscillation. The general solution for the motion of the over-damped system may be written in terms of the damping ratio as follows:

$$y(t) = A \exp\{-[\zeta - (\zeta^2 - 1)^{1/2}]\omega_n t\} + B \exp\{-[\zeta + (\zeta^2 - 1)^{1/2}]\omega_n t\} \qquad (9\text{-}31)$$

If the initial displacement is $y(0) = y_o$ and the initial velocity is $v(0) = v_o$, we may find the following expressions for the constants of integration for the over-damped system, $\zeta > 1$:

$$A = -\frac{v_o + y_o[\zeta - (\zeta^2 - 1)^{1/2}]\omega_n}{[2(\zeta^2 - 1)^{1/2}]\omega_n} \qquad (9\text{-}32)$$

$$B = \frac{v_o + y_o[\zeta + (\zeta^2 - 1)^{1/2}]\omega_n}{[2(\zeta^2 - 1)^{1/2}]\omega_n} \qquad (9\text{-}33)$$

The motion of the over-damped system is not oscillatory, and the system moves toward the equilibrium position more slowly than is the case for the critically damped system.

9.2.3 Under-Damped System, $\zeta < 1$

For this case, the second term in Eqs (9-22) and (9-23) is imaginary. The factors may be written in the following form for the under-damped system:

$$s_1 = -[\zeta - j(1 - \zeta^2)^{1/2}]\omega_n \qquad (9\text{-}34)$$

$$s_2 = -[\zeta + j(1 - \zeta^2)^{1/2}]\omega_n \qquad (9\text{-}35)$$

Let us define the *damped natural frequency* ω_d by the following expression:

$$\omega_d = \omega_n (1 - \zeta^2)^{1/2} \qquad (9\text{-}36)$$

Vibration Isolation for Noise Control

If we make this substitution into Eqs (9-34) and (9-35), we obtain the following result:

$$s_1 = -\zeta\omega_n - j\omega_d \tag{9-37}$$

$$s_2 = -\zeta\omega_n + j\omega_d \tag{9-38}$$

The general solution for the motion of the mass for the under-damped case may be written in the following form:

$$y(t) = e^{-\zeta\omega_n t}(A\cos\omega_d t + B\sin\omega_d t) \tag{9-39}$$

The constants of integration may be written in terms of the initial displacement and initial velocity as follows:

$$A = y_o \quad \text{and} \quad B = (v_o + \zeta\omega_n y_o)/\omega_d \tag{9-40}$$

The motion of the under-damped system may also be written in the following form:

$$y(t) = C e^{-\zeta\omega_n t}\cos(\omega_d t + \phi) \tag{9-41}$$

The constant C and the phase angle ϕ may be expressed in terms of the initial displacement y_o and initial velocity v_o:

$$C = y_o(1 + \tan^2\phi)^{1/2} \tag{9-42}$$

$$\tan\phi = -\frac{v_o + \zeta\omega_n y_o}{\omega_d y_o} \tag{9-43}$$

We note from Eq. (9-41) that the amplitude of the vibratory motion for the under-damped system is not constant but decays exponentially with time. The motion for the three cases is illustrated in Fig. 9-3.

9.3 DAMPING FACTORS

In Sec. 9.2, we considered the damping as being produced by a linear viscous damper, in which the force of the damper is directly proportional to the relative velocity of the ends of the damper element. There are many other types of damping or energy dissipation elements, and many of these elements are nonlinear. It is usually possible to define an "equivalent damping coefficient" or mechanical resistance for these elements, so the general results that we have developed are valid.

In analogy with the Helmholtz resonator analysis, one alternative measure of the effect of damping may be expressed through the *mechanical quality factor* Q_M (Kinsler et al., 1982), which is defined in a form similar to the acoustic quality factor in Eq. (8-48):

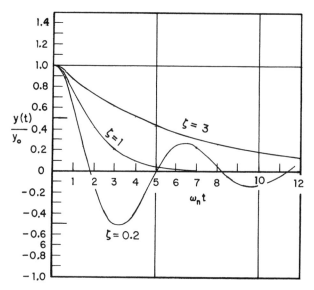

FIGURE 9-3 Vibratory motion for various values of the damping ratio ζ. The curves are plotted for an initial displacement $y(0) = y_o$ and an initial velocity $v(0) = 0$.

$$Q_M = \frac{\omega_n M}{R_M} = \frac{2\pi f_n M}{R_M} \tag{9-44}$$

From Eq. (9-24), we find:

$$\omega_n M = \tfrac{1}{2} R_{M,cr} \tag{9-45}$$

Using this result, the mechanical quality factor may be written in terms of the damping ratio $\zeta = R_M/R_{M,cr}$:

$$Q_M = \frac{R_{M,cr}}{2R_M} = \frac{1}{2\zeta} \tag{9-46}$$

The mechanical quality factor or damping ratio may be measured experimentally by measuring the frequencies at which the power dissipated in the damper element is one-half of the power dissipated at resonance. An expression similar to Eq. (8-60) may be used to relate the half-power frequencies (f_1 and f_2) to the mechanical quality factor:

$$\frac{f_2 - f_1}{f_n} = \frac{1}{Q_M} = 2\zeta \tag{9-47}$$

One of the oldest measures of damping in mechanical systems is the *logarithmic decrement* δ (Thomson and Dahleh, 1998), which is defined by

Vibration Isolation for Noise Control

the natural logarithm of the ratio of the peak amplitudes N cycles apart, divided by the number of cycles:

$$\delta = \frac{1}{N} \ln\left[\frac{y_{\max}(t_o)}{y_{\max}(t_N)}\right] \tag{9-48}$$

The quantities $y_{\max}(t_o)$ and $y_{\max}(t_N)$ are the maximum or peak amplitudes of the motion at time t_o and t_N, respectively. The quantity N is the number of cycles during the time interval between t_o and t_N:

$$N = \frac{\omega_d(t_N - t_o)}{2\pi} \tag{9-49}$$

For small damping (or for $\zeta \leq 0.3$), Eq. (9-36) indicates that the undamped natural frequency ω_n and the damped natural frequency ω_d are practically equal, with less than a 5% error. With this approximation, Eq. (9-49) may be written in the following form:

$$N = \frac{\omega_n(t_N - t_o)}{2\pi} \tag{9-50}$$

At the peak amplitude, the cosine function in Eq. (9-41) is unity, so the ratio of the peak amplitudes may be written in the following form, with the help of Eq. (9-50):

$$\frac{y(t_o)}{y(t_N)} = \exp[-\zeta\omega_n(t_o - t_N)] = \exp[2\pi\zeta N] \tag{9-51}$$

Taking the natural logarithm of both sides of Eq. (9-51), we obtain the following relationship:

$$\ln\left[\frac{y(t_o)}{y(t_N)}\right] = 2\pi\zeta N \tag{9-52}$$

By comparing Eqs (9-48) and (9-52), we see that the logarithmic decrement is directly related to the damping ratio:

$$\delta = 2\pi\zeta \tag{9-53}$$

The logarithmic decrement may be conveniently measured by displaying the motion (on an oscilloscope, for example) and measuring the amplitude ratio directly.

The peak amplitude $y_{\max}$ may be expressed in "level" form, where the displacement level is defined by the following relationship:

$$L_d = 20 \log_{10}(y_{\max}/y_{\text{ref}}) \tag{9-54}$$

The reference displacement is:

$$y_{\text{ref}} = 10\,\text{pm} = 10 \times 10^{-12}\,\text{m}$$

Another measure of the damping is the *decay rate* Δ (Plunkett, 1959), defined as the change in the peak displacement level with time, in units of dB/s:

$$\Delta = -\frac{dL_d}{dt} = -20(\log_{10} e)\frac{(dy_{max}/dt)}{y_{max}} \tag{9-55}$$

According to Eq. (9-41), the peak amplitude for the system with a viscous damper is given by the following relationship:

$$y_{max}(t) = C e^{-\zeta\omega_n t} \tag{9-56}$$

Using this expression in Eq. (9-55), we find the following relationship for the decay rate (dB/s):

$$\Delta = (-20)(0.43429)(-\zeta\omega_n) = 8.6859\zeta\omega_n = 54.575\zeta f_n \tag{9-57}$$

In analogy with the acoustic concepts presented in Chapter 7, *the reverberation time* T_{60} may be defined as the time required for the displacement level to decrease by 60 dB:

$$T_{60} = \frac{60\,\text{dB}}{\Delta\,\text{dB/s}} = \frac{1.0994}{\zeta f_n} \tag{9-58}$$

Finally, the *loss factor* or *energy dissipation factor* η (Ungar and Kerwin, 1962) may be defined as the ratio of the average energy dissipated per radian (energy dissipated per cycle E_{diss} divided by 2π) to the total energy (kinetic energy plus potential energy) of the system E_{tot}:

$$\eta = \frac{E_{diss}/2\pi}{E_{tot}} \tag{9-59}$$

The energy dissipated per cycle may be evaluated from the following expression, involving an integration over one cycle of vibration:

$$E_{diss} = \int F_d\, dy \tag{9-60}$$

For a viscous damper, the force is $F_d = R_M v(t)$ and the displacement $dy = v(t)\,dt$. If we write the displacement in the form $y(t) = y_{max}\cos(\omega t)$, where y_{max} is the peak amplitude, the velocity of the system may be evaluated as follows:

$$v(t) = \frac{dy(t)}{dt} = -\omega y_{max}\sin(\omega t) \tag{9-61}$$

Making the substitutions from Eq. (9-61) into Eq. (9-60), we obtain the following integral:

Vibration Isolation for Noise Control 417

$$E_{diss} = R_M \omega^2 y_{max}^2 \int_0^{2\pi/\omega} \cos^2(\omega t)\, dt \qquad (9\text{-}62)$$

The following result for the energy dissipated per cycle is obtained after carrying out the integration:

$$E_{diss} = \pi R_M \omega y_{max}^2 \qquad (9\text{-}63)$$

The total energy stored in the system is equal to the potential energy stored in the spring at the peak displacement (the point at which the kinetic energy of the mass is zero):

$$E_{tot} = \int F_S\, dy = K_S \int_0^{y_{max}} y\, dy = \tfrac{1}{2} K_S y_{max}^2 \qquad (9\text{-}64)$$

If we make the substitutions from Eqs (9-63) and (9-64) into the expression for the loss coefficient, Eq. (9-59), the following result is obtained:

$$\eta = \frac{R_M \omega}{K_S} \qquad (9\text{-}65)$$

The loss factor is often measured in systems involving freely decaying vibrations, which occur at the resonant frequency, $\omega = \omega_n$. We note from Eqs (9-15) and (9-24) that:

$$\frac{2K_S}{\omega_n} = 2M\omega_n = R_{M,cr} \qquad (9\text{-}66)$$

Using this result in Eq. (9-65), we find that the loss factor and the damping ratio are related by the following expression:

$$\eta = 2\zeta \qquad (9\text{-}67)$$

The relationships between the various measures of damping are summarized in the following expression:

$$\zeta = \tfrac{1}{2}\eta = \frac{1}{2Q_M} = \frac{\delta}{2\pi} = \frac{D}{54.575 f_n} = \frac{1.0994}{f_n T_{60}} \qquad (9\text{-}68)$$

Representative values for the loss factor η for materials at room temperatures are given in Table 9-1. In general, the loss factor is strongly temperature-dependent, as illustrated in Table 9-2 for an acoustic absorbing foam material.

Example 9-2. A machine has a mass of 50 kg (110.2 lb$_m$). The spring constant for the support is 49.35 kN/m (281.8 lb$_f$/in) and the undamped natural frequency is 5 Hz. It is desired to select the damping for the support such that the damping ratio is 0.100. Determine the damping coefficient R_M and the other measures of damping capacity for the system.

TABLE 9-1 Typical Values of the Loss Factor η for Materials at Room Temperature

Material	η	Material	η
Aluminum	0.0010	Masonary blocks	0.006
Brass; bronze	0.0010	Plaster	0.005
Brick	0.015	Plexiglas®	0.020
Concrete	0.015	Plywood	0.030
Copper	0.002	Sand (dry)	0.90
Cork	0.150	Steel; iron	0.0013
Glass	0.0013	Tin	0.002
Gypsum board	0.018	Wood, oak	0.008
Lead	0.015	Zinc	0.0003

Source: Beranek (1971).

TABLE 9-2 Variation of the Loss Factor η for a Typical Acoustic Foam Material Used in Composite Panels[a]

Temperature, °C	Loss factor, η
0	0.087
5	0.116
10	0.168
15	0.28
20	0.48
25	0.72
30	0.79
35	0.66
40	0.47
45	0.30
50	0.19

[a]The material has a density of 32 kg/m^3 (2 lb$_m$/ft^3). The temperature range of application for the material is between $-40°$C ($-40°$F) and $+70°$C (160°F).

Vibration Isolation for Noise Control

The damping ratio is defined by Eq. (9-25). The damping coefficient for a damping ratio of $\zeta = 0.100$ is found as follows:

$$R_M = 2\zeta(K_S M)^{1/2} = (2)(0.100)[(49.35)(10^3)(50)]^{1/2}$$

$$R_M = 314.2 \text{ N-s/m}$$

The mechanical quality factor is given by Eq. (9-46):

$$Q_M = 1/2\zeta = 1/[(2)(0.100)] = 5.00$$

The logarithmic decrement is given by Eq. (9-53):

$$\delta = 2\pi\zeta = (2\pi)(0.100) = 0.6283$$

The decay rate for the system is given by Eq. (9-57):

$$\Delta = 54.575\zeta f_n = (54.575)(0.100)(5) = 27.3 \text{ dB/s}$$

The reverberation time is given by Eq. (9-58):

$$T_{60} = 60/\Delta = (60)/(27.3) = 2.20 \text{ sec}$$

Finally, the loss factor is given by Eq. (9-67):

$$\eta = 2\zeta = (2)(0.100) = 0.200$$

9.4 FORCED VIBRATION

If an external driving force is applied to the spring–mass–damper system, as shown in Fig. 9-4, the governing equation of motion for the system may be found from Newton's second law of motion:

$$M\frac{d^2 y}{dt^2} + R_M \frac{dy}{dt} + K_S y = F(t) \tag{9-69}$$

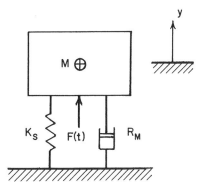

FIGURE 9-4 Forced vibratory system.

If we introduce the undamped natural frequency from Eq. (9-2) and the damping ratio from Eq. (9-25), we obtain the following form for the equation of motion, Eq. (9-69):

$$\frac{d^2y}{dt^2} + 2\zeta\omega_n \frac{dy}{dt} + \omega_n^2 y = \frac{\omega_n^2 F(t)}{K_S} \tag{9-70}$$

We find, in general, that the solution of Eq. (9-70) for the motion of the system involves two components: (a) the *transient* part, which is the solution of the homogeneous equation, $F(t) = 0$, and (b) the *steady-state* part, which is a particular solution of the complete equation. The transient part is the same as that obtained for free motion, given by Eq. (9-21).

As we note from Eqs (9-30), (9-31), or (9-41), the transient portion of the solution involves negative exponentials of the form $\exp(-\zeta\omega_n t)$. These terms generally decay to negligible values after a few cycles, unless the damping is identically zero. Because all physical systems involve some level of damping or energy dissipation, we see that the transient component will approach zero for all mechanical vibrating systems. For example, if the undamped natural frequency is $f_n = 5\,\text{Hz}$ and the damping ratio is $\zeta = 0.10$, we find the following numerical value for the argument of the exponential:

$$\zeta\omega_n t = (0.10)(2\pi)(5)(t) = 3.14t$$

The exponential term is less than 0.010 for an argument of -4.71 or larger (absolute value). For a time of 1.50 seconds, we find the following value:

$$\zeta\omega_n t = (3.14)(1.50) = 4.712$$

$$\exp(-\zeta\omega_n t) = e^{-4.712} = 0.0090$$

Thus, for the conditions given in this example, the homogeneous solution or transient component of the vibratory displacement has become negligible only 1.50 seconds after the driving force has been applied, and only the particular solution or steady-state component is of importance.

Let us examine the case in which the driving force is sinusoidal, or:

$$F(t) = F_o \cos(\omega t) \tag{9-71}$$

We may also write the driving force in complex notation, keeping in mind that only the real part has "real" physical significance:

$$F(t) = F_o e^{j\omega t} \tag{9-72}$$

The quantity ω is the frequency for the applied force.

Because the right-hand side of Eq. (9-70) involves an exponential function, in this case, the particular solution (steady-state solution) will

Vibration Isolation for Noise Control

also involve exponential functions. Let us consider the steady-state solution in the form:

$$y(t) = C e^{j(\omega t - \phi)} = y_{max} e^{j(\omega t - \phi)} \quad (9\text{-}73)$$

The solution may also be written in the trigonometric form:

$$y(t) = C \cos(\omega t - \phi) \quad (9\text{-}74)$$

The quantity ϕ is the phase angle between the applied force and the displacement of the system. We note that the forced-vibration system will oscillate at the same frequency as that of the applied force, but the force and displacement will not be exactly in-phase, in general.

If we make the substitution from Eq. (9-73) for $y(t)$ into Eq. (9-70), the governing equation of motion, the following result is obtained:

$$-C\omega^2 e^{j(\omega t - \phi)} + j2C\zeta\omega_n\omega e^{j(\omega t - \phi)} + C\omega_n^2 e^{j(\omega t - \phi)} = \frac{\omega_n^2 F_o e^{j\omega t}}{K_S} \quad (9\text{-}75)$$

Let us define the frequency ratio as follows:

$$r \equiv \omega/\omega_n = f/f_n \quad (9\text{-}76)$$

If we divide both sides of Eq. (9-75) by $[C\omega_n^2 e^{j(\omega t - \phi)}]$ and introduce the frequency ratio, we obtain the following result:

$$-r^2 + j2\zeta r + 1 = \frac{F_o e^{j\phi}}{C K_S} = \frac{F_o e^{j\phi}}{y_{max} K_S} \quad (9\text{-}77)$$

We may define the magnification factor (MF) as follows:

$$\text{MF} = \frac{y_{max}}{(F_o/K_S)} \quad (9\text{-}78)$$

The magnification factor is the ratio of the maximum amplitude of vibration of the system to the static displacement of the system if the force F_o acts as a static force. The quantity F_o is the maximum amplitude of the applied force. We note that *the peak-to-peak amplitude* of motion $y_p = 2y_{max}$. The *rms amplitude* of motion is $y_{rms} = y_{max}/(2^{1/2})$.

We may write the left-hand side of Eq. (9-77) in the form involving a magnitude $[\text{Re}^2 + \text{Im}^2]^{1/2}$ and a phase angle $\tan\phi = \text{Im}/\text{Re}$:

$$[(1 - r^2)^2 + (2\zeta r)^2]^{1/2} e^{j\phi} = \frac{e^{j\phi}}{\text{MF}} \quad (9\text{-}79)$$

The phase angle between the displacement and the applied force is given by the following:

$$\tan\phi = \frac{2\zeta r}{1 - r^2} \quad (9\text{-}80)$$

The magnification factor may be obtained from Eq. (9-79):

$$\text{MF} = \frac{1}{[(1-r^2)^2 + (2\zeta r)^2]^{1/2}} \qquad (9\text{-}81)$$

A plot of the magnification factor MF as a function of frequency ratio $r = f/f_n$ for several different damping ratios ζ is shown in Fig. 9-5.

There are several important observations that we can make from Fig. 9-5. First, if the forcing frequency f is less than the undamped natural frequency f_n, so that $r < 1$, the magnification ratio is always greater than

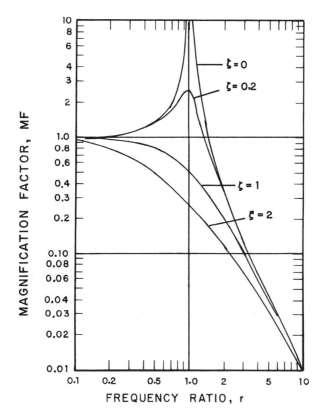

FIGURE 9-5 Plot of the magnification factor $\text{MF} = K_S y_{\max}/F_o$ vs. frequency ratio $r = f/f_n$ for an SDOF spring–mass–damper system with various values of the damping ratio $\zeta = R_M/R_{M,\text{cr}}$.

Vibration Isolation for Noise Control

unity for damping ratios $\zeta < \frac{1}{2}$. If $r < 1$, for a damping ratio $\zeta = \frac{1}{2}$, we see that the denominator of Eq. (9-81) is as follows:

$$[(1-r^2)^2 + (2\zeta r)^2]^{1/2} = [1 - r^2 + r^4]^{1/2} < 1$$

and MF > 1 for this case.

Secondly, we observe that the magnification factor is always less than unity for all frequencies if the damping ratio is larger than $2^{-1/2} = 0.707$. For a damping ratio $\zeta = 1/2^{1/2}$, the denominator of Eq. (9-81) is as follows:

$$[(1-r^2)^2 + (2\zeta r)^2]^{1/2} = [1 + r^4]^{1/2} > 1$$

and MF < 1 for this case. If we wish to reduce the amplitude of motion of the mass in the low-frequency region ($0 < r < 1$), we must select that damping ratio to be 0.707 or larger.

Finally, we note that the magnification factor is always less than unity for any value of damping ratio if the frequency ratio $r > 2^{1/2} = 1.414$. For a frequency ratio $r = 2^{1/2}$, the denominator of Eq. (9-81) is as follows:

$$[(1-r^2)^2 + (2\zeta r)^2]^{1/2} = [1 + 8\zeta^2]^{1/2} \geq 1 \quad \text{for all } \zeta \geq 0$$

and MF $\leq$ 1 for this case. This means that, if we are really serious about reducing the motion significantly for a given frequency f, we must adjust the undamped natural frequency of the system such that $f > 2^{1/2} f_n = 1.414 f_n$. This feat may be accomplished for a given mass M by selecting the proper value of the spring constant K_S for the system, as illustrated by Eq. (9-15).

Example 9-3. A machine having a mass of 50 kg (110.2 lb$_m$) is driven by a sinusoidal force having a maximum amplitude of 40 N (8.99 lb$_f$) and a frequency of 10 Hz. The support system has a spring constant of 50 kN/m (285 lb$_f$/in) and a damping coefficient of 600 N-s/m (3.43 lb$_f$-sec/in). Determine the maximum amplitude of vibration for the system.

The undamped natural frequency for the system is determined from Eq. (9-15):

$$f_n = \frac{(K_S/M)^{1/2}}{2\pi} = \frac{(50{,}000/50)^{1/2}}{(2\pi)} = 5.033 \text{ Hz}$$

The frequency ratio is:

$$r = f/f_n = (10)/(5.033) = 1.987$$

The damping ratio may be found from Eq. (9-25):

$$\zeta = \frac{R_M}{2(K_S M)^{1/2}} = \frac{(600)}{(2)[(50{,}000)(50)]^{1/2}} = \frac{(600)}{(3162.3)} = 0.1897$$

The magnification factor may be found from Eq. (9-81):

$$MF = \frac{1}{\{[1-1.987^2]^2 + [(2)(0.1897)(1.987)]^2\}^{1/2}} = 0.3287 = \frac{K_S y_{max}}{F_o}$$

The maximum amplitude of motion for the system may be calculated:

$$y_{max} = \frac{(0.3287)(40)}{(50{,}000)} = 0.263 \times 10^{-3}\,\text{m} = 0.263\,\text{mm} \quad (0.0104\,\text{in})$$

The peak-to-peak amplitude of motion is as follows:

$$y_p = 2y_{max} = (2)(0.263) = 0.526\,\text{mm} \quad (0.0207\,\text{in})$$

9.5 MECHANICAL IMPEDANCE AND MOBILITY

There are many cases in which the velocity of the mass is an important vibration function to be controlled, instead of the displacement. In this case, the *mechanical impedance* Z_M may be utilized. The mechanical impedance gives a measure of how strongly the system resists applied forces (or moments). The mechanical impedance is defined as the ratio of the applied force to the resulting velocity of the system:

$$Z_M = \frac{F(t)}{v(t)} \tag{9-82}$$

Let us take the displacement of the system from Eq. (9-73) or (9-74) and take the derivative with respect to time to obtain the velocity of the system:

$$v(t) = \frac{dy(t)}{dt} = j\omega y_{max}\, e^{j(\omega t - \phi)} \tag{9-83}$$

We note that we may write $j = e^{j\pi/2}$, so Eq. (9-83) may be written in the following alternative form:

$$v(t) = \omega y_{max}\, e^{j(\omega t - \phi + \pi/2)} = \omega y_{max}\, e^{j(\omega t - \theta)} = v_{max}\, e^{j(\omega t - \theta)} \tag{9-84}$$

The quantity θ is related to the displacement phase angle ϕ as follows:

$$\theta = \phi - \pi/2\,\text{(radians)} = \phi - 90°\,\text{(degrees)} \tag{9-85}$$

The complex representation of the mechanical impedance may be obtained by combining Eqs (9-84) and (9-82):

$$Z_M = \frac{F_o\, e^{j\theta}}{\omega y_{max}} = |Z_M|\, e^{j\theta} \tag{9-86}$$

Vibration Isolation for Noise Control

The magnitude of the mechanical impedance may be expressed in terms of the magnification factor (MF) by using Eq. (9-78), defining the factor:

$$|Z_M| = \frac{F_o}{\omega y_{max}} = \frac{K_S(F_o/K_S)}{\omega y_{max}} = \frac{K_S}{\omega \text{MF}} \quad (9\text{-}87)$$

If we introduce the expression for the magnification factor from Eq. (9-81) into Eq. (9-87), we obtain the following result for the magnitude of the mechanical impedance:

$$|Z_M| = (K_S/\omega)[(1 - r^2)^2 + (2\zeta r)^2]^{1/2} \quad (9\text{-}88)$$

The expression may be further simplified by using the expression for the undamped natural frequency, Eq. (9-2), $\omega_n^2 = K_S/M$, and the damping factor relationship, Eq. (9-25), $2\zeta r = \omega R_M/K_S$.

$$|Z_M| = R_M\{1 + (1/2\zeta)^2[r - (1/r)]^2\}^{1/2} \quad (9\text{-}89a)$$

The expression may also be written in terms of the mechanical quality factor Q_M by using Eq. (9-46):

$$|Z_M| = R_M\{1 + Q_M^2[r - (1/r)]^2\}^{1/2} \quad (9\text{-}89b)$$

The variation of the mechanical impedance in the various limiting regions may be noted. First, in the low-frequency region ($r \ll 1$), we note that r is negligible compared with $(1/r)$, and $(1/r)$ is much larger than unity. In this region, the magnitude of the mechanical impedance approaches the following limiting value:

$$|Z_M| \sim \frac{R_M}{2\zeta r} = \frac{K_S}{\omega} = \frac{K_S}{2\pi f} \quad (9\text{-}90)$$

For low frequencies of vibration (or, for practical purposes, when $r < 0.16$), the motion is governed by the stiffness or the spring constant of the system. Thus, Region I ($r < 0.16$) could be denoted as the *stiffness-controlled region* of vibration. The mechanical impedance is inversely proportional to the frequency in this region.

Secondly, for the frequency region around the undamped natural frequency ($r \approx 1$), the term $[r - (1/r)]$ is small. In this region, the magnitude of the mechanical impedance approaches the following limiting value:

$$|Z_M| \sim R_M \quad (9\text{-}91)$$

For frequencies around the undamped natural (or, for practical purposes, when $[1 - 0.22\zeta] < r < [1 + 0.22\zeta]$), the motion is governed by the damping of the system. Thus, Region II could be denoted as the *damping-controlled region* of vibration.

Finally, in the high-frequency region ($r \gg 1$), we note that ($1/r$) is negligible compared with r. In this region, the magnitude of the mechanical impedance approaches the following limiting value:

$$|Z_M| \sim \frac{R_M r}{2\zeta} = \omega M = 2\pi f M \qquad (9\text{-}92)$$

For high frequencies of vibration (or, for practical purposes, when $r > 6$), the motion is governed by the inertia or the mass of the system. Thus, Region III ($r > 6$) could be denoted by the *mass-controlled region* of vibration. The mechanical impedance is directly proportional to the frequency in this region.

When analyzing electromechanical systems (combinations of electric and mechanical components), it is convenient to use the *mechanical admittance* Y_M, which is defined by the following expression:

$$Y_M = \frac{v(t)}{F(t)} = \frac{1}{Z_M} = |Y_M| e^{-j\theta} \qquad (9\text{-}93)$$

The mechanical admittance is also called the *mechanical mobility* of the system. The magnitude of the mechanical admittance or mobility may be expressed in terms of the magnification factor MF by using Eq. (9-78):

$$|Y_M| = \frac{v_{max}}{F_o} = \frac{\omega y_{max}}{F_o} = \frac{\omega y_{max}}{K_S(F_o/K_S)} = \frac{\omega \text{MF}}{K_S} \qquad (9\text{-}94)$$

If we introduce the expression for the magnification factor from Eq. (9-81) into Eq. (9-94), we obtain the following result for the mechanical admittance or mobility:

$$|Y_M| = \frac{r\omega_n}{K_S[(1-r^2)^2 + (2\zeta r)^2]^{1/2}} \qquad (9\text{-}95)$$

The undamped natural frequency may be eliminated from the expression as follows:

$$\frac{\omega_n}{K_S} = \frac{(K_S/M)^{1/2}}{K_S} = \frac{1}{(K_S M)^{1/2}} = \frac{2}{R_{M,cr}} = \frac{2\zeta}{R_M} \qquad (9\text{-}96)$$

$$|Y_M| = \frac{2r}{R_{M,cr}[(1-r^2)^2 + (2\zeta r)^2]^{1/2}} = \frac{2\zeta r}{R_M[(1-r^2)^2 + (2\zeta r)^2]^{1/2}} \qquad (9\text{-}97)$$

A plot of the mobility as a function of the frequency ratio is shown in Fig. 9-6.

Example 9-4. Determine the magnitude of the maximum velocity for the system given in Example 9-3.

Vibration Isolation for Noise Control

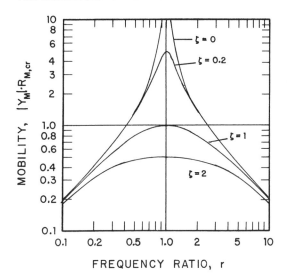

FIGURE 9-6 Plot of the dimensionless mechanical mobility $|Y_M| \times R_{m,cr} = R_{M,cr} v_{max}/F_o$ vs. frequency ratio $r = f/f_n$ for an SDOF spirng–mass–damper system with various values of the damping ratio $\zeta = R_M/R_{M,cr}$.

The magnitude of the mobility for the system is given by Eq. (9-94):

$$|Y_M| = \frac{v_{max}}{F_o} = \frac{2\pi f \, MF}{K_S} = \frac{(2\pi)(10)(0.3287)}{(50,000)} = 0.413 \times 10^{-3} \, \text{m/N-s}$$

$$|Y_M| = 0.413 \, \text{mm/N-s} \quad (0.0723 \, \text{in/lb}_f\text{-sec})$$

The maximum amplitude of the velocity for the system is as follows:

$$v_{max} = |Y_M| F_o = (0.413)(40) = 16.52 \, \text{mm/s} \quad (0.650 \, \text{in/sec})$$

9.6 TRANSMISSIBILITY

One of the important factors in design for vibration isolation is reduction of the force transmitted to the base or support of the system. Generally, the objective of vibration isolation is to reduce the transmitted force to an acceptable value. The force transmitted to the base is equal to the sum of the spring force and the damper force, as illustrated in Fig. 9-7:

$$F_T(t) = F_S + F_d = K_S y(t) + R_M v(t) \tag{9-98}$$

The displacement is given by Eq. (9-73), and the velocity is given by Eq. (9-83):

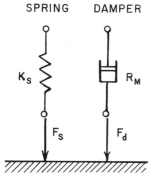

FIGURE 9-7 The force transmitted to the foundation is the sum of the spring force and the damper force.

$$F_T(t) = K_S y_{max} e^{j(\omega t - \phi)} + j\omega R_M y_{max} e^{j(\omega t - \phi)} \quad (9\text{-}99)$$

The *transmissibility* (Tr) is defined as the ratio of the transmitted force to the applied force for the system.

$$\text{Tr} = \frac{F_T(t)}{F(t)} = \frac{K_S y_{max}}{F_o}\left[1 + j\frac{\omega R_M}{K_S}\right] e^{-j\phi} \quad (9\text{-}100)$$

If we use Eq. (9-76) for the frequency ratio, Eq. (9-78) for the magnification factor, and Eq. (9-96), we may write the transmissibility in the following form:

$$\text{Tr} = \text{MF}[1 + j(2\zeta r)]e^{-j\phi} = |\text{Tr}|e^{-j(\phi - \tau)} \quad (9\text{-}101)$$

The magnitude of the transmissibility may be found from the real and imaginary parts of Eq. (9-101):

$$|\text{Tr}| = \text{MF}[1 + (2\zeta r)^2]^{1/2} \quad (9\text{-}102)$$

The transmissibility may be written in the following from by substituting for the magnification factor from Eq. (9-81):

$$|\text{Tr}| = \left[\frac{1 + (2\zeta r)^2}{(1 - r^2)^2 + (2\zeta r)^2}\right]^{1/2} \quad (9\text{-}103)$$

A plot of this function is shown in Fig. 9-8. The quantity ζ is the damping ratio and $r = f/f_n$ is the frequency ratio.

The transmissibility may also be expressed in "level" form. The *transmissibility level* L_{Tr} is defined by the following expression:

$$L_{Tr} = 20\log_{10}|\text{Tr}| \quad (9\text{-}104)$$

Vibration Isolation for Noise Control

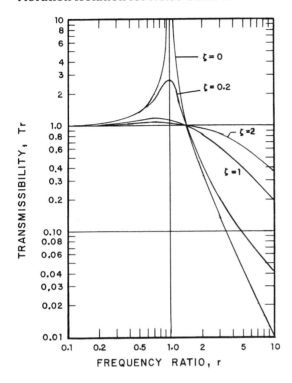

FIGURE 9-8 Plot of the transmissibility $\text{Tr} = F_T/F_o$ vs. frequency ratio $r = f/f_n$ for an SDOF spring–mass–damper system with various values of the damping ratio $\zeta = R_M/R_{M,\text{cr}}$.

We note that the transmissibility level may be positive (if $|\text{Tr}| > 1$) or negative (if $|\text{Tr}| < 1$).

The phase angle between the transmitted force and the applied force may be determined as follows. The angle τ may be found from the real and imaginary components of the term in brackets in Eq. (9-101) or, $\tan \tau = \text{Im}/\text{Re}$.

$$\tan \tau = 2\zeta r \tag{9-105}$$

The phase angle is $(\phi - \tau)$.

There are several observations that we may make from Fig. 9-8. First, the transmissibility approaches unity as the frequency ratio becomes small (less than about 0.2) for all values of the damping ratio. This means that, if we wish to reduce the force transmitted, the natural frequency of the system should not be large compared with the forcing frequency.

Secondly, if the frequency ratio r is less than $\sqrt{2}$ (i.e., 1.414), the transmissibility is always greater than unity for all values of the damping ratio. In this range of frequencies, the effect of damping is to reduce the transmitted forces below that which would occur with zero damping; however, the transmitted force is still larger than the exciting or applied force.

Finally, the transmissibility is always less than unity for the frequency ratio larger than 1.414. The effect of damping is to increase the transmitted force above that which would occur with zero damping; however, the transmitted force is smaller than the applied force, because some of the effort is expended in accelerating the mass at the higher frequencies. This means that, if we wish to isolate the foundation from the vibrating mass, we should select a spring constant for the support system such that $f_n < 0.707f$, or $r > 1.414$, and use as little damping as is practical.

For the case of zero damping ($\zeta = 0$), the magnitude of the transmissibility reduces to the following expression:

$$|\text{Tr}| = \frac{1}{|r^2 - 1|} \tag{9-106}$$

In many cases of vibration isolation design, we need to determine the frequency ratio required to achieve a given transmissibility. Using Eq. (9-103), we may write the following:

$$\text{Tr}^2[(1 - r^2)^2 + (2\zeta r)^2] = 1 + (2\zeta r)^2 \tag{9-107}$$

This expression may be simplified as follows:

$$r^4 - 2\beta r^2 - \frac{1 - \text{Tr}^2}{\text{Tr}^2} = 0 \tag{9-108}$$

The quantity β is defined by the following expression:

$$\beta = 1 + \frac{2\zeta^2(1 - \text{Tr}^2)}{\text{Tr}^2} \tag{9-109}$$

For a known value of the damping ratio ζ, the frequency ratio r required to achieve a given value of the transmissibility Tr may be determined. Because Eq. (9-108) is a quadratic equation in r^2, two solutions for r^2 are obtained; however, only the *positive* solution has a physical meaning.

Example 9-5. A machine has a mass of 50 kg (110.2 lb$_m$). The damping ratio for the support system is $\zeta = 0.10$. The driving force acting on the mass has a maximum amplitude of 5.00 kN (1124 lb$_f$), and the frequency of the driving force is 35 Hz. Determine the spring constant of the support such that the transmissibility is 0.020 or the transmissibility level is -34 dB.

The value of the parameter β may be found from Eq. (9-109):

Vibration Isolation for Noise Control

$$\beta = 1 + \frac{(2)(0.10)^2(1 - 0.020^2)}{(0.020)^2} = 1 + 49.98 = 50.98$$

The required frequency ratio may be found by solving Eq. (9-108):

$$(r^2)^2 - (2)(50.98)r^2 - (1 - 0.020^2)/(0.020)^2 = 0$$

$$r^2 = 50.98 + [(50.98)^2 + 2499]^{1/2} = 50.98 + 71.40 = 122.38$$

$$r = (122.38)^{1/2} = 11.06 = f/f_n$$

The required undamped natural frequency may be determined:

$$f_n = \frac{35}{11.06} = 3.164\,\text{Hz} = \frac{(K_S/M)^{1/2}}{2\pi}$$

The required spring constant for the support system may now be found:

$$K_S = [(2\pi)(3.164)]^2(50) = 19{,}759\,\text{N/m} = 19.76\,\text{kN/m} \quad (112.8\,\text{lb}_f/\text{in})$$

The magnitude of the transmitted force may be calculated from the definition of the transmissibility:

$$F_T = F_o|Tr| = (5000)(0.020) = 100\,\text{N} \quad (22.5\,\text{lb}_f)$$

The phase angle between the transmitted force and the applied force may be found from Eqs (9-80) and (9-105):

$$\tan\phi = \frac{(2)(0.10)(11.06)}{1 - (11.06)^2} = -0.01823$$

$$\phi = -0.0182\,\text{rad} = -1.04°$$

$$\tan\tau = (2)(0.10)(11.06) = 2.213$$

$$\tau = 1.146\,\text{rad} = 65.68°$$

The phase angle between the transmitted and applied forces is as follows:

$$(\phi = -\tau) = -1.04° - 65.68° = -66.72° = -1.164\,\text{rad}$$

Note that, because of the small damping ratio, the phase angle ϕ between the displacement and the applied force is almost zero.

9.7 ROTATING UNBALANCE

Many vibrating problems arise because of rotating unbalance in the piece of machinery. The unbalance may be expressed in terms of the product of an equivalent unbalanced mass m and the eccentricity of the mass ε. The total mass of the system is denoted by M.

Let us consider the system shown in Fig. 9-9. The distance $y(t)$ is the displacement of the center of mass of the system from the static equilibrium position, and $y_1(t)$ is the displacement of the unbalanced mass with respect to the same reference. The displacement of the unbalanced mass may be written in the following form:

$$y_1(t) = y(t) + \varepsilon\, e^{j\omega t} \tag{9-110}$$

The quantity ω is the rotational frequency for the unbalanced mass.

If we apply Newton's second law of motion to the system, we obtain the following expression:

$$(M - m)\frac{d^2 y}{dt^2} + m\frac{d^2 y_1}{dt^2} + R_M \frac{dy}{dt} + K_S y = 0 \tag{9-111}$$

We may eliminate the term involving the displacement of the unbalanced mass by using Eq. (9-110):

$$\frac{d^2 y_1}{dt^2} = \frac{d^2 y}{dt^2} - \varepsilon\omega^2\, e^{j\omega t} \tag{9-112}$$

If we make the substitution from Eq. (9-112) into Eq. (9-111), the following result is obtained:

$$M\frac{d^2 y}{dt^2} + R_M \frac{dy}{dt} + K_S y = m\varepsilon\omega^2\, e^{j\omega t} = F_{eq}\, e^{j\omega t} \tag{9-113}$$

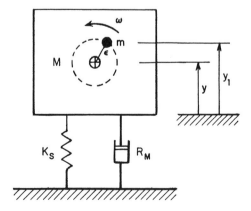

FIGURE 9-9 Mechanical system with a rotating unbalance.

Vibration Isolation for Noise Control

We observe that Eq. (9-113) for the system driven by an unbalanced mass is exactly the same as Eq. (9-69) for an external applied force, if we use an *equivalent force* F_{eq}, given by the following:

$$F_{eq} = m\varepsilon\omega^2 \qquad (9\text{-}114)$$

Based on this observation, we may write the following expression for the magnitude of the maximum displacement of the system in terms of the magnification factor MF:

$$y_{max} = \frac{F_{eq}\text{MF}}{K_S} = \frac{m\varepsilon\omega^2\text{MF}}{K_S} \qquad (9\text{-}115)$$

If we introduce the undamped natural frequency from $K_S = M\omega_n^2$ and the damping ratio $r = \omega/\omega_n$ into Eq. (9-115), we obtain the following result:

$$y_{max} = (m\varepsilon/M)r^2(\text{MF}) \qquad (9\text{-}116)$$

The magnitude of the force transmitted to the foundation may be found from the definition of the transmissibility, using the equivalent force as the driving force for the system:

$$F_T = F_{eq}\text{Tr} = m\varepsilon\omega^2\text{Tr} = (m\varepsilon\omega_n^2)r^2(\text{Tr}) \qquad (9\text{-}117)$$

In the low-frequency region (the stiffness-controlled region), the magnification factor and the transmissibility each approach unity. For small frequency ratios (or for practical purposes, $r < 0.2$), the maximum amplitude and the transmitted force are proportional to the square of the frequency:

$$y_{max} \sim (m\varepsilon/M)r^2 \qquad (r \ll 1) \qquad (9\text{-}118a)$$

$$F_T \sim (m\varepsilon\omega_n^2)r^2 \qquad (r \ll 1) \qquad (9\text{-}118b)$$

In the high-frequency region (the mass-controlled region), the magnification factor is almost inversely proportional to the square of the frequency ratio. For large frequency ratios (or for practical purposes, $r > 6$), the maximum amplitude and transmitted force approach the following limiting values:

$$y_{max} \sim m\varepsilon/M \qquad (r \gg 1) \qquad (9\text{-}119a)$$

$$F_T \sim (m\varepsilon\omega_n^2)(1 + 4\zeta^2 r^2)^{1/2} \qquad (r \gg 1) \qquad (9\text{-}119b)$$

If the speed of rotation for a system with rotational (or translational) unbalance is constant, the magnitude of the equivalent driving force is constant. On the other hand, for situations in which the speed of rotation is not constant, such as during start-up or change in operating conditions for the machine, the equivalent force is not constant, but varies directly proportional to the square of the frequency. At low speeds, the equivalent force is small, and consequently, the amplitude of vibration and the transmitted force are both relatively small. However, at high speeds, the equivalent force is large, and the amplitude of the transmitted force may be quite large, depending on the value of the damping ratio. We note from Eq. (9-119b) that the transmitted force is directly proportional to the damping ratio at high frequencies, so it is desirable to have as little damping as practical (ζ less than 0.15) for vibration isolation systems used to isolate vibration due to a rotary machine with variable speed.

Some subjective reactions to vibration of machines are presented in Table 9-3. The velocity values are maximum or *peak* vibration velocity values, v_{max}.

Example 9-6. A reciprocating air compressor has a total mass of 700 kg (1543 lb$_m$) and operates at a speed of 1800 rpm. The mass of the unbalanced reciprocating parts is 12 kg (26.5 lb$_m$), and the eccentricity is 100 mm (3.94 in). The damping ratio for the support system is 0.050. Determine

TABLE 9-3 Subjective Response to Machine Vibration—the Velocity Values are Peak or Maximum Vibrational Velocities[a]

Subjective impression	v_{max} range, mm/s	L_v range, dB(p)
Very rough	>16	>124
Rough	8–16	118–124
Slightly rough	4–8	112–118
Fair	2–4	106–112
Good	1–2	100–106
Very good	0.50–1	94–100
Smooth	0.25–0.50	88–94
Very smooth	0.125–0.25	82–88
Extremely smooth	<0.125	<82

[a]The velocity is related to the displacement and frequency by the following expression for sinusoidal vibration:

$v_{max} = 2\pi f y_{max}$

The velocity level is defined by the following expression:

$L_v = 20 \log_{10}(v/v_{ref})$

The reference velocity level is $v_{ref} = 10$ nm/s. The designation dB(p) denotes peak or maximum velocity levels.
Source: Fox (1971).

Vibration Isolation for Noise Control 435

the required spring constant for the support to achieve a transmissibility level of $-20\,\text{dB}$.

The value of the transmissibility is as follows:

$$\text{Tr} = 10^{-20/20} = 0.100$$

The value of the parameter β may be calculated from its definition, Eq. (9-109):

$$\beta = 1 + \frac{(2)(0.050)^2(1 - 0.100^2)}{(0.100)^2} = 1 + 0.495 = 1.495$$

The frequency ratio may be found from the value of the transmissibility and Eq. (9-108):

$$(r^2)^2 - (2)(1.495)r^2 - \frac{1 - 0.100^2}{0.100^2} = 0$$

$$r^2 = 1.495 + (1.495^2 + 99)^{1/2} = 11.557$$

$$r = (11.557)^{1/2} = 3.40 = f/f_\text{n}$$

The required undamped natural frequency for the system may now be found:

$$f_\text{n} = \frac{(1800/60)}{(3.40)} = 8.825\,\text{Hz} = \frac{(K_\text{S}/M)^{1/2}}{2\pi}$$

The required spring constant for the supports may be calculated from the value of the undamped natural frequency:

$$K_\text{S} = (4\pi^2)(8.825)^2(700) = 2.152 \times 10^6\,\text{N/m}$$
$$= 2.152\,\text{MN/m} \quad (12{,}290\,\text{lb}_\text{f}/\text{in})$$

If we use four springs (one at each corner) to support the compressor, the spring constant for each spring would have the following value:

$$K_\text{S1} = K_\text{S}/4 = (2152)/(4) = 538\,\text{kN/m} \quad (3072\,\text{lb}_\text{f}/\text{in})$$

Let us determine the other parameters for the vibrating system. The equivalent force may be found from Eq. (9-114):

$$F_\text{eq} = m\varepsilon\omega_\text{n}^2 = (12)(0.100)(4\pi^2)(1800/60)^2 = 42.64 \times 10^3\,\text{N}$$

$$F_\text{eq} = 42.64\,\text{kN} \quad (9585\,\text{lb}_\text{f})$$

The magnitude of the force transmitted into the foundation is found from the transmissibility:

$$F_\text{T} = F_\text{eq}\text{Tr} = (42.64)(0.100) = 4.264\,\text{kN} \quad (958.5\,\text{lb}_\text{f})$$

The magnification factor may be calculated from Eq. (9-81):

$$\mathrm{MF} = \frac{1}{\{[1-3.40^2]^2+[(2)(0.050)(3.40)]^2\}^{1/2}} = 0.0947 = \frac{K_S y_{max}}{F_{eq}}$$

The maximum amplitude of vibration for the system is as follows:

$$y_{max} = \frac{(0.0947)(42.64)(10^3)}{(2.152)(10^6)} = 0.00188\,\mathrm{m} = 1.88\,\mathrm{mm} \quad (0.074\,\mathrm{in})$$

The damping coefficient for the system may be determined from Eq. (9-25):

$$R_M = 2(K_S M)^{1/2}\zeta = (2)[(2.152)(10^6)(700)]^{1/2}(0.050)$$

$$R_M = 3881\,\mathrm{N\text{-}s/m} = 3.381\,\mathrm{kN\text{-}s/m}$$

The magnitude of the mechanical impedance may be determined from Eq. (9-89):

$$|Z_M| = (3881)\{1+[1/(2)(0.050)]^2[3.40-(1/3.40)]^2\}^{1/2}$$

$$|Z_M| = (3881)(31.07) = 120.6 \times 10^3\,\mathrm{N\text{-}s/m} = 120.6\,\mathrm{kN\text{-}s/m}$$

The mechanical admittance or mobility is equal to the reciprocal of the mechanical impedance:

$$|Y_M| = \frac{1}{|Z_M|} = \frac{1}{(120.6)(10^3)} = 8.292 \times 10^{-6}\,\mathrm{m/s\text{-}N} = 8.292\,\mathrm{\mu m/s\text{-}N}$$

The maximum amplitude of the vibratory velocity of the system may be found from the definition of the mechanical mobility:

$$v_{max} = F_{eq}|Y_M| = (42.64)(10^3)(8.292)(10^{-6})$$
$$= 0.3536\,\mathrm{m/s} \quad (13.92\,\mathrm{in/sec})$$

This level of vibration corresponds to "smooth" operation, as suggested by the data in Table 9-3.

The maximum vibratory acceleration of the machine may be calculated from the following:

$$a_{max} = 2\pi f v_{max} = (2\pi)(1800/60)(0.3536) = 66.65\,\mathrm{m/s^2}$$

$$a_{max}/g = (66.65)/(9.806) = 6.80g$$

9.8 DISPLACEMENT EXCITATION

In many cases, the designer seeks to isolate a device from external vibrations or motion of the support, as illustrated in Fig. 9-10. If we apply Newton's

Vibration Isolation for Noise Control

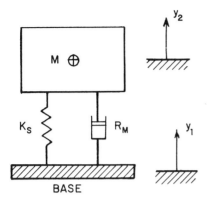

FIGURE 9-10 An SDOF spring–mass–damper system excited by the displacement motion of the base.

second law of motion to the system, the following differential equation is obtained:

$$M \frac{d^2 y_2}{dt^2} + R_M \left(\frac{dy_2}{dt} - \frac{dy_1}{dt} \right) + K_S (y_2 - y_1) = 0 \qquad (9\text{-}120)$$

The quantity y_1 is the displacement of the base, and y_2 is the displacement of the main system. If we introduce the undamped natural frequency from Eq. (9-2) and the damping ratio from Eq. (9-25), we obtain the following expression for the equation of motion of the system:

$$\frac{d^2 y_2}{dt^2} + 2\zeta\omega_n \frac{dy_2}{dt} + \omega_n^2 y_2 = 2\zeta\omega_n \frac{dy_1}{dt} + \omega_n^2 y_1 \qquad (9\text{-}121)$$

Let us consider the case for which the displacement of the base is sinusoidal:

$$y_1(t) = y_{1m} e^{j\omega t} \qquad (9\text{-}122)$$

The right side of Eq. (9-121) may then be written in the following form:

$$2\zeta\omega_n \frac{dy_1}{dt} + \omega_n^2 y_1 = (1 + j2\zeta r)\omega_n^2 y_{1m} e^{j\omega t} \qquad (9\text{-}123)$$

The quantity $r = f/f_n$ is the frequency ratio. Because the right side of Eq. (9-121) is sinusoidal for this case, the steady-state response (particular solution) or the motion of the main system will also be sinusoidal:

$$y_2(t) = y_{2m} e^{j(\omega t - \psi)} \qquad (9\text{-}124)$$

If we make the substitution from Eqs (9-123) and (9-124) into the equation of motion, Eq. (9-121), the following result is obtained:

$$[(1-r^2)+j2\zeta r]y_{2m}e^{-j\psi} = (1+j2\zeta r)y_{1m} \tag{9-125}$$

The result may also be written in the following form:

$$[(1-r^2)^2+(2\zeta r)^2]^{1/2} y_{2m} e^{-j\phi_2} = [1+(2\zeta r)^2]^{1/2} y_{1m} e^{-j\phi_1} \tag{9-126}$$

The phase angles are given as follows:

$$\psi = \phi_2 - \phi_1 \tag{9-127}$$

$$\tan\phi_1 = \frac{2\zeta r}{1-r^2} \tag{9-128}$$

$$\tan\phi_2 = 2\zeta r \tag{9-129}$$

The ratio of the system displacement to the displacement of the base can be obtained from Eq. (9-126):

$$\frac{y_{2m}}{y_{1m}} = \frac{[1+(2\zeta r)^2]^{1/2} e^{j\psi}}{[(1-r^2)^2+(2\zeta r)^2]^{1/2}} \tag{9-130}$$

The magnitude of this ratio is exactly the same as that found for the transmissibility, as given by Eq. (9-103):

$$|y_{2m}| = |\text{Tr}||y_{1m}| \tag{9-131}$$

The same principles, as discussed previously for reduction of the force transmitted to the base for a system excited by an external force, may be applied for reduction of the displacement induced by motion of the base of a system.

Example 9-7. An instrument package having a mass of 10 kg (22.0 lb$_m$) is supported by a spring–damper system having a damping ratio of $\zeta = 0.060$. The maximum displacement of the base to which the package is attached is 12 mm (0.472 in), and the frequency of vibration of the base is 15 Hz. Determine the spring constant for the support to limit the displacement amplitude of the instrument package to 0.60 mm (0.024 in).

The transmissibility for the system may be found from Eq. (9-131):

$$\text{Tr} = \frac{y_{2m}}{y_{1m}} = \frac{0.60}{12.0} = 0.050$$

Let us calculate the parameter from Eq. (9-109):

$$\beta = 1 + \frac{(2)(0.060)^2(1-0.050^2)}{(0.050)^2} = 1 + 2.873 = 3.873$$

Vibration Isolation for Noise Control

The required frequency ratio may be found from Eq. (9-108):

$$(r^2)^2 - (2)(3.873)r^2 - \frac{(1 - 0.050^2)}{(0.050)^2} = 0$$

$$r^2 = 3.873 + (3.873^2 + 399)^{1/2} = 24.22$$

$$r = 4.921 = f/f_n$$

The undamped natural frequency for the system is as follows:

$$f_n = (15)/(4.921) = 3.046 \text{ Hz} = (K_S/M)^{1/2}/2\pi$$

The required spring constant for the support is found from the undamped natural frequency:

$$K_S = (4\pi^2)(3.046)^2(10) = 3668 \text{ N/m} = 3.668 \text{ kN/m} \quad (20.94 \text{ lb}_f/\text{in})$$

The required damping coefficient for the system may be calculated from Eq. (9-25):

$$R_M = 2(K_S M)^{1/2}\zeta = (2)[(3668)(10)]^{1/2}(0.060)$$

$$R_M = 22.98 \text{ N-s/m} \quad (0.1312 \text{ lb}_f\text{-sec/in})$$

The maximum velocity of the system is found as follows:

$$v_{max} = 2\pi f y_{2m} = (2\pi)(15)(0.60) = 56.4 \text{ mm/s} \quad (2.23 \text{ in/sec})$$

The maximum acceleration of the system is as follows:

$$a_{max} = 4\pi^2 f^2 y_{2m} = (4\pi^2)(15)^2(0.60)(10^{-3}) = 5.33 \text{ m/s}^2 \quad (17.49 \text{ ft/sec}^2)$$

$$a_{max}/g = (5.33)/(9.806) = 0.544g$$

9.9 DYNAMIC VIBRATION ISOLATOR

There are some vibration isolation situations in which the machine may operate at or near the resonant frequency of the system. This may occur if the support system is a flexible or resilient floor. Other cases may arise such that conventional vibration isolation techniques, such as those discussed in previous sections, are not practical when the system operates near the resonant frequency. In these problems, one solution may be to add an additional mass connected through a spring and damper such that the additional mass opposes the motion and practically cancels out the motion of the main mass. The additional mass, spring, and damper system is called a *dynamic absorber*. A typical system is shown in Fig. 9-11.

Let us denote the mass, spring constant, and damping coefficient of the main mass by M, K_S, and R_M, respectively. The corresponding proper-

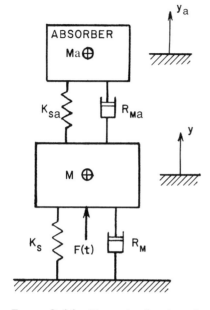

FIGURE 9-11 Dynamic vibration absorber system.

ties of the dynamic absorber will be denoted by M_a, K_{Sa}, and R_{Ma}, respectively. If we apply Newton's second law of motion to each mass, we obtain the following equations:

$$M \frac{d^2 y}{dt^2} + R_M \frac{dy}{dt} + R_{Ma}\left(\frac{dy}{dt} - \frac{dy_a}{dt}\right) + K_S y + K_{Sa}(y - y_a) = F(t) \tag{9-132}$$

$$M_a \frac{d^2 y_a}{dt^2} + R_{Ma}\left(\frac{dy_a}{dt} - \frac{dy}{dt}\right) + K_{Sa}(y_a - y) = 0 \tag{9-133}$$

The variable $y(t)$ is the displacement of the main mass and $y_a(t)$ is the displacement of the additional mass.

We may introduce the following variables into Eqs (9-132) and (9-133):

$$\omega_n^2 = K_S/M \quad \text{and} \quad \omega_a^2 = K_{Sa}/M_a \tag{9-134}$$

$$2\zeta\omega_n = R_M/M \quad \text{and} \quad 2\zeta_a\omega_a = R_{Ma}/M_a \tag{9-135}$$

Vibration Isolation for Noise Control

The following result is obtained:

$$\frac{d^2y}{dt^2} + 2\zeta\omega_n \frac{dy}{dt} + 2\zeta_a\omega_a\left(\frac{dy}{dt} - \frac{dy_a}{dt}\right)(M_a/M)$$
$$+ \omega_n^2 y + \omega_a^2(M_a/M)(y - y_a) = \frac{\omega_n^2 F(t)}{K_S} \quad (9\text{-}136)$$

$$\frac{d^2 y_a}{dt^2} + 2\zeta_a\omega_a\left(\frac{dy_a}{dt} - \frac{dy}{dt}\right) + \omega_a^2(y_a - y) = 0 \quad (9\text{-}137)$$

The steady-state solution for Eqs (9-136) and (9-137) is quite lengthy (Reynolds, 1981). Dynamic absorbers are often designed such that the following relationships are valid:

$$\omega_n = \omega_a \quad \text{and} \quad \zeta = \zeta_a \quad (9\text{-}138)$$

We may define the frequency ratio $r = \omega/\omega_n$, where ω is the frequency of the applied force and the magnification factor for the main mass $\mathrm{MF} = y_{\max}/(F_o/K_S)$. If we have a sinusoidal force applied to the main mass and the conditions of Eq. (9-138) are valid, we may solve for the magnification factor for the main mass:

$$\mathrm{MF} = \frac{y_{\max}}{F_o/K_S} = \frac{[(1-r^2)^2 + (2\zeta r)^2]^{1/2}}{(A^2 + 4B^2)^{1/2}} \quad (9\text{-}139)$$

The quantities A and B are defined as follows:

$$A = r^4 - (2 + \mu + 4\zeta^2)r^2 + 1 \quad (9\text{-}140)$$

$$B = 2\zeta r - \zeta(2 + \mu)r^3 \quad (9\text{-}141)$$

The quantity μ is the ratio of the absorber mass to the main mass, $\mu = M_a/M$.

A similar expression may be obtained for the maximum amplitude of motion for the absorber mass, $y_{a,\max}$:

$$\mathrm{MF}_a = \frac{y_{a,\max}}{F_o/K_{Sa}} = \frac{[1 + (2\zeta r)^2]^{1/2}}{(A^2 + 4B^2)^{1/2}} \quad (9\text{-}142)$$

Note that Eq. (9-142) is valid only under the conditions given by Eq. (9-138).

The transmissivity for the dynamic absorber with $\zeta = \zeta_a$ and $\omega_n = \omega_a$ may be determined from the displacement expression and the fact that the force transmitted is given by $F_T(t) = K_S y_1(t) + R_M(dy_1/dt)$.

$$\mathrm{Tr} = \frac{F_T}{F_o} = \frac{\{[1 + (2\zeta r)^2][(1-r^2)^2 + (2\zeta r)^2]\}^{1/2}}{(A^2 + 4B^2)^{1/2}} \quad (9\text{-}143)$$

The addition of the dynamic absorber to an SDOF system results in a two-degree-of-freedom system, so there are two resonant frequencies for the combination at which the displacements are large, even though the displacement may be small at the resonant frequency for the SDOF system. For the special case of an undamped dynamic absorber ($\zeta = \zeta_a = 0$) and for $\Omega = \omega_a/\omega_n$ not necessarily equal to unity, the magnification factor for the main mass has been determined (Rao, 1986):

$$\text{MF} = \frac{y_{\max}}{(F_o/K_S)} = \frac{\Omega^2 - r^2}{(1 + \Omega^2\mu - r^2)(\Omega^2 - r^2) - \Omega^4\mu} \tag{9-144}$$

The magnification factor for the absorber mass has also been determined for this case:

$$\text{MF}_a = \frac{y_{a,\max}}{(F_o/K_{Sa})} = \frac{(1/\mu)}{(1 + \Omega^2\mu - r^2)(\Omega^2 - r^2) - \Omega^4\mu} \tag{9-145}$$

We note that the expressions for magnification factor, Eqs (9-144) and (9-145), become infinite when the denominator achieves a value of zero:

$$(1 + \Omega^2\mu - r^2)(\Omega^2 - r^2) - \Omega^4\mu = 0$$

$$r^4 - [1 + \Omega^2(1 + \mu)]r^2 + \Omega^2 = 0 \tag{9-146}$$

The solution for the two frequencies from Eq. (9-146) is as follows:

$$r_1^2 = \tfrac{1}{2}[1 + \Omega^2(1 + \mu)] - \{\tfrac{1}{4}[1 + \Omega^2(1 + \mu)]^2 - \Omega^2\}^{1/2} \tag{9-147a}$$

$$r_2^2 = \tfrac{1}{2}[1 + \Omega^2(1 + \mu)] + \{\tfrac{1}{4}[1 + \Omega^2(1 + \mu)]^2 - \Omega^2\}^{1/2} \tag{91-47b}$$

The operating frequency of the system should not be equal to either of the frequencies $f_1 = f_n r_1$ or $f_2 = f_n r_2$. These frequencies are functions of the mass ratio, $\mu = M_a/M$, and the natural frequency ratio, $\Omega = \omega_a/\omega_n = f_a/f_n = [(K_{Sa}/K_S)(M/M_a)]^{1/2}$.

Example 9-8. A large electric motor has an effective mass of 300 kg (661 lb$_m$). The frequency of the driving force causing vibration of the motor is 100 Hz, and the effective force is 250 N (56.2 lb$_f$). The motor is attached to a concrete floor having dimensions of 3.00 m (9.843 ft) × 3.00 m × 150 mm (5.91 in) thick. The damping ratio for the support is $\zeta = 0.06$. It is desired to attach a dynamic absorber to the motor to limit the vibratory motion of the motor. The undamped natural frequency for the absorber and motor are equal, and the damping ratios are the same for the absorber support and the motor support. Determine the required absorber mass and the maximum amplitude of motion for the motor.

Vibration Isolation for Noise Control

The spring constant for the support of the motor (the concrete floor) may be found from the following expression, which is valid for a force applied at the center of a rectangular plate rigidly attached (clamped) along all four edges (Timoshenko and Woinowsky-Krieger, 1959). The plate dimensions are a (shorter length), b (longer length), and h (thickenss).

$$K_S = \frac{C_K E h^3}{(1-\sigma^2)a^2} \quad (9\text{-}148)$$

The quantities E and σ are the Young's modulus and Poisson's ratio for the plate material, respectively. The coefficient C_K depends on the aspect ratio (b/a) for the plate. Numerical values for C_K are listed in Table 9-4. The aspect ratio for the floor slab in this problem is $b/a = 1$. From Appendix C,

TABLE 9-4 Coefficients C_K in the Spring Constant Expression for a Rectangular Plate Having Dimensions a (Shorter Length), b (Longer Length), and h (thickness)[a]

Aspect ratio, b/a	Coefficient, C_K
1.0	14.88
1.1	13.66
1.2	12.88
1.3	12.39
1.4	12.06
1.5	11.85
1.6	11.70
1.8	11.57
2.0	11.54
3.0	11.53
4.0	11.52
∞	11.49

[a]The plate has the driving force applied at the center of the plate, and all four edges of the plate are fixed or clamped. E is Young's modulus, and σ is Poisson's ratio for the plate material.

$$K_S = \frac{C_K E h^3}{(1-\sigma^2)a^2}$$

we find Young's modulus $E = 20.7$ GPa (3.00×10^6 psi) and Poisson's ratio $\sigma = 0.13$ for concrete. Therefore:

$$K_S = \frac{(14.88)(20.7)(10^9)(0.150)^3}{(1 - 0.13^2)(3.00)^2}$$

$K_S = 117.49 \times 10^6$ N/m $= 116.49$ MN/m   (670,900 lb$_f$/in)

The undamped natural frequency for the motor system is as follows:

$\omega_n = (K_S/M)^{1/2} = [(117.49)(10^6)/(300)]^{1/2} = 625.8$ rad/s

$f_n = (625.8)/(2\pi) = 99.6$ Hz $= f_a$

The frequency ratio is practically unity in this case:

$r = f/f_n = (100)/(99.6) = 1.004$

The magnification ratio for the absorber is given by Eq. (9-142):

$$\mathrm{MF}_a = \frac{y_{a,\max}}{(F_o/K_{Sa})} = \frac{y_{a,\max} M_a \omega_a^2}{F_o} = \frac{y_{a,\max} \mu M \omega_n^2}{F_o}$$

$$\mathrm{MF}_a = \frac{(0.003)(\mu)(300)(625.8)^2}{(250)} = 1409.9\,\mu$$

From Eqs (9-140) and (9-141), we find the following values:

$A = (1.004)^4 - [1 + \mu + (4)(0.06)^2](1.004)^2 + 1 = -2.01445 - 1.00802\mu$

$B = (2)(0.06)(1.004) - (0.06)(2 + \mu)(1.004)^3 = -0.0009658 - 0.06072\mu$

The magnification ratio for the absorber may also be written as follows:

$$\mathrm{MF}_a = \frac{\{1 + [(2)(0.06)(1.004)]^2\}^{1/2}}{(A^2 + 4B^2)^{1/2}} = \frac{1.00723}{(A^2 + 4B^2)^{1/2}} = 1409.9\,\mu$$

The expression for the magnification factor of the absorber is a function of the mass ratio μ. By iteration, we find the mass ratio as follows:

$\mu = M_a/M = 0.0203$

The corresponding mass of the absorber may be found:

$M_a = (0.0203)(300) = 6.09$ kg   (13.43 lb$_m$)

The magnification factor for the absorber is as follows:

$$\mathrm{MF}_a = (1409.9)(0.0203) = 28.62 = \frac{y_{a,\max}}{(F_o/K_{Sa})}$$

Vibration Isolation for Noise Control

The required spring constant for the absorber is as follows:

$$K_{Sa} = \frac{(28.62)(250)}{(0.0030)} = 2.385 \times 10^6 \, \text{N/m} = 2.385 \, \text{MN/m} \quad (13{,}620 \, \text{lb}_f/\text{in})$$

The magnification factor for the motor may be determined from Eq. (9-139):

$$\text{MF} = \frac{\{(1 - 1.0042)^2 + [(2)(0.06)(1.004)]^2\}^{1/2}}{[(-0.03491)^2 + (4)(-0.002198)^2]^{1/2}} = 3.431$$

The maximum amplitude of motion of the motor may be found from the definition of the magnification factor:

$$y_{\text{max}} = \frac{F_o \text{MF}}{K_S} = \frac{(250)(3.431)}{(117.49)(10^6)} = 7.30 \times 10^{-6} \, \text{m} = 0.0073 \, \text{mm}$$

The deflection of the floor under the weight (Mg) for the motor may be found as follows:

$$d_{\text{st}} = \frac{Mg}{K_S} = \frac{(300)(9.806)}{(117.49)(10^6)} = 25 \times 10^{-6} \, \text{m} = 0.025 \, \text{mm}$$

The amplitude of the vibration is $(0.0073/0.025) = 0.292$ or 29% of the static deflection.

The undamped resonant frequencies for the motor–absorber system may be found from Eq. (9-147) with $\Omega = 1$ and $\mu = 0.0203$:

$$r_1^2 = \tfrac{1}{2}(2 + \mu) - [\tfrac{1}{4}(2 + \mu)^2 - 1]^{1/2} = 0.8673 \quad \text{and} \quad r_1 = 0.9313$$

$$r_2^2 = \tfrac{1}{2}(2 + \mu) + [\tfrac{1}{4}(2 + \mu)^2 - 1]^{1/2} = 1.1530 \quad \text{and} \quad r_2 = 1.0738$$

$$f_1 = f_n r_1 = (99.6)(0.9313) = 92.8 \, \text{Hz}$$

$$f_2 = f_n r_2 = (99.6)(1.0738) = 106.9 \, \text{Hz}$$

The second harmonic of the forcing frequency $(2)(100) = 200 \, \text{Hz}$ is far removed from these two frequencies.

Let us determine the vibration amplitude for the case in which no dynamic absorber is used. The magnification factor for the basic vibrating system is given by Eq. (9-81):

$$\text{MF} = \frac{1}{\{[1 - 1.004^2]^2 + [(2)(0.060)(1.004)]^2\}^{1/2}} = 8.282$$

The maximum amplitude of vibration of the motor without the dynamic absorber attached is as follows:

$$y_{max} = \frac{MFF_o}{K_S} = \frac{(8.282)(250)}{(117.49)(10^6)} = 17.6 \times 10^{-6}\,m$$
$$= 0.0176\,mm \quad (0.00069\,in)$$

The use of the dynamic absorber reduces the amplitude of vibration of the motor in this example by a factor of $(0.0073/0.0176) = 0.414 = 1/2.41$.

9.10 VIBRATION ISOLATION MATERIALS

There are several commonly used materials for vibration isolation applications, including felt, cork, rubber or elastomers, and metal springs. The characteristics of these materials are presented in this section.

9.10.1 Cork and Felt Resilient Materials

Cork and felt materials are used in vibration isolation applications in which relatively low surface pressures (less than about 200 kPa or 30 psi) can be attained, and applications where the required static deflection is in the range of 0.25–1.8 mm (0.01–0.07 in). In applications where cork and felt materials are used, care should be exercised to prevent contaminants, such as oil and water, from coming into contact with the material. Exposure to oil and water over an extended period of time will cause the material to deteriorate. Cork and felt materials are well-suited for applications in which high-frequency vibration must be damped.

Felt may be composed of wool fibers, and synthetic fibers may also be mixed with the wool fibers. Because the felt has no strong binder for the fibers, felt material should be used only in the form of pads loaded in compression.

When felt is used as a vibration isolation material, the lowest transmissibility is obtained by using a small surface area, a soft felt (one having a low density), and a large thickness. Felt material generally has a damping ratio around $\zeta = 0.060$. The uncompressed thickness of the felt pad is usually selected in practice as 30 or more times the static deflection. The surface pressure for felt should generally be limited to values less than about 140–200 kPa (20–30 psi).

An empirical relationship (Crede, 1951) between the natural frequency and the surface pressure for a felt pad is as follows:

$$f_n\{Hz\} = \left\{ C_1 + \frac{14.93[1 - (0.148)/(\rho_f/\rho_{ref})]}{(P_s/P_{ref})} \right\} \psi(h) \qquad (9\text{-}149)$$

Vibration Isolation for Noise Control

The quantity P_s is the surface pressure on the pad, and the quantity C_1 is a function of the density of the felt ρ_f:

$$C_1 = \frac{14.29}{1 - 1.068(\rho_f/\rho_{ref})} \tag{9-150}$$

The quantity $\psi(h)$ is a function of the unloaded thickness of the felt pad h:

$$\psi(h) = 1 - 0.25 \ln(h/h_{ref}) \tag{9-151}$$

The reference surface pressure $P_{ref} = 101.325 \text{ kPa}$ (14.696 psi), the reference density $\rho_{ref} = 1000 \text{ kg/m}^3$ (62.43 lb_m/ft^3), and the reference thickness is $h_{ref} = 25.4 \text{ mm}$ (1.00 in). The empirical relationship is valid for the following range of the variables:

$$180 \text{ kg/m}^3 \leq \rho \leq 400 \text{ kg/m}^3$$
$$30 \text{ kPa} \leq P_s \leq 250 \text{ kPa}$$
$$5 \text{ mm} \leq h \leq 200 \text{ mm}$$

Cork is one of the oldest materials used for vibration isolation. It is made from the elastic outer layer of bark of the cork oak tree. Like felt, cork is used only in the form of pads under compressive loading. The slope of the load–deflection curve for cork is not constant, but tends to increase as the load on the cork is increased. Typical thickness of cork used for pads is between about 25 mm (1 in) and 150 mm (6 in). The allowable static deflection for cork should be in the range from 0.4 mm (0.015 in) to 2.0 mm (0.079 in). The damping ratio for cork is approximately $\zeta = 0.075$.

An empirical expression (Crede, 1951) for the undamped natural frequency as a function of the surface pressure for cork material is as follows:

$$f_n\{\text{Hz}\} = C_2(P_s/P_{ref})^{-1/3}\psi(h) \tag{9-152}$$

It is recommended that the *design* natural frequency be selected as about 1.5 times the value given by Eq. (9-152). The quantity P_s is the surface pressure on the pad and the quantity C_2 is a function of the density of the cork ρ_c:

$$C_2 = \frac{19.75}{1 - 1.091(\rho_c/\rho_{ref})} \tag{9-153}$$

The quantity $\psi(h)$ is given by Eq. (9-151), and the reference surface pressure, density, and thickness are the same as those used in the expressions for the natural frequency for felt.

The recommended maximum allowable surface pressure for cork is a function of the density of the cork:

$$P_{max}/P_{ref} = 13(\rho_c/\rho_{ref}) \tag{9-154}$$

The empirical relationships are valid for the following range of the variables:

$$200 \text{ kg/m}^3 \leq \rho \leq 350 \text{ kg/m}^3$$
$$70 \text{ kPa} \leq P_s \leq 400 \text{ kPa}$$
$$25 \text{ mm} \leq h \leq 150 \text{ mm}$$

Cork material is sometimes sold in units of board feet, where 1 board foot is equal to $144 \text{ in}^3 = \frac{1}{12} \text{ ft}^3 = 2.36 \text{ dm}^3$. The conversion factor for density expressed in units of $\text{lb}_m/\text{bd ft}$ is as follows:

$$192.22 \, (\text{kg/m}^3)/(\text{lb}_m/\text{bd ft}) = 12.00 \, (\text{lb}_m/\text{ft}^3)/(\text{lb}_m/\text{bd ft})$$

Example 9-9. A machine having a total weight of 1800 N (405 lb$_f$) operates at a speed of 6000 rpm or 100 Hz. Four felt pads (damping ratio, $\zeta = 0.060$) are used to support the machine. The pads have a density of 260 kg/m^3 ($16.2 \text{ lb}_m/\text{ft}^3$) and a thickness of 40 mm (1.57 in). The maximum transmissibility for the system is to be 0.100. Determine the size of the felt pads.

First, let us determine the frequency ratio. The β parameter may be calculated from Eq. (9-109):

$$\beta = 1 + \frac{(2)(0.060)^2(1 - 0.10^2)}{(0.10)^2} = 1 + 0.7128 = 1.7128$$

The frequency ratio is found from Eq. (9-108):

$$r^4 - (2)(1.7128)r^2 - \frac{1 - 0.10^2}{0.10^2} = 0$$

$$r^2 = 1.7128 + [(1.7128)^2 + 99]^{1/2} = 11.809$$

$$r = 3.436 = f/f_n$$

The undamped natural frequency for the system is as follows:

$$f_n = (100)/(3.436) = 29.1 \text{ Hz}$$

Let us check the static deflection for the system, using Eq. (9-17):

$$d = \frac{g}{4\pi^2 f_n^2} = \frac{(9.806)}{(4\pi^2)(29.1)^2} = 0.293 \times 10^{-3} \text{ m} = 0.293 \text{ mm} \quad (0.0115 \text{ in})$$

$$h/d = (40)/(0.293) = 137 > 30$$

Thus, the static deflection is satisfactory in this case.

The surface pressure may now be calculated. The coefficient C_1 may be calculated from Eq. (9-150):

Vibration Isolation for Noise Control

$$C_1 = \frac{(14.29)}{1-(1.068)(0.260)} = 19.78$$

The thickness parameter is found from Eq. (9-151):

$$\psi(h) = 1 - (0.25)\ln(40/25.4) = 0.886$$

The surface pressure may be found from Eq. (9-149):

$$f_n = 29.1\,\text{Hz} = \left\{19.78 + \frac{(14.93)[1-(0.148)/(0.260)]}{(P_s/P_{\text{ref}})}\right\}(0.886)$$

$$P_s/P_{\text{ref}} = \frac{(6.431)}{[(29.1)/(0.886)]-19.78} = 0.4941$$

$$P_s = (0.4941)(101.325) = 50.0\,\text{kPa} \quad (7.25\,\text{psi})$$

The total required load-bearing area for the pads is as follows:

$$S_f = \frac{(1800)}{(50.0)(10^3)} = 0.0360\,\text{m}^2 = 360\,\text{cm}^2 \quad (55.8\,\text{in}^2)$$

The area per pad (four pads are used) is as follows:

$$\tfrac{1}{4}S_f = (360)/(4) = 90.0\,\text{cm}^2 = 9000\,\text{mm}^2 \quad (13.95\,\text{in}^2)$$

If the cross-sectional area of the pad is square, the edge dimensions of the pad are as follows:

$$a = (9000)^{1/2} = 94.9\,\text{mm} \quad (3.72\,\text{in})$$

Let us determine the size of cork pads having a density of $250\,\text{kg/m}^3$ ($15.61\,\text{lb}_m/\text{ft}^3$ or $1.30\,\text{lb}_m/\text{bd ft}$) and a thickness of 50 mm (1.969 in) for the same application. The damping ratio for cork is $\zeta = 0.075$.

Let us determine the design natural frequency:

$$\beta = 1 + \frac{(2)(0.075)^2(1-0.10^2)}{(0.10)^2} = 1 + 1.1138 = 2.1138$$

The frequency ratio is found from Eq. (9-108):

$$r^4 - (2)(2.1138)r^2 - \frac{1-0.10^2}{0.10^2} = 0$$

$$r^2 = 2.1138 + [(2.1138)^2 + 99]^{1/2} = 12.286$$

$$r = 3.505 = f/f_n$$

The design undamped natural frequency for the system is as follows:

$$f_n(\text{design}) = (100)/(3.505) = 28.53\,\text{Hz}$$

According to the recommended design procedure, the design frequency should be about 1.50 times the frequency given by Eq. (9-152):

f_n[Eq. (9-152)] = (28.53)/(1.50) = 19.02 Hz

The pad surface pressure may be determined from Eq. (9-152). The coefficient C_2 is found from Eq. (9-153):

$$C_2 = \frac{(19.75)}{1 - (1.091)(0.250)} = 27.157$$

The thickness function is found from Eq. (9-151):

$\psi(h) = 1 - (0.25)\ln(50/25.4) = 0.8307$

$$f_n = 19.02 \text{ Hz} = \frac{(27.157)(0.8307)}{(P_s/P_{ref})^{1/3}}$$

$P_s/P_{ref} = (1.1861)^3 = 1.668$

$P_s = (1.668)(101.325) = 169.1 \text{ kPa}$ (24.5 psi)

The maximum recommended surface pressure for cork is found from Eq. (9-154):

$P_{max}/P_{ref} = (13)(0.250) = 3.25$

$P_{max} = (3.25)(101.325) = 329 \text{ kPa} > P_s = 169 \text{ kPa}$

The required load-bearing area of the cork is as follows:

$$S_c = \frac{(1800)}{(169.1)(10^3)} = 0.01065 \text{ m}^2 = 106.5 \text{ cm}^2 \quad (16.51 \text{ in}^2)$$

The area per pad (four pads are used) is as follows:

$\frac{1}{4} S_c = (106.5)/(4) = 26.6 \text{ cm}^2 = 2660 \text{ mm}^2$ (4.13 in^2)

Suppose the cork pad is cylindrical. The diameter of one pad may be found as follows:

$D_c = [(4/\pi)(2660)]^{1/2} = 58.2 \text{ mm}$ (2.29 in)

It is noted that both the felt pads and the cork pads provide a practical solution to the vibration isolation problem in this case.

9.10.2 Rubber and Elastomer Vibration Isolators

Rubber and elastomer (such as neoprene) materials are used in vibration isolation applications in which the static deflection on the order of 10–15 mm (0.40–0.60 in) is required and where moderate surface pressures are encountered. Some of the properties of rubber materials are given in Table

Vibration Isolation for Noise Control

9-5. It is noted that the physical properties of rubber are strongly dependent on the hardness of the material.

The design of rubber isolators is best carried out by consulting the data from manufacturers' catalogs, because there are a wide variety of shapes and sizes of mount available. For compression loading, the static deflection should be limited to about 10–15% of the unloaded thickness. For shear loading, the static deflection should be limited to about 20–30%. The design or maximum operating compressive stress for rubber isolators should generally not exceed about 600 kPa (90 psi), and the design shear stress for shear-loaded isolators should be limited to about 200–275 kPa (30–40 psi). The manfacturers' catalog data generally states the maximum allowable load for the isolator. The normal operating temperatures for rubber mounts is usually below 60°C (140°F), although temperatures as high as 75°C (170°F) may be used for some mounts without seriously deteriorating service performance.

The spring constant for a rubber mounting loaded in compression is strongly dependent on the friction between the mount and the support for unbonded mounts. An approximate empirical relationship for the spring constant for compression-loaded rubber isolators, similar to that shown in Fig. 9-12, is as follows:

$$K_S = \frac{SE_c}{h} \left[\frac{S^{1/2}}{(h_o h^2 a^2 / b^2)^{1/3}} \right] \tag{9-155}$$

The quantity a is the length (longer dimension, $a \geq b$), b is the width, and h is the thickness of the rubber slab. The surface area $S = ab$, and the quantity

TABLE 9-5 Properties of Rubber at 21°C (70°F)

Durometer hardness, Shore A	Density, kg/m³	Static modulus, MPa[a]				Damping ratio, ζ
		Compression	Tension	Shear	Bulk	
30	1,010	1.28	1.21	0.345	2,030	0.015
40	1,060	1.86	1.59	0.483	2,220	0.020
50	1,110	2.59	2.10	0.655	2,380	0.045
60	1,180	3.79	3.10	0.965	2,550	0.075
70	1,250	5.17	4.21	1.34	2,870	0.105
80	1,310	8.27	7.07	1.76	3,130	0.140

[a]To convert (MPa) to (psi), the conversion factor is 145.038 psi/MPa.
Source: U.S. Rubber (1941).

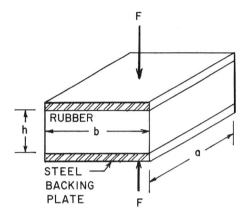

FIGURE 9-12 Rubber or elastomer vibration isolator loaded in compression. The dimension a is the larger dimension, i.e., $a \geq b$.

$h_o = 25.4$ mm (1.00 in). The quantity E_c is Young's modulus in compression for the rubber.

For the case of a mount loaded in shear, consisting of two rubber pads bonded between three steel plates, as shown in Fig. 9-13, the following expression applies for the spring constant:

$$K_S = \frac{2SG}{h} \qquad (9\text{-}156)$$

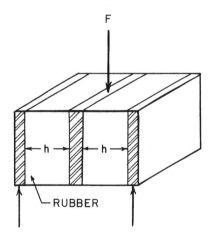

FIGURE 9-13 Rubber or elastomer vibration isolator loaded in shear.

Vibration Isolation for Noise Control

The quantity S is the surface area of the pad on one side, G is the modulus of elasticity in shear, and h is the thickness of one rubber pad. If a single side is used (one rubber pad between two steel plates), the factor 2 is omitted from Eq. (9-156).

For a shear-loaded mount, consisting of a cylinder of rubber mounted between two cylinders of steel, as shown in Fig. 9-14, the spring constant is given by the following expression:

$$K_S = \frac{2\pi h G}{\ln(D_o/D_i)} \qquad (9\text{-}157)$$

The quantities D_o and D_i are the outside diameter and inside diameter of the rubber, respectively, and h is the length of the cylinder. The quantity G is the modulus of elasticity in shear for the rubber pad.

Shear-loaded rubber mounts are usually used in more applications than compression-loaded mounts because the spring constant for the shear-loaded mounts is generally much smaller than that for compression-loaded mounts. The smaller spring constant results in a smaller undamped natural frequency and a larger frequency ratio, which produces a smaller transmissibility.

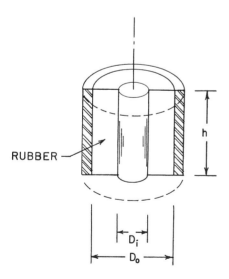

FIGURE 9-14 Cylindrical rubber or elastomer vibration isolator loaded in shear. The load is applied to the ends of the inner and outer steel cylinders.

Compression-loaded rubber mounts are used in applications in which greater load-carrying capacity per unit volume of rubber is required. Rubber in compression is also used where large deflections are not required.

Example 9-10. A machine having a total mass of 1400 kg (3086 lb$_m$) is supported by four rubber isolators loaded in compression. The rubber has a hardness of 50 Durometer. The dimensions of each pad are 76.2 mm (3.00 in) × 76.2 mm × 25.4 mm (1.00 in) thick. The effective dynamic force on the system has an amplitude of 1.60 kN (360 lb$_f$) and a frequency of 45 Hz. Determine the maximum amplitude of motion of the machine and the force transmitted to the foundation.

The surface area of the pad is as follows:

$$S = ab = (0.0762)(0.0762) = 58.06 \times 10^{-4}\,\text{m}^2 = 58.06\,\text{cm}^2 \quad (9.00\,\text{in}^2)$$

The Young's modulus in compression for the rubber is found from Table 9-5 to be $E_c = 2.59$ MPa. The spring constant for one compression-loaded pad may be found from Eq. (9-155):

$$K_S = \frac{(58.06)(10^{-4})(2.59)(10^6)}{(0.0254)} \left\{ \frac{[(58.06)(10^{-4})]^{1/2}}{[(0.0254)(0.0254)^2(1)^2]^{1/3}} \right\}$$

$$K_S = (0.5920)(10^6)(3.00) = 1.776 \times 10^6\,\text{N/m}$$
$$= 1.776\,\text{MN/m} \quad (10{,}140\,\text{lb}_f/\text{in})$$

The spring constant for the system (four pads) is as follows:

$$4K_s = (4)(1.776) = 7.104\,\text{MN/m}$$

We may check the static deflection at this point. The total supported weight of the machine is as follows:

$$Mg = (1400)(9.806) = 13{,}730\,\text{N} = 13.73\,\text{kN} \quad (3086\,\text{lb}_f)$$

The static deflection is found from Eq. (9-16):

$$d = \frac{(13{,}730)}{(7.104)(10^6)} = 0.001932\,\text{m} = 1.932\,\text{mm} \quad (0.076\,\text{in})$$

The static deflection is $(1.832/25.4) = 0.072 = 7.2\%$ of the unloaded thickness, which is satisfactory (maximum allowable is between 10 and 15%).

The surface pressure due to the static load (weight of the machine) is as follows:

$$P_s = (13{,}730/(4)(58.06 \times 10^{-4})) = 591.2 \times 10^3\,\text{Pa}$$
$$= 591.2\,\text{kPa} \quad (85.7\,\text{psi})$$

Vibration Isolation for Noise Control

The value is satisfactory, because the surface pressure should be limited to about 600 kPa (90 psi).

The undamped natural frequency for the system may be found from Eq. (9-15):

$$f_n = \frac{[(7.104)(10^6)/(1400)]^{1/2}}{(2\pi)} = 11.34 \text{ Hz}$$

The frequency ratio $r = f/f_n$ is as follows:

$$r = (45)/(11.34) = 3.969$$

From Table 9-5, we find the damping ratio for the rubber to be $\zeta = 0.045$. The magnification factor is found from Eq. (9-81):

$$\text{MF} = \frac{1}{\{[1-(3.969)^2]^2 + [(2)(0.045)(3.969)]^2\}^{1/2}} = 0.06776 = \frac{K_S y_{max}}{F_o}$$

The maximum amplitude of vibration of the machine is as follows:

$$y_{max} = (0.06776)(1600)/(7.104)(10^6) = 0.0153 \times 10^{-3} \text{ m} = 0.0153 \text{ mm}$$

The transmissibility may be calculated from Eq. (9-103):

$$\text{Tr} = \left\{ \frac{1+[(2)(0.045)(3.969)]^2}{[1-(3.969)^2]^2 + [(2)(0.045)(3.969)]^2} \right\}^{1/2} = 0.07195$$

The transmissibility level is found from Eq. (9-104):

$$L_{Tr} = 20 \log_{10}(0.07195) = -22.9 \text{ dB}$$

The force transmitted to the foundation is as follows:

$$F_T = F_o \text{Tr} = (1600)(0.07195) = 115.1 \text{ N} \quad (25.9 \text{ lb}_f)$$

Example 9-11. An instrument package having a total mass of 20 kg (44.1 lb$_m$) is connected to a structure through four single-shear rubber isolator mounts, as shown in Fig. 9-15. The hardness of the rubber isolator material is 40 Durometer, and the diameter of the cylindrical isolator is 25 mm (0.984 in). The structure vibrates with a maximum amplitude of 3.6 mm (0.142 in) with a frequency of 62 Hz. Determine the thickness of the isolator such that the maximum amplitude of vibration for the instrument package is limited to 0.06 mm (0.0024 in).

The shear force on each isolator is as follows:

$$\tfrac{1}{4} Mg = (\tfrac{1}{4})(20)(9.806) = 49.03 \text{ N} \quad (11.02 \text{ lb}_f)$$

The shear area is as follows:

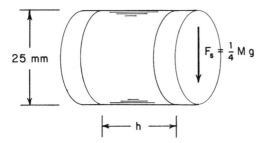

FIGURE 9-15 Diagram for the single-shear rubber isolator mount in Example 9-11.

$$S = (\pi/4)(0.025)^2 = 4.909 \times 10^{-4}\,\text{m}^2 = 4.909\,\text{cm}^2 \quad (0.0761\,\text{in}^2)$$

The shear stress on each isolator is as follows:

$$s_s = (49.03)/(4.909 \times 10^{-4}) = 99.9\,\text{kPa} \quad (14.5\,\text{psi})$$

This value is satisfactory, because the shear stress for shear-loaded isolators should be limited to about 200–275 kPa (30–40 psi).

The required transmissibility for the support system is determined from Eq. (9-131):

$$\text{Tr} = \frac{|y_{2m}|}{|y_{1m}|} = \frac{0.06}{3.60} = 0.01667$$

The damping ratio for 40 Durometer rubber is $\zeta = 0.020$, as given in Table 9-5. The β-parameter is found from Eq. (9-109):

$$\beta = 1 + \frac{(2)(0.020)^2(1 - 0.01667^2)}{(0.01667)^2} = 1 + 2.8792 = 3.8792$$

The frequency ratio is found from Eq. (9-108):

$$r^4 - (2)(3.8792)r^2 - \frac{(1 - 0.01667^2)}{(0.01667)^2} = 0$$

$$r^2 = 3.8792 + [(3.8792)^2 + 3599]^{1/2} = 64.00$$

$$r = 8.00 = f/f_n$$

The required undamped natural frequency for the system is as follows:

$$f_n = \frac{(62)}{(8.00)} = 7.75\,\text{Hz} = \frac{(K_S/M)^{1/2}}{2\pi}$$

The spring constant for the system is as follows:

$$K_S = (4\pi^2)(7.75)^2(20.0) = 47{,}420 \text{ N/m}$$

The spring constant for one of the four individual isolators is as follows:

$$\tfrac{1}{4} K_S = (\tfrac{1}{4})(47{,}420) = 11{,}860 \text{ N/m} = 11.86 \text{ kN/m} \quad (67.7 \text{ lb}_f/\text{in})$$

The shear modulus for the rubber material is $G = 0.483$ MPa from Table 9-5. The thickness for the isolator is as follows:

$$h = \frac{SG}{(\tfrac{1}{4} K_S)} = \frac{(4.909)(10^{-4})(0.483)(10^6)}{(11{,}860)} = 0.0200 \text{ m}$$
$$= 20.0 \text{ mm} \quad (0.787 \text{ in})$$

Let us check the static deflection for each isolator:

$$d = \frac{Mg}{K_S} = \frac{(\tfrac{1}{4} M)g}{(\tfrac{1}{4} K_S)} = \frac{(20)(9.806)}{(47{,}420)} = 0.00414 \text{ m} = 4.14 \text{ mm} \quad (0.163 \text{ in})$$

This value is satisfactory, because the static deflection is $(4.14/20.0) = 0.207 = 20.7\%$ of the unloaded thickness. This is within the range of 20–30%. The maximum dynamic deflection of the isolator is as follows:

$$\delta = d + (y_{2m} + y_{1m}) = 4.14 + (3.60 + 0.06) = 7.80 \text{ mm} \quad (0.307 \text{ in})$$

9.10.3 Metal Spring Isolators

Metal springs are commonly used elements in vibration isolation, especially for applications in which the required undamped natural frequency is less than 5 Hz and large (up to 125 mm or 5 in) static deflections are encountered. Metal springs have been used (Beranek, 1971) to isolate small delicate instrument packages and have been used to isolate masses as large as 400 Mg (400 metric tons or 900,000 lb$_m$). Metal springs have the advantage that spring materials that are not adversely affected by oil and water can be selected.

In many cases, pads of neoprene or other elastomers are mounted in series with the spring (between the spring and the supporting structure, as shown in Fig. 9-16) to prevent high-frequency waves from traveling through the spring into the support structure.

Metal springs have been constructed of several materials, including spring steel, 304 stainless steel, spring brass, phosphor bronze, and beryllium copper. The pertinent physical properties of these materials are listed in Table 9-6. The standard size (wire gauge) for ferrous wire, excluding music wire, is the Washburn and Moen gauge (W&M). The Music Wire

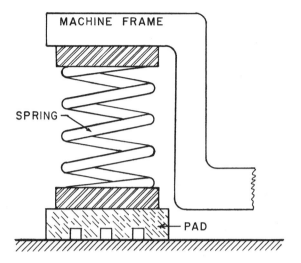

FIGURE 9-16 Metal spring support with damping pad or neoprene or other elastomer.

gauge is used for music wire sizes. For non-ferrous metals, the Brown and Sharp gauge (B&S) or the American wire gauge (AWG) are used (Avallone and Baumeister, 1987).

There are several types of metal springs, including helical springs, leaf springs, Belleville springs (coned disk springs), and torsion springs. In this section, we will concentrate on helical compression springs.

Helical compression springs may be used as freestanding springs (unrestrained springs) or as housed or restrained springs. For freestanding springs, care must be taken to avoid sideways (lateral) instability or buckling. The unrestrained compression spring will always be stable if the following condition is valid:

$$\xi \equiv \left(\frac{1+2\sigma}{2+\sigma}\right)\left(\frac{\pi D}{H_o}\right)^2 \geq 1 \qquad (9\text{-}158)$$

For a value of Poisson's ratio $\sigma = 0.3$, Eq. (9-158) reduces to the following:

$$D/H_o \geq 0.382 \qquad (9\text{-}159)$$

If the value of the ξ ratio is less than 1, the spring will be stable if the ratio of the total deflection Δy to the free height H_o (spring height when unloaded) meets the following criterion (Timoshenko and Gere, 1961):

Vibration Isolation for Noise Control

TABLE 9-6 Properties of Metal Spring Materials

Material	Density, ρ, kg/m^3	Young's modulus, E, GPa	Shear modulus, G, GPa	Poisson's ratio, σ	Shear yield strength, s_{ys}, MPa
Spring steel	7,830	203.4	79.3	0.287	a
304 stainless steel	7,820	190.3	73.1	0.305	179
Spring brass	8,550	106.0	40.1	0.324	200
Phosphor bronze	8,800	111.0	41.4	0.349	315
Beryllium copper	8,230	124.0	48.3	0.285	675

[a]Note that the strength properties are strongly dependent on the heat treatment, cold working, etc. The shear yield strength of spring steels is also dependent on the wire size. The shear yield strength may be approximated by the following expression for small sizes:

$$s_{ys} = s_{ys1}(d_{ref}/d_w)^n$$

where $d_{ref} = 1$ mm and s_{ys1} and n are as follows:

	Size range, mm	s_{ys1}, MPa	Exponent, n
Music wire	0.10–6.5	940	0.146
	>6.5	715	0
Oil-tempered wire	0.50–12	814	0.186
	>12	513	0
Hard-drawn wire	0.70–12	758	0.192
	>12	470	0

$$\frac{\Delta y}{H_o} < \left(\frac{1+\sigma}{1+2\sigma}\right)\left\{1 - \left[1 - \left(\frac{1+2\sigma}{2+\sigma}\right)\left(\frac{\pi D}{H_o}\right)^2\right]^{1/2}\right\} \quad (9\text{-}160)$$

The quantity σ is Poisson's ratio, D is the mean coil diameter for the spring, and H_o is the free height for a spring that is not clamped at the ends. If both ends of the spring are clamped, use $H_o = (2 \times$ free height for spring). If Poisson's ratio for the spring material is $\sigma = 0.3$, Eq. (9-160) reduces to the following:

$$(\Delta y/H_o) < 0.8125\{1 - [1 - 6.87(D/H_o)^2]^{1/2}\} \quad (9\text{-}161)$$

The spring constant for axial compression of a helical spring is given by the following expression:

$$K_S = \frac{G d_w^4}{8 D^3 N_c} \quad (9\text{-}162)$$

The quantity G is the shear modulus, d_w is the wire diameter, D is the mean diameter of the wire coil, and N_c is the number of *active* coils in the spring. The ratio D/d_w is called the *spring index*, and usually has values in the range between 6 and 12 (Shigley and Mischke, 1989).

The number of active coils for a spring depends on the treatment of the ends of the spring wire. A spring with *plain ends* has no special treatment of the ends; the ends are the same as if a spring had been cut to make two shorter springs. For the case of *plain and ground ends*, the last coil on the end of the spring has the wire ground with a flat surface so that approximately half of the coil is in direct contact with the supporting surface. For a *squared* or *closed end*, the end coil is deformed to a zero degree helix angle such that the entire coil touches the supporting surface. For a *squared and ground end*, the end coil is squared, then the wire is ground with a flat surface such that practically all of the end coil is in direct contact with the supporting surface. The number of active coils for the various end treatments is summarized in Table 9-7. Unless other factors indicate otherwise, the ends of the springs should be both squared and ground because better transfer of the load on the spring is achieved for this end treatment.

It is obvious that the spring should not be compressed solid (i.e., with the coils in contact with the adjacent coils) during operation of the spring. The expressions for the solid height of spring with various end treatments are also given in Table 9-7. For a helical compression spring, the spring height under maximum deflection conditions should not be less than about 1.20 times the solid height.

The shear stress in a helical compression spring is a function of the force applied F, which includes both the supported weight and the dynamic force, and the dimensions of the spring:

$$s_s = \frac{8FDk_{sh}}{\pi d_w^3} \tag{9-163}$$

TABLE 9-7 Characteristics of Helical Coil Springs

End treatment	Active coils, N_c	Free height, H_o	Solid height, H_s	Spring pitch, p_s[b]
Plain	N_t[a]	$p_s N_t + d_w$	$d_w(N_t + 1)$	$(H_o - d_w)/N_t$
Plain and ground	$N_t - 1$	$p_s N_t$	$d_w N_t$	H_o/N_t
Squared	$N_t - 2$	$p_s N_c + 3d_w$	$d_w(N_t + 1)$	$(H_o - 3d_w)/N_c$
Squared and ground	$N_t - 2$	$p_s N_c + 2d_w$	$d_w N_t$	$(H_o - 2d_w)/N_c$

[a] N_t is the total number of coils for the spring.
[b] p_s is the spring pitch (reciprocal of the number of coils per unit height of the spring).

Vibration Isolation for Noise Control

The quantity k_{sh} is a *shear-stress correction factor*, given by the following expression:

$$k_{sh} = \frac{2(D/d_w) + 1}{2(D/d_w)} \qquad (9\text{-}164)$$

Springs that support machinery are often subjected to loads in the lateral direction (perpendicular to the axis of the spring). The spring constant in the lateral direction K_{lat} is related to the spring constant in the axial direction K_S by the following expression:

$$\frac{K_{lat}}{K_S} = \frac{2(1+\sigma)}{1 + 4(2+\sigma)(H_o/D)^2} \qquad (9\text{-}165)$$

For the special case of Poisson's ratio $\sigma = 0.3$, Eq. (9-165) reduces to the following:

$$\frac{K_{lat}}{K_S} = \frac{2.60}{1 + 9.20(H_o/D)^2} \qquad (9\text{-}166)$$

Another factor that must be considered in spring design is the problem of *spring surge*. If one end of a helical spring is forced to oscillate, a wave will travel from the moving end to the fixed end of the spring, where the wave will be reflected back to the other end. The critical or surge frequency for a spring that has one end against a flat plate and the other end driven by an oscillatory force is given by (Wolford and Smith, 1976):

$$f_s = \frac{d_w(G/2\rho)^{1/2}}{2\pi D^2 N_c} \qquad (9\text{-}167)$$

The quantity d_w is the diameter of the spring wire, G is the shear modulus, ρ is the density of the spring material, D is the mean diameter of the spring coil, and N_c is the number of active coils for the spring. The surge frequency for the spring should be at least 15 times the forcing frequency for the system to avoid problems with resonance in the spring. The surge frequency may be increased by using a larger spring wire diameter or a smaller spring coil diameter (or a smaller spring index, D/d_w).

Example 9-12. A machine having a mass of 80 kg (176.4 lb$_m$) is to be supported by four metal springs. The springs are to be made of hard-drawn steel wire and have squared and ground ends. The damping ratio for the springs is $\zeta = 0.050$, and the required transmissibility is 0.05 or -26 dB. The driving force for the machine has a maximum amplitude of 5.00 kN and a frequency of 36 Hz. Determine the dimensions of the spring.

First, let us determine the required frequency ratio. The parameter β is found from Eq. (9-109):

$$\beta = 1 + \frac{(2)(0.050)^2(1 - 0.05^2)}{(0.05)^2} = 1 + 1.995 = 2.995$$

The frequency ratio is as follows:

$$r^4 - (2)(2.995)r^2 - \frac{(1 - 0.05^2)}{(0.05)^2} = 0$$

$$r^2 = 2.995 + (2.995^2 + 399)^{1/2} = 23.193$$

$$r = 4.816 = f/f_n$$

The undamped natural frequency for the system is as follows:

$$f_n = (36)/(4.816) = 7.475 \text{ Hz}$$

The required spring constant for one spring, supporting a mass of $(\frac{1}{4}M) = 20$ kg may now be found:

$$K_S = (4\pi^2)(7.475)^2(20) = 44.12 \times 10^3 \text{ N/m} = 44.12 \text{ kN/m} \quad (252 \text{ lb}_f/\text{in})$$

Let us try a spring with a spring index $(D/d_w) \approx 6$ and $N_c = 5$ active coils. The spring wire diameter may be found from Eq. (9-162) with a shear modulus of 79.3 GPa:

$$d_w = \frac{(8)(44.12)(10^3)(6)^3(5)}{(79.3)(10^9)} = 4.807 \times 10^{-3} \text{ m} = 4.807 \text{ mm} \quad (0.1893 \text{ in})$$

The next larger standard gauge is #6 W&M gauge wire, with a diameter of $d_w = 0.1920$ in $= 4.877$ mm. Let us try this size wire for the spring.

The actual mean diameter of the spring may be found from Eq. (9-162):

$$D = \left[\frac{(79.3)(10^9)(0.004877)^4}{(8)(44.12)(10^3)(5)} \right]^{1/3} = 0.02940 \text{ m} = 29.40 \text{ mm} \quad (1.157 \text{ in})$$

The actual spring index is as follows:

$$D/d_w = (29.4)/(4.877) = 6.029$$

The outside diameter of the spring is as follows:

$$D_o = D + d_w = 29.40 + 4.877 = 34.28 \text{ mm} \quad (1.349 \text{ in})$$

Let us check the static shear stress in the spring. The shear correction factor is found from Eq. (9-164):

$$k_{sh} = \frac{(2)(6.029) + 1}{(2)(6.029)} = 1.041$$

Vibration Isolation for Noise Control

The static shear stress for the spring is found from Eq. (9-163):

$$s_s = \frac{(8)(1.041)(20)(9.806)(0.02940)}{(\pi)(0.004877)^3} = 131.8 \times 10^6 \, \text{Pa}$$
$$= 131.8 \, \text{MPa} \quad (19{,}120 \, \text{psi})$$

The shear yield strength for a hard-drawn wire with a diameter of 4.877 mm is found from the data in Table 9-6.

$$s_{ys} = (758)(1/4.877)^{0.192} = 559.2 \, \text{MPa} \quad (81{,}100 \, \text{psi})$$

The static factor of safety for the spring is as follows:

$$\text{FS} = \frac{s_{ys}}{s_s} = \frac{559.2}{131.8} = 4.24 > 3$$

The static shear stress level is satisfactory.

The total number of coils for the spring with squared and ground ends is as follows:

$$N_t = N_c + 2 = 5 + 2 = 7 \text{ coils total}$$

The solid height of the spring is as follows:

$$H_s = (4.877)(7) = 34.14 \, \text{mm} \quad (1.344 \, \text{in})$$

The static deflection for the spring is found from Eq. (9-16):

$$d = \frac{(20)(9.806)}{(44{,}120)} = 0.004445 \, \text{m} = 4.445 \, \text{mm} \quad (0.175 \, \text{in})$$

The magnification factor is found from Eq. (9-81):

$$\text{MF} = \frac{1}{\{[1 - (4.816)^2]^2 + [(2)(0.050)(4.816)]^2\}^{1/2}} = 0.04505 = \frac{K_S y_{\max}}{F_o}$$

The maximum amplitude of vibration for the system is as follows:

$$y_{\max} = \frac{(0.04505)(5000)}{(44{,}120)} = 0.005105 \, \text{m} = 5.105 \, \text{mm} \quad (0.201 \, \text{in})$$

The maximum deflection of the spring is the sum of the static and dynamic displacements:

$$d_{\max} = d + y_{\max} = 4.445 + 5.105 = 9.55 \, \text{mm} \quad (0.376 \, \text{in})$$

To ensure that the spring will not be compressed solid, let us take the design maximum deflection as follows:

$$d_{\max}(\text{design}) = 1.25 d_{\max} = (1.25)(9.55) = 11.94 \, \text{mm} \quad (0.470 \, \text{in})$$

The design free height of the spring may now be determined as the sum of the solid height and the design maximum deflection:

$$H_o = H_s + d_{max}(\text{design}) = 34.14 + 11.94 = 46.08 \text{ mm} \quad (1.814 \text{ in})$$

The pitch of the spring may be determined from the data in Table 9-7.

$$p_s = \frac{H_o - 2d_w}{N_c} = \frac{46.08 - (2)(4.877)}{(5)} = 7.27 \text{ mm} \quad (0.286 \text{ in})$$

The pitch is the center-to-center spacing of the wire in adjacent coils of the spring. There are $(1/7.27) = 0.1376$ coils/mm $= 1.376$ coils/cm or 3.50 coils/inch height of the spring.

Let us check the buckling stability of the spring. The ξ parameter may be found from Eq. (9-158):

$$\xi = \frac{[1 + (2)(0.287)][(\pi)(29.40)/(46.08)]^2}{(2 + 0.287)} = 2.765 > 1$$

The spring is quite stable and buckling will not be a problem.

Finally, let us check the surge frequency from Eq. (9-167):

$$f_s = \frac{(0.00487)}{(2\pi)(0.02940)^2(5)} \left[\frac{(79.3)(10^9)}{(2)(7830)} \right]^{1/2} = 404 \text{ Hz}$$

The ratio of the surge frequency to the driving force frequency is as follows:

$$f_s/f = (404)/(36) = 11.2$$

Although the surge frequency is not greater than 15 times the forcing frequency, surging probably would to be a serious problem, in this case, because of the damping in the support system.

A summary of the spring characteristics is as follows:

spring wire diameter, d_w	4.877 mm (0.1920 in), #6 W&M gauge wire
spring mean diameter, D	29.40 mm (1.157 in)
spring outside diameter, D_o	34.28 mm (1.349 in)
number of coils	7 total coils; 5 active coils
spring pitch, p_s	7.27 mm (0.286 in)
free height, H_o	46.08 mm (1.814 in)
solid height, H_s	34.14 mm (1.344 in)

9.11 EFFECTS OF VIBRATION ON HUMANS

The human body is a relatively complex vibratory system, because it contains both linear and nonlinear "springs" and "dampers." As in the case of

Vibration Isolation for Noise Control

hearing damage studies, it is difficult (and unethical, in extreme cases) to conduct research on vibratory damage on living human subjects. As a consequence of this difficulty, much of the research data on vibratory effects on humans have been obtained from experiments on animals or by simulation.

For the frequency range below about 40 Hz, the human body can be modeled approximately by a system of masses (the head, upper torso, hips, legs, and arms), spring elements, and damping elements (Coermann et al., 1960).

Generally, exposure to vibration at the workplace is more severe than vibration exposure at home, in terms of both levels of vibration and duration of vibration exposure. Most of the work-related whole-body vibration exposure arises from forces transmitted through the person's feet while standing, or the buttocks while seated (Von Gierke and Goldman, 1988). Hand–arm vibration exposure may also occur while holding tools.

There are two important frequency regions as far as vibration of the whole human body is concerned: (a) from 3 Hz to 6 Hz, where resonance of the thorax–abdomen system occurs, and (b) from 20 Hz to 30 Hz, where resonance of the head–neck–shoulder system occurs. The resonance of the thorax–abdomen system is expecially important, because this resonance places stringent requirements on the vibration isolation of a sitting or standing person. For example, at a frequency of 4 Hz, the acceleration of the hip region of a standing person is approximately 1.8 times the acceleration of the surface on which the person is standing. For a person seated, the acceleration of the head–shoulder region is about 3.5 times the acceleration of the surface on which the person is seated, for a frequency of 30 Hz.

In the frequency region between 60 Hz and 90 Hz, resonance in the eyeballs occurs. There is a resonant effect in the lower jaw–skull system in the frequency range between 100 Hz and 200 Hz. Resonance within the skull occurs in the frequency region between 300 Hz and 400 Hz. Human response to vibration at frequencies above about 100 Hz is influenced significantly by the clothing or shoes at the point of application of the vibratory force.

Vibration at frequencies below about 1 Hz affects the inner ear and produces annoyance, such as *cinerosis* (motion sickness). For frequencies greater than about 100 Hz, the perception of vibration is noticed mainly on the skin, and depends on the specific body region affected and on the clothing, shoes, etc., that the person is wearing.

Criteria for acceptable vibration exposure have been developed by national (ANSI, 1979) and international (ISO, 1985) standards organizations. The rms acceleration levels corresponding to fatigue-induced decrease in work proficiency are given by the following relationships. If a person is exposed to rms acceleration levels that exceed the values given by the following relationships, the person will generally experience noticeable fatigue and decreased job proficiency in most tasks:

for $1\,\text{Hz} \le f < 4\,\text{Hz}$

$$L_a = 90 - 10\log_{10}(f/4) + \text{CF}_t \qquad (9\text{-}168)$$

for $4\,\text{Hz} \le f \le 8\,\text{Hz}$

$$L_a = 90\,\text{dB} + \text{CF}_t \qquad (9\text{-}169)$$

for $8\,\text{Hz} < f \le 80\,\text{Hz}$

$$L_a = 90 + 20\log_{10}(f/8) + \text{CF}_t \qquad (9\text{-}170)$$

The rms acceleration level must not exceed $L_a(\text{max}) = 116.8\,\text{dB}$, which corresponds to an acceleration of $0.707g$ or $6.94\,\text{m/s}^2$ ($22.75\,\text{ft/sec}^2$).

The factor CF_t is a correction for the duration of the acceleration exposure, and may be estimated by the following relationships:

for $t \le 8\,\text{hours}$

$$\text{CF}_t = 20[1 - (t/8)^{1/2}] \qquad (9\text{-}171)$$

for $8 < t \le 16\,\text{hours}$

$$\text{CF}_t = 20[(8/t)^{1/2} - 1] \qquad (9\text{-}172)$$

The acceleration limits for a condition of "reduced comfort" due to the vibration may be found by subtracting 10 dB from the values given by Eqs (9-168), (9-169), or (9-170). The upper bound of allowable acceleration exposure, which represents a hazard to the person's health if exceeded, is found by adding 6 dB to the values given by Eqs (9-168), (9-169), or (9-170).

The acceleration level is defined as follows:

$$L_a = 20\log_{10}(a/a_{\text{ref}}) \qquad (9\text{-}173)$$

The reference acceleration, as given in Table 2-1, is $a_{\text{ref}} = 10\,\mu\text{m/s}^2$ ($0.00039\,\text{in/sec}^2$). An acceleration of $1g$ ($g = 9.806\,\text{m/s}^2 = 32.174\,\text{ft/sec}^2 = 386.1\,\text{in/sec}^2$) corresponds to an acceleration level of the following:

$$L_a = 20\log_{10}(9.806/10 \times 10^{-6}) = 119.8\,\text{dB} \approx 120\,\text{dB}$$

If the vibrational displacement is sinusoidal, the rms acceleration is related to the maximum or peak acceleration a_{max} by:

$$a_{\text{rms}} = a_{\text{max}}/2^{1/2} = 0.707 a_{\text{max}} \qquad (9\text{-}174)$$

For a vibrational displacement $y(t)$ given by the following sinusoidal relationship, we may determine the relationship between the acceleration and displacement:

Vibration Isolation for Noise Control

$$y(t) = y_{max} e^{j\omega t} \qquad (9\text{-}175)$$

$$a(t) = \frac{dy}{dt} = -\omega^2 y_{max} e^{j\omega t} = \omega^2 y_{max} e^{j(\omega t + \pi)} \qquad (9\text{-}176)$$

The acceleration of the mass is π radians or $180°$ out of phase with the displacement. The maximum or peak acceleration is related to the maximum displacement by the following expression:

$$a_{max} = \omega^2 y_{max} = 4\pi^2 f^2 y_{max} \qquad (9\text{-}177)$$

If we differentiate the expression for the vibration of a mass subjected to displacement excitation, given by Eqs (9-124) and (9-130), we obtain the following relationship for the maximum acceleration of a mass subjected to displacement excitation:

$$a_{2,max} = \omega^2 y_{1,max} \text{Tr} = (K_S/M) r^2 y_{1,max} \text{Tr} \qquad (9\text{-}178)$$

The quantity r is the frequency ratio, $r = \omega/\omega_n = f/f_n$, and Tr is the transmissibility.

If we substitute for the transmissibility given by Eq. (9-103), we obtain the following dimensionless relationships for the maximum acceleration of a mass subjected to displacement excitation:

$$\frac{(a_{2,max}/g)}{(K_S y_{1,max}/Mg)} = r^2 \text{Tr} = r^2 \left[\frac{1 + (2\zeta r)^2}{(1-r^2)^2 + (2\zeta r)^2}\right]^{1/2} \qquad (9\text{-}179)$$

If the spring constant K_S is the design variable that we are seeking, the following form is more convenient to use:

$$\frac{(a_{2,max}/g)}{(4\pi^2 f^2 y_{1,max}/g)} = \text{Tr} = \left[\frac{1 + (2\zeta r)^2}{(1-r^2)^2 + (2\zeta r)^2}\right]^{1/2} \qquad (9\text{-}180)$$

Example 9-13. A person is seated in a seat that is supported by a spring–damper system. The mass of the seat and the person is 80 kg (176.4 lb$_m$), and the damping ratio for the support system is $\zeta = 0.060$. The maximum amplitude of motion for the foundation to which the support system is attached is 5 mm (0.197 in), and the vibration frequency for the foundation is 10 Hz. The time that the person will be seated is 6 hours per day. Determine the spring constant for the support such that the person would experience little fatigue-induced decrease in work proficiency.

The correction for time of vibration exposure may be found from Eq. (9-171):

$$CF_t = 20[1 - (6/8)^{1/2}] = 2.68 \text{ dB}$$

The rms acceleration level at the fatigue-induced proficiency limit is found from Eq. (9-170) for a frequency of 10 Hz:

$$L_a = 90 + 20 \log_{10}(10/8) + 2.68 = 90 + 1.94 + 2.68 = 94.62 \, \text{dB}$$

$$a_{\text{rms}} = (10)(10^{-6})(10^{94.62/20}) = 0.5383 \, \text{m/s}^2 \quad (0.1472 \, \text{in/sec}^2)$$

For design purposes, let us use an acceleration that is 80% of the limiting value:

$$a_{\text{rms}} = (0.80)(0.5383) = 0.4306 \, \text{m/s}^2 \quad (0.1177 \, \text{in/sec}^2)$$

The peak acceleration, assuming sinusoidal excitation, is as follows:

$$a_{2,\text{max}} = (2)^{1/2}(0.4306) = 0.6090 \, \text{m/s}^2 \quad (0.1665 \, \text{in/sec}^2)$$

$$a_{2,\text{max}}/g = (0.6090)/(9.806) = 0.06210$$

Let us calculate the parameter in Eq. (9-180):

$$\frac{4\pi^2 f^2 y_{1,\text{max}}}{g} = \frac{(4\pi^2)(10)^2(0.0050)}{(9.806)} = 2.0130$$

The required transmissibility for the support system is found from Eq. (9-180):

$$\text{Tr} = \frac{0.06210}{2.0130} = 0.03085 \quad (L_{\text{Tr}} = -30.2 \, \text{dB})$$

The parameter β is as follows:

$$\beta = 1 + \frac{(2)(0.06)^2(1 - 0.03085)^2}{(0.03085)^2} = 1 + 7.557 = 8.557$$

The required frequency ratio is found from Eq. (9-108):

$$r^2 = 8.557 + \left[(8.557)^2 + \frac{(1 - 0.03085^2)}{(0.03085)^2}\right]^{1/2} = 42.066$$

$$r = (42.066)^{1/2} = 6.486 = f/f_n$$

The required undamped natural frequency for the support system is as follows:

$$f_n = (10)/(6.486) = 1.542 \, \text{Hz}$$

The required spring constant for the support may now be calculated:

$$K_S = (4\pi^2)(1.542)^2(80) = 7508 \, \text{N/m} = 7.508 \, \text{kN/m} \quad (42.87 \, \text{lb}_f/\text{in})$$

Note that the static displacement under the weight of the person and the seat is as follows:

Vibration Isolation for Noise Control

$$d = \frac{Mg}{K_S} = \frac{(80)(9.806)}{(7508)} = 0.1045 = 104.5\,\text{mm} \quad (4.11\,\text{in})$$

The person may experience some problems when sitting in a seat for which the supports deflect this much, however.

The maximum amplitude of vibration for the seated person during vibration is as follows:

$$y_{2,\text{max}}/y_{1,\text{max}} = \text{Tr} = 0.03085$$

$$y_{2,\text{max}} = (0.03085)(5.00) = 0.154\,\text{mm} \quad (0.0061\,\text{in})$$

The maximum or peak velocity for the foundation is as follows:

$$v_{1,\text{max}} = 2\pi f y_{1,\text{max}} = (2\pi)(10)(0.0050) = 0.314\,\text{m/s}$$
$$= 314\,\text{mm/s} \quad (12.4\,\text{in/sec})$$

According to the data in Table 9-3, this degree of vibration corresponds to a very rough machine.

PROBLEMS

9-1. A machine has a mass of 75 kg (165.3 lb$_m$), and the static deflection of the support system is 10 mm (0.394 in). Determine the undamped natural frequency for the support system and the value of the spring constant.

9-2. A spring–mass system has a mass of 35 kg (77.2 lb$_m$) and a spring constant of 200 kN/m (1142 lb$_f$/in). The system is located on the surface of the moon, where the local acceleration due to gravity is 1.70 m/s^2 (5.58 ft/sec^2 or 66.9 in/sec^2). Determine the undamped natural frequency and the static deflection for the system.

9-3. A machine has a mass of 75 kg (165.3 lb$_m$), and the undamped natural frequency for the support system is 4 Hz. Determine the damping coefficient R_M such that the damping ratio is $\zeta = 0.080$. Determine the mechanical quality factor Q_M, the decay rate Δ, the reverberation time T_{60}, and the loss factor η for the support system.

9-4. A machine has a mass of 22.5 kg (49.6 lb$_m$) and is subjected to a sinusoidal force having a maximum amplitude of 1.25 kN (281 lb$_f$) and a frequency of 12 Hz. The damping ratio for the support system is $\zeta = 0.040$. The spring system supporting the machine deflects 27.6 mm (1.087 in) when the weight of the machine is applied. Determine the magnitude of the dynamic deflection and velocity of the machine.

9-5. A machine has a mass of 115 kg (253.5 lb$_m$), and the frequency of the exciting force on the machine is 10 Hz. The maximum amplitude of the exciting force is 4.50 kN (1012 lb$_f$). The suspension system for the machine has a damping ratio of $\zeta = 0.10$. It is desired to select a spring support such that the magnification factor for the machine motion is 0.0250. Determine the required undamped natural frequency for the system, the spring constant for the support, and the maximum amplitude of the displacement of the machine.

9-6. A compressor having a mass of 15 kg (33.1 lb$_m$) is placed directly on the floor of a room. The deflection of the floor caused by the weight of the compressor was measured to be 0.621 mm (0.0244 in). The damping coefficient for the floor is 0.008, and the effect of the floor mass may be neglected. The exciting force has a magnitude of 150 N (33.7 lb$_f$) and a frequency of 20 Hz. Determine the magnitude of the displacement of the compressor.

9-7. A device has a mass of 2.00 kg (4.41 lb$_m$). The device is supported by a spring–damper system with a spring constant of $K_S = 4.00$ kN/m (22.8 lb$_f$/in) and a damping coefficient of $R_M = 36$ N-s/m (0.206 lb$_f$-sec/in). The system is excited by a driving force having a maximum amplitude of 8.00 N (1.80 lb$_f$). Determine the maximum amplitude of motion y_{max} for the mass if the driving force has a frequency of (a) the undamped natural frequency of the system, (b) 1.50 Hz, and (c) 30 Hz.

9-8. For the device described in Problem 9-7, determine the mechanical impedance, mechanical mobility, and magnitude of the maximum velocity for the system at the following frequencies of the driving force: (a) the undamped natural frequency of the system, (b) 1.50 Hz, and (c) 30 Hz.

9-9. A machine has a mass of 25 kg (55.1 lb$_m$). The spring constant for the support is 60 kN/m (378 lb$_f$/in), and the support damping coefficient is 245 N-s/m (1.45 lb$_f$-sec/in). The exciting force on the machine has a magnitude of 625 N (140.5 lb$_f$) and a frequency of 60 Hz. Determine the magnitude of the force transmitted to the foundation and the phase angle between the transmitted force and the excitation force.

9-10. A stamping machine has a total mass of 50 kg (110.2 lb$_m$). The machine is supported by springs having a total spring constant of 640 kN/m (3654 lb$_f$/in) and the damping coefficient for the supports is 28 N-s/m (0.143 lb$_f$-sec/in). The applied force on the stamping machine has a magnitude of 1.00 kN (224.8 lb$_f$) and a frequency of 72 Hz. Determine the transmissibility level (dB) and the magnitude of the force transmitted into the foundation.

9-11. A refrigerator having a mass of 30 kg (66.1 lb$_m$) is supported by a spring–damper system. The damping ratio for the system is 0.200, and the frequency of the exciting force is 60 Hz. Determine the

Vibration Isolation for Noise Control

required spring constant for the support system such that the transmissibility is 0.100 or the transmissibility level is $-20\,\text{dB}$.

9-12. A washing machine has a mass of 25 kg (55.1 lb$_\text{m}$) and is supported by a spring–damper system having a damping ratio $\zeta = 0.050$. The washer is subjected to an external sinusoidal force having a maximum amplitude of 250 N (56.2 lb$_\text{f}$) and a frequency of 36 Hz. It is desired to limit the force transmitted through the support system to the floor to 2.73 N (0.614 lb$_\text{f}$). Determine the required spring constant for the support system. What is the maximum amplitude of the velocity of the washer under this condition?

9-13. A machine has a total mass of 180 kg (397 lb$_\text{m}$). The moving element on the machine has a mass of 16 kg (35.3 lb$_\text{m}$) and an eccentricity of 90 mm (3.543 in). The system is mounted on a spring–damper support, with a spring constant of 64 kN/m (365 lb$_\text{f}$/in) and a damping coefficient of 4.50 kN-s/m (25.7 lb$_\text{f}$-sec/in). The operating speed of the machine is 900 rpm (15 Hz). Determine the maximum amplitude of the vibratory motion (displacement and velocity) of the machine.

9-14. A machine having a total mass of 100 kg (220 lb$_\text{m}$) has a 20 kg (44.1 lb$_\text{m}$) rotor with a 6 mm (0.236 in) eccentricity. The operating speed of the machine is 600 rpm (10 Hz), and the unit is constrained to move in the vertical direction only. The damping ratio for the support system is 0.010. It is desired to limit the maximum amplitude of the vibratory motion to 0.25 mm (0.0098 in). Determine the value of the spring constant for the support system.

9-15. A motor is rigidly attached to a seismic mass, and the total mass of the system is 1200 kg (2646 lb$_\text{m}$)). The damping ratio for the spring–damper support system is 0.100. The machine operates at a frequency of 3600 rpm (60 Hz). There is a rotating unbalance of 24 kg (52.9 lb$_\text{m}$) with an eccentricity of 2.00 mm (0.0787 in). Determine the value of the spring constant such that the maximum force transmitted to the floor is 400 N (89.9 lb$_\text{f}$). Determine the maximum amplitude of the vibratory velocity for the motor system.

9-16. An indicator system having a mass of 50 kg (110.2 lb$_\text{m}$) is supported by a spring–damper system. The spring constant for the support is 68.5 kN/m (391 lb$_\text{f}$/in) and the damping ratio is $\zeta = 0.050$. The system is attached to a panel that vibrates with a maximum amplitude of 9.00 mm (0.354 in) at a frequency of 30 Hz. Determine the maximum amplitude of motion (displacement) for the indicator system.

9-17. A small temperature transducer having a mass of 150 g (0.331 lb$_\text{m}$) is attached to a panel, which vibrates with a maximum amplitude of 20 mm (0.787 in). The frequency of vibration of the panel is 167.3 Hz. The damping ratio for the spring–damper support of the transducer is $\zeta = 0.020$. Determine the spring constant for the support such that the maximum amplitude of vibration of the transducer is limited to 0.20 mm (0.00787 in).

9-18. A compressor has a mass of 350 kg (771.6 lb$_m$). The compressor is supported by a spring–damper system with a spring constant of 5.527 MN/m (31,560 lb$_f$/in) and a damping ratio of $\zeta = 0.050$. The exciting force has a maximum amplitude of 11.054 kN (2485 lb$_f$) and a frequency of 20 Hz. Determine the maximum amplitude of the displacement of the compressor. To limit the motion of the compressor, a dynamic vibration absorber is added to the system described. The mass of the absorber is 280 kg (617.3 lb$_m$), the spring constant is 4.422 MN/m (25,250 lb$_f$/in), and the damping ratio is 0.050. Determine the maximum amplitude of the displacement of the compressor with the dynamic absorber attached. Determine the maximum amplitude of displacement of the absorber mass.

9-19. A machine that weighs 3600 N (1601 lb$_f$) is to be supported by four cork vibration absorbers, each having a thickness of 100 mm (3.937 in) and a damping ratio of 0.075. The density of the cork is 250 kg/m^3 (15.61 lb$_m$/ft^3). The frequency of the exciting force on the machine is 60.4 Hz. The required transmissibility is 0.20 (transmissibility level of -14 dB). If the cross section of each cork pad is square, determine the design dimensions of the cork pads.

9-20. A machine is to be supported by four felt pads having a thickness of 50 mm (1.969 in), and the pads have a square cross section. The damping ratio for the felt material is 0.060, and the density is 340 kg/m^3 (21.23 lb$_m$/ft^3). The weight of the machine is 11.30 kN (2540 lb$_f$). It is desired that the maximum force transmitted to the foundation is 1200 N (270 lb$_f$). The frequency of the exciting force on the machine is 90 Hz, and the magnitude of the exciting force is $F_o = 6.00$ kN (1349 lb$_f$). Determine the dimensions of each felt pad.

9-21. A rubber mounting is designed to be loaded in compression. The mounting is made of 40 Durometer rubber with dimensions of 75 mm (2.953 in) × 100 mm (3.937 in) × 60 mm (2.362 in) thick. The static deflection of the mount is 5.00 mm (0.197 in). Determine the undamped natural frequency for the system and the value of the static load supported.

9-22. A shear-loaded mount, as shown in Fig. 9-14, is made of 60 Durometer rubber. The dimensions of the mount cross section are 28.5 mm (1.122 in) inside diameter × 53.4 mm (2.102 in) outside diameter. The spring constant for the mount is 965 kN/m (5510 lb$_f$/in), and the weight supported by the mount is 1930 N (434 lb$_f$).

Vibration Isolation for Noise Control

Determine the length of the mount and undamped natural frequency for the system.

9-23. A helical spring is made of oil-tempered #8 W&M gauge (wire diameter, 4.115 mm or 0.1620 in) spring steel wire. The mean diameter of the spring coil is 30 mm (1.181 in), and the free (unloaded) height of the spring is 90 mm (3.543 in). The ends are squared and ground, and the number of active coils is 10. Determine the spring constant and solid height for the spring. If a compressive load of 260 N (26.51 lb$_f$) is applied to the spring, determine the shear stress in the spring. If the spring is unconstrained, determine whether the spring is stable under compressive loads.

9-24. A helical spring is made of 304 stainless steel and has squared and ground ends. The spring has 8 active coils and the spring index $(D/d_w) = 9.00$. The spring constant for the spring is $K_S = 10.45$ kN/m (59.7 lb$_f$/in). Determine the wire diameter, mean coil diameter, and outside diameter for the spring. If the static shear stress is limited to 70 MPa, determine the maximum static load to which the spring may be subjected. If the height of the spring under this load is equal to 1.25 times the solid height of the spring, determine the free (unloaded) length of the spring. Is the unconstrained spring stable under the maximum compressive load?

9-25. A helical spring is made of 304 stainless steel and has 5 active coils. The ends are squared and ground. The spring index $(D/d_w) = 6.50$, and the spring wire is #14 W&M gauge, with a diameter of 2.032 mm (0.0800 in). Determine the surge frequency for the spring.

9-26. A person is seated on a support for 4 hours. The frequency of vibration of the support is 20 Hz. Determine the rms acceleration and maximum velocity amplitude to which the person may be exposed to produce fatigue-induced decrease in work proficiency. Determine the rms acceleration and maximum velocity amplitude to which the person may be exposed to produce a feeling of "reduced comfort."

9-27. A person stands on a vibrating platform for 2 hours during the work activities each day. The mass of the person and the platform is 90 kg (198.4 lb$_m$), and the damping ratio for the support of the platform is $\zeta = 0.080$. The maximum amplitude of vibration of the foundation on which the platform is attached is 2.50 mm (0.984 in), and the frequency of vibration of the foundation is 20 Hz. Determine the undamped natural frequency and spring constant for the support such that the vibration is at the limit of the fatigue-induced decrease in work proficiency condition.

REFERENCES

American National Standards Institute (ANSI). 1979. Guide for the evaluation of human exposure to whole-body vibration, ANSI S3.18-1979. Acoustical Society of American, New York.

Avallone, E. A. and Baumeister III, T. 1987. *Marks' Standard Handbook for Mechanical Engineers*, 9th edn, pp. 6–45. McGraw-Hill, New York.

Beranek, L. L. 1971. *Noise and Vibration Control*, pp. 422, 453. McGraw-Hill, New York.

Coermann, R. R., Ziegenruecker, G. H., Witter, A. L. and Von Gierke, H. E. 1960. The passive dynamic mechanical properties of the human thorax–abdomen system and of the whole body. *Aerospace Med.* 31(60): 443–455.

Crede, C. E. 1951. *Vibration and Shock Isolation*. John Wiley and Sons, New York.

Fox, R. L. 1971. Machinery vibration monitoring and analysis techniques, *Sound and Vibration* 5: 35–40.

ISO. 1985. Evaluation of human exposure to whole-body vibration. Part 1: General requirements, ISO 2631/1-1985. International Standarization Organization, Geneva, Switzerland.

Kinsler, L. E., Frey, A. R., Copens, A. B., and Sanders, J. V. 1982. *Fundamentals of Acoustics*, 3rd edn, p. 16. John Wiley and Sons, New York.

Plunkett, R. 1959. Measurement of damping. In: *Structural Damping*, J. F. Ruzicka (Ed.). American Society of Mechanical Engineers, New York.

Rao, S. S. 1986. *Mechanical Vibrations*, pp. 447–454. Addison-Wesley, Reading, MA.

Reynolds, D. D. 1981. *Engineering Principles of Acoustics*, pp. 107–114. Allyn and Bacon, Boston.

Shigley, J. E. and Mischke, C. R. 1989. *Mechanical Engineering Design*, 5th edn, p. 414. McGraw-Hill, New York.

Thomson, W. T. and Dahleh, M. D. 1998. *Theory of Vibrations with Applications*. Prentice Hall, Upper Saddle River, NJ.

Timoshenko, S. P. and Gere, J. M. 1961. *Theory of Elastic Instability*, 2nd edn, pp. 142–144. McGraw-Hill, New York.

Timoshenko, S. P. and Woinowsky-Krieger, S. 1959. *Theory of Plates and Shells*, 2nd edn, pp. 203–206. McGraw-Hill, New York.

Tongue, B. H. 1996. *Principles of Vibrations*. Oxford University Press, New York.

Ungar, E. E. and Kerwin, Jr. E. M. 1962. Loss factors of viscoelastic systems in terms of energy concepts. *J. Acoust. Soc. Am.* 34(7): 954–957.

U.S. Rubber Company. 1941. *Some Physical Properties of Rubber*. U.S. Rubber Co., Cleveland, Ohio.

Von Gierke, H. E. and Goldman, D. E. 1988. Effects of shock and vibration on man. In: *Shock and Vibration Handbook*, 3rd edn, C. M. Harris (Ed.). McGraw-Hill, New York.

Wolford, J. C. and Smith, G. M. 1976. Surge of helical springs. *Mechn. Eng. News* 13(1): 4–9.

10
Case Studies in Noise Control

10.1 INTRODUCTION

We will present some case studies illustrating the application of noise control procedures that have been discussed in the previous chapters of this book. Additional case studies may be found in the literature (Beranek, 1960; Salmon et al., 1975; Faulkner, 1976).

The case studies include a statement of the noise problem and a description of the acoustic measures taken to solve the problem. The measurements taken before the noise control procedures were applied generally indicate that the noise levels were in violation of some acoustic criterion—in many cases the Occupation Safety and Health Administration (OSHA) criterion. A short analysis is presented to identify the primary sources of noise causing the problem. Some noise control procedures are suggested, and the degree of noise reduction achieved by the application of these procedures is described. In most cases, the cost of the application of the noise control procedures is important. Finally, some of the pitfalls or potential problems to avoid when attempting to reduce the noise level are noted.

The specific noise control procedures illustrated by the case studies are as follows:

1. Acoustic barriers—folding carton packing station with an air-hammer noise source

2. Equipment enclosure—metal cut-off saw noise
3. Operator enclosure—paper machine wet end
4. Sound absorption material—air scrap handling duct
5. Silencer—air-operated hoist motor
6. Vibration isolation—blanking press
7. Acoustic wall treatment—small meeting room.

10.2 FOLDING CARTON PACKING STATION NOISE

In manufacturing folding cartons, such as those used for soft drink bottles, the individual cartons are cut and stacked on a pallet (Salmon et al., 1975). The cartons are held together for transfer by a nick or uncut part of the carton. The individual cartons are separated by an air-driven chisel, which breaks the nicks and frees the entire stack of cartons. When the operations are completed, the stacks of cartons are packed in cases for shipment. A schematic of the layout of the stripper and packer line is shown in Fig. 10-1.

The air hammer or chisel produces noise that has not been practical to eliminate by system design. Because of this characteristic, the stripper was required to wear hearing protection while working with the air chisel. The cartons are transferred from a conveyor belt to a skid for shipment. The packer is located about 4.27 m (14 ft) from the air chisel. The purpose of this noise control study was to develop a means for reduction of the noise experienced by the packer at the end of the conveyor.

10.2.1 Analysis

The noise generated by the air chisel is broadband, with no significant peaks in the frequency spectrum, as shown in Fig. 10-2 (Plunkett, 1955). The A-weighted sound level at the packer's location with no noise treatment is 95 dBA, which exceeds the OSHA limit for an 8-hour daily noise exposure. A barrier would solve the noise control problem if the direct field were found to be significant, compared with the reverberant field, as discussed in Chapter 7.

The room constant R for the space without the barrier may be estimated from measurements of the reverberation time T_r and the total surface area of the space S_o. The number of absorption units a is defined by Eq. (7-34):

$$\frac{a}{S_o} = \frac{55.26 V}{T_r c S_o} \qquad (10\text{-}1)$$

Case Studies in Noise Control

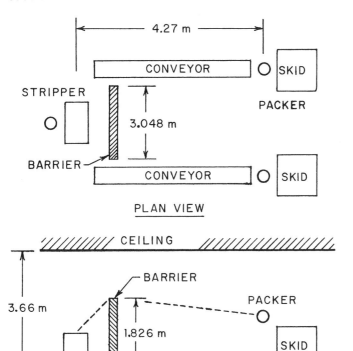

FIGURE 10-1 Air-hammer stripper and packer line layout.

The average surface absorption coefficient $\bar{\alpha}$ may be found from Eq. (7-30):

$$1 - \bar{\alpha} = \exp(-a/S_o) \qquad (10\text{-}2)$$

The room constant may be determined from Eq. (7-13):

$$R = \frac{\bar{\alpha} S_o}{1 - \bar{\alpha}} = \frac{(1 - e^{-a/S_o})S_o}{e^{-a/S_o}} = (e^{a/S_o} - 1)S_o \qquad (10\text{-}3)$$

The ceiling height of the room in which the system was located was 3.66 m (12 ft) and the room constant was relatively large. For example, for the 1000 Hz octave band, the room constant was approximately $R = 3360$ m^2 (36,170 ft^2). The direct distance between the air chisel and the packer's ear was $r = 4.27$ m (14 ft), and the directivity factor for the chisel was approximately $Q = 2$. The contributions of the reverberant and the direct sound fields may be found as follows for the 1000 Hz octave band:

Reverberant field: $4/R = (4)/(3360) = 0.001190$ m^{-2}
Direct field: $Q/4\pi r^2 = (2)/[(4\pi)(4.27)^2] = 0.00873$ m^{-2}

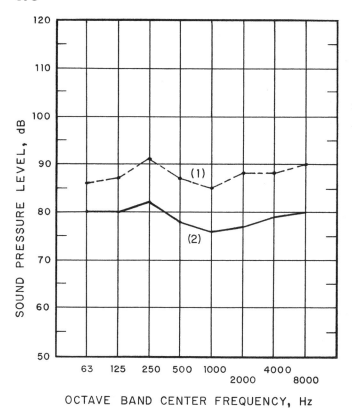

FIGURE 10-2 Sound pressure level spectrum for the air-hammer noise at the packer's location (1) before installation of the barrier, $L_A = 95\,\text{dBA}$, and (2) after installation of the barrier, $L_A = 85\,\text{dBA}$.

The contribution of the direct field is about eight times that of the reverberant field, so a barrier would be effective in reducing the noise experienced by the packer. If we combine Eq. (7-18) for the sound pressure level without the barrier (L_p^o) with Eq. (7-96) for the sound pressure level with the barrier in place (L_p) we obtain the sound pressure level reduction:

$$L_p^o - L_p = \Delta L_p = 10 \log_{10} \left[\frac{\dfrac{4}{R} + \dfrac{Q}{4\pi r^2}}{\dfrac{4}{R_b} + \dfrac{Q(a_b + a_t)}{4\pi(A + B)^2}} \right] \qquad (10\text{-}4)$$

The room constant with the barrier in place and the room constant without the barrier are practically the same, $R_b \approx R$, and the transmission

Case Studies in Noise Control

coefficient is generally negligible compared with the barrier coefficient, $a_t \ll a_b$. For a barrier coefficient of $a_b = 0.02$ at 1000 Hz and $(A + B) = 4.724$ m (15.5 ft), the anticipated reduction in sound pressure level (in the 1000 Hz octave band) by using a barrier is as follows:

$$\Delta L_p = 10 \log_{10} \left[\frac{\frac{4}{3360} + \frac{2}{(4\pi)(4.267)^2}}{\frac{4}{3360} + \frac{(2)(0.02)}{(4\pi)(4.724)^2}} \right] = 10 \log_{10} \left[\frac{0.00993}{0.001333} \right]$$

$$\Delta L_p = 8.7 \, \text{dB}$$

Because the noise level reduction that is needed is about $(95 - 90) = 5$ dBA or more, this magnitude of sound pressure level should be satisfactory.

10.2.2 Control Approach Chosen

A barrier wall was selected as the noise control measure in this case. The wall was 3.048 m (10 ft) long and 1.829 m (6 ft) high. The air chisel was located about 1.219 m (4 ft) behind the barrier and about 1.143 m (3 ft 9 in) below the top of the barrier. The distance from the barrier to the packer's ear was about 3.035 m (9 ft 11.5 in) from the barrier and about 0.305 m (1 ft) below the top of the barrier.

The barrier was constructed of $\frac{1}{4}$-inch (6.4 mm) thick plywood attached on both sides of a frame constructed of 2 × 4's. The barrier was simple to construct and was quite sturdy. No sound-absorbing materials were needed on the plywood surface.

The sound level spectrum at the packer's location with the barrier in place is shown in Fig. 10-2. With no barrier, the A-weighted sound level was 95 dBA, and the sound level was 85 dBA with the barrier in place. The overall sound pressure levels (measured on the C-scale) were 97 dB with no barrier and 88 dB with the barrier in place. The addition of the barrier reduced the sound level such that the packers did not need hearing protection.

10.2.3 Cost

The material and labor costs for the barrier were as follows. Five sheets of $\frac{1}{4}$-inch plywood, 4 ft × 8 ft, had a total cost of $85.00. (Note: Cost values used throughout this chapter are US dollars at year 2000.) A total of 60 ft of 2 × 4's were used, for a total cost of $20.00. The in-plant labor cost for construction of the barrier was $205.00. Therefore, the total cost for the barrier, which reduced the noise level by about 10 dBA, was $310.

10.2.4 Pitfalls

In this installation, the room size was large and the sound radiated directly from the air chisel to the packer's ears was a significant portion of the total noise. The barrier would not have given satisfactory results for an application in which the room was small and the walls had very low surface absorption coefficients (acoustically "hard" surfaces).

10.3 METAL CUT-OFF SAW NOISE

One of the common problems in industrial settings is that of protecting workers from the effects of noise generated by machines that the worker must guide or monitor directly. One example is a cut-off saw used on metal shapes (Handley, 1973). The noise generated by the sawing operation originates from two primary sources: (a) the saw blade and (b) the workpiece being sawed.

The metal cut-off saw in this case study was actuated downward into the workpiece by a lever attached to the hinged and counterbalanced saw and motor. It was necessary that the worker observe the cutting operation. In addition to the visual clues, vibration and opposing forces transmitted to the worker through the lever arm furnished feedback on the cutting operation. The problem was to reduce the noise that the worker received in front of the saw, with little interference with the workflow, visibility, or with operation of the lever arm.

10.3.1 Analysis

Because of the constraints placed on the situation, an enclosure with clear, transparent front doors was considered as a solution to the noise problem. The operator's location would be directly in front of the doors, so the important noise path would be from the saw through the doors to the operator.

The change in sound pressure level due to the insertion of the enclosure would be approximately the same as the difference in sound power level radiated from the saw without an enclosure and the sound power level radiated from the enclosure with the saw inside. The sound power level difference is given by Eq. (7-85), expressed in "level" form:

$$\Delta L_p = L_p^o - L_p \approx 10 \log_{10}(W/W_{out}) = 10 \log_{10}\left[1 + \frac{\Sigma S_j \alpha_j}{\Sigma S_j a_{tj}}\right] \quad (10\text{-}5)$$

Let us consider the 500 Hz octave band and determine the thickness needed for a sound pressure level reduction of 12 dB. Because the primary

Case Studies in Noise Control

element of interest in the preliminary analysis is noise transmitted through the door, let us consider this element only. The surface absorption coefficient for Plexigas ($\alpha = 0.05$ at 500 Hz) may be found in Table 7-4:

$$\Delta L_p = 10\,\text{dB} = 10\log_{10}\left[1 + \frac{0.05}{a_t}\right]$$

$$(0.05/a_t) = 10 - 1 = 9.00$$

The required sound transmission coefficient for the doors is as follows:

$$a_t = (0.05)/(9.00) = 0.00556$$

The corresponding transmission loss is found from the following:

$$\text{TL} = 10\log_{10}(1/a_t) = 10\log_{10}(1/0.00556) = 22.6\,\text{dB}$$

The surface mass of the Plexiglas $M_S = \rho_w h$, may be found from Eq. (4-171):

$$\text{TL} = 22.6\,\text{dB} = 20\log_{10} M_S + 20\log_{10}(500\,\text{Hz}) - 47.3$$

$$20\log_{10} M_S = 22.6 + 47.3 - 54.0 = 15.9\,\text{dB}$$

$$M_S = 10^{15.9/20} = 6.24\,\text{kg/m}^2$$

Using the density of Plexiglas from Appendix C, $\rho_w = 1150\,\text{kg/m}^3$, the following estimate of the thickness h of the doors is obtained:

$$h = (6.24)/(1150) = 0.00542\,\text{m} = 5.42\,\text{mm} \quad (0.214\,\text{in})$$

Based on the preliminary analysis, an enclosure with transparent doors having a thickness on the order of $\frac{1}{4}$ inch (6.4 mm) would solve the noise control problem satisfactorily.

10.3.2 Control Approach Chosen

The enclosure chosen in this case is illustrated in Fig. 10-3. Workpieces were fed into a slot in one side of the enclosure and exited through another slot on the opposite side. Flaps of leaded vinyl covered the slot openings to reduce noise transmitted through the slots. The front of the enclosure was closed by two doors constructed of $\frac{1}{4}$-inch thick clear plastic (polymethylmethacrylate, Plexiglas^{TM} or Lexan^{TM}). The plastic allowed the operator to see clearly the piece being cut. The doors closed with a gap having a width slightly greater than the width of the control lever. Each door had a flat strip of leaded vinyl approximately 76 mm (3 in) wide to close the gap. The saw operating lever pushed aside the flaps only in the place where it protruded through the door opening.

The sound pressure level spectrum at the worker position before and after the enclosure was installed is shown in Fig. 10-4. Before the enclosure

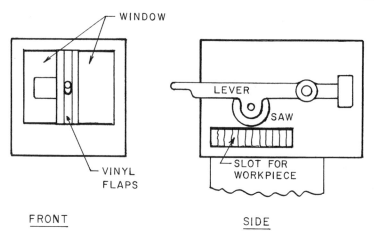

FIGURE 10-3 Schematic of the metal cut-off saw and enclosure.

was applied, the A-weighted sound level at the operator's location was 97 dB, whereas the sound level was 84 dBA after the enclosure was installed.

10.3.3 Cost

The cost of a commercially available acoustic enclosure, as described in the previous section, was approximately $5000.

10.3.4 Pitfalls

As is the case with any acoustic enclosure, it is important to seal openings as effectively as possible. The slots through which material was introduced and withdrawn and the opening through which the operating lever protruded were sealed with leaded vinyl strips in this case.

One possible improvement of the system would be to offset the operating lever of the saw such that the operator's head is not directly in front of the gap between the two doors. In addition to moving the region of noise leakage further from the operator's ears, the operator's view of the workpiece would be improved.

10.4 PAPER MACHINE WET END

In a paper machine, the fluid pulp is introduced at one end (the "wet end") of the machine, where the fluid flows across the *couch roll* to begin the drying process, as shown in Fig. 10-5. The wet paper moves over a screen

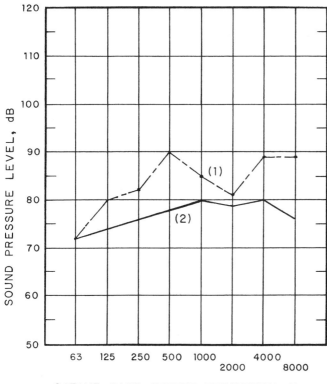

FIGURE 10-4 Sound pressure level spectrum for the metal cut-off saw noise at the operator's location (1) before installation of the enclosure, $L_A = 97\,\text{dBA}$, and (2) after installation of the enclosure, $L_A = 84\,\text{dBA}$.

to the suction rolls, and the drying and sizing process is continued down the machine (Salmon et al., 1975).

The major source of noise around the wet end of the paper machine is the couch roll suction air movement, the pumps, and the whipper roll. The whipper roll provides a mechanical beating action on the felt of the paper machine to keep the web felt clean. The air around the wet end of the paper machine has a very high humidity, so it is not practical to provide an acoustic enclosure for the machine itself.

10.4.1 Analysis

The noise level at the wet end of the operator aisle in front of the machine was in the range from 92 dBA and 94 dBA. The sound pressure level spec-

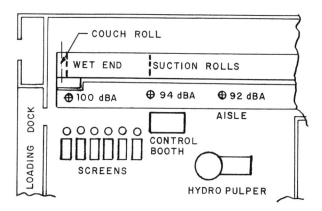

FIGURE 10-5 Layout of the area around the wet end of the paper machine.

trum around the wet end is shown in Fig. 10-6. Higher levels of approximately 100 dBA were measured around the couch roll.

The operator usually spent about 1 hour making adjustments around the couch roll area ($L_A = 100$ dBA) and about 2 hours making general observations in other areas around the machine, where the sound level was 92 dBA. The remainder of the 8-hour day was spent at the control station, where the sound level was about 94 dBA. The allowable exposure time, according to the OSHA standards discussed in Chapter 6, may be calculated from Eq. (6-2):

$L_{A1} = 100$ dBA, $T_1 = 2$ hours

$L_{A2} = 92$ dBA, $T_2 = 6.063$ hours

$L_{A3} = 94$ dBA, $T_3 = 4.595$ hours

The corresponding noise exposure dose (NED) for the situation if no noise control measures were implemented may be determined from Eq. (6-3):

$$\text{NED} = \frac{1}{2} + \frac{2}{6.063} + \frac{5}{4.595} = 0.500 + 0.330 + 1.088 = 1.918 > 1$$

This noise exposure dose is not in OSHA compliance.

If it is impractical to enclose the noise source (the paper machine wet end), then an alternative approach would be to enclose the operator. A personnel booth may be used to house the operator and the operating controls. The operator would need to go outside the enclosure to make adjustments around the couch roll and to make general observations along the machine. If the operator can spend 5 hours each day in the booth, where the

Case Studies in Noise Control

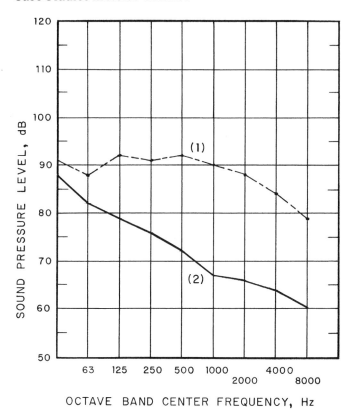

FIGURE 10-6 Sound pressure level spectrum (1) around the wet end of the paper machine, $L_A = 94\,\text{dBA}$, $L_p = 99\,\text{dB}$, and (2) within the personnel enclosure, $L_A = 75\,\text{dBA}$, $L_p = 85\,\text{dB}$.

noise level is less than 85 dBA, the corresponding noise exposure would be in OSHA compliance:

$$\text{NED} = 0.500 + 0.330 + 0 = 0.830 < 1$$

If a personnel acoustic enclosure were selected to reduce the operator's noise exposure, the required transmission loss for the wall facing the machine may be estimated as follows. The sound pressure level outside the enclosure may be approximated by Eq. (7-73) with the direct sound field term, $Q/4\pi r_1^2$, neglected:

$$L_{p1} \approx L_W - 10\log_{10}(R_1/4) + 0.1 = L_W - 10\log_{10} R_1 + 6.1\,\text{dB} \quad (10\text{-}6)$$

The term R_1 is the room constant for the space outside the enclosure and L_W is the power level of the source of noise. The sound pressure level within the enclosure may be estimated from Eq. (7-71):

$$L_{p2} = [L_W - 10\log_{10} R_1] + 10\log_{10}[(4S_w/R_2) + 1] - \text{TL} + 0.1 \quad (10\text{-}7)$$

The quantity S_w is the surface area of the enclosure wall facing the noise source. Note that, for a wall having dimensions of about 4.27 m (14 ft) × 2.45 m (8 ft), the parameter $(S_w/2\pi)^{1/2} = 1.287$ m (4.22 ft). It is likely that the operator would be located within 1.3 m of the wall; therefore, the condition of Eq. (7-71) would apply.

The pressure level difference between inside and outside of the enclosure may be found by combining Eqs (10-6) and (10-7):

$$\Delta L_p = L_{p1} - L_{p2} = -10\log_{10}[(4S_w/R_2) + 1] + \text{TL} + 6.0\,\text{dB} \quad (10\text{-}8)$$

By using acoustic treatment inside the enclosure, an average surface absorption coefficient of $\bar{\alpha} \approx 0.30$ could be achieved. For preliminary design, let us try a ratio of wall area (wall facing the paper machine) S_w to total wall surface area S_o (i.e., S_w/S_o) of 0.15:

$$\frac{4S_w}{R_2} = \frac{4(S_w/S_o)(1-\bar{\alpha})}{\bar{\alpha}} = \frac{(4)(0.15)(1-0.30)}{(0.30)} = 1.40$$

Let us consider a sound pressure level reduction of $\Delta L_p = 15\,\text{dB}$. Making these substitutions into Eq. (10-8), we obtain the following estimate of the required transmission loss for the wall facing the paper machine:

$$\text{TL} = 15 - 6 + 10\log_{10}(1.40 + 1) = 13\,\text{dB}$$

Let us suppose that the wall facing the paper machine is a 2 × 4 frame construction with $\frac{1}{2}$-inch plywood sheets attached to both sides of the frame. In the 1000 Hz octave band, the transmission loss is approximately 43 dB ($a_t = 50.1 \times 10^{-6}$) for the wall (Reynolds, 1981). Let us suppose the wall has two windows, each 0.914 m (3 ft) × 1.524 m (5 ft). The windows are aluminum frame windows, with double glazing and a 12.7 mm ($\frac{1}{2}$ in) thick air space between the panes. The transmission loss in the 1000 Hz octave band for this window is approximately 40 dB ($a_t = 100 \times 10^{-6}$) (Reynolds, 1981).

The overall transmission loss for the partition may be found from Eq. (4-173). The total surface area of the windows is as follows:

$$S_1 = (2)(0.914)(1.524) = 2.786\,\text{m}^2 \quad (30\,\text{ft}^2)$$

Suppose the partition dimensions are 4.267 m (14 ft) × 2.438 m (8 ft) high. The wall portion has a surface area as follows:

$$S_2 = (4.267)(2.438) - 2.786 = 10.403 - 2.786 = 7.617\,\text{m}^2 \quad (82\,\text{ft}^2)$$

Case Studies in Noise Control

The overall transmission loss for the wall facing the paper machine may be calculated:

$$\bar{a}_t = \frac{\Sigma S_j a_{t,j}}{S_w} = \frac{[(2.786)(100) + (7.617)(50.1)](10^{-6})}{(10.403)} = 63.46 \times 10^{-6}$$

The wall transmission loss is as follows:

$$\text{TL} = 10\log_{10}(1/\bar{a}_t) = 10\log_{10}(63.46 \times 10^{-6}) = 42\,\text{dB} > 13\,\text{dB}$$

Therefore, the proposed enclosure wall construction should yield satisfactory acoustical results.

10.4.2 Control Approach Chosen

To protect the paper machine operator from excessive noise exposure at the wet end, an operator enclosure was provided. The paper machine operating controls and main instruments were placed within the booth, and double-glazed viewing windows were provided on the wall of the booth facing the machine to allow the operator to observe the machine operation continuously. A solid wood core door with gaskets and drop closure seals was provided in one of the side walls for entry into the enclosure.

The operator's booth was constructed with 2 × 4-inch framing with $\frac{1}{2}$-inch (12.7 mm) thick plywood sheathing on the inside and outside. The dimensions of the room were 4.27 m (14 ft) × 3.05 m (10 ft) × 2.44 m (8 ft) high. Two double-glazed windows, 3 ft × 5 ft (0.914 m × 1.524 m), were provided for operator observation of the machine operation. The ceiling and upper one-half of the walls were covered with acoustic tile to reduce internal noise levels. The room was provided with lighting, heating and air conditioning for operator comfort.

The sound pressure level spectrum inside the booth is shown in Fig. 10-6. The A-weighted sound level was 75 dBA, and the overall sound pressure level was 85 dB, with the main contribution at the lower frequencies (below 250 Hz). This sound level was well below the 85 dBA required to meet OSHA criteria.

10.4.3 Cost

The personnel enclosure was constructed in-plant for a cost of approximately $7500, including materials and in-plant labor. Although greater noise attenuation could be achieved by purchasing commercial acoustic enclosures (NIOSH, 1975), the higher attenuation was not required in this application. A typical commercially available personnel enclosure of the size needed for this application would cost approximately $15,000.

10.4.4 Pitfalls

Most of the problems with the use of the personnel enclosure tend to be non-acoustical. For example, it is essential that the operator has a clear and unobstructed view of the paper machine for monitoring purposes. This consideration places a restriction on the location of the booth and on the size and location of the windows in the booth.

It is important that the enclosure door be sealed effectively to prevent noise "leakage" around the door. The windows should be double-glazed to provide the largest attenuation of sound passing through the windows.

10.5 AIR SCRAP HANDLING DUCT NOISE

In a facility for the manufacture of corrugated boxes, the sheets of corrugated paper were trimmed with circular-blade cutters. The side-trim scrap was removed from the conveyor by an air jet. The trim was passed through a trim blower fan with extra thick blades to cut the strips of trim into smaller pieces. The smaller pieces of scrap trim were conveyed through 305-mm (12-inch) diameter ducts to bins where the scrap was baled. The ductwork was suspended from the ceiling at a level of about 3 m (10 ft) from the floor and passed across a 12 m (40 ft) distance through a work room to the bins in the baler room (Salmon et al., 1975). The duct cross section is shown in Fig. 10-7.

The trim was moved through the duct by an air stream having a nominal velocity of about 30 m/s (100 ft/sec). The major source of noise was the impact of the trim against the duct wall, especially at bends in the duct. The sound level spectrum at the worker's ear level (approximately 1.5 m or 5 ft from the duct) is shown in Fig. 10-8. The A-weighted sound level with no acoustic treatment was 93 dBA, which exceeded the 8-hour allowable noise exposure for OSHA compliance (90 dBA). In addition, the noise level was such that it was difficult for workers to communicate in the area under the duct.

10.5.1 Analysis

The speech interference level (L_{SIL}) may be determined from the average of the octave band sound level readings in the 500 Hz, 1000 Hz, 2000 Hz, and 4000 Hz octave bands, as discussed in Sec. 6.4:

$$L_{SIL} = \tfrac{1}{4}(86 + 86 + 86 + 85) = 85.75 \, \text{dB} \quad \text{or} \quad L_{SIL} = 86 \, \text{dB}$$

Case Studies in Noise Control

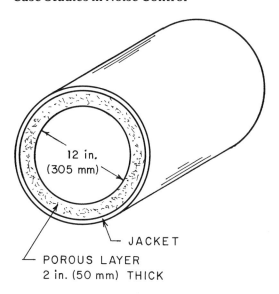

FIGURE 10-7 Cross-section of the scrap-handling duct with the acoustic wrapping in place.

According to the data given in Table 6-4, the expected voice level for face-to-face communication for this value of SIL would be "shouting" ($77 \leq L_{SIL} \leq 91$ dB, women; $80 \leq L_{SIL} \leq 99$ dB, men).

Because of the presence of other noise sources in the room, it would probably be impractical to try to reduce the SIL to values such that conversation in a "normal voice" could be carried out. Let us consider the situation where the worker would communicate in a "loud voice." The SIL range for this condition is given in Table 6-4:

$61 \leq L_{SIL} \leq 71$ dB, women; $63 \leq L_{SIL} \leq 77$ dB, men

The reduction in the SIL is as follows:

$16 \leq \Delta L_{SIL} \leq 20$ dB, women; $17 \leq \Delta L_{SIL} \leq 22$ dB, men

The midrange of the required SIL reduction is $18 \leq \Delta L_{SIL} \leq 20$ dB, or about 19 dB. If we consider the 1000 Hz octave band only, the reduction in sound pressure level required to achieve the SIL reduction of 19 dB would also be approximately 19 dB.

Let us examine the feasibility of using an acoustic pipe wrapping to reduce the noise radiated from the scrap-handling duct. Pipe wrappings usually consist of a resilient layer of porous material (fiberglass, mineral wool, etc.) placed directly on the pipe, and the resilient blanket is covered

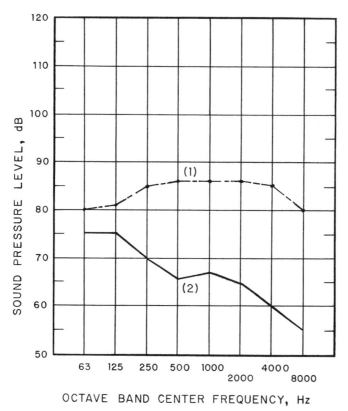

FIGURE 10-8 Sound pressure levels around the scrap duct: (1) before, $L_A = 93$ dBA, and (2) after application of the duct wrapping, $L_A = 72$ dBA.

with an impervious jacket (plastic, galvanized steel, etc.). The following empirical expression has been developed for the insertion loss (IL) for pipe wrapping (Michelsen et al., 1980). The *insertion loss* is defined as the difference between the sound pressure level before the noise control measure has been applied and the sound pressure level after the noise control measure has been applied:

$$\text{IL} = \frac{40}{1 + (D_{\text{ref}}/D)} \{[\log_{10}(f/f_o)] - 0.342\} \quad (10\text{-}9)$$

This expression is valid for frequencies greater than about twice f_o, or for $f \geq 2f_o$. The quantity D is the outside diameter of the pipe, and D_{ref} is a reference diameter, $D_{\text{ref}} = 120$ mm. The quantity f is the frequency, and the frequency f_o is given by the following expression:

Case Studies in Noise Control

$$f_o = \frac{1}{2\pi}\left[\frac{\rho_o c^2}{\rho_s h_s h}\right]^{1/2} \qquad (10\text{-}10)$$

The quantities ρ_o and c are the density and sonic velocity, respectively, of the air within the porous layer. The quantities ρ_s and h_s are the density and thickness, respectively, of the solid jacket material. The quantity h is the thickness of the resilient material over the pipe.

The thickness of the solid jacket h_s may be estimated as follows. Let us try a resilient material thickness, $h = 50.8$ mm (2 in). The density and speed of sound for air at atmospheric pressure and 300K (27°C or 80°F) are $\rho_o = 1.177$ kg/m³ (0.0735 lb$_m$/ft³) and $c = 347.2$ m/s (1139 fps), respectively. For the jacket material, let us try building paper having a density of $\rho_s = 1120$ kg/m³ (70 lb$_m$/ft³). For an insertion loss of IL $= 19$ dB at a frequency $f = 1000$ Hz and a duct diameter of $D = 304.8$ mm (12 in), Eq. (10-9) may be used to find the jacket thickness:

$$\text{IL} = 19\,\text{dB} = \frac{(40)}{1+(120/304.8)}\{[\log_{10}(f/f_o)] - 0.342\}$$

$$f/f_o = 10.09 > 2$$

The frequency f_o is as follows:

$$f_o = (1000)/(10.09) = 99.08\,\text{Hz}$$

The required jacket thickness may be found from Eq. (10-10):

$$\rho_s h_s = \frac{(1.177)(347.2)^2}{(4\pi^2)(99.08)^2(0.0508)} = 7.21\,\text{kg/m}^2$$

$$h_s = (7.21)/(1120) = 0.00643\,\text{m} = 6.43\,\text{mm} \quad (0.253\,\text{in})$$

The thickness of heavy building paper is approximately 3.2 mm (1/8 in); therefore, two layers of building paper could be used for the jacket over the resilient material.

10.5.2 Control Approach Chosen

The noise control approach chosen for this application was to wrap the ducts with 50 mm (2 in) of mineral wool building insulation material to provide the resilient and sound-absorbing layer. The jacket of the wrapping was two impervious layers of heavy building paper ("tar" paper), spirally wrapped over the mineral wool with 50% overlap between layers.

The sound pressure level spectrum in the room with the pipe wrapping applied is shown in Fig. 10-8. The A-weighted sound level was reduced to

72 dBA, which is well within the OSHA limits for daily worker noise exposure. The speech interference level with the wrapping in place is as follows:

$$L_{SIL} = \tfrac{1}{4}(66 + 67 + 65 + 60) = 64.5 \, dB \quad \text{or} \quad L_{SIL} = 65 \, dB$$

This SIL value falls in the range for communication in a "loud voice" for either men or women. If men were to carry on a face-to-face conversation with a background SIL of 65 dB, the expected separation of the two people could be determined from Eq. (6-5), with $K = 60$:

$$L_{SIL} = 65 = 60 - 20 \log_{10} r$$

$$r = 10^{-5/20} = 0.562 \, m \quad (22.1 \, in)$$

The standard building materials selected for the pipe wrapping were economical; however, special acoustic pipe coverings with leaded vinyl sheeting could also have been used if higher attenuation or insertion loss were required.

10.5.3 Cost

The material costs were relatively small. The cost for building paper to cover two 40-ft sections of 12-in diameter duct was $20, and the cost of the mineral wool was $70 for the job. In-house labor costs were approximately $240, so the total cost for the noise reduction system was $330.

10.5.4 Pitfalls

The case study presented in this section illustrates that common inexpensive building materials may be used to achieve a modest attenuation of sound in some cases. For pipe wraps, it is important to eliminate any leakage at seams of the wrap. In this case, the building paper was overlapped by 50% to eliminate leakage and reduction of attenuation.

10.6 AIR-OPERATED HOIST MOTOR

There are many noise sources associated with discharge of shop air through vents that cause noise problems in industry. One such situation is that of air-operated hoist motors used in industrial materials-handling systems (Salmon et al., 1975).

The sound pressure level spectrum at the operator's location around the air exhaust from a hoist motor is shown in Fig. 10-9. The A-weighted sound level is 115 dBA, at which the allowable daily exposure time is only 15 min, according to the OSHA criteria.

Case Studies in Noise Control

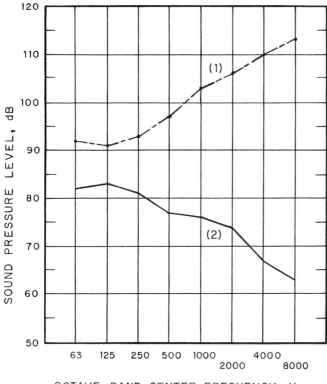

FIGURE 10-9 Sound pressure levels around the hoist: (1) before, $L_A = 115\,\text{dBA}$, and (2) after installation of the muffler on the air exhaust of the hoist motor, $L_A = 81\,\text{dBA}$.

The octave band sound pressure level increases as the frequency is increased, which is characteristic of flow-induced noise at frequencies below the peak frequency. The frequency f_p at which the peak sound power level occurs for turbulent air jet noise (Beranek, 1971) is given by the following expression:

$$\frac{f_p D}{u} = 0.2 \tag{10-11}$$

The quantity D is the diameter of the jet outlet and u is the exit velocity of the fluid. For an air exhaust pipe with a diameter of $D = 12.7\,\text{mm}$ (0.500 in) and an exit air velocity equal to the sonic velocity ($c = 347.2\,\text{m/s}$ at 300K), the frequency at which the peak acoustic energy is radiated from the exhaust is as follows:

$$f_p = \frac{(0.2)(347.2)}{(0.0127)} = 5468 \, \text{Hz}$$

This frequency lies in the 4000 Hz octave band, but it is near the edge of the 8000 Hz octave band.

A muffler or silencer would provide a straightforward solution to the noise control problem. The octave band sound level spectrum given in Table 10-1 would result in an A-weighted sound level of 85 dBA. This spectrum is obtained as follows. Since there are eight octave bands given in the data, we may calculate the following parameter:

$$L_p(\text{OB}) + \text{CFA} = 10 \log_{10}[(1/8)(10^{85/10})] = 76 \, \text{dB}$$

$$L_p(\text{OB}) = 76 - \text{CFA}$$

The conversion factor for the A-scale weighting is given in Table 2-4.

An off-the-shelf muffler was selected as the noise control procedure in this case, because the major source of noise was the exhaust air noise from the air-operated hoist motor. The sound pressure level spectrum with the muffler installed is given in Fig. 10-9. The A-weighted sound level with the muffler in place was $L_A = 81$ dBA, which is well below the OSHA limit for daily noise exposure.

10.7 BLANKING PRESS NOISE

A typical blanking press is a massive machine, with a mass on the order of 125,000 kg (125t or 275,000 lb$_m$ or 137.5 tons). The unit in this case study was mounted on four footings set on heavy concrete piers (Salmon et al., 1975). The press produced automobile chassis steel sections of $\frac{1}{4}$-inch (6.4 mm) thick steel with a width of approximately 254 mm (10 in) and lengths between 2.4 m (8 ft) and 3.0 m (10 ft). The normal operation involved 30 strokes per minute.

TABLE 10-1 Transmission Loss for the Muffler Discussed in Sec. 10.6

	Octave band center frequency, Hz							
	63	125	250	500	1,000	2,000	4,000	8,000
L_p for L_A of 85 dBA	102	92	85	79	76	75	75	77
TL for L_A of 85 dBA	—	—	8	18	27	31	35	36
TL provided, dB	10	8	12	20	27	32	43	50

Case Studies in Noise Control

10.7.1 Analysis

The vertical acceleration levels for the press pier as a function of frequency in the acoustic frequency range are shown in Fig. 10-10. These values are peak acceleration levels averaged over an 0.6-second sampling time. The peak values are less than the maximum acceleration values because of the averaging over the sampling time.

The sound pressure level spectrum at the operator's location is shown in Fig. 10-11. The operator was located approximately 1.2 m (4 ft) from the press. These octave band sound pressure level readings were obtained from measurements when the press made a single stroke, so the measurements corresponded to the peak values averaged over an 0.6-second sampling time, as was the case for the acceleration data. The sound pressure level indication

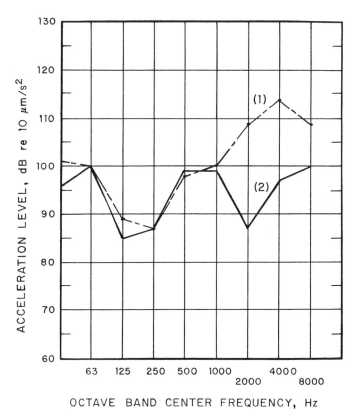

FIGURE 10-10 Vertical acceleration levels referenced to $10\,\mu m/s^2$ on the support pier: (1) before and (2) after installation of vibration isolators.

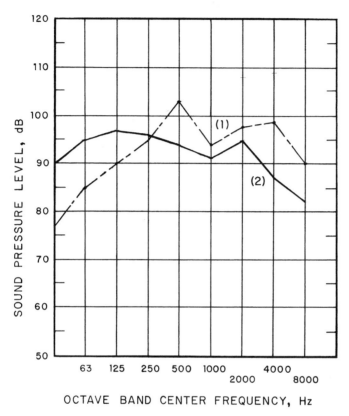

FIGURE 10-11 Sound pressure level spectrum (peak values averaged over 0.6 second) for the blanking press noise at the operator's location: (1) before installation of the vibration isolators and (2) after installation of the vibration isolators.

on the "slow" setting on the sound level meter was approximately 10 dB lower than the peak values shown in Fig. 10-11. The A-weighted sound level calculated from the data in Fig. 10-11 was 105 dBA, whereas the data measured on the "slow" setting of the sound level meter was 95 dBA before any noise control procedure was applied. This value corresponds to an allowable exposure time of 4 hours according to the OSHA criteria.

The estimation of the acoustic energy radiated from a vibrating surface of complicated shape due to impact loading must generally be carried out through a numerical analysis (Beranek and Vér, 1992). Reduction of the vibration transmitted from the machine to the foundation will usually result in some reduction of the noise radiated by the foundation, however.

Case Studies in Noise Control

10.7.2 Control Approach Chosen

Based on the speed of the press and its mass, special vibration isolators were designed for placement under the four feet of the press. The vertical peak acceleration levels (averaged over 0.6 seconds) on the pier after the vibration isolators were installed are shown in Fig. 10-10. There was no large reduction in the vibration levels in the octave bands from 31.5 Hz to 1000 Hz; however, there was significant attenuation in the octave bands above 1000 Hz.

The sound pressure level spectrum at the operator's location after the vibration isolators were installed is shown in Fig. 10-11. The octave band sound pressure levels actually increased somewhat for the range from 31.5 Hz to 250 Hz; however, the octave band sound pressure level was reduced in the higher octave bands. Sound in the octave bands from about 250 Hz to 8000 Hz is more significant, as far as damage to the human ear, than noise in the lower frequency range, as discussed in Sec. 6.2. The A-weighted peak sound level calculated from the data in Fig. 10-11 was 99 dBA, and the A-weighted rms sound level measured with the "slow" setting on the sound level meter was approximately 89 dBA.

Although the noise generated by the press after vibration isolators were installed was less than the OSHA limit for 8-hour exposure (90 dBA), noise radiated by adjacent presses contribute to the operator's work noise exposure. These other operational noise sources for the press would be controlled separately from the vibration isolation.

The impact of the blanking press produces vibration in the floor of the building and in the press structure. Often the background noise around a press that has no vibration isolation is a result of induced vibration of the building structure, which is probably caused by the anchor bolt aftershock. Vibration isolators can act to reduce the building structure vibration.

10.7.3 Cost

The vibration isolators selected for this problem were not stock or off-the-shelf items, but were specifically designed for this case. The cost of the isolators was approximately $5000. The cost for in-plant labor to install the vibration isolators was approximately $3000.

10.7.4 Pitfalls

One of the major problems in using vibration isolators to reduce noise radiated from a foundation of complicated shape is that the prediction of the amount of noise reduction is usually quite difficult and/or expensive. Although some reduction in support-generated noise will be achieved by vibration isolation, the actual noise levels are best determined after the vibration isolators have been installed.

Some degree of vibration isolation is usually recommended for such items of machinery as presses. The life of the dies is usually increased and maintenance problems are usually decreased after the machine has been treated to reduce vibration. Foundation failures, anchor bolt breakage, and fracture of the press feet are usually significantly reduced by vibration isolation of the press.

10.8 NOISE IN A SMALL MEETING ROOM

In a small meeting room in a university building, the noise level was considered to be somewhat high; however, the main complaint of people using the room was that there was an "echo" in the room when someone was speaking. The floor plan for the meeting room is shown in Fig. 10-12. The sound pressure level spectrum for the background noise and for the room in the original condition with eight people present in the room is shown in Fig. 10-13. The overall sound pressure level (measured on the C-scale of the

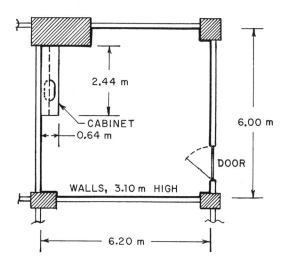

FIGURE 10-12 Meeting room floor plan. The walls are 3.10 m (10 ft 2 in) high. The floor is carpet on concrete, the ceiling is acoustic tile, the walls are plaster on metal lath, and the cabinet is constructed of $\frac{3}{4}$-in (20-mm) thick plywood. The top cabinet height is 800 mm ($31\frac{1}{2}$ in), and the lower cabinet height is 940 mm (37 in). The door is 1.20 m × 2.20 m high (3 ft 11 in × 7 ft $2\frac{1}{2}$ in high), with a thickness of 45 mm ($1\frac{3}{4}$ in). The door has a glass insert, 640 mm × 910 mm ($25\frac{1}{8}$ in × $35\frac{5}{8}$ in). There are 16 wooden chairs in the room.

Case Studies in Noise Control

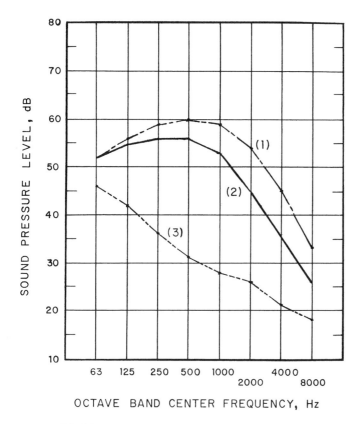

Figure 10-13 Sound pressure level spectrum for the small meeting room: (1) before installation of the acoustic material on the walls and (2) after installation of the acoustic material. The background noise level spectrum (3) is also shown

sound level meter) and the A-weighted sound levels are presented in Table 10-2.

10.8.1 Analysis

The speech interference level for the meeting room before acoustic treatment was applied may be calculated from the data in Fig. 10-12:

$$(L_{SIL})^o = \tfrac{1}{4}(60 + 59 + 54 + 45) = 54.5 \quad \text{or} \quad (L_{SIL})^o = 55\,\text{dB}$$

According to the data presented in Table 6-4, a SIL of 55 dB corresponds to face-to-face communication with a "raised voice" for both men and women.

TABLE 10-2 Sound Pressure Levels for the Background Noise and the Noise Before and after Acoustic Treatment of the Meeting Room Shown in Figure 10-12

	L_A, dBA	L_p, dB	L_{SIL}, dB
Background noise	35	48	27
Before acoustic treatment of walls	63	65	55
After acoustic treatment of walls	57	62	48

there was a need for reduction of the noise level in the room, because the room was used for meetings of small groups of people.

Because the reverberation noise seemed to be the major problem with the acoustic environment of the meeting room, the reverberation time was estimated by using the Fitzroy expression, Eq. (7-35). The total surface area of the room was $S_o = 147.1\,\text{m}^2$ (1583.6 ft^2). The total surface area for the individual surfaces was as follows: (a) side walls, $S_x = 37.20\,\text{m}^2$ (400.4 ft^2); end walls, $S_y = 38.44\,\text{m}^2$ (413.8 ft^2); and floor–ceiling combination, $S_z = 71.48\,\text{m}^2$ (769.4 ft^2). The room volume was $V = 110.8\,\text{m}^3$ (3913 ft^3).

The average surface absorption coefficient for the three sets of interior surfaces in the 500 Hz octave band was estimated as follows:

Side walls: $\alpha_x = 0.058$; $S_x/S_o = 0.2529$
End walls: $\alpha_y = 0.060$; $S_y/S_o = 0.2612$
Floor–ceiling: $\alpha_z = 0.445$; $S_z/S_o = 0.4859$

The number of absorption units for 16 wooden chairs in the 500 Hz octave band was estimated as $\Sigma(\alpha S) = 0.32\,\text{m}^2$. It was observed that the surface absorption of the walls of the room was much smaller than that for the floor and ceiling. This characteristic would allow sound waves traveling horizontally (from wall to wall) to decay at a slower rate than those waves traveling vertically (floor to ceiling and back). It was assumed that this phenomenon was the source of the "echoes" about which people had complained.

Using Eq. (7-35), the number of absorption units, exclusive of the chairs, was calculated:

$$\frac{1}{a_o} = -\frac{1}{(147.1)}\left[\frac{(0.2549)}{\ln(1-0.058)} + \frac{(0.2612)}{\ln(1-0.060)} + \frac{(0.4859)}{\ln(1-0.445)}\right]$$

$$\frac{1}{a_o} = -\frac{(-4.233 - 4.221 - 0.825)}{(147.1)} = 0.06307\,\text{m}^{-2}$$

$$a_o = 15.85\,\text{m}^2$$

Case Studies in Noise Control

The total number of absorption units, including the 16 chairs, was as follows, according to Eq. (7-36):

$$a = 15.85 + 0.32 = 16.17 \, \text{m}^2$$

The reverberation time in the 500 Hz octave band before acoustic treatment was applied was found from Eq. (7-34), using the speed of sound in air at 22°C (72°F) as 344.4 m/s (1130 ft/sec):

$$T_{r,o} = \frac{(55.26)(110.8)}{(344.4)(16.17)} = 1.10 \, \text{seconds}$$

This value of reverberation time was in agreement with measurements for the empty room.

The optimum reverberation time for a conference room was found from Eq. (7-38), with $a = -0.101$ and $b = 0.3070$ from Table 7-3.

$$T_{r,opt} = -0.101 + 0.3070 \log_{10}(110.8) = 0.53 \, \text{seconds}$$

The optimum reverberation time was approximately half that of the room before acoustic treatment was applied.

To achieve a reverberation time of 0.53 s for the empty room, including the 16 chairs, the required number of absorption units was calculated as follows:

$$a = \frac{(55.26)(110.8)}{(344.4)(0.53)} = 33.544 \, \text{m}^2$$

The average surface absorption coefficients for the side walls and the end walls were approximately the same; therefore, the required surface absorption coefficient was estimated by using the same value for both sets of surfaces:

$$\frac{S_o}{a - 0.32} = \frac{(147.1)}{(33.544 - 0.32)} = 4.428 = -\frac{(S_x/S_o) + (S_y/S_o)}{\ln(1 - \alpha_{x,y})} + 0.825$$

$$-\ln(1 - \alpha_{x,y}) = \frac{(0.2529 + 0.2612)}{(4.428 - 0.825)} = 0.1426$$

$$\alpha_{x,y} = 0.133$$

If we let $x =$ fraction of the side-wall surface area covered with acoustic material and $\alpha_m =$ surface absorption coefficient for the acoustic material, the following relationship is valid for the absorption coefficient with the acoustic material applied:

$$\alpha_{x,y} = \frac{\alpha_x S_x (1 - x) + \alpha_m x S_x}{S_x} \tag{10-12}$$

The acoustic material selected was an acoustic foam, 25-mm (1-in) thick, with a decorative surface coating. The surface absorption coefficient for the acoustic material in the 500 Hz octave band was $\alpha_m = 0.51$. The fraction of the side-wall surface area that should be covered with acoustic material was determined as follows:

$$x = \frac{0.133 - 0.058}{0.51 - 0.058} = 0.1657$$

The surface area for both side walls that was to be covered with the acoustic material was as follows:

$$S_{m,x} = (0.1657)(37.20) = 6.164 \, \text{m}^2 \quad (66.35 \, \text{ft}^2)$$

For each side wall, the covered surface area was $(\frac{1}{2})(6.164) = 3.082 \, \text{m}^2$ ($33.17 \, \text{ft}^2$).

The fraction of the end walls y to be covered with the acoustic material was calculated in a similar manner:

$$y = \frac{0.133 - 0.060}{0.51 - 0.060} = 0.1620$$

$$S_{m,y} = (0.1620)(38.44) = 6.223 \, \text{m}^2 \quad (67.03 \, \text{ft}^2)$$

By using this amount of wall coverage, the reverberation time for the empty room should be reduced to approximately 0.5 second.

10.8.2 Control Approach Chosen

Because the walls of the room had a fairly small value of surface absorption coefficient, it was decided to cover the side walls with a total of $7.15 \, \text{m}^2$ ($77 \, \text{ft}^2$) of acoustic material, and the end walls were covered with a total of $7.71 \, \text{m}^2$ ($83 \, \text{ft}^2$) of acoustic material.

The acoustic material selected was an acoustic foam, 25-mm (1-in) thick, with a decorative surface coating. The acoustic material was available in the form of 305-mm (12-in) squares, and the material was attached to the plaster wall using an adhesive compatible with the acoustic material. The squares were distributed over the wall surface in an aesthetically pleasing pattern (at least, pleasing to the engineers who used the room).

The overall sound pressure level (measured on the C-scale of the sound level meter) was 62 dB (a 3 dB reduction) with eight people in the room after the acoustic treatment was applied, as given in Table 10-2. The A-weighted sound level after the acoustic material was applied was 57 dBA (a reduction of 6 dBA) with eight people in the room. The speech interference level after the acoustic material was applied was found from the octave band sound pressure level measurements given in Fig. 10-13. The SIL after acoustic treatment of the room was $L_{SIL} = 48\,dB$, which corresponded to the situation in which conversation in a "normal" voice should be possible in the room.

The reverberation time for the room after the acoustic treatment was applied was 0.49 seconds, which is slightly lower than the calculated optimum value. The problem with annoying "echoes" in the conference room was eliminated by the use of the acoustic treatment, however.

10.8.3 Cost

The cost for the acoustic material was $140 for one carton containing $1.49\,m^2$ ($16\,ft^2$) of material. Ten cartons were used, for a total cost of $1400 for the acoustic material. The cost of the adhesive required to attach the material to the wall was $26, and shipping and handling costs were $49. The total material cost for the wall treatment of the conference room was $1475. University maintenance personnel were used to install the acoustic material, and the resulting labor cost was approximately $385. The corresponding total cost for the acoustic project was $1860.

10.8.4 Pitfalls

For the type of acoustic treatment used in this application to be most effective, it was important to distribute the acoustic material over the walls. The acoustic material was not concentrated in one area on the walls.

It was noted that the steady-state sound pressure level reduction was modest (about 6 dB total). If a reduction in steady-state sound pressure level much more than about 8–10 dB were required, noise control measures other than acoustic treatment of the walls should be considered. In addition, if the direct sound field were predominant, the acoustic treament of the room surface would be ineffective in reducing the noise received directly from the source.

PROBLEMS

There are many acoustic design problems or projects that could be suggested. It is advantageous to undertake projects in which acoustic measurements may be taken on an existing piece of equipment or in an environment where noise is causing a problem. An analysis may be carried out, using the principles presented in previous chapters, to recommend an engineering solution of the noise problem. The final test of the effectiveness of the solution involves implementing the noise control procedure and making measurements on the piece of equipment or environment with the noise control procedure in place. The following projects are suggested as guidelines for possible acoustic design projects.

10-1. Many power tools used around residences emit significant noise, and the user often does not use hearing protection when the equipment is operated. The purpose of this proposed design project is to design a noise control system to reduce the A-weighted sound level produced by a portable leaf blower. The goal of the design is to reduce the sound level at the ear level of the operator by at least 10 dBA below the sound level in the as-purchased condition. Because the leaf blower is portable, the noise control system should be lightweight, it should not interfere with the thermal (cooling) requirements of the blower motor, and the cost of the system should be minimized.

10-2. The noise generated by an industrial cooling tower near an office building can be annoying when the cooling tower fan is operating. The purpose of this proposed design project is to design a noise control system to reduce the environmental noise experienced by office personnel near a large industrial cooling tower. The design goal is to reduce the sound level at a location 6 m (20 ft) from the edge of the cooling tower to a value 10 dBA below the existing operating level. The cost of the acoustic treatment should be minimized, and the noise control procedure must not interfere with the thermal operation of the tower.

10-3. The noise generated by a small shop air compressor is excessive for effective conversation in the area near the compressor. The purpose of this proposed design project is to design a noise control system to reduce the noise in the large shop room in which the compressor is located. The design goal is to reduce the sound level at a location 6 m (20 ft) from the edge of the compressor to a value of 55 dBA or less. The cost of the acoustic treatment should be minimized, and the noise control procedure must not interfere with the operation and maintenance of the compressor.

REFERENCES

Beranek, L. L. 1960. *Noise Reduction*, pp. 571–643. McGraw-Hill, New York.
Beranek, L. L. 1971. *Noise and Vibration Control*, pp. 515–518. McGraw-Hill, New York.
Beranek, L. L. and Vér, I. L. 1992. *Noise and Vibration Control Engineering*, pp. 328–337. McGraw-Hill, New York.
Faulkner, L. L. 1976. *Handbook of Industrial Noise Control*, pp. 506–555. Industrial Press, New York.

Handley, J. M. 1973. Noise—the third pollution, IAC Bulletin 6.0011.0. Industrial Acoustsics Co., Inc., Bronx, NY.

Michelsen, R., Fritz, K. R., and Sazenhofan, C. V. 1980. Effectiveness of acoustic pipe wrappings [in German], Proceedings of the DAGA, pp. 301–304. VDE-Verlag, Berlin.

NIOSH. 1975. *Compendium of Materials for Noise Control*, HEW Publication no. (NIOSH) 75–165, pp. 49–52. National Institute for Occupational Safety and Health. U.S. Government Printing Office, Washington, DC.

Plunkett, R. 1955. Noise reduction of pneumatic hammers. *Noise Control* 1(1): 78.

Reynolds, D. D. 1981. *Engineering Principles of Acoustics*, pp. 605–617. Allyn and Bacon, Inc., Boston.

Salmon, V., Mills, J. S., and Petersen, A. C. 1975. *Industrial Noise Control Manual*, HEW Publication No. (NIOSH) 75-183. U.S. Government Printing Office, Washington, DC.

Appendix A: Preferred Prefixes in SI

APPENDIX A Preferred Prefixes in SI (International System of Units[a])

Prefix	Abbr.	Multiplier	Example	Name	Value
atto	a	E−18	aW	attowatt	10^{-18} W
femto	f	E−15	fW	femtowatt	10^{-15} W
pico	p	E−12	pW	picowatt	10^{-12} W
nano	n	E−09	nW	nanowatt	10^{-9} W
micro	μ	E−06	μW	microwatt	10^{-6} W
milli	m	E−03	mW	milliwatt	10^{-3} W
Base unit			W	watt	
kilo	k	E+03	kW	kilowatt	10^{3} W
mega	M	E+06	MW	megawatt	10^{6} W
giga	G	E+09	GW	gigawatt	10^{9} W
tera	T	E+12	TW	terawatt	10^{12} W
peta	P	E+15	PW	petawatt	10^{15} W
exa	E	E+18	EW	exawatt	10^{18} W

[a]The prefixes centi (c) multiplier of 10^{-2} and deci (d) multiplier 10^{-1} are recommended only for area and volume units (cm^2 and dm^3, for example). According to SI usage, the prefixes should be applied only to a unit in the numerator of a set of units (MN/m^2, for example, and *not* N/mm^2). When an exponent is involved with a unit with a prefix, the exponent applies to the entire unit. For example, mm^2 is a square millimeter, or $10^{-6} m^2$, and not a milli-(square meter).

Appendix B: Properties of Gases, Liquids, and Solids

APPENDIX B Properties of Gases, Liquids, and Solids

Material	Density ρ, kg/m^3	Speed of sound c, m/s	Characteristic impedance Z_0, rayl
Gases at 25°C (77°F) and 1 atm.			
Air	1.184	346.1	409.8
Ammonia	0.696	434.5	302.4
Carbon dioxide	1.799	269.5	484.7
Helium	0.1636	1,016.1	166.2
Hydrogen	0.0824	1,316.4	108.5
Methane	0.666	448.1	293.7
Nitrogen	1.145	352.0	403.0
Oxygen	1.308	328.5	429.6
Steam at 100°C	0.5978	472.8	282.6
Liquids:			
Ethyl alcohol (25°C)	787	1,144	0.900×10^6
Ethylene glycol (25°C)	1,100	1,644	1.808×10^6
Gasoline (25°C)	700	1,171	0.820×10^6
Kerosene (25°C)	823	1,320	1.086×10^6
Sea water (20°C)	1,026	1,500	1.539×10^6
Water (15°C or 59°F)	999.1	1,462.7	1.461×10^6
Water (20°C or 68°F)	998.2	1,483.2	1.481×10^6
Water (25°C or 77°F)	997.0	1,494.5	1.490×10^6
Water (30°C or 86°F)	995.6	1,505.8	1.499×10^6

APPENDIX B (Continued)

Material	Density ρ, kg/m^3	Speed of sound c, m/s	Characteristic impedance Z_0, rayl
Solids:			
Aluminum (pure)	2,700	6,400	17.28×10^6
Brass	8,700	4,570	39.76×10^6
Brick (common)	1,750	4,270	7.47×10^6
Concrete	2,400	3,100	7.44×10^6
Copper	8,910	4,880	43.48×10^6
Glass (window)	2,500	6,000	15.00×10^6
Glass (Pyrex)	2,300	5,200	11.96×10^6
Ice	920	3,200	2.94×10^6
Lead	11,300	1,980	22.47×10^6
Lucite	1,200	1,800	2.16×10^6
Polyethylene	935	1,980	1.85×10^6
Steel (C1020)	7,700	5,790	44.58×10^6
Wood (oak)	770	4,300	3.31×10^6
Wood (pine)	640	4,750	3.04×10^6
Zinc	7,140	4,270	30.49×10^6

Appendix C: Plate Properties of Solids

APPENDIX C Plate Properties of Solids[a]

Material	c_L, m/s	ρ_w, kg/m^3	$M_S f_c$, Hz-kg/m^2	η	E, GPa	σ
Aluminum (2014)	5,420	2,800	34,090	0.001	73.1	0.33
Brass (red)	3,710	8,710	155,200	0.001	103.4	0.37
Brick	3,800	1,800	31,250	0.015	25.0	0.20
Chipboard	675	750	73,400	0.020	0.340	0.08
Concrete	2,960	2,400	50,200	0.020	20.7	0.13
Glass	5,450	2,500	30,300	0.0013	71.0	0.21
Granite	4,413	2,690	40,270	0.001	48.3	0.28
Gypsum board	6,790	650	6,320	0.018	29.5	0.13
Lead	1,206	11,300	819,000	0.015	13.8	0.40
Lexan^{TM}	1,450	1,200	54,650	0.015	2.12	0.40
Marble	4,600	2,800	40,200	0.001	55.2	0.26
Masonry block (6 in)	3,120	1,100	23,300	0.007	10.6	0.10
Plaster	4,550	1,700	24,700	0.005	32.0	0.30
Plexiglas$^{TM.}$	2,035	1,150	37,300	0.020	4.00	0.40
Plywood	3,100	600	12,780	0.030	4.86	0.40
Polyethylene	765	935	80,700	0.010	0.48	0.35
Pyrex	5,350	2,300	28,400	0.004	62.0	0.24
Rubber (hard)	1,700	950	36,900	0.080	2.30	0.40
Steel (C1020)	5,100	7,700	99,700	0.0013	200.0	0.27
Wood (oak)	3,860	770	11,900	0.008	11.2	0.15
Wood (pine)	4,680	640	8,160	0.020	13.7	0.15

[a] c_L is the longitudinal speed of sound; ρ_w is the material density; $M_S = \rho_w h$ = surface density; f_c is the critical or wave coincidence frequency, η is the damping coefficient; E is Young's modulus; and σ is Poisson's ratio.

Appendix D: Surface Absorption Coefficients

APPENDIX D Surface Absorption Coefficients α for Various Materials[a]

Material	Octave band center frequency, Hz					
	125	250	500	1,000	2,000	4,000
Walls and ceilings:						
Brick, unglazed	0.03	0.03	0.03	0.04	0.05	0.07
Brick, unglazed and painted	0.01	0.01	0.02	0.02	0.02	0.03
Concrete block, unpainted	0.36	0.44	0.31	0.29	0.39	0.25
Concrete block, painted	0.10	0.05	0.06	0.07	0.09	0.08
Door, solid wood panel	0.10	0.07	0.05	0.04	0.04	0.04
Marble or terrazzo	0.01	0.01	0.015	0.02	0.02	0.02
Plaster, gypsum or lime, on tile/brick	0.013	0.015	0.02	0.03	0.04	0.05
Plaster, smooth finish on lath	0.14	0.10	0.06	0.04	0.04	0.03
Plaster, fibrous	0.35	0.30	0.20	0.15	0.10	0.04
Plaster, on wood wool	0.40	0.30	0.20	0.15	0.10	0.10
Poured concrete, unpainted	0.01	0.01	0.02	0.02	0.02	0.03
Poured concrete, painted	0.01	0.01	0.01	0.02	0.02	0.02
Sprayed-on cellulose fibers:						
$\frac{5}{8}$-inch thick on solid backing	0.05	0.16	0.44	0.79	0.90	0.91
1-in thick on solid backing	0.08	0.29	0.75	0.98	0.93	0.76
1-in thick on metal lath with air space	0.47	0.90	0.99	0.99	0.99	0.99

Surface Absorption Coefficients

Material	\multicolumn{6}{c}{Octave band center frequency, Hz}
	125	250	500	1,000	2,000	4,000
Open-cell polyurethane foam:						
$1\frac{1}{2}$-in thick	0.05	0.16	0.66	0.99	0.99	0.92
$1\frac{3}{4}$-in thick	0.10	0.25	0.52	0.89	0.99	0.95
2-in thick	0.16	0.25	0.57	0.82	0.86	0.86
$2\frac{1}{4}$-in thick	0.23	0.41	0.75	0.93	0.85	0.77
Fiberglass formboard:						
1-in thick	0.18	0.34	0.79	0.99	0.93	0.90
$1\frac{1}{2}$-in thick	0.25	0.49	0.98	0.99	0.91	0.85
2-in thick	0.33	0.67	0.99	0.99	0.94	0.90
Ceiling board, fiberglass cloth faced:						
1-in linear	0.07	0.24	0.66	0.95	0.99	0.95
1-in nubby surface	0.06	0.25	0.68	0.97	0.99	0.92
1-in textured surface	0.10	0.27	0.75	0.99	0.99	0.84
Plywood panels:						
$\frac{1}{8}$-in with $1\frac{1}{4}$-in air space	0.15	0.25	0.12	0.08	0.08	0.08
$\frac{1}{8}$-in with $2\frac{1}{4}$-in air space	0.28	0.20	0.10	0.10	0.08	0.08
$\frac{3}{16}$-in with 2-in air space	0.38	0.24	0.17	0.10	0.08	0.05
$\frac{3}{16}$-in with 2-in air space filled with fibrous insulation	0.42	0.36	0.19	0.10	0.08	0.05
$\frac{1}{4}$-in with small air space	0.30	0.25	0.15	0.10	0.10	0.10
$\frac{3}{4}$-in with small air space	0.20	0.18	0.15	0.12	0.10	0.10
Plywood panelling, $\frac{3}{8}$-in	0.28	0.22	0.17	0.09	0.10	0.11
Gypsum board, $1\frac{1}{2}$-in on studs	0.29	0.10	0.06	0.05	0.04	0.04
Sound-absorbing masonry blocks:						
4-in thick, unpainted, 2 unfilled cavities	0.19	0.83	0.41	0.38	0.42	0.40
4-in, painted, 2 insulation-filled cavities	0.20	0.88	0.63	0.65	0.52	0.43
6-in, painted, 2 unfilled cavities	0.62	0.84	0.36	0.43	0.27	0.50
6-in, painted, 2 insulation-filled cavities	0.39	0.99	0.65	0.58	0.43	0.45
8-in, painted, 2 unfilled cavities	0.97	0.44	0.38	0.39	0.50	0.60
8-in, painted, 2 insulation-filled cavities	0.33	0.94	0.62	0.60	0.57	0.49
8-in, painted, 3 insulation-filled cavities	0.61	0.91	0.65	0.65	0.42	0.49
Perforated acoustic ceiling tile:						
$\frac{1}{2}$-in thick, hard backing	0.08	0.17	0.55	0.73	0.72	0.67
$\frac{3}{4}$-in thick, hard backing	0.10	0.22	0.72	0.88	0.75	0.66

APPENDIX D (Continued)

Material	\multicolumn{6}{c}{Octave band center frequency, Hz}
	125	250	500	1,000	2,000	4,000
1-in thick, hard backing	0.12	0.31	0.93	0.90	0.72	0.63
$\frac{3}{4}$-in, furring backing	0.12	0.42	0.71	0.88	0.75	0.65
$\frac{3}{4}$-in, mech. suspension	0.30	0.51	0.80	0.85	0.78	0.66
Acoustic foam:						
$\frac{1}{4}$-in thick	0.08	0.10	0.20	0.30	0.60	0.93
$\frac{1}{2}$-in thick	0.12	0.21	0.36	0.54	0.92	0.98
$\frac{3}{4}$-in thick	0.14	0.25	0.44	0.70	0.98	0.99
1-in thick	0.16	0.28	0.51	0.78	0.99	0.99
2-in thick	0.27	0.48	0.80	0.99	0.99	0.99
Ventilating grille	0.30	0.40	0.50	0.50	0.50	0.50
Floors:						
Carpet:						
44 oz $\frac{1}{4}$-in thick wool woven pile, uncoated backing, on 40 oz hair pad	0.17	0.35	0.66	0.71	0.70	0.65
44 oz $\frac{1}{4}$-in thick wool woven pile, coated backing, on 40 oz pad	0.17	0.35	0.46	0.36	0.40	0.45
$\frac{3}{8}$-in thick wool pile on concrete; no pad	0.09	0.08	0.21	0.26	0.27	0.37
$\frac{5}{8}$-in thick wool pile, with pad	0.20	0.25	0.35	0.40	0.50	0.75
Indoor–outdoor carpet	0.01	0.05	0.10	0.20	0.45	0.65
Linoleum/rubber tile on concrete	0.02	0.03	0.03	0.03	0.03	0.02
Varnished wood joist floor	0.15	0.11	0.10	0.07	0.06	0.07
Windows:						
Large panes, plate glass	0.18	0.06	0.04	0.03	0.02	0.02
Small panes, ordinary glass	0.04	0.04	0.03	0.03	0.02	0.02
Draperies:						
Cotton fabric, draped to half area, 14 oz/yd^2	0.07	0.31	0.49	0.81	0.66	0.54
Velour, straight, 10 oz/yd^2	0.04	0.05	0.11	0.18	0.30	0.35
Velour, straight, 14 oz/yd^2	0.05	0.07	0.13	0.22	0.32	0.35
Velour, straight, 18 oz/yd^2	0.05	0.12	0.35	0.48	0.38	0.36

Surface Absorption Coefficients

Material	Octave band center frequency, Hz					
	125	250	500	1,000	2,000	4,000
Velour, draped to half area, 14 oz/yd^2	0.07	0.31	0.49	0.75	0.70	0.60
Velour, draped to half area, 18 oz/yd^2	0.14	0.35	0.53	0.75	0.70	0.60
Natural outdoor materials:						
Grass, 2 in high	0.11	0.26	0.60	0.69	0.92	0.99
Gravel soil, loose and moist	0.25	0.60	0.65	0.70	0.75	0.80
Snow, 4 in deep	0.45	0.75	0.90	0.95	0.95	0.95
Water surface	0.01	0.01	0.01	0.02	0.02	0.03
People and furniture: Values of (αS), m^2 per person or item						
Person in upholstered chair	0.30	0.315	0.35	0.42	0.42	0.39
Upholstered chair, empty	0.23	0.26	0.30	0.325	0.325	0.30
Person in wood theater seat	0.23	0.28	0.325	0.35	0.37	0.35
Wood theater seat, empty	0.01	0.02	0.02	0.03	0.04	0.04
Adult or high school student in desk	0.20	0.28	0.31	0.37	0.41	0.42
Child or elementary school student in desk	0.17	0.21	0.26	0.30	0.325	0.37
Empty desk	0.10	0.13	0.14	0.17	0.18	0.15
Person standing	0.19	0.325	0.44	0.42	0.46	0.37

[a] The values have been selected from manufacturer's catalog data.

Appendix E: Nomenclature

a	acceleration, m/s² or ft/sec²
	dimension of a panel, m or ft, Eq. (4-152)
	number of absorption units, m² or ft², Eq. (7-30)
	tube diameter, m or ft, Eq. (8-5)
a_b	barrier coefficient, Eq. (7-90)
a_{max}	maximum vibratory acceleration, m/s² or in/sec²
a_r	sound power reflection coefficient, Eq. (4-91)
a_t	sound power transmission coefficient, Eq. (4-88)
A	constant of integration, Eq. (4-38)
	cross-sectional area, m² or ft², Example 4-9
	constant in Eq. (5-57)
	distance for barrier calculation, m or ft, Eq. (7-89)
	function defined by Eq. (9-140)
A_Δ	conversion function, dB, Eq. (3-7)
b	dimension of a panel, m or ft, Eq. (4-152)
B	isothermal bulk modulus, Pa or lb$_f$/ft², Eq. (2-2)
	constant of integration, Eq. (4-38)
	flexural rigidity, N-m or lb$_f$-ft
	constant in Eq. (5-57)
	distance for barrier calculation, m or ft, Eq. (7-89)
	function defined by Eq. (9-141)
B_T	blade tone component of fan noise, dB, Table 5-1
c	speed of sound, m/s or ft/sec
c_L	speed of longitudinal sound waves, m/s or ft/sec, Eq. (4-156)

Nomenclature

c_p	specific heat at constant pressure, J/kg-K or Btu/lb$_m$-°R
c_v	specific heat at constant volume, J/kg-K or Btu/lb$_m$-°R
$C_1 = C_g/C_V$	coefficient ratio
C_A	acoustic compliance, m^3/Pa or ft^5/lb$_f$
C_D	dissipation coefficient, Pa-s
	grille pressure drop function, Eq. (5-70)
C_E	electrical capacitance, F
C_g	valve-sizing coefficient for gas flow, Eq. (5-43)
C_k	coefficient in Eq. (9-148)
C_m	regression constant in Eq. (8-181)
$C_M = 1/K_S$	mechanical compliance, m/N or in/lb$_f$
C_S	specific mechanical compliance, m^3/N or ft^3/lb$_f$, Eq. (4-138)
C_V	valve-sizing coefficient for liquids
C_w	volume compliance of a panel, m^5/N or m^3/Pa
CF	correction factor for composite noise rating, dB, Eq. (6-6)
$CF_1, CF_2, \ldots, CF_7$	factors to convert to octave band sound pressure levels, dB
CF_{DN}	correction factor for day–night level, dBA, Eq. (6-9)
CF_g	factor to convert to octave band sound pressure levels for a grille, dB
CFA	A-weighting conversion factors, Table 2-4
CFC	C-weighting conversion factors, Table 2-4
d	spacing between panels, m or ft, Eq. (4-175)
	center-to-center spacing of ribs, m or ft, Eq. (4-186)
	inside diameter of a vent tube, m or ft, Eq. (5-38)
	mean free path for sound in a room, m or ft, Eq. (7-23)
	slant distance in a plenum chamber, m or ft, Eq. (8-185)
	static deflection, m or ft, Eq. (9-16)
d_w	spring wire diameter, m or in
D	acoustic energy density, J/m^3
	diameter of a cylinder, m or ft
	mean diameter of a spring, m or in
D_D	acoustic energy density for the direct field, J/m^3, Eq. (7-3)
D_e	equivalent diameter, m or ft
D_E	equivalent distance for traffice noise, m or ft, Eq. (5-74)

D_R	acoustic energy density for the reverberant field, J/m^3, Eq. (7-4)
D_t	tower diameter, m or ft
DI	directivity index, dB, Eq. (2-28)
Δe	voltage drop, V
E	energy, J
f	frequency, Hz
f_b	blade pass frequency for compressor, Hz, Eq. (5-20)
f_B	blade pass frequency for a fan, Hz, Eq. (5-12)
f_c	critical or wave coincidence frequency, Hz
f_n	undamped natural frequency, Hz
f_o	octave band center frequency, Hz
	peak frequency for a gas jet, Hz, Eq. (5-40)
	resonant frequency for a Helmholtz resonator, Hz
Δf_P	width of the TL plateau, Hz, Table 4-1
F	force, N or lb_f
	pressure function, Eq. (5-57)
	temperature function, Eq. (4-229)
F_1, F_2	quantities defined by Eqs (8-142) and (8-143)
F_d	damper force, N or lb_f
F_s	noise spectrum function, dB, Eq. (5-64)
F_S	spring force, N or lb_f
F_{sb}	side-branch noise spectrum function, dB, Eq. (5-66)
F_T	transmitted force, N or lb_f
g	local acceleration due to gravity, m/s^2 or ft/sec^2
g_c	units conversion factor, 1 kg-m/N-s^2 or 32.174 lb_m-ft/lb_f-sec^2
G	shear modulus, Pa or lb_f/ft^2
G_1, G_2	quantities defined by Eqs (8-146) and (8-147)
h	thickness of a panel, m or ft
	fraction of molecules that are H_2O, Eq. (4-227)
	distance water falls in a tower, m or ft, Eq. (5-32)
h_o	distance between the bottom of packing and pond surface, m or ft, Eq. (5-32)
h_p	depth of packing below tower ring beam, m or ft, Eq. (5-32)
h_r	height of ribs on a panel, m or ft, Eq. (4-186)
H	dimension of plenum chamber, m or ft, Eq. (8-185)
$H(\theta, \varphi)$	pressure distribution function, Eq. (2-29)
H_o	free height of a spring, m or in
H_s	solid height of a spring, m or in
i	electric current, A

Nomenclature

I	acoustic intensity, W/m²
	area moment of inertia, m⁴ or ft⁴, Eq. (4-189)
IL	insertion loss, dB, Eq. (7-76)
Im	imaginary part of a complex number, Eq. (4-21)
$j = \sqrt{-1}$	imaginary number
k	wavenumber, m⁻¹ or ft⁻¹
k_t	thermal conductivity, W/m-K or Btu/hr-ft-°F
K_f	greatest common factor, Eq. (5-20)
K_L	pressure function, Eq. (5-51)
K_o	constant, dB, Eq. (5-18)
K_S	quantity defined by Eq. (4-143)
	spring constant, N/m or lb_f/in
L	quantity defined by Eq. (4-95), m or ft
	thickness of a wall, m or ft, Eq. (4-115)
	dimension used in predicting NEF contours, Sec. 6.9
	tube length, m or ft, Eq. (8-6)
	expansion chamber muffler length, m or ft
L°	A-weighted sound level for stationary locomotive, dBA, Eq. (5-82)
L_1	A-weighted sound level for train passby, dBA, Eq. (5-84)
L_{10}, L_{50}, L_{90}	A-weighted sound levels that are exceeded 10%, 50%, or 90% of the time, respectively
L_a	vibratory acceleration level, dB
L_A	A-weighted sound level, dBA
L_{CNR}	corrected composite noise rating, dB, Eq. (6-6)
L_d	displacement level, dB
L_D	acoustic energy density level, dB
	energy-equivalent pressure level during the daytime, dBA, Eq. (6-8
L_{DN}	day–night sound level, dBA (DN)
L_e	equivalent length, m or ft, Eq. (8-12)
L_E	mutual inductance, H
L_{EPN}	effective perceived noise level, dB(PN), Eq. (6-14)
L_{eq}	energy-equivalent sound level, dBA, Eq. (6-7)
L_F	vibratory force level, dB
L_G	sound pressure level gain for a resonator, dB, Eq. (8-62)
L_{Go}	sound pressure level gain at resonance for a resonator, dB, Eq. (8-67)
L_I	acoustic intensity level, dB

L_N	energy-equivalent pressure level during the nighttime, dBA, Eq. (6-8)
L_{NP}	noise pollution level, dBa (NP), Eq. (6-10)
L_p	sound pressure level, dB
L_{SIL}	speech interference level, dB
L_{Tr}	transmissibility level, dB, Eq. (9-104)
L_v	acoustic velocity level, dB
L_W	sound power level, dB
$\Delta L_1, \Delta L_2$	additional equivalent lengths, m or ft, Eqs (8-8) and (8-10)
m	mass flow rate of cooling water, kg/s or lb_m/sec, Eq. (5-32)
	mass flow rate of gas, kg/s, Eq. (5-57)
	mass being accelerated, kg or lb_m, Eq. (8-1)
	unbalance mass, kg. or lb_m, Eq. (9-111)
$m = S_2/S_1$	muffler area ratio, Eq. (8-123)
$m = 2\alpha$	energy attenuation coefficient, Np/m
M	molecular weight, kg/mol or lb_m/lbmole, Eq. (5-38)
	mass of vibrating system, kg or lb_m
M_a	molecular weight of air, kg/mol or lb_m/lbmole, Eq. (5-38)
	mass of vibration absorber, kg or lb_m, Eq. (9-133)
M_A	acoustic mass, kg/m^4 or lb_m/ft^4
$M_S = \rho_w L$	specific (surface) mass, kg/m^2 or lb_m/ft^2
MF	magnification factor, Eq. (9-78)
n	integer
	number of sound wave reflections, Eq. (7-26)
	exponent in Eq. (8-181)
n_r	rotational speed, rev/sec or rpm
N	Fresnel number, Eq. (7-91)
	number of cycles, Eq. (9-50)
N_b	number of blades for a fan, Eq. (5-12)
N_d	number of train passbyes during the daytime
N_D	number if airplane flights during the daytime, Eq. (6-18)
N_e	effective noy value, Eq. (6-13)
N_{EF}	number of effective airplane flights, Eq. (6-18)
N_{max}	largest noy value, Eq. (6-13)
N_n	number of train passbyes during the nighttime
N_N	number of airplane flights during the nighttime, Eq. (6-18)
N_o	base composite noise rating, dB, Table 6-10

Nomenclature

N_r	number of rotating blades
N_s	number of stationary blades
N_S	Strouhal number, Eq. (5-65)
N_t	number of tubes, Eq. (8-83)
NCB	balanced noise criterion
NED	noise exposure dosage, Eq. (6-3)
NEF	noise exposure forecast, dBA
NRC	noise reduction coefficient, Sec. 7.1.2
$p(x, t)$	instantaneous acoustic pressure, Pa or lb_f/ft^2
p_m	peak amplitude of the acoustic pressure, Pa or lb_f/ft^2
p_{max}	amplitude of the acoustic pressure, Pa or lb_f/ft^2
p_{rms} or p	root-mean-square acoustic pressure, Pa or lb_f/ft^2
p_s	pitch of spring coils, m or in
P	pressure rise across a fan, Pa or in H_2O, Eq. (5-11)
P_1	absolute pressure at the valve inlet, psia, Eq. (5-44)
P_o	ambient pressure, Pa or lb_f/ft^2
P_s	surface pressure, Pa or psi
P_v	vapor pressure, Pa or psia, Eq. (5-52)
P_W	wetted perimeter, m or ft, Eq. (5-61)
$Pr = \mu c_p / k_t$	Prandtl number
ΔP	pressure drop across a valve, Pa or psi, Eq. (5-43)
Q	directivity factor, Eq. (2-13)
	volumetric flow rate, m^3/s or ft^3/min, Eq. (5-11)
Q_A	acoustic quality factor, Eq. (8-48)
Q_g	gas flow rate, scfh (standard cubic feet per hour), Eq. (5-48)
Q_M	mechanical quality factor
r	radial distance, m or ft
$r = f/f_n$	frequency ratio, Eq. (9-76)
r^*	characteristic distance, m or ft, Eq. (5-34)
$r^* = (S_w/2\pi)^{1/2}$	characteristic distance, m or ft, Sec. 7.5
r_o	reference distance for traffic and train noise, (30 m)
r_v	boundary layer ratio, Eq. (8-32)
R	specific gas constant, J/kg-K or $ft\text{-}lb_f/lb_m\text{-}°R$
	room constant, m^2 or ft^2, Eq. (5-7)
R_1	specific flow resistance per unit thickness, rayl/m, Eq. (8-165)
R_A	acoustic resistance, $Pa\text{-}s/m^3$ or $lb_f sec/ft^5$
R_b	room constant including a barrier, m^2 or ft^2, Eq. (7-95)
R_e	effective flow resistance per unit thickness, rayl/m, Eq. (8-165)

R_E	electrical resistance, Ω
R_M	mechanical resistance or damping coefficient, N-s/m or lb$_f$-sec/ft
R_S	specific acoustic resistance, rayl = Pa-s/m or lb$_f$-sec/ft^3
R_{S1}	specific acoustic resistance for one screen, rayl = Pa-s/m or lb$_f$-sec/ft^3
Re	real part of a complex quantity, Eq. (4-21)
RH	relative humidity, Eq. (4-230)
S	surface area, m^2 or ft^2
	vehicle speed, km/h, Eq. (5-75)
S_{50}	values of TL for the STC-50 curve, dB, Table 4-4
S_b	surface area of one side of a barrier, m^2 or ft^2, Eq. (7-95)
S_F	floor area, m^2 or ft^2, Eq. (7-37)
S_L	surface area of lining material in a plenum chamber, m^2 or ft^2
S_o	total surface area of a room, m^2 or ft^2
S_w	surface area of a wall, m^2 or ft^2
t	time, s
	thickness of pipe wall, m or ft, Eq. (5-43)
	fraction of the time that a given noise level occurs, Eq. (6-11)
t_1	time between sound wave reflections, s, Eq. (7-22)
T	absolute temperature, K or °R
	OSHA permissible time of exposure, h, Eq. (6-2)
T_1	absolute temperature at the valve inlet, K or °R, Eq. (5-48)
T_{60}	vibrational reverberation time, s, Eq. (6-58)
T_L	total length of railroad cars, m, Eq. (5-81)
T_r	reverberation time, s
T_t	total length of the train, m, Eq. (5-88)
TL	transmission loss, dB, Eq. (4-90)
TL$_n$	transmission loss for normal incidence, dB
TL$_P$	plateau transmission loss, dB, Table 4-1
Tr	transmissibility, Eq. (9-100)
u	rms acoustic velocity, m/s or ft/sec
	velocity, m/s or ft/sec, Eq. (5-64)
$u(x, t)$	instantaneous acoustic velocity, m/s or ft/sec
$U = Su$	acoustic volume velocity, m^3/s or ft^3/sec
U_t	blade tip speed, m/s or ft/sec, Eq. (5-21)
v	velocity, m/s or ft/sec

Nomenclature

V	volume, m³ or ft³
	vehicle volume, vehicles/h, Eq. (5-75)
	train speed, m/s, Eq. (5-81)
$V(t)$	velocity of a panel, m/s or ft/sec, Eq. (4-137)
W	acoustic power, W
	dimension used in predicting the NEF contours, Sec. 6.9
x	linear coordinate or distance, m or ft
X	equivalent number of passbyes, Eq. (5-86)
$X(t)$	displacement of a wall or panel, m or ft
X_A	acoustic reactance, Pa-s/m³, Eq. (8-15)
X_m	peak amplitude of motion, m or ft
y	linear coordinate or distance, m or ft
y_j	mole fraction of the jth component in a mixture, Eq. (4-226)
y_{max}	maximum amplitude of vibratory motion, m or in
y_P	peak-to-peak amplitude of vibratory motion, m or in
Y	porosity
$Y_M = 1/Z_M$	mechanical admittance or mobility, m/N-s or in/lb$_f$-sec
z	linear coordinate or distance, m or ft
	complex number, Eq. (4-21)
z^*	complex conjugate, Eq. (4-65)
Z_A	acoustic impedance, Pa-s/m³
Z_{Ab}	acoustic impedance of side branch, Pa-s/m³
$Z_o = \rho_o c$	characteristic impedance, rayl = Pa-s/m
Z_M	mechanical impedance, N-s/m or lb$_f$-sec/in
$Z_s = p/u$	specific acoustic impedance, rayl = Pa-s/m

Greek letters:

α	surface absorption coefficient
	attenuation coefficient Np/m, Eq. (4-207)
$\bar{\alpha}$	average surface absorption coefficient
β	function in Eq. (8-87)
	function defined by Eq. (9-109)
$\beta = a/b$	panel aspect ratio
γ	function in Eq. (8-94)
$\gamma = c_p/c_v$	specific heat ratio
$\delta = (L_{10} - L_{90})$	difference, dBA, Eq. (6-12)
δ	logarithmic decrement, Eq. (9-48)
	dynamic deflection, m or in, Example 9-11
δ_f	diameter of fibers, μm, Eq. (8-180)
$\Delta = S_{50} - TL$	difference, dB, Eq. (4-193)

Δ	decay rate, dB/s, Eq. (9-55)
$\Delta_1, \Delta_2, \Delta_3$	vehicle noise adjustment factors, dB, Eqs (5-79) and (5-80)
Δ_{tc}	turbocharger adjustment factor, Eq. (5-82)
ε	eccentricity, m or in, Eq. (9-110)
ζ	damping ratio, Eq. (9-25)
η	damping coefficient or energy dissipation factor
θ	angular coordinate
	function in Eq. (5-48)
	phase angle, Eq. (9-85)
θ_{cr}	critical angle of incidence, Eq. (4-105)
κ	effective elasticity coefficient, Pa or lb_f/ft^2
λ	wavelength, m or ft
λ_b	wavelength of bending waves, m or ft, Eq. (7-74)
μ	viscosity, Pa-s or lb_m/ft-sec
$\mu = M_a/M$	mass ratio for a vibration absorber, Eqs (9-140) and (9-141)
ν	Poisson's ratio, Eq. (2-3)
$\nu = S_3/S_1$	muffler area ratio, Eq. (8-122)
ξ	acoustic particle displacement, m or ft
ρ	density, kg/m^3 or lb_m/ft^3
ρ_c	bulk density of cork, kg/m^3 or lb_m/ft^3
ρ_f	bulk density of felt, kg/m^3 or lb_m/ft^3
ρ_L	density of liquid in Eq. (5-56)
ρ_m	bulk density of acoustic material, kg/m^3 or lb_m/ft^3
ρ_o	density at atmospheric pressure, kg/m^3 or lb_m/ft^3
ρ_w	density of the solid wall, kg/m^3 or lb_m/ft^3
	density of water in Eq. (5-57)
σ	Poisson's ratio, Eq. (4-152)
	standard deviation, dBA, Eq. (6-11)
	dissipative muffler attenuation coefficient, Np/m
$\tau = 1/f$	period, s
$\tau = C_D/\rho_o c^2$	relaxation time, s, Eq. (4-201)
τ	phase angle for transmissibility, Eq. (9-105)
ϕ	phase angle
	pressure ratio, Eq. (5-52)
ϕ_s	structure factor, Eq. (8-169)
χ	distance from the interface to the overall neutral axis, Eq. (4-180)
ψ	ratio defined by Eq. (8-165)
	phase angle, Eq. (9-127)
ψ_1	ratio defined by Eq. (8-171)

Nomenclature

$\psi(h)$	quantity defined by Eq. (9-151)
ψ_o	ratio defined by Eq. (8-167)
$\psi(x), \psi(r)$	amplitude function, m or ft
$\omega = 2\pi f$	angular frequency, rad/s
ω_a	undamped natural frequency for vibration absorber, rad/s
ω_d	damped natural frequency, rad/s, Eq. (9-36)
ω_n	undamped natural frequency, rad/s
$\Omega = \omega_a/\omega_n$	frequency ratio, Eq. (9-144)

Subscripts:

in	denotes incident quantity
n	denotes normal incidence
ref	denotes a reference quantity, p_{ref} = reference acoustic pressure
tr	denotes transmitted quantity

Index

Absorber, dynamic vibration, 439–446
Absorption coefficient, 55, 141, 143–153, 269–270, 274, 289–293, 318, 394, 510–513
Absorption in air, 145–153, 289–293
Absorption units, 283, 476, 500
Acceleration level, 29, 465, 495
Acceleration limits, 466
Acceptable noise levels (*see* Acoustic criteria)
Acoustic analyzers, 50
Acoustic barriers:
  located indoors, 317–321, 476–480
  located outdoors, 313–317
Acoustic calibrator, 44
Acoustic compliance, 335–337
Acoustic criteria:
  for aircraft noise, 255–262
  for auditoriums, 241
  CHABA, 232
  EPA, 247–253, 267
  HUD, 253–255, 258–260
  for interior spaces, 238–243

[Acoustic criteria]
  for locations outdoors, 243–253
  OSHA, 231–235
  for offices, 241
  for speech interference, 235–238
  for telephone communication, 236–237
Acoustic enclosures:
  for automatic press, 299
  design practice for, 311
  door seals for, 312
  large, 304–311
  openings in, 305–307
  for saws, 300, 480–482
  small, 300–304
  transmission paths for, 304
Acoustic energy density:
  absorption of, 289–293
  definition of, 19
  of direct field, 275
  for plane waves, 20
  for spherical waves, 20, 22, 91
Acoustic filters (*see* Silencers)

Index

Acoustic impedance:
  characteristic, 17
  for Helmholtz resonators, 343
  specific, 16, 22
  for spherical wave, 22, 89
Acoustic intensity:
  definition of, 17
  measurement of, 46–49
Acoustic levels, 27–31
Acoustic mass, 332–335
Acoustic ohm, 339
Acoustic particle velocity:
  definition of, 16
  for plane waves, 79, 87
  for spherical waves, 23–24
Acoustic power, 17, 51–66, 277, 305, 344
Acoustic pressure:
  definition of, 15
  for plane waves, 87
  rms, 16
  for spherical waves, 88
Acoustic quality factor, 343
Acoustic resistance:
  of orifices, 339
  of tubes, 339
  of wire screens, 353
Acoustic sources (*see* Sound sources)
Acoustic wrapping, 489–492
Admittance, mechanical, 426
Aircraft noise, 255–262
Air distribution system noise, 192–207
Air, properties of, 145, 507
Air space, 122–124
Aluminum, properties of, 119, 508–509
Ambient noise, 71, 244
Ammonia, properties of, 507
Analogy, 332, 341
Anechoic room, 58–62
Annoyance due to noise, 247, 249
Appliance noise, 185–186
A-scale, 34
Attenuation:
  in argon gas, 145
  in atmospheric air, 145–153, 289–293

[Attenuation]
  by barriers, 313–319
  coefficient, 141, 143–153, 379, 382–389
  in ducts, 193–195
  in elbows, 196
  by enclosures, 300–311
  in helium gas, 145
  in mufflers, 377–381
Auditory canal, 226
Auditory threshold, 229
Automobile noise, 207–211

Background noise, 71, 239–241, 244, 499
Baffle-type silencer, 392
Balanced noise criterion (NCB), 239–243
Band filters:
  octave, 49
  one-third octave, 49
Bandwidth, half-power, 344–346, 414
Barrier coefficient, 313
Barriers:
  located indoors, 317–321
  located outdoors, 313–317
Bel, 28
Bending waves, 300
Biederbecke, Bix, 38
Blade-pass frequency, 167, 174
Blade tone, 167
Blanking press noise, 494–498
Boundary layer, 339
Brass, properties of, 508–509
Brick, properties of, 119, 508–510
B-scale, 34
Bulk modulus, 13

Calibrated sound source, 54, 62
Calibrator, 44
Carpet, properties of, 512
Cars:
  passenger, noise of, 207–211
  railroad, 211–217
Cavitation, 187

Index 527

Characteristic impedance, 17, 87, 507–508
Cinerosis, 465
Closed tube, 361–365
Cochlea, 226–228
Coincidence frequency, 113, 128
Combination of levels, 31–33
Communication:
  face-to-face, 236–238
  telephone, 236–238
Community response to noise, 245–253
Complex numbers, 83, 89
Compliance:
  acoustic, 335–337
  mechanical, 109, 111
  volume, 301
Composite walls, 122–124
Compressibility, 91
Compressor noise, 173–177
Concrete, properties of, 119, 508–510
Condensation, 81
Construction machinery noise, 185–186
Cooling tower noise, 178–182
Copper, properties of, 508–509
Cork, 446–450
Criteria for noise (*see* Acoustic criteria)
Criteria for vibration, 464–469
Critical angle, 99
Critical damping, 411
Critical frequency, 113
Crysippus, 3
C-scale, 34
Cut-off frequency, 375
Cut-off saw, 480–482
Cylindrical waves, 153–154

Damage to hearing, 229–231
Damping coefficient, 114, 128, 410, 507
Damping ratio, 411, 446–447
Day-night level (L_{DN}), 215, 247–248, 317
dB (*see* Decibels)
Dead room, 274

Decay rate, 416
Decibels:
  A-weighted, 34
  combining, 31–33
  definition of, 27
  reference quantities for, 28–29
Decrement, logarithmic, 414
Deflection, static, 409
Density of materials, 507–508
Design:
  of expansion chamber mufflers, 371–372
  of side branch mufflers, 357–358
Directivity factor:
  definition of, 19, 24
  of gas vents, 184
  measurement of, 66–69
Directivity index:
  definition of, 19, 25
  of gas vents, 184
  measurement of, 66–69
Direct sound field, 52, 275–278, 312, 477
Displacement excitation, 436–439
Displacement level, 29, 415
Displacement, particle, 79, 82, 86, 226
Dissipation coefficient, 139
Dissipation, viscous, 410
Dissipative muffler, 377–381
Dosimeter, 50
Double-chamber plenum, 395
Double walls, 122–124
Draperies, properties of, 512
Ducts, noise in, 193–195, 488–492
Dynamic vibration absorber, 439–446

Ear:
  cross-section of, 227
  damage to, 229–231
  parts of, 226
Eccentricity, 431
Echo noise, 498
Elasticity coefficient, 384
Elastic moduli, 459
Elastomers, 450–455
Elbows, attenuation in, 196

Electric motor noise, 169–171
Enclosures:
 acoustic, 299–311
 for automatic press, 299
 design practice for, 311
 door seals for, 312
 large, 304–311
 openings in, 305–307
 for saw, 300, 480–482
 small, 300–304
End effects, 334, 387
Energy absorption, 289–293
Energy density:
 definition of, 19
 level, 18
 for plane waves, 20
 reference, 29
 for spherical waves, 91
Energy dissipation factor, 416–418, 509
Energy level, 29
Environmental noise, 243–255
EPA criteria, 247–253, 267
Equipment noise, 185–186
Expansion chamber mufflers:
 double-chamber, 373–376
 single-chamber, 368–373
Exposure level, vibration, 464–469

Face-to-face communication, 236–238
Fan noise, 164–169
Fans, types of, 164–166
Far-field, 22, 181
Fatigue due to vibration, 466
Felt, 446–450
Filters, octave band, 49–50
Fitzroy absorption units, 284, 500
Flanking path, 124
Flashing of liquids, 187–188
Flexural rigidity, 128, 132, 302
Flow resistance, 385–386
Folding carton packer, 476–480
Forced vibration, 419–424
Force level, 29
Free height of springs, 460
Free vibration, 407–419

Frequency:
 of bending waves, 300
 blade pass, 167, 174
 of closed tube, 361
 coincident, 113, 128
 cut-off, 375
 definition of, 14
 natural, undamped, 407, 446–447
 resonant, 111, 342, 361
 surge, 461
Fresnel number, 314

Gain, sound pressure level, 348–350
Galileo, 4
Gas compressor noise, 173–177
Gases, properties of, 507
Gasoline, properties of, 507
Gassendi, 4
Gas valve noise, 188–190
Gas vent noise, 182–185
Glass, properties of, 119, 306, 508–509
Granite, properties of, 509
Grass, properties of, 513
Gravel, properties of, 513
Grille coefficient, 199, 205
Grille noise, 198–199, 205–206
Guericke, Otto von, 4
Gypsum board, properties of, 509, 511

Half-power bandwidth, 344–346, 414
Hearing:
 effects of age, 230–231
 loss, 229–231
 protection, 10
 threshold of, 229
Heat conduction, 143
Helium gas, properties of, 145, 507
Helmholtz resonator, 341–350
Hertz, H. R., 14
Hiss noise, 239–243
HUD:
 noise criteria, 253–255, 258–260, 268
 sound transmission class, 139

Index

Human response:
  to noise, 245–253
  to vibration, 464–467
Humidity, effects of, 144–148
Hydrogen gas, properties of, 507

Ice, properties of, 508
Impedance:
  characteristic, 17, 87, 507–508
  for Helmholtz resonator, 343
  mechanical, 424
  for plane waves, 87
  specific acoustic, 16
  for spherical waves, 89
Incident wave, 91, 96, 101, 110, 352
Incus, 226
Infrasound, 14
Insertion loss, 301, 331, 490
Intensity, acoustic:
  definition of, 17
  level, 28
  meters, 46–49
  of plane waves, 19
  of spherical waves, 19, 21, 90
Interference:
  with sleep, 244
  with speech, 235–238, 488–489, 499–500
Isolators (see Vibration isolators)

Kerosene, properties of, 507
Kinetic energy, 19–20

Lagging of ducts, 194, 489–492
Laminate, 127–131
Land use compatibility, 259
Lead, properties of, 119, 508–509
Levels:
  acceleration, 29, 465
  A-weighted, 34–37
  combination of, 31–33
  day-night, 215, 247–248, 317
  definition of, 28
  displacement, 29
  energy density, 28
  force, 29

[Levels]
  intensity, 28
  measurement of, 41–73
  noise pollution, 253
  perceived noise, 256
  power, 28
  pressure, 29
  reference, 28–29
  speech interference, 235–238, 488–489, 499–500
  transmissibility, 428
  velocity, 29
Line source, 215
Liquids, properties of, 507
Live room, 274
Locomotive noise, 213–214
Logarithmic decrement, 414–415
Loss factor, 416–418
Loudness, 256
Lucite, properties of, 508

Machine noise prediction (see Sound sources)
Magnification factor, 421–422, 441
Malleus, 226
Masking, 235–236
Masonry block, properties of, 119, 511
Mass:
  acoustic, 332–335
  mechanical, 407
  specific (surface), 104, 112
Mass law, 104, 112
Mean free path, 282
Measurement:
  of acoustic intensity, 46–49
  of acoustic power, 51–66
  of directivity factor, 66–69
  procedures, 69–73
  of sound pressure, 42–45
Meatus, 226
Mechanical impedance, 424
Mersenne, 4, 10
Meters:
  intensity level, 46–49
  sound level, 42–45

Methane gas, properties of, 507
Mobility, 426–427
Motion sickness, 465
Motor noise, electric, 169–171
Mufflers (see Silencers)
Multiple sources, 31–33

Natural frequency:
 damped, 412
 for plates, 111–112
 undamped, 407
Near-field, 22, 181
Neper, 143
Newton, Isaac, 5, 10
Nitrogen gas, properties of, 507
Noise:
 background, 72
 of blanking press, 494–498
 criteria (see Acoustic criteria)
 definition of, 12
 of equipment, 185–186
 measurement procedures, 69–73
 of paper machines, 482–488
 paths, 7
 perceived, 256–257
 rating, composite, 243–247
 reduction, 331
 of saws, 480–482
 sources of (see Sound sources)
 structure-borne, 4
 vehicular, 207–211
Noise exposure dosage (NED), 233, 484
Noise exposure forecast (NEF), 257–261
Noise induced permanent threshold shift (NIPTS), 230
Noise levels, 27–31
Noise pollution level (L_{NP}), 253
Noise reduction coefficient (NRC), 270, 321
Noisiness, 256
Normal incidence, 91–96
Norris-Eyring reverberation time, 284
Noy, 256

Oblique incidence, 96–101
OSHA criteria, 10, 231–235, 484–485
Octave, 3, 33
Octave band filters, 49–50
Octave bands, 33–34, 49
Ohm, acoustic, 339
Optimum reverberation time, 285–286, 501
Organ of Corti, 228
Orifices, 339, 365–368
Outdoor noise criteria, 243–253
Overall levels, 36
Oxygen gas, properties of, 507

Panel absorber, 272, 392
Panels (see Plates and Walls)
Paper machine noise, 482–488
Particle displacement, acoustic, 79, 82, 86
Particle velocity, acoustic, 16, 79, 87
Pass bands for mufflers, 374, 376
Path identification, 7, 124
People, absorptivity of, 270, 513
Perceived noise level (L_{PN}), 256–257
Period, 14
Personnel booth, 484–488
Phase angle:
 in complex notation, 83
 for Helmholtz resonator, 344
 for plane waves, 86
 for spherical waves, 22, 89
 for vibration, 421, 429, 438
 through walls, 103
Pinna, 226
Plane waves:
 energy density for, 20
 equation for, 82
 impedance for, 87
 intensity for, 19
Plaster, properties of, 119, 511
Plateau, 118
Plates:
 composite, 122–124
 laminated, 127–131
 rib-stiffened, 131–134
Plenum chambers, 391–397

Index

Plexiglas, 306, 481
Plywood:
  panels, 272
  properties of, 119, 511
Poisson's ratio, 13, 111, 459
Polyurethane foam, properties of, 511
Porosity, 382
Porous materials, 271–272, 382–386
Potential energy, 20
Power, acoustic, 17, 51, 277, 305
Power level, 28, 51
Prandtl number, 362
Presbycusis, 230
Press noise, 494–498
Pressure level, 28
Propagation of sound:
  indoors, 164
  outdoors, 163
  over barriers, 313–321
Pump noise, 171–173
Pythagoras, 3

Quadrupole source, 38
Quality factor:
  acoustic, 343
  mechanical, 413–414

Rail noise, 211–213
Random incidence, 110, 386–387
Rayleigh, W. S., 5, 10, 405
Rayl unit, 17, 353
Receiver, 9
Reflected wave, 91, 96, 101
Reflection coefficient, 93
Reference quantities, 28–29
Reference sound source, 54, 62
Relaxation effects, 143–148
Relaxation time, 140
Residential areas, noise levels in, 246
Resistance:
  acoustic, 338–339
  mechanical, 339
  of orifices, 339
Resonant frequency, 111–112, 342, 361, 465

Reverberant room, 52–58
Reverberant sound field, 52, 275–278, 477
Reverberation time, 55, 281–285, 416, 476, 501
Rib-stiffened panels, 131–134
Rigidity, flexural, 128, 132, 302
Room constant, 55, 164, 277, 289–290, 297, 318, 477
Rooms:
  anechoic, 58–62
  noise in, 498–503
  reverberant, 52–58
Root-mean-square (rms) pressure, 16
Rotating unbalance, 431–436
Rubber, properties of, 450–457, 509
Rumble noise, 240–241

Sabine, W. C., 5, 269, 328
Sabin unit, 283
Saw noise, 480–482
Scale, decibel, 28
Screens:
  acoustic (*see* Acoustic barriers)
  wind, 45
  wire, 353
Shear modulus, 459
Silencers:
  for air-operated hoist, 494
  baffle-type, 392
  commercial, 389–391, 494
  design procedures for, 357–358, 371–372
  design requirements for, 330–332
  dissipative, 377–381
  double-chamber, 373–376
  expansion chamber, 368–373
  plenum, 391–397
  side-branch, 350–361
  types of, 330
Snell's law, 97
Snow, properties of, 513
Solid height for springs, 459–460
Solids, properties of, 508–513
Sonar dome, 105
Sone, 256–257

Sound:
  absorption, 139–143, 289–293
  from adjacent room, 295–297
  attenuation coefficient, 141,
    143–153, 289–293
  barriers, 312–321
  definition of, 12
  elbows, 195–198
  energy density level, 28
  indoors, 164, 274–281
  intensity level, 28
  levels, weighted, 32–37
  mean free path, 282
  outdoors, 162–163
  power level, 28
  power measurement, 51–66
  power reflection coefficient, 93
  power transmission coefficient, 93,
    99, 103, 352
  pressure
    definition, 15
    level, 28
    measurements, 42–45
    root-mean-square (rms), 16
  pressure level gain, 348–350
  speed of, 12–13, 507–508
    in ideal gases, 12, 507
    in liquids, 13, 507
    in solids, 13, 508
  steady-state, 274–281
  transmission, 162–164
  wavelength of, 14
  waves,
    cylindrical, 153–154
    plane, 17
    spherical, 18
Sound level meters, 42–45
Sound sources:
  in adjacent room, 295–298
  air distribution systems, 192–207
  appliances, 185–186
  calibrated, 54, 62
  combination of, 31–33
  compressors, 173–177
  cooling towers, 178–182
  electric motors, 169–171

[Sound sources]
  fans, 164–169
  gas vents, 182–185
  grilles, 198–199
  modification of, 8
  paths, 7, 9, 166
  pumps, 171–173
  quadrupole, 38
  traffic, 207–211
  trains, 211–217
  transformers, 177–178
  valves, 186–192
  walls, 293–295
Sound transmission class (STC),
  134–139
Source (*see* Sound sources)
Specific acoustic impedance, 16, 22, 89
Specific (surface) mass, 104, 112
Speech interference level (SIL),
  235–238, 488–489, 499–500
Speed of sound, 12–13, 507–509
Spherical waves:
  acoustic impedance for, 22, 89
  energy density for, 20, 91
  intensity for, 19, 21, 90
Spring constant:
  definition of, 335, 407
  for metal springs, 459
  for plates, 443
  for rubber isolators, 451–453
Spring index, 460
Springs, metal, 457–464
Stability, lateral, 458–459
Stapes, 122
Static deflection, 409, 447, 451, 457
Steam, properties of, 507
Steel, properties of, 119, 459, 508–509
Stethoscope, 6
Strouhal number, 196–197
Structure-borne noise, 4
Structure factor, 383
Surface absorption coefficient, 269–270

Taylor, Brook, 4
Telephone communication, 236–237
Temporary threshold shift (TTS), 230

Index

TL (see Transmission loss)
Train noise, 211–217
Traffic noise, 207–211
Transfer matrix, 339–340
Transformer noise, 177–178
Transmissibility, 428, 441
Transmissibility level, 428
Transmission coefficient, 93, 99, 103, 107–117
Transmission loss:
   approximate method for determination of, 117–120
   for barriers, 314
   for composite walls, 120–131
   definition of, 93
   for dissipative silencers, 377–381
   for expansion chamber silencers, 369–371
   mass law for, 104, 112
   for mufflers, 369–381
   normal incidence, 91–96
   oblique incidence, 96–101
   for parallel elements, 121–122
   for plenum chambers, 391–397
   for rib-stiffened panels, 131–134
   for side-branch, 351–355
   for walls, 107–117
Transmission of sound, 162–164
Transmitted wave, 92, 96, 101, 352
Truck noise, 207–211
Tube:
   closed, 361–365
   open, 365–368
Tympanic membrane, 226

Ultrasound, 14
Unbalance, rotating, 431–436
Units:
   acoustic ohm, 339
   board feet, 448
   mechanical ohm, 410
   metric (SI), 506
   neper, 143
   rayl, 17, 353
   sabin, 283

Valve noise, 186–192
Velocity level, 29, 434
Velocity, particle, 16, 79, 226
Ventilation system noise, 192–207
Vent noise, 182–185
Vibration:
   blanking press, 494–498
   damped, 410–413
   displacement excitation, 436–439
   effects on humans, 464–469
   exposure criteria, 465–466
   forced, 419–424
   free, 407–419
   isolators,
     for blanking press, 497
     cork, 446–450
     dynamic, 439–446
     felt, 446–450
     rubber, 450–457
     shear mount, 453
     spring, 457–464
   magnification factor, 421–422
   natural frequency of, 407
   subjective response to, 434
   transmissibility, 428
   with unbalanced mass, 431–436
   undamped, 407–410
Vibration isolation materials:
   cork, 446–450
   felt, 446–450
   metal springs, 457–464
   rubber, 450–457
Vibration levels:
   acceleration level, 29, 466
   displacement level, 29, 415
   force level, 29
   velocity level, 29, 434
Vinyl curtain, 306
Viscosity, effective, 362
Viscous damping, 410
Viscous effects, 143
Voice communication, 236–238
Vortex noise, 199

Walls:
   composite, 122–124

[Walls]
  double, 122–124
  laminated, 127–131
  sound transmission class (STC) for, 134–139
  as a source of sound, 293–295
  transmission loss for, 107–117
Walsh-Healy Act, 1, 6, 232
Water, properties of, 507
Wave equation, 78
Wavelength:
  for bending waves, 300
  definition of, 14
Wave number, 14, 20, 362
Waves:
  cylindrical, 153–154

[Waves]
  equation for, 78
  plane, 17
  spherical, 18, 21–24, 88–91
Weighting networks, 34
Windows, properties of, 512
Wind screen, 45
Wire screen, 353
Wire, spring, 459
Wood, properties of, 508–509
Wrapping, acoustic, 489–492

Yield strength for metals, 459
Young's modulus, 13, 111, 459

Zinc, properties of, 508